Quantum Mechanics

2nd Edition

Quantum Mechanics

2nd Edition

Fayyazuddin
Riazuddin

National Centre for Physics, Pakistan

W **World Scientific**

NEW JERSEY · LONDON · SINGAPORE · BEIJING · SHANGHAI · HONG KONG · TAIPEI · CHENNAI

Published by

World Scientific Publishing Co. Pte. Ltd.

5 Toh Tuck Link, Singapore 596224

USA office: 27 Warren Street, Suite 401-402, Hackensack, NJ 07601

UK office: 57 Shelton Street, Covent Garden, London WC2H 9HE

British Library Cataloguing-in-Publication Data
A catalogue record for this book is available from the British Library.

QUANTUM MECHANICS
2nd Edition

ISBN 978-981-4412-90-2

Printed in Singapore by B & Jo Enterprise Pte Ltd

To the memory of our *parents*, brother 𝔄*bdul* 𝔎*arim* and
sister 𝔷*ohra*

Preface

Preface for the 2nd Edition

Since the publication of the first edition of our book on Quantum Mechanics 22 years ago, physics has progressed in many directions. To catch the flavor of these developments where quantum mechanics has found application, we have added three new chapters. One is on the Two state problem, where a unified treatment of Rabi oscillations, Particle mixing and Neutrino oscillations is given, providing a beautiful illustration of quantum mechanical phenomena of superposition and interferometry. The second is on Quantum computing which is again an application of two state physics. The third one is a beautiful application of the Dirac equation in (1+2) dimensions to Graphene, where there is now evidence for the existence of massless quasi particles obeying the Dirac equations. This is of topical interest as currently a lot of work, both experimental and theoretical, is going on in this field.

The other major changes we have made are as follows: In chapter 10, a new section on the Path integral formulation of quantum mechanics was added. In chapter 20 (chapter 18 in the first edition) on the Dirac equation, we have adopted a more commonly used metric and notation for gamma matrices. We have also expanded the chapter so that it now contains a section on the Lorentz group and its representations. In chapter 9 (operators), a section on supersymmetric oscillators was added, as an example of a supersymmetric algebra. In chapter 12 (Time independent perturbation theory), a section on the "variational principle" and its applications was added. Some portions of the chapters, namely 5 (Harmonic oscillator), 8 (Collision theory), 9 (Operators), 17 (old 15, Interaction of radiation with matter) have been rewritten for more simplicity and greater clarity. Another important feature of the second edition is the addition of a substantial number of new problems.

In this and the first edition, we have followed what Freeman J Dyson[1] has called a third legacy of Dirac (not so well known to people) namely "Dirac's disengagement from verbal dispute about the meaning of quantum mechanics. When expressed in mathematical equations, the laws of quantum mechanics are clear and unambiguous." This is what is required of a theory.

Finally we wish to express our deep appreciation to Dr. Farhan Saif for writing the chapter on Quantum computing for us, and to Dr. Kashif Sabeeh for critically reading chapter 21, on Graphene. We wish to express our deep thanks to Ishtiaq Ahmed (our graduate student) for managing and doing the typing, drawing figures, and preparing the index. Thanks are also due to M. Tahir Iftikhar, Muhammad Usman, Faisal Munir, Ghulam Farid, Bilal Tariq and Saadi Ishaq for assistance with typing the manuscript.

Fayyazuddin

Riazuddin

Preface for the 1st Edition

Quantum Mechanics was discovered in 1920's to explain the stability of atoms. It is undoubtedly one of the greatest achievements in theoretical physics in the century. It is the basic theory for atomic and subatomic phenomena.

The concepts of quantum theory are radically different from the classical theory, which describes the everyday phenomena successfully. The quantum mechanical concepts are described in mathematical language. This is the approach we have followed in this book. We, however, do not assume any advanced knowledge of mathematics. The knowledge of differential and integral calculus and familiarity with matrices are sufficient to understand this book. The mathematics needed beyond this is developed in the text.

We have tried to keep the presentation well motivated and to provide sufficient details in order to facilitate the understanding of the subject. Our emphasis is on the basic theory rather than on specific applications in atomic, molecular, solid state and nuclear physics.

The book could be divided into 3 semester courses. Chapters 1–7, Chapter 8 (Secs. 1–7) and Chapter 12 (Secs. 1–2) could form a one semester undergraduate course. Chapter 8 (Secs. 8–13), Chapter 9, Chapter 10

---

[1]F. J. Dyson *Silent Quantum Genius*.
New York Review of Books: *The Strangest Man*:
The Hidden Life of Paul Dirac, Mystic of the Atom, by G. Farmelo.

(Secs. 1–3), Chapter 11 (Secs. 1–5), Chapter 12 (Sec. 4), Chapters 13 and 14 should be suitable for the second semester undergraduate course. The rest of the sections and chapters could form a one semester graduate course.

This book is based on a course of lectures which we have given at the Punjab University, Lahore, The Quaid-I-Azam University, Islamabad, The King Fahd University of Petroleum and Minerals, Dhahran(R), The King Saud University, Riyadh(F) and the Ummal-Qura University, Makkah Al-Mukarramah(F) at various times. In fact we have been encouraged by our students to write these lectures in a book form. We would like to express our thanks to them and acknowledge the respective universities for their support.

In particular we are grateful to our former students Dr. M. M. Ilyas and especially Dr. Sajjad Mahmood for help in preparing this book for publication. We are grateful to our colleagues Dr. Fahim Hussain and Dr. Pervaiz Hoodbhouy for reading the first draft of the manuscript and for some useful suggestions.

We also wish to express our thanks to Mr. Shbahat Ullah Khan for typing the first draft of the manuscript.

We were first introduced to this subject by Prof. Abdus Salam. We would like to take this opportunity to express our deep sense of gratitude to him for the encouragement throughout our careers.

Fayyazuddin
Riazuddin
March 4, 1999

Contents

Chapter 1

Breakdown of Classical Concepts

1.1 Introduction

Quantum mechanics is the theory which describes phenomena on the atomic and molecular scale. An event in this domain is not visible to the human eye. The concepts of classical physics which have been developed to describe phenomena on the macroscopic scale may not be applicable for processes on the microscopic scale of dimensions 10^{-6} to 10^{-13} cm. There are, however, some macroscopic quantum systems, e.g. superfluids, superconductors, transistors and main sequence stars.

We first outline the concepts of classical theory and then describe how, for the microphysical world, the necessity for departure from classical physics is clearly shown by experimental results.

In classical physics, matter is treated in terms of particles of definite mass, and radiation is described as wave motion. The two great disciplines of classical physics are Newtonian Mechanics and Maxwell's theory. The former describes motion of material particles according to Newton's Laws. Classical mechanics successfully describes electrically neutral macroscopic systems. Energy E and momentum $\mathbf{p}$ of a particle are two important dynamical variables.

The electric and magnetic phenomena are described in terms of electric and magnetic fields $\mathbf{E}$ and $\mathbf{B}$ which satisfy Maxwell's equations

$$\operatorname{div} \mathbf{E} = 4\pi\rho \,, \tag{1.1a}$$

$$\operatorname{curl} \mathbf{E} = -\frac{1}{c}\frac{\partial \mathbf{B}}{\partial t} \,, \tag{1.1b}$$

$$\operatorname{curl} \mathbf{B} = \frac{4\pi}{c}\mathbf{j} + \frac{1}{c}\frac{\partial \mathbf{E}}{\partial t} \,, \tag{1.1c}$$

$$\operatorname{div} \mathbf{B} = 0. \tag{1.1d}$$

Here ρ is the charge density and $\mathbf{j}$ is the electric current and they satisfy the continuity equation

$$\frac{\partial \rho}{\partial t} + \operatorname{div} \mathbf{j} = 0. \tag{1.2}$$

In free space, $\mathbf{E}$ and $\mathbf{B}$ satisfy the wave equation

$$\left(\frac{1}{c}\frac{\partial^2}{\partial t^2} - \nabla^2\right)\mathbf{E}, \mathbf{B} = 0. \tag{1.3}$$

A solution of this equation shows that $\mathbf{E}$ and $\mathbf{B}$ propagate through space as waves with speed c. For appropriate frequencies, these waves should be identified with visible light. The whole spectrum of radiation from extremely long wave length region of radio waves, through visible range, to extremely small wave length region of X-rays and γ-rays is described in terms of electromagnetic waves as given by Maxwell's theory.

As we have seen, electromagnetic radiation is regarded as consisting of waves which propagate through space with velocity c. A typical wave in x-direction is expressed as:

$$\psi(x,t) = A e^{i\left(\frac{2\pi}{\lambda}x - 2\pi\nu t\right)},$$

$$\lambda : \text{wave length},$$

$$\tau = 1/\nu : \text{periodic time } (\nu \text{ frequency}), \tag{1.4}$$

$$k = \frac{2\pi}{\lambda} \quad : \text{wave number},$$

$$\omega = 2\pi\nu : \text{angular frequency}.$$

k can be regarded as vector. A wave in 3 dimensions can be written as

$$\psi(\mathbf{r},t) = A e^{i(\mathbf{k}\cdot\mathbf{r} - \omega t)}. \tag{1.5}$$

We define a phase $\theta \equiv \mathbf{k}\cdot\mathbf{r} - \omega t$. A surface of constant phase is called a wave front. The velocity with which this surface moves is called the phase velocity. To calculate the phase velocity we note that θ is constant on this surface:

$$\frac{d\theta}{dt} = 0$$

or

$$\mathbf{k}\cdot\frac{d\mathbf{r}}{dt} - \omega = 0.$$

This gives the phase velocity $[\mathbf{k} = |\mathbf{k}|\mathbf{n}]$

$$\mathbf{n}\cdot\mathbf{u} = \frac{\omega}{|\mathbf{k}|} = \nu\lambda. \tag{1.6}$$

We cannot send a signal in the form of a monochromatic wave. However, what we do, in practice, is to send a signal in the form of a wave packet or group of waves. The only velocity which can be experimentally measured is the group velocity which we define below. The wave packet can be generated by superposition of a number of simple harmonic waves with wave numbers centered round the mean wave number. Consider first the superposition of two waves

$$\psi_1 = Ae^{i(kx-\omega t)}$$

$$\psi_2 = Ae^{i[(k+\Delta k)x-(\omega+\Delta\omega)t]}$$

$$\psi = \psi_1 + \psi_2$$

$$= A\exp\left[i\left(k+\frac{\Delta k}{2}\right)x - \left(\omega+\frac{\Delta\omega}{2}\right)t\right]$$

$$\times \left[\exp\left[-i\left(\frac{\Delta k}{2}x - \frac{\Delta\omega}{2}t\right)\right] + \exp\left[i\left(\frac{\Delta k}{2}x - \frac{\Delta\omega}{2}t\right)\right]\right]$$

$$\approx Ae^{i(kx-\omega t)}2\cos\left(\frac{\Delta k}{2}x - \frac{\Delta\omega}{2}t\right) \tag{1.7}$$

where we have put

$$k + \frac{\Delta k}{2} \approx k,$$

$$\omega + \frac{\Delta\omega}{2} \approx \omega.$$

The first term in the last line of Eq. (1.7) is carrier wave of frequency $\frac{\omega}{2\pi}$ and wave length $\frac{2\pi}{k}$. The second term gives the modulation of the amplitude of the carrier wave.

The phase velocity as before is given by

$$u = \frac{\omega}{k} = \nu\lambda.$$

On the other hand, the maximum of the amplitude moves with velocity

$$v_g = \frac{\Delta\omega}{\Delta k} = \frac{\Delta\nu}{\Delta(1/\lambda)}. \tag{1.8}$$

This is the group velocity.

The wave length of modulation is given by

$$\lambda_m = \frac{2\pi}{\frac{1}{2}\Delta k} = \frac{4\pi}{\Delta k}. \tag{1.9}$$

The plot of Eq. (1.7) is shown in Fig. 1.1. The width of the wave packet is evidently given by (half the wave length of the modulation)

$$\Delta x = \frac{1}{2}\lambda_m = \frac{2\pi}{\Delta k}$$

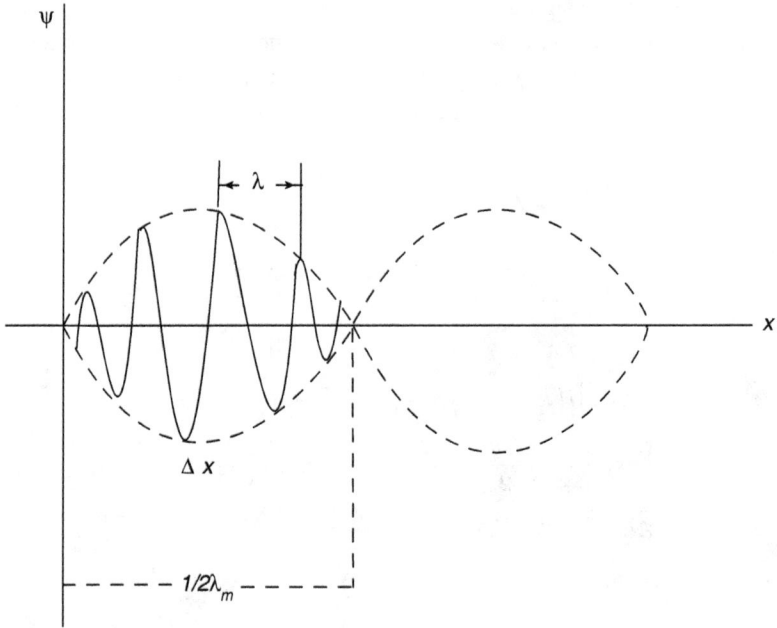

Fig. 1.1   The superposition of two simple harmonic waves of slightly different frequencies and wave numbers.

or

$$\Delta k \Delta x = 2\pi$$

or

$$\Delta k \Delta x > 1 \, . \tag{1.10}$$

In general, we can represent a wave packet (which is a superposition of monochromatic waves with wave numbers centered around the mean value k_0) as:

$$\psi(x,t) = \int_{-\infty}^{\infty} A(k) e^{i(kx - \omega(k)t)} dk \, , \tag{1.11a}$$

$$\psi(x,0) = \int_{-\infty}^{\infty} A(k) e^{ikx} dk \, , \tag{1.11b}$$

$$A(k) = \frac{1}{2\pi} \int_{-\infty}^{\infty} \psi(x) e^{-ikx} dx \, . \tag{1.11c}$$

The wave packet is localised within a distance Δx and has a spread Δk in wave number as shown in Figs. 1.2 and 1.3. It can be shown by Fourier analysis that Δx and Δk are such that

$$\Delta x \Delta k \geq 1 \, .$$

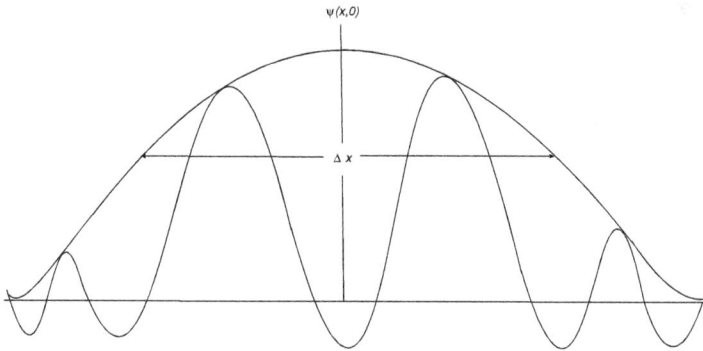

Fig. 1.2   A wave packet pictured at $t = 0$

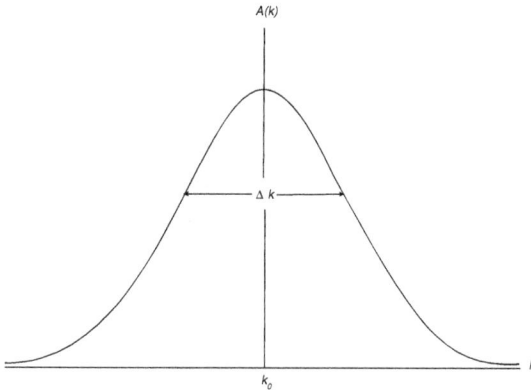

Fig. 1.3   Picture of $A(\mathbf{k})$

Since ω is a function of k, i.e.

$$\omega = \omega(k) \ , \tag{1.12}$$

in general there will be dispersion in the wave, the wave packet spreading out as it moves along. The range over which $A(k)$ is appreciably different from zero is

$$k_0 - \frac{1}{2}\Delta k \le k \le k_0 + \frac{1}{2}\Delta k \ . \tag{1.13}$$

To reduce the dispersion, we take

$$\Delta k \ll k_0 \ ,$$

and expand Eq. (1.12) around k_0:

$$\omega(k) = \omega(k_0) + \left(\frac{\partial \omega}{\partial k}\right)_{k_0} (k - k_0) + \dots \quad . \tag{1.14}$$

Now let us rewrite Eq. (1.11b) in the following way

$$
\begin{aligned}
\psi(x,0) &= \int_{-\infty}^{\infty} A(k) e^{ik_0 x + i(k-k_0)x} dk \\
&= e^{ik_0 x} \int_{-\infty}^{\infty} A(k) e^{i(k-k_0)x} dk
\end{aligned}
\tag{1.15}
$$

or

$$\psi(x) = e^{ik_0 x} X(x) , \tag{1.16}$$

where

$$X(x) = \int_{-\infty}^{\infty} A(k) e^{i(k-k_0)x} dk \tag{1.17}$$

is appreciably different from zero only in the range Δx. Now using Eq. (1.14) in Eq. (1.11a)

$$
\begin{aligned}
\psi(x,t) &\approx \int_{-\infty}^{\infty} A(k) e^{ik_0 x + i(k-k_0)x} e^{-i(\omega_0 t + v_g(k-k_0)t)} dk \\
&= e^{i(k_0 x - \omega_0 t)} \int_{-\infty}^{\infty} A(k) e^{i(k-k_0)(x - v_g t)} dk \\
&= e^{i(k_0 x - \omega_0 t)} X(x - v_g t) .
\end{aligned}
\tag{1.18}
$$

The wave packet is composed of two factors: the first factor represents a wave of frequency $\omega_0/2\pi$ and wave length $2\pi/k_0$; the last factor describes a modulation of the amplitude of this wave. This modulation moves with velocity v_g, so that Eq. (1.18) describes a group of waves (wave packet) which moves without a change in shape with group velocity v_g.

In deriving Eq. (1.18), we have neglected the term

$$\frac{1}{2} it \left(\frac{d^2 \omega}{dk^2}\right)_{k_0} (k - k_0)^2 \tag{1.19}$$

in the exponential. This is justified only if

$$t \left(\frac{d^2 \omega}{dk^2}\right)_{k_0} (\Delta k)^2 \ll 1$$

or

$$t \left(\frac{d^2 \omega}{dk^2}\right)_{k_0} \Delta k \ll \frac{1}{\Delta k} \approx \Delta x$$

or

$$t \left(\frac{d^2 \omega}{dk^2}\right)_{k_0} \Delta k \ll \Delta x . \tag{1.20}$$

The wave picture of radiation satisfactorily explains the phenomena of interference and diffraction of light.

1.2 Breakdown of Classical Concepts

Can the classical concepts outlined above, viz, matter consisting of point particles and radiation consisting of waves provide a framework for physical phenomena on a microscopic scale, i.e. when point particles are electrons and protons, each carrying mass and electric charge and interacting through the fundamental electromagnetic force? As we shall see, the classical concepts were found to be quite inadequate in describing the motion of electrons and their interaction with radiation or other particles. Also the emission and absorption of radiation could not be explained by the classical concept of the wave nature of radiation.

(a) Dual Behaviour of Radiation

(i) Phenomena of interference and diffraction, which are characteristic of waves.
(ii) Phenomena of emission and absorption of radiation which cannot be explained by the wave nature of radiation. Examples of phenomena indicating particle aspects of radiation are as follows.

(1) Blackbody spectrum

This concerns the thermodynamics of the exchange of energy between radiation and matter. It was shown by Planck that the correct thermodynamical formula, namely

$$E(\omega) = \frac{8\pi\hbar\omega^3}{(8\pi c)^3}(e^{(\hbar\omega)/kT} - 1)^{-1} , \tag{1.21}$$

where $E(\omega)$ is the energy per unit volume of wave, with angular frequency ω and k is the Boltzmann constant, is obtained only if radiation of frequency ω exchanges energy with matter in discrete units of $\hbar\omega$. Here as well as in the formula (1.21), $\hbar = \frac{h}{2\pi}$, h being a universal constant, called Planck's constant, has dimensions of $[E][t]$ and a numerical value

$$\hbar = 1.055 \times 10^{-27} \text{erg } s = 6.582 \times 10^{-22} \text{ MeV} s . \tag{1.22}$$

Incidently the Planck's formula (1.21) reduces to the classical Rayleigh-Jean's law for low frequencies $\hbar\omega \ll kT$ while the total emitted energy

$$E = \int_0^\infty E(\omega)d\omega ,$$

remains finite and gives the Stefan's law

$$E = \frac{4\sigma}{c} T^4 \qquad (1.23)$$

with $\sigma = \frac{2\pi^5 k^4}{15 h^3 c^2}$.

Thus, according to Planck's hypothesis, the only available energies for the wave of frequency ω will be $0, \hbar\omega, 2\hbar\omega, \ldots$, i.e. radiation of frequency ω behaves like a stream of particles (photons) of energy $E = \hbar\omega$.

Now according to the special theory of relativity

$$E^2/c^2 = p^2 + m^2 c^2 . \qquad (1.24)$$

Since photons travel with the velocity of light c, their rest mass should be zero, so that

$$E^2 = p^2 c^2 = \hbar^2 \omega^2$$

or

$$\mathbf{p} = \hbar \mathbf{k} . \qquad (1.25)$$

These equations clearly show the relation between the particle parameters $(E, \mathbf{p})$ of the photon, and the parameters $(\omega, \mathbf{k})$ of the corresponding wave.

(2) Photoelectric effect

This quantum idea was later used by Einstein to explain the photoelectric effect. If a beam of light of frequency ω is incident on the surface of a metal, electrons may be emitted. Note that when electrons are emitted, their energy T does not depend on the intensity of radiation, but only on its frequency. The more intense the beam the more electrons are emitted, but their energy will depend on the frequency. This cannot be explained in the classical picture of continuous exchange of energy. However, it is easily explained by Planck's hypothesis. When the incident energy of the photon $\hbar\omega > W$, where W is the work required to free the electron from the attractive potential of the metal, electrons are emitted, and carry the energy T given by

$$\hbar\omega = W + T . \qquad (1.26)$$

Otherwise, i.e. for $\hbar\omega < W$, no electron is emitted. Since the number of photons are proportional to intensity of light, the more intense the light, the more electrons are emitted if $\hbar\omega > W$.

(3) Compton effect

The particle aspect of radiation is most clearly shown in the Compton effect, the scattering of X-rays by electrons bound in atoms. The binding energy which is in eV range can be neglected as X-rays are in the MeV range. The change of wave length $\Delta\lambda$ observed by Compton depends on the angle of scattering and not on its wavelength. This cannot be explained classically since in terms of radiation waves one expects $\Delta\lambda$ proportional to λ. This can be easily explained as an elastic scattering of a photon of energy $E = \hbar\omega$ and momentum $\mathbf{p} = \hbar\mathbf{k}$ with an electron at rest as shown in Fig. 1.4. It is easy to see from energy-momentum conservation that

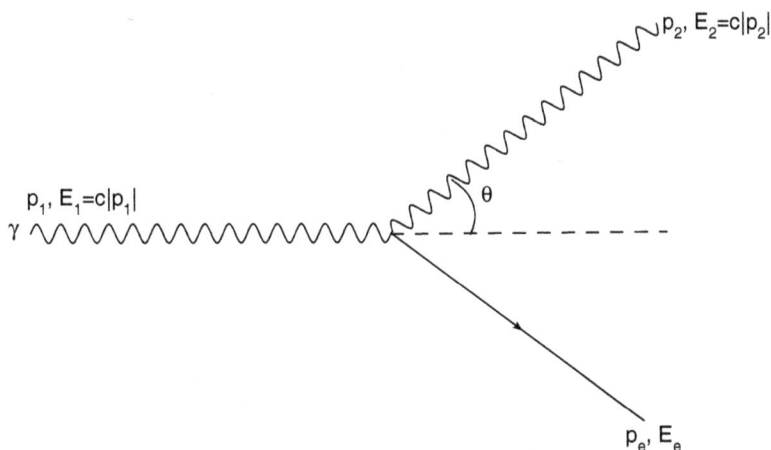

Fig. 1.4   Collision of a photon with electron at rest in Compton effect.

$$\Delta\frac{\lambda}{2\pi} \equiv (\frac{\lambda_2}{2\pi} - \frac{\lambda_1}{2\pi}) = \left(\frac{\hbar}{p_2} - \frac{\hbar}{p_1}\right) = \frac{2\hbar}{mc}\sin^2\frac{\theta}{2} ,$$

$$= 2\frac{\lambda_e}{2\pi}\sin^2\frac{\theta}{2} , \qquad (1.27)$$

where

$$\frac{\lambda_e}{2\pi} \equiv \frac{\hbar}{mc} \approx 3.86 \times 10^{-11} \text{ cm} \qquad (1.28)$$

is called Compton wave length of the electron.

(b) Dual Nature of Matter

Later it was found that this dual behaviour was not confined to radiation, but holds for particle as well. This was experimentally discovered by Davisson and Germer and by G. P. Thomson. They showed that when a homogeneous beam of electrons passes through a crystal, the emergent beam exhibits a pattern of alternate maxima and minima of intensity, wholly similar to the diffraction pattern observed in the diffraction of electromagnetic radiation. The beam of electrons thus behaves as though it were a train of waves of wave length

$$\lambda = h/p \tag{1.29}$$

or

$$p = h/\lambda = \hbar k \ ,$$

where p is the momentum of each electron.

This was predicted, however, by de Broglie before it was discovered experimentally. Thus an electron of given energy E and momentum p was associated with a de Broglie wave such that

$$E = h\nu = \hbar\omega \ ,$$
$$p = \tfrac{h}{\lambda} = \hbar k \ . \tag{1.30}$$

Now

$$E = \frac{mc^2}{\sqrt{1 - v^2/c^2}} \ , \qquad p = \frac{mv}{\sqrt{1 - v^2/c^2}} \ . \tag{1.31}$$

Therefore, phase velocity for a de Brogile wave is given by

$$u = \frac{\omega}{k} = \frac{E}{p} = \frac{c^2}{v} > c \ . \tag{1.32}$$

For group velocity we have

$$v_g = \frac{d\omega}{dk} = \frac{dE}{dp} \ . \tag{1.33}$$

Now

$$E^2 = p^2 c^2 + m^2 c^4$$

or

$$2E\frac{dE}{dp} = 2c^2 p \ .$$

Therefore

$$v_g = \frac{dE}{dp} = \frac{pc^2}{E} = v \ . \tag{1.34}$$

The relationship thus brought out is very attractive; we associate a wave packet

$$\psi(x,t) = \int A(k)e^{i(kx-\omega(k)t)}dk$$

with a particle. For a free particle

$$\psi(x,t) = \int A(k)e^{i[kx-(\hbar k^2/2m)t]}dk \ ,$$

since for a free non-relativistic particle

$$\omega(k) = \frac{E}{\hbar} = \frac{p^2}{2m\hbar} = \frac{\hbar k^2}{2m} \ . \tag{1.35}$$

For a wave packet, we have shown that

$$\Delta k \Delta x \geq 1 \ .$$

Since $p = \hbar k$, we have

$$\Delta p \Delta x \geq \hbar \ . \tag{1.36}$$

This is an example of the Heisenberg uncertainty principle, which states that uncertainty Δx in the position of a particle and uncertainty Δp in its momentum is given by the above relation. The association of a wave packet with a particle implies that it is not possible to determine simultaneously the position and momentum of a particle precisely.

As we have discussed previously $A(k)$ is appreciably different from zero in the range $(\Delta k \ll k)$

$$k_0 - \frac{1}{2}\Delta k \leq k \leq k_0 + \frac{1}{2}\Delta k \ .$$

Now $\omega(k)$ can be written in the form

$$\omega(k) = \frac{\hbar}{2m}[k_0^2 + 2k_0(k - k_0) + (k - k_0)^2]; \Delta = (k - k_0) \ .$$

Thus the wave packet $\psi(x,t)$:

$$\psi(x,t) = \int A(k)e^{i(k_0+k-k_0)x}e^{\frac{-i\hbar}{2m}[k_0^2+2k_0(k-k_0)+(k-k_0)^2]}$$

$$\approx e^{i(k_0 x - \frac{\hbar k_0^2}{2m})}X(x - vt - \frac{\hbar}{m}\Delta kt)$$

where

$$X\left(x - vt - \frac{\hbar}{m}\Delta k\right) = \int_{-\infty}^{\infty} A(k)e^{i(kx-\omega(k)t)[x-vt-\frac{\hbar}{m}\Delta kt]}dk. \quad (1.37)$$

Thus, we see that the wave packet propagates without appreciable spread with the velocity v, the particle velocity. This is true only if according to Eq. (1.20)

$$t\left(\frac{\hbar}{m}\right)\Delta k \ll \Delta x \;,$$

or

$$t\Delta v \ll \Delta x \;. \qquad (1.38)$$

Now $t\Delta v$ gives uncertainty in position of the particle at time t over and beyond the initial uncertainty Δx. Thus the wave packet spreads as it propagates, this spread is negligible provided that

$$t\Delta v \ll \Delta x \;.$$

To sum up, what are conventionally labeled as particles have wave properties, which can be exhibited under suitable conditions and conversely light is as much an elementary particle as an electron, a proton or a neutron.

Thus, for phenomena on the atomic scale, some basic change was needed in our ideas. In the new theory, essentially non-classical features of particle aspects of radiation and wave aspects of particles should appear naturally.

(c) Hydrogen Spectrum; The Bohr atom

The spectrum of hydrogen is very simple. It has discrete frequencies and the corresponding wave lengths satisfy the general law

$$\frac{1}{\lambda} = R\left(\frac{1}{n_1^2} - \frac{1}{n_2^2}\right) , \qquad (1.39)$$

where $n_2 = 3, 4, 5, \ldots, n_1 = 2$ gives the Balmer series and R is the Rydberg constant given by

$$R = \frac{m_e e^4}{4\pi\hbar^3 c} = \frac{m_e c}{4\pi\hbar}(\alpha^2) = 109,677,581 \text{ cm}^{-1}(= 13.595 \text{ volts}) ,$$

where $\alpha = (e^2/\hbar c) = \frac{1}{137}$ is a dimensionless constant and is known as the fine structure constant.

According to Rutherford, an atom consists of a heavy nucleus with a charge Ze, about which Z electrons revolve. According to Maxwell's theory, an accelerated charge emits electromagnetic radiation. An orbiting electron constitutes a rapidly accelerating charge and should emit radiation continuously and thus lose energy. As a result its orbit would become smaller and smaller and it would ultimately be drawn into the nucleus. The classical theory has two important consequences:

(1) The atom should be very unstable.

(2) It should radiate energy over a continuous range of frequencies.

Both of these consequences completely contradict experiment. The atom is stable and as we have seen above, the spectrum of the hydrogen atom has discrete frequencies.

In order to explain the hydrogen spectrum and the stability of the atom, Bohr postulated the following rules:

(i) The magnitude of the angular momentum L of the electron is an integral multiple of $\hbar$

$$L = n\hbar, \qquad n = 1, 2, \ldots .$$

(ii) The radiation is emitted or absorbed when electron makes a transition (quantum jump) from an orbit of energy E_n to one of energy E'_n, say, and the resulting angular frequency $\omega_{nn'}$ is given by

$$\hbar\omega_{nn'} = |E_n - E_{n'}| . \tag{1.40}$$

Let us apply these rules to the hydrogen atom. For a circular orbit of atom a, we get by equating Coulomb force to centripetal force:

$$e^2/a^2 = ma\omega^2 .$$

Now

$$L = ma^2\omega = n\hbar .$$

This gives discrete values for a and ω:

$$a_n = \left(\frac{\hbar^2}{me^2}\right)n^2 = a_0 n^2$$

$$\omega_n = \frac{e^4 m}{\hbar^3}\frac{1}{n^3} , \tag{1.41}$$

where

$$a_0 \equiv \frac{\hbar^2}{me^2} = 5.29 \times 10^{-11} \text{ m} \tag{1.42}$$

is called the first Bohr atom.

Now

$$E_n = \frac{1}{2}m(a_n\omega_n)^2 - \frac{e^2}{a_n} = \left(-\frac{1}{2}\frac{e^2}{a_0}\right)\frac{1}{n^2} , \tag{1.43}$$

and from Eq. (1.40)

$$\omega_{nn'} = \frac{e^2}{2\hbar a_0}\left(\frac{1}{n^2} - \frac{1}{n'^2}\right).$$ (1.44)

This explains the discrete spectrum of the hydrogen atom; in particular we get the Balmer series with $n = 2$ and $n' = 3, 4, 5, \ldots$.

The Bohr atom is stable, since no further radiation of energy is possible once it has reached the lowest level (ground state) $E_1 = \left(-\frac{1}{2}e^2/a_0^2\right)$.

The success of the Bohr theory for hydrogen-like atoms is impressive but clearly unsatisfactory in the sense that the motion of the electron is described in classical terms by the superposition of additional arbitary rules.

To summarise, the non-classical features which have emerged are:

(i) Particle aspect of radiation.

(ii) Wave aspect of particles.

(iii) Some dynamical variables such as energy and angular momentum having discrete values. These features should appear as natural consequences of the new theory — quantum mechanics.

1.3 Problems

1.1 Find the energy and momentum of an X-ray photon whose frequency is 5×10^{18} cycles sec^{-1}.

1.2 (a) The velocity of ripples on a liquid surface is $\sqrt{2\pi S/\lambda\rho}$, where S is the surface tension and ρ the density of the liquid. Find the group velocity of these waves.

(b) Velocity of ocean waves is $\sqrt{g\lambda/2\pi}$, where g is the acceleration due to gravity. Find the group velocity of ocean waves.

1.3 Given that $A(k) = \left(\frac{1}{2K}\right)^{\frac{1}{2}}$, $|k| \leq K$, $A(k) = 0$, $|k| > K$. Find $\psi(x)$. Make a sketch of $A(k)$ and $\psi(x)$ and hence show that $\Delta k\Delta x > 1$.

1.4 For a Guassian wave packet

$$\psi(x,t) = Ne^{-x^2/2\delta^2}e^{i(k_0 x - \omega_0 t)}.$$

Show that

$$A(k) = \frac{N}{\sqrt{2\pi}}\delta e^{-\delta^2(k-k_0)^2/2}.$$

Make a sketch of $\psi(x) = \psi(x,0)$ and $A(k)$ and hence show that

$$\Delta x\Delta k = 1.$$ (1.45)

1.5 Light of wave length $\lambda = 10$ meters falls on the surface of a metal. If the work function of metal is 3.64 eV, is an electron emitted or not?

1.6 Find the velocity of a proton whose de Broglie wave length is equal to that of a 1 keV X-ray.

1.7 A photon whose initial frequency was 1.5×10^{19} sec^{-1} emerges from a collision with an electron with a frequency 1.2×10^{19} sec^{-1}. How much kinetic energy was imparted to electron?

1.8 Derive a formula expressing de Broglie wave length (in Å) of an electron in terms of potential difference V (in volts) through which it is accelerated.

1.9 Show that velocity v of the electron in the first Bohr orbit of hydrogen atom is given by

$$v/c = e^2/\hbar c = 1/137 \ .$$

1.10 A beam of 100 MeV electrons travels a distance of 10 meters. If the width Δx of initial packet is 10^{-2} cm, calculate the spread in the wave packet in traveling this distance and show that this spread is much less than Δx.

1.11 Find the de Broglie wave length of an electron accelerated through a 10 Volts potential difference. What is its velocity?

1.12 A light of wave length $2500 Å$ falls on a surface of metal whose work function is 3.64 eV. What is the kinetic energy of the electrons emitted from the surface? If the intensity of light is 4.0 W/m^2, find the average number of photons per unit time per unit area that strike the surface.

Chapter 2

Quantum Mechanical Concepts

2.1 Uncertainty Principle

Classical mechanics breaks down when applied to small systems, provided they are small enough, and science is concerned with observable phenomena. For classical systems, an assumption is tacitly made that the operation of observation does not appreciably disturb the system and the disturbance which may be caused by measurement can be corrected for exactly, at least in principle. On the other hand, we can observe phenomena on atomic scale only by using microscopic objects (atoms, nuclei or photons) of the same dimensions so that the process of measurement disturbs the system to be observed and for small enough objects this disturbance is not negligible. If for example, the momentum of an electron is known initially and we try to measure its position, then at least one photon must be scattered by the electron to come through the microscope onto the screen. This would disturb the electron by an amount which cannot be predicted so that we cannot observe an electron without disturbing it. As Dirac has put it "there is a limit to the fineness of our powers of observation and the smallness of the accompanying disturbance — a limit which is inherent in the nature of things and can never be surpassed by improved technique or increased skill on the part of the observer. If the object under observation is such that the unavoidable limiting disturbance is negligible, then the object is big and we may apply classical mechanics to it. If on the other hand, the limiting disturbance is not negligible, then the object is small and we have to apply a new theory, i.e. quantum mechanics to it". (P. M. A. Dirac, the Principles of Quantum Mechanics, Oxford University Press, 4th Edition, p. 3.)

The limit to the fineness of our powers of observation and the smallness of the accompanying disturbance is expressed by the Heisenberg Uncertainty Principle.

Suppose we wish to look at an electron. To do so we must touch it with something else, e.g. shine light on it. Let the wave length of light be λ. Each photon of this light has momentum $Q = h/\lambda$. In order to "see" the electron, one of these photons must bounce off the electron, thus disturbing the electron and changing the electron's original momentum. The exact amount of change Δp cannot be predicted, but it cannot be greater than the photon momentum h/λ. Hence

$$\Delta p \leq h/\lambda.$$

The longer the wave length, the smaller the uncertainty in momentum. Because light is a wave phenomenon, we cannot expect to determine the electron's position with perfect accuracy; however the uncertainty Δx cannot be less than λ, i.e.

$$\Delta x \geq \lambda$$

the smaller the wave length, the smaller the uncertainty in position. Hence if we use light of short wave length to increase the accuracy of position measurement, there will be a corresponding decrease in the accuracy of momentum measurement, while light of long wave length will decrease the accuracy of position measurement with the corresponding increase in momentum measurement. Combining the above two estimates, we get

$$\Delta x \Delta p \geq h \qquad (2.1)$$

which is one form of the uncertainty principle.

Thus if we make a measurement on any object, and we determine the x-component of its momentum with an uncertainty Δp_x, we cannot at the same time know its x position more accurately than $\Delta x = h/\Delta p_x$. The uncertainties $\Delta x, \Delta p_x$ in the position and momentum of a particle at any instant must satisfy the relation

$$\Delta x \Delta p_x \geq \hbar \qquad (2.2)$$

$\hbar$ is so small that the uncertainty principle is important only on the atomic scale where laws of motion have to be altered to include it.

Hence quantum mechanics cannot make completely definite predictions concerning the future behaviour of, for instance, electrons. For a given initial state of the electron, a subsequent measurement can give various results. The typical problem in quantum mechanics consists of determining the probability of obtaining various results on performing this measurement.

Thus only probability relations exist between present and future. However, there is a casual relation between probabilities. Quantum mechanics enables us to formulate these relations in a general set of equations that will supply us with a description of atomic phenomena.

To sum up we no longer have determinacy as is the case on the macroscopic scale, where classical mechanics is applicable and we have to replace determinacy by probability. Quantum mechanics enables us to calculate the probability.

2.2 Illustration of Heisenberg Uncertainty Principle

Consider an experiment in which the momentum P of an electron in the x-direction is accurately known at the beginning and the position is measured. We look at the electron through a microscope by means of light of wave length λ. Each photon of this light has momentum $P = h/\lambda = h\nu/c$. In order to 'see' it, one of the photons must be scattered from the electron, thus disturbing the electron and changing its original momentum.

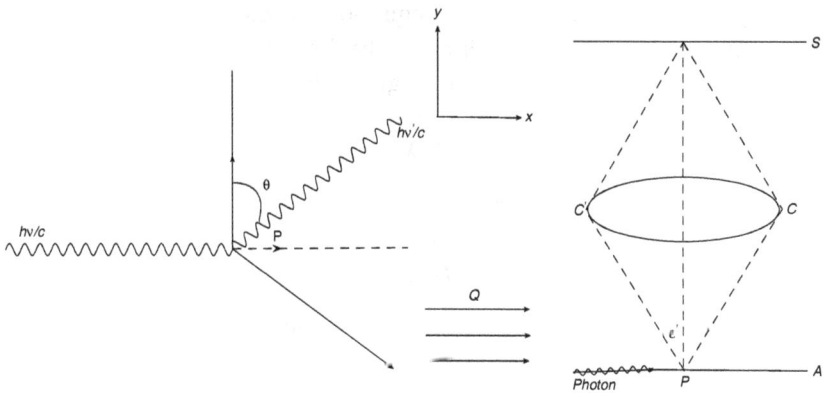

Fig. 2.1   Microscopic experiment

From conservation of momentum (see Fig. 2.1)

$$P + \frac{h\nu}{c} = \frac{h\nu'}{c} \sin\theta + p_x \tag{2.3}$$

where P is the original momentum of electron, p_x its momentum in the x-direction, θ the angle of scattering, ν and ν' are the frequencies of the photon before and after scattering because of Compton effect. Now $\theta \leq \varepsilon$ in order to observe the electron (i.e. light goes through the microscope). Now we can determine the position of the electron with accuracy (given by the resolving power of the microscope)

$$\Delta x \approx \frac{\lambda'}{\sin \varepsilon}. \tag{2.4}$$

Now the quantum may have scattered anywhere within the angle $\pm \epsilon$, therefore from Eq. (2.3) we see that p_x the momentum of the electron in x-direction has uncertainty

$$\Delta p_x = 2\frac{h}{\lambda'} \sin \varepsilon. \tag{2.5}$$

From Eqs. (2.4) and (2.5), we have

$$\Delta p_x \Delta x \approx 2h > h \ .$$

2.3   Schrödinger Equation

Since wave aspects of particles are not a result of classical mechanics, it is not possible to derive the basic equations of quantum mechanics from there. We are here dealing with a new field of physics. We shall however, use the de Broglie rule as a guide to guess the basic equation of quantum mechanics. The de Broglie rule is that with each particle of momentum p is associated a wave of wave length λ given by

$$\lambda = \frac{2\pi}{k} = \frac{h}{p} = \frac{h}{mv} = \frac{2\pi}{k}. \tag{2.6}$$

The Einstein relation for energy E is

$$E = h\nu = \hbar\omega \qquad (\text{wave aspect}). \tag{2.7}$$

For a free non-relativistic particle

$$E = \frac{p^2}{2m} \qquad (\text{particle aspect}). \tag{2.8}$$

Thus,

$$\omega = \frac{E}{\hbar} = \frac{p^2}{2m\hbar} = \frac{\hbar k^2}{2m} \tag{2.9}$$

so that

$$v_g = \frac{d\omega}{dk} = \frac{\hbar k}{m} = \frac{p}{m} = v \ .$$

Consider for simplicity a particle moving in the x-direction. We now introduce a quantity $\psi(x,t)$, called the wave function which specifies all that can be known about the system it represents. We put two requirements: (i) to find the physical interpretation of ψ, i.e. to have a set of rules which will enable us, knowing ψ, to predict what results a certain measurement will give, and (ii) to find an equation which ψ satisfies so that this equation represents the motion of the system which ψ represents. We deal first with (ii) and come to (i) later. In analogy with other fields of physics, we look for a differential equation for $\psi(x,t)$. For this purpose, we describe a particle with a wave packet,

$$\psi(x,t) = \int_{-\infty}^{\infty} A(k)e^{i[kx-\omega(k)t]}dk \tag{2.10}$$

where ω is given by Eq. (2.9). Hence the wave packet associated with the particle[1] is

$$\psi(x,t) = \int_{-\infty}^{\infty} A(k)e^{i[kx-(\hbar k^2/2m)t]}dk. \tag{2.11}$$

For the wave function given by Eq. (2.11), the following relation holds [c.f. Eq. (1.11d)]

$$\Delta k \Delta x \geq 1, \tag{2.12}$$

which on using relation (2.6) gives

$$\Delta p \Delta x \geq \hbar. \tag{2.13}$$

Thus our description of a particle by wave packet satisfies the basic tenet of quantum mechanics.

We now proceed to find a differential equation having (2.10) as its solution. We have from Eq. (2.11)

$$i\hbar \frac{\partial \psi(x,t)}{\partial t} = \int_{-\infty}^{\infty} A(k)\frac{\hbar^2 k^2}{2m}e^{i[kx-(\hbar k^2/2m)t]}dk$$

$$-(\hbar^2/2m)\frac{\partial^2 \psi}{\partial x^2} = \int_{-\infty}^{\infty} A(k)\frac{\hbar^2 k^2}{2m}e^{i[kx-(\hbar k^2/2m)t]}dk$$

so that the differential equation

$$i\hbar \frac{\partial \psi}{\partial t} = -\frac{\hbar^2}{2m}\frac{\partial^2 \psi}{\partial x^2}, \tag{2.14}$$

follows. This is the simplest form of the Schrödinger equation. This is the basic equation of quantum mechanics and describes phenomena on the

---

[1] Note the important difference. In quantum mechanics, the wave function $\psi(x,t)$ is in general complex, whereas in classical electromagnetism, $\psi(x)$ represents a real wave.

atomic scale. The above equation is for a free particle, i.e. a particle on which no force is acting and for which

$$\text{Total energy} = \text{Kinetic energy} = \frac{p^2}{2m}. \tag{2.15}$$

It may be seen that Eq. (2.14) can be obtained from Eq. (2.15) by regarding energy and momentum as differential operators

$$E \longrightarrow i\hbar\frac{\partial}{\partial t} \; ; \quad p \longrightarrow -i\hbar\frac{\partial}{\partial x} \tag{2.16}$$

that act on the wave function ψ to give

$$i\hbar\frac{\partial \psi}{\partial t} = H\psi$$

where

$$H = -\frac{\hbar^2}{2m}\frac{\partial^2}{\partial x^2} \tag{2.17}$$

is the Hamiltonian operator for a free particle.

This is an example of the correspondence principle:

"The essentially definitive relations between physical variables in classical mechanics, which do not involve derivatives, are also satisfied by the corresponding quantum operators."

For a particle moving in a field of force given by the potential $V(x)$, the total energy is given by

$$E = \frac{p^2}{2m} + V(x). \tag{2.18}$$

The rule given in Eq (2.16), gives the differential equation:

$$i\hbar\frac{\partial \psi(x,t)}{\partial t} = \left(-\frac{\hbar^2}{2m}\frac{\partial^2}{\partial x^2} + V(x)\right)\psi(x,t) = H\psi(x,t) \tag{2.19}$$

where the Hamiltonian H is:

$$H = T + V = \frac{\hat{p}^2}{2m} + V(x) = -\frac{\hbar^2}{2m}\frac{\partial^2}{\partial x^2} + V(x). \tag{2.20}$$

In general we denote the momentum and position operators by $\hat{p}$ and $\hat{x}$. The important representation of these operators in x-space is provided by Eq. (2.16) which is known as the Schrödinger representation.

The correspondence principle will ensure that in the limit of the observed system becoming large and the disturbance becoming negligible, it must go over into classical mechanics.

2.4 Physical Interpretation of Wave Function

The wave funciton $\psi(x,t)$ has a statistical interpretation. It is related to the probability, $P(x,t)$ of finding the particle at the position x and time t. This is because particle is associated with a wave packet which has a spread. The probability is positive definite : $P(x,t) \geq 0$. Therefore $P(x,t)$ should be proportional to $\psi^*(x,t)\psi(x,t)$. We take it

$$P(x,t) = \psi^*(x,t)\psi(x,t) = |\psi(x,t)|^2. \tag{2.21}$$

Now

$$\int_{-\infty}^{+\infty} P(x,t)dx = \int_{-\infty}^{+\infty} |\psi(x,t)|^2 dx = 1 \tag{2.22}$$

because the particle must be somewhere between $-\infty$ and $+\infty$. When it is satisfied, the wave function is said to be normalised. Thus, in order to normalise the wave function, all we have to do is to adjust the coefficient of $\psi(x,t)$ in such a way that the integral over all space is one.

Eq (2.20) implies that $\int_{-\infty}^{+\infty} P(x,t)dx$ is independent of time i.e.

$$\frac{d}{dt}\int_{-\infty}^{\infty} P(x,t)dx = 0. \tag{2.23}$$

But

$$\frac{d}{dt}\int_{-\infty}^{\infty} P(x,t)dx = \int_{-\infty}^{\infty} \left(\frac{\partial \psi^*(x,t)}{\partial t}\psi(x,t) + \psi^*(x,t)\frac{\partial \psi(x,t)}{\partial t} \right)dx.$$

From Schrödinger's equation, we have

$$i\hbar\frac{\partial \psi}{\partial t} = \left(-\frac{\hbar^2}{2m}\frac{\partial^2}{\partial x^2} + V(x) \right)\psi. \tag{2.24}$$

Taking its complex conjugate

$$-i\hbar\frac{\partial \psi^*}{\partial t} = \left(-\frac{\hbar^2}{2m}\frac{\partial^2}{\partial x^2} + V(x) \right)\psi^*. \tag{2.25}$$

From these two equations, we get

$$\frac{\partial \psi^*}{\partial t}\psi + \psi^*\frac{\partial \psi}{\partial t} = \frac{\hbar}{2mi}\left(\psi\frac{\partial^2 \psi^*}{\partial x^2} - \psi^*\frac{\partial^2 \psi}{\partial x^2} \right)$$

$$= \frac{\hbar}{2mi}\frac{\partial}{\partial x}\left(\psi\frac{\partial \psi^*}{\partial x} - \psi^*\frac{\partial \psi}{\partial x} \right). \tag{2.26}$$

Hence

$$\frac{d}{dt}\int_{-\infty}^{\infty} P(x,t)dx = \frac{\hbar}{2mi}\int_{-\infty}^{\infty} \frac{\partial}{\partial x}\left(\psi\frac{\partial \psi^*}{\partial x} - \psi^*\frac{\partial \psi}{\partial x} \right)dx$$

$$= \frac{\hbar}{2mi}\left(\psi\frac{\partial \psi^*}{\partial x} - \psi^*\frac{\partial \psi}{\partial x} \right)_{-\infty}^{+\infty}$$

$$= 0 \tag{2.27}$$

if $\psi(x,t) \to 0$ as $x \to \infty$. Hence the probability is conserved if we use definition (2.21) of the probability.

Next, we define a quantity called probability current density denoted by $S(x,t)$:

$$S(x,t) = \frac{\hbar}{2mi}\left(\psi^*\frac{\partial\psi}{\partial x} - \psi\frac{\partial\psi^*}{\partial x}\right). \tag{2.28}$$

Then from Eq. (2.30), we have

$$\frac{\partial P(x,t)}{\partial t} + \frac{\partial}{\partial x}S(x,t) = 0 \tag{2.29}$$

which is analogous to the equation of continuity viz.,

$$\frac{\partial\rho}{\partial t} + \nabla\cdot\mathbf{j} = 0 \tag{2.30}$$

of electricity. It is because of this reason that $S(x,t)$ given by Eq. (2.31) is called Probability current density.

Remarks:

1. We have shown that the probability density

$$P(x,t) = |\psi(x,t)|^2 \tag{2.31}$$

and that

$$\frac{d}{dt}\int_{-\infty}^{\infty} P(x,t)dx = 0 \tag{2.32}$$

provided that

$$\psi(x,t) \longrightarrow 0 \quad \text{as} \quad x \longrightarrow \infty.$$

This is the boundary condition satisfied when the wave function $\psi(x,t)$ represents a bound state, for example, an electron bound in the hydrogen atom, or a particle in a potential well. For a bound state, the particle can never go to infinity, i.e. $\psi(x,t) = 0$ at $x \to \infty$. Therefore $\int|\psi(x,t)|^2 dx$ is convergent and can be normalised.

There is another kind of wave function encountered in collision problems in which case a particle arrives from infinity (great distance), is scattered by a field of force and goes off to infinity (i.e. to a great distance from the centre of force) again. For such cases the wave function remains finite even as $x \to \infty$ and $\int|\psi|^2 dx$, diverges. Here $|\psi|^2$ does not directly determine the probability and must be regarded as a quantity proportional to this probability. For example, for a wave function of type

$$\psi = \exp[i(kx - \omega t)] \tag{2.33}$$

the normalisation integral $\int_{-\infty}^{\infty} |\psi|^2 dx$ when taken over the interval diverges but in this case ψ can be defined in a finite interval so that it vanishes or has a periodic structure at the boundary.

2. In the Schrödinger picture we have considered, x is an algebraic variable but momentum p is a differential operator

$$\widehat{x} = x, \quad \widehat{p} = -i\hbar \frac{\partial}{\partial x}.$$

In general the two operators $\widehat{A}$ and $\widehat{B}$ do not commute, i.e.

$$\widehat{A}\widehat{B} \neq \widehat{B}\widehat{A} \tag{2.34}$$

(cf ordinary numbers which commute). As an example of illustrating this property of operators, we take the operators $\widehat{x}, \widehat{p}$. Then in Schrodinger picture

$$\widehat{x}\widehat{p} = x\left(-i\hbar\frac{\partial}{\partial x}\right)$$

$$\widehat{p}\widehat{x} = \left(-i\hbar\frac{\partial}{\partial x}\right)x. \tag{2.35}$$

Therefore

$$\widehat{x}\widehat{p}\psi(x,t) = x\left(-i\hbar\frac{\partial}{\partial x}\right)\psi(x,t)$$

$$\widehat{p}\widehat{x}\psi(x,t) = -i\hbar\frac{\partial}{\partial x}\Big(x\psi(x,t)\Big)$$

$$= -i\hbar\psi(x,t) - i\hbar x\frac{\partial\psi(x,t)}{\partial x}. \tag{2.36}$$

From these equations we see that

$$\widehat{x}\widehat{p} \neq \widehat{p}\widehat{x}. \tag{2.37}$$

Note whenever we say $\widehat{A} = \widehat{B}$, we mean that for all functions $f(x)$

$$\widehat{A}f(x) = \widehat{B}f(x).$$

But for brevity we write $\widehat{A} = \widehat{B}$. In order to see how badly two operators do not commute, we define a quantity called the "commutator":

$$[\widehat{x}, \widehat{p}] \equiv (\widehat{x}\widehat{p} - \widehat{p}\widehat{x}). \tag{2.38}$$

From Eq. (2.36), we get

$$[\widehat{x}, \widehat{p}]\psi(x,t) = i\hbar\psi(x,t)$$

or

$$[\widehat{x}, \widehat{p}] = i\hbar. \tag{2.39}$$

This is known as the quantum condition. It simply means that x and p cannot be measured precisely and simultaneously. In fact it can be shown that this quantum condition is completely equivalent to Heisenberg's uncertainty principle. Note that as $\hbar \to 0$, $\widehat{x}$ and $\widehat{p}$ commute as in classical mechanics and in this limit quantum mechanics reduces to classical mechanics.

3. The foregoing treatment has been for a one-dimensional case. It may be readily extended to three dimensions. Then

$$\mathbf{p} = \hbar \mathbf{k}; \quad E = \frac{\mathbf{p}^2}{2m} + V(\mathbf{r}). \tag{2.40}$$

The corresponding operators are

$$\mathbf{p} \longrightarrow -i\hbar\nabla \tag{2.41}$$

$$E \longrightarrow i\hbar\frac{\partial}{\partial t} \tag{2.42}$$

$$H \longrightarrow -\frac{\hbar^2}{2m}\nabla^2 + V(\mathbf{r}) \tag{2.43}$$

so that the Schrödinger equation

$$i\hbar\frac{\partial\psi(\mathbf{r},t)}{\partial t} = H\psi(\mathbf{r},t) \tag{2.44}$$

can now be written

$$i\hbar\frac{\partial\psi(\mathbf{r},t)}{\partial t} = \left(-\frac{\hbar^2}{2m}\nabla^2 + V(\mathbf{r})\right)\psi(\mathbf{r},t). \tag{2.45}$$

The probability density

$$P(\mathbf{r},t) = |\psi(\mathbf{r},t)|^2 \tag{2.46}$$

and probability current density is

$$\mathbf{S}(\mathbf{r},t) = \frac{\hbar}{2mi}[\psi^*\nabla\psi - \psi\nabla\psi^*] \tag{2.47}$$

and the equation of continuity is

$$\frac{\partial P(\mathbf{r},t)}{\partial t} + \nabla \cdot \mathbf{S}(\mathbf{r},t) = 0. \tag{2.48}$$

Now

$$\mathbf{p} \equiv (p_x, p_y, p_z) \equiv (p_1, p_2, p_3)$$
$$= p_i \quad i = 1,2,3, \tag{2.49}$$

$$\mathbf{r} \equiv (x, y, z) \equiv (x_1, x_2, x_3)$$
$$= x_i \quad i = 1,2,3. \tag{2.50}$$

In this case the quantum conditions are given by

$$[\widehat{x}_i, \widehat{p}_j] = i\hbar\delta_{ij}. \tag{2.51}$$

2.5   Stationary States

By stationary state of a system we mean one where energy is constant in time. Now we have the Schrödinger equation

$$i\hbar \frac{\partial \psi(x,t)}{\partial t} = H\psi(x,t). \tag{2.52}$$

We may look for a particular solution of the form

$$\psi(x,t) = u(x)f(t). \tag{2.53}$$

Substituting into Eq. (2.43), we have

$$i\hbar u(x)\frac{\partial}{\partial t}f(t) = Hu(x)f(t) \tag{2.54}$$

where H Hamiltonian is of the form

$$H\left(x, -i\hbar\frac{\partial}{\partial x}\right) = -\frac{\hbar^2}{2m}\frac{\partial^2}{\partial x^2} + V(x). \tag{2.55}$$

Note that it does not contain t, so that

$$Hu(x)f(t) = f(t)Hu(x). \tag{2.56}$$

Therefore dividing both sides of Eq. (2.53) by $u(x)f(t)$, we have

$$i\hbar \frac{\frac{\partial}{\partial t}f(t)}{f(t)} = \frac{H\left(x, -i\hbar\frac{\partial}{\partial x}\right)u(x)}{u(x)}. \tag{2.57}$$

Since left-hand side depends only on t, and right-hand side only on x, and these variables can be varied independently, each side must in fact be equal to a constant. We call this constant E. Thus,

$$\frac{i\hbar\frac{\partial}{\partial t}f(t)}{f(t)} = E \tag{2.58}$$

$$H\left(x, -i\hbar\frac{\partial}{\partial x}\right)u(x) = Eu(x). \tag{2.59}$$

From the Correspondence principle, it follows that the number E appearing in Eq. (2.58) may be interpreted as the energy.

Equation (2.58) is such that Hamiltonian operator H acting on u does not alter u apart from multiplying it by a number E. Such an equation is an example of what is called an eigenvalue equation. Here E is the eigenvalue of the Hamiltonian operator or simply energy eigenvalue, while $u(x)$ is called the eigenfunction belonging to the eigenvalue E and is often written as $u_E(x)$. Thus we write Eq. (2.58) as

$$H\left(x, -i\hbar\frac{\partial}{\partial x}\right)u_E(x) = Eu_E(x). \tag{2.60}$$

With the expression of H as given in Eq. (2.55), Eq. (2.60) becomes

$$\left(-\frac{\hbar^2}{2m}\frac{\partial^2}{\partial x^2} + V(x)\right)u_E(x) = Eu_E(x). \tag{2.61}$$

This is the Schrödinger equation for stationary states for which

$$\psi_E(x,t) = u_E(x)e^{-iEt/\hbar}, \tag{2.62}$$

since solution of Eq. (2.58) is

$$f(t) = e^{-iEt/\hbar}. \tag{2.63}$$

Solution of Eq. (2.60) or Eq. (2.61) will give us energy eigenvalues or energy levels and corresponding eigenfunctions. Note that for a stationary state

$$P(x,t) = |\psi_E(x,t)|^2$$
$$= |u_E(x)|^2, \tag{2.64}$$

i.e. the probability of finding the particle in the interval dx is constant in time for a stationary state.

We can develop certain continuity properties of the wave function $u(x)$, which will help to solve the Schrödinger equation (2.61). Let the potential energy $V(x)$ have only finite discontinuities. Since energy is finite and we want the wave function $u(x)$ to be finite (otherwise there would be trouble with the probability interpretation of the wave function), therefore, it follows from the Schrödinger equation (2.61) that $\frac{\partial^2 u}{\partial x^2}$ must be finite. Hence $\frac{\partial u}{\partial x}$ and u are differentiable. As a result of the two wave functions shown in Fig. 2.2, (1) is permissible while (2) is not.

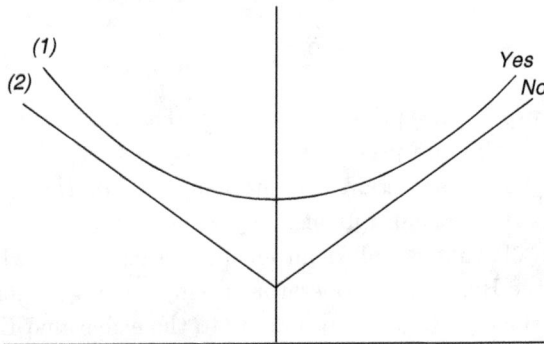

Fig. 2.2   Illustration of permissible and not permissible wave functions.

2.6 Eigenvalues and Eigenfunctions

An operator $\widehat{A}$ is a mathematical entity which, in general acts on any function of x and can be written as $\widehat{A}\left(x, \frac{\partial}{\partial x}\right)$. A simple example is the operation of differentiation, $\widehat{A}$ being any function of $\frac{\partial}{\partial x}$

$$\widehat{A} \equiv \widehat{A}\left(\frac{\partial}{\partial x}\right) \quad \text{e.g.} \quad \widehat{A} = \widehat{p} = -i\hbar\frac{\partial}{\partial x}. \tag{2.65}$$

Another example is

$$\widehat{A} \equiv \widehat{A}\left(x, \frac{\partial}{\partial x}\right) = \frac{\partial}{\partial x}x$$

which gives

$$\left(\frac{\partial\psi(x)}{\partial x}x\right)\psi(x) = \psi(x) + x\frac{\partial}{\partial x}$$
$$= \left(1 + x\frac{\partial}{\partial x}\right)\psi(x).$$

However, if to each operator $\widehat{A}\left(x, \frac{\partial}{\partial x}\right)$ there belongs a set of numbers a_n and a set of functions such that

$$A\left(x, \frac{\partial}{\partial x}\right)u_n(x) = a_n u_n(x) \tag{2.66}$$

then a_n is called an eigenvalue of $\widehat{A}$ and $u_n(x)$ is called an eigenfunction of $\widehat{A}$ belonging to the eigenvalue a_n. The eigenfunctions of an operator are thus those special functions which remain unaltered under the operation of the operator, apart from a multiplication by the eigenvlaue. Equation (2.66) is called the eigenvalue equation for the operator $\widehat{A}$.

Examples:

$$\widehat{A} = \frac{d^2}{dx^2}.$$

Eigenvalue equation becomes the differential equation

$$\frac{d^2}{dx^2}u(x) - au(x) = 0. \tag{2.67}$$

We want to find a and $u(x)$ so as to satisfy Eq. (2.67) subject to the conditions

$$u(0) = 0,$$
$$u(\pi) = 0 \quad \text{(Boundary conditions)}.$$

Solution of Eq. (2.67) is

$$u(x) = C_1 \sin(\sqrt{-a}\,x) + C_2 \cos(\sqrt{-a}\,x),$$
$$u(0) = 0 \Longrightarrow C_2 = 0,$$
$$u(\pi) = 0 \Longrightarrow 0 = C_1 \sin(\sqrt{-a}\,\pi).$$

This is only true if $\sqrt{-a} = n$, n being an integer:

$$a = -n^2.$$

Therefore the only eigenvalues of the operator are

$$a_n = -n^2$$

and eigenvalue equation becomes

$$\frac{d^2}{dx^2} u_n(x) = -n^2 u_n(x)$$

where eigenfunctions are

$$f_n(x) = C_1 \sin(nx).$$

Note that the eigenvalues depend very much on the boundary conditions imposed on the solution to the eigenvalue equation (2.66). It is not necessary that eigenvalues should always be discrete. They may be whole continuum.

2.7   Problems

2.1 If the excited level in an atom lasts for about 10^{-10} sec., what is the order of magnitude of electron energy spread measured in eV?

2.2 What is the approximate momentum imparted to a proton initially at rest by a measurement which locates its position within 10^{-11} meters?

2.3 A particle has an uncertainty in its position $\Delta x = 2a$. The magnitude of the momentum must be at least as large as the uncertainty in the momentum. Hence find an estimate of its energy.

2.4 Show that

$$u(x) = e^{-x^2/2} \quad \text{and} \quad u(x) = 2xe^{-\frac{1}{2}x^2}$$

are eigenfunctions of the operator

$$\hat{A}\left(x, \frac{\partial}{\partial x}\right) \equiv -\frac{d^2}{dx^2} + x^2.$$

Find the corresponding eigenvalues.

2.5 Show that for a wave function represented by a plane wave

$$\psi(\mathbf{r}) = e^{i\mathbf{p}\cdot\mathbf{r}/\hbar} \tag{2.68}$$

the probability current density **S** is given by

$$\mathbf{S}(\mathbf{r}, t) = \frac{\mathbf{p}}{m} = \mathbf{v},$$

where m is the mass of the particle.

2.6 In β-decay, electrons are emitted from the nuclei with energy of a few MeV. Take the energy to be 1 MeV and size of the nucleus 10^{-13} cm. Use the uncertainty principle to show that electrons cannot be contained in the nucleus before the decay.

Chapter 3

Basic Postulates of Quantum Mechanics

3.1 Basic Postulates of Quantum Mechanics

We have introduced two concepts: (i) the state of a particle or a quantum mechanical system is described by a wave function or state function $\psi(\mathbf{r}, t)$ which satisfies Schrödinger equation; (ii) the dynamical variables like momentum $\mathbf{p}$ and energy E are operators.

Now important properties of a physical system are quantities like $\mathbf{p}, E$ which can be measured or observed; such quantities are called observables. There must be a means of predicting the values of observables from the state function and the procedure for doing this is given by following set of postulates:

(1) To every observable there corresponds an operator $\widehat{A}$.

(2) The possible result of a measurement of an observable is one of the eigenvalues a_n of $\widehat{A}$ given by the equation

$$\widehat{A}u_n = a_n u_n \tag{3.1}$$

where $\widehat{A}$ is an operator and a_n is eigenvalue corresponding to eigenfunction u_n.

(3) A measurement of $\widehat{A}$ on a system in an eigenstate certainly leads to the result a_n, the eigenvalue.

(4) The average value of a large number of measurements of an observable on a system described by an arbitrary state ψ is given by

$$\langle \widehat{A} \rangle \equiv \bar{a}_\psi \equiv \bar{a} = \int \psi^* \widehat{A} \psi d\mathbf{r} \tag{3.2}$$

provided that $\int \psi^* \psi d\mathbf{r} = 1$, and there exist suitable boundary conditions. Thus for example, the average value of momentum in x-direction for a state

$\psi(x,t)$ is given by

$$\bar{p} = \int \psi^*(x)\left(-i\hbar\frac{\partial}{\partial x}\right)\psi(x)dx. \tag{3.3}$$

Applying to the special case of $\widehat{A} = \widehat{x}$, we get the average value of position

$$\bar{x} = \int \psi^*(x)\widehat{x}\psi(x)dx$$

$$= \int x\psi^*\psi dx = \int x|\psi(x)|^2 dx. \tag{3.4}$$

This is consistent with our interpretation of $\psi(x)$, that $|\psi(x)|^2$ determines the probability density $P(x)$ of the particle in space, since $|\psi(x)|^2$ appears as the weighing factor appropriate to x in the calculation of the average position.

It is convenient to introduce a compact notation for the matrix element $(\phi|\widehat{A}\psi)$ or $\langle\phi|\widehat{A}|\psi\rangle$ to mean $\int \phi^*\widehat{A}\psi d\mathbf{r}$. Furthermore, the integration may not always be over space and it may be necessary to imply integration over other continuous or discontinuous variables.

3.2 Formal Properties of Quantum Mechanical Operators

The quantum mechanical operators (observables) possess certain properties which are important and we discuss some of them briefly.

(a) They are linear, i.e. if (C_n are numbers)

$$\psi = \sum_n C_n\psi_n \tag{3.5}$$

then we have

$$\widehat{A}\psi = \sum_n C_n\widehat{A}\psi_n \tag{3.6}$$

(b) They obey the laws of association and distribution. Thus if $\widehat{A}, \widehat{B}$ and $\widehat{C}$ are three operators, we have

$$\widehat{A}(\widehat{B}\widehat{C}) = (\widehat{A}\widehat{B})\widehat{C}, \tag{3.7}$$

$$\widehat{A}(\widehat{B} + \widehat{C}) = \widehat{A}\widehat{B} + \widehat{A}\widehat{C}. \tag{3.8}$$

(c) An observable corresponds to a hermitian operator. We define the hermitian conjugate or adjoint $\widehat{A}^\dagger$ of an operator $\widehat{A}$ by the equation

$$(\psi|\widehat{A}\phi) = (\phi|\widehat{A}^\dagger\psi)^* = (\widehat{A}^\dagger\psi|\phi) \tag{3.9}$$

i.e.

$$\int \psi^* \widehat{A}\phi d\mathbf{r} = \int (\widehat{A}^\dagger \psi)^* \phi d\mathbf{r}.$$

An operator is said to hermitian if $\widehat{A}^\dagger = \widehat{A}$, i.e.

$$(\psi|\widehat{A}\phi) = (\phi|\widehat{A}\psi)^* = (\widehat{A}\psi|\phi). \tag{3.10}$$

Theorems:

(i) The eigenvalues of a hermitian operator are real.

(ii) The eigenfunctions of a hermitian operator corresponding to different eigenvalues are orthogonal:

We have eigenvalue equation

$$\widehat{A}u_n = a_n u_n \tag{3.11}$$

and then

$$\langle u_m|\widehat{A}u_n\rangle^* = a_n^*\langle u_m|u_n\rangle^* = a_n^*\langle u_n|u_m\rangle. \tag{3.12}$$

But from Eq (3.9):

$$\langle u_m|\widehat{A}u_n\rangle^* = \langle u_m|\widehat{A}^\dagger u_n\rangle = \langle u_n|\widehat{A}u_m\rangle$$

$$= a_m\langle u_n|u_m\rangle. \tag{3.13}$$

Hence from Eqs (3.12) and (3.13):

$$(a_m - a_n^*)\langle u_n|u_m\rangle = 0.$$

Therefore if

$$\text{i) } m = n, \quad a_n = a_n^*, \quad \text{since} \quad \langle u_n|u_n\rangle \neq 0 \tag{3.14}$$

$$\text{ii) } m \neq n, \quad \langle u_n|u_m\rangle = \int u_n^* u_m d\vec{r} = 0. \tag{3.15}$$

If the eigenfunctions are normalized,

$$\langle u_n|u_n\rangle = \int u_n^* u_n d\vec{r} = 1. \tag{3.16}$$

Hence Eqs (3.15) and (3.16) can be written in a compact form :

$$\langle u_n|u_m\rangle = \int u_n^* u_m d\vec{r} = \delta_{mn} \tag{3.17}$$

where

$$\begin{cases} \delta_{mn} = 0 & m \neq n, \\ \delta_{mn} = 1 & m = n. \end{cases} \tag{3.18}$$

δ_{mn} is called the Kronecker delta. The eigenfunctions are then said to be orthonormal.

(d) The eigenfunctions u_n of a hermitian operator (observable) form a complete orthonormal set so that any arbitrary state function ψ can be expanded in terms of them, i.e.

$$\psi = \sum_n C_n u_n .\qquad(3.19)$$

This is what we mean by a complete set. The Eq. (3.19) is called to the superposition principle, a basic ingredient of quantum mechanics.

Now in analogy with vector analysis where we express a vector in terms of basis vectors $\mathbf{i}, \mathbf{j}, \mathbf{k}$, the u_n's are called basis vectors and C_n the corresponding coordinates. It follows that

$$\int u_m^* \psi d\mathbf{r} = (u_m|\psi)$$
$$= \sum_n C_n (u_m|u_n)$$
$$= \sum_n C_n \delta_{mn}$$
$$= C_m ,\qquad(3.20)$$

where C_m is related to the probability of finding the system described by state ψ in an eigenstate u_m. Thus

$$|C_m|^2 = |\int u_m^* \psi d\mathbf{r}|^2\qquad(3.21)$$

gives the probability of operator $\hat{A}$ having the eigenvalue a_m, when the system is described by a state ψ. To see this we note that the average value of the operator $\hat{A}$ in state ψ is given

$$\bar{a} = \int \psi^* \hat{A} \psi d\mathbf{r}$$
$$= \sum_m \sum_n C_m^* C_n \int u_m^* \hat{A} u_n d\mathbf{r}$$
$$= \sum_m \sum_n C_m^* C_n a_n \int u_m^* u_n d\mathbf{r}$$
$$= \sum_m \sum_n C_m^* C_n a_n \delta_{mn}$$
$$= \sum_m |C_m|^2 a_m ,\qquad(3.22)$$

The weighing factor $|C_m|^2$ above gives the probability of finding the Eigenvalue a_m.

(e) The operators do not necessarily obey a commutative law, i.e. two operators $\widehat{A}$ and $\widehat{B}$ need not give $\widehat{A}\widehat{B} = \widehat{B}\widehat{A}$. If $\widehat{A}\widehat{B}$ equals $\widehat{B}\widehat{A}$, the operators are said to commute, i.e.

$$[\widehat{A}, \widehat{B}] = \widehat{A}\widehat{B} - \widehat{B}\widehat{A} = 0 \tag{3.23}$$

$[\widehat{A}, \widehat{B}]$ is called the "commutator" of two operators.

(f) If two observables commute, then it is possible to find a set of functions which are simultaneously eigenfunctions of $\widehat{A}$ and $\widehat{B}$. If they do not commute, i.e. $[\widehat{A}, \widehat{B}] \neq 0$, this cannot be done except for a state ψ which has $[\widehat{A}, \widehat{B}]\psi = 0$.

We now show that if u_n is a simultaneous eigenfunction of $\widehat{A}$ and $\widehat{B}$ corresponding to eigenvalues a_n and b_n, then

$$[\widehat{A}, \widehat{B}]u_n = 0 . \tag{3.24}$$

Now

$$\widehat{A}u_n = a_n u_n , \tag{3.25a}$$
$$\widehat{B}u_n = b_n u_n . \tag{3.25b}$$

Further

$$\widehat{A}\widehat{B}u_n = \widehat{A}b_n u_n$$
$$= b_n \widehat{A}u_n$$
$$= b_n a_n u_n , \tag{3.26a}$$

$$\widehat{B}\widehat{A}u_n = \widehat{B}a_n u_n$$
$$= a_n \widehat{B}u_n$$
$$= a_n b_n u_n . \tag{3.26b}$$

Therefore

$$(\widehat{A}\widehat{B} - \widehat{B}\widehat{A})u_n = (b_n a_n - a_n b_n)u_n = 0 ,$$

i.e.

$$[\widehat{A}, \widehat{B}]u_n = 0 . \tag{3.27}$$

The implication of this result is as follows:

Since u_n form a complete set so that an arbitrary function ψ can be expanded in terms of them, it follows that

$$[\widehat{A}, \widehat{B}]\psi = 0 .$$

Since ψ is arbitrary, $[\widehat{A}, \widehat{B}] = 0$. Thus if a set of simultaneous eigenfunctions of two observables $\widehat{A}$ and $\widehat{B}$ exist, then $\widehat{A}$ and $\widehat{B}$ commute.

If $[\widehat{A}, \widehat{B}] = 0$ and $[\widehat{A}, \widehat{C}] = 0$ but $[\widehat{B}, \widehat{C}] \neq 0$, then it is not possible to find functions which are simultaneously eigenfunctions of $\widehat{A}, \widehat{B}$ and $\widehat{C}$. It is only possible to find eigenfunctions for $\widehat{A}$ and $\widehat{B}$ or for $\widehat{A}$ and $\widehat{C}$. Corresponding to this, we can only make simultaneous measurements of the observables corresponding to the pair of operators $\widehat{A}$ and $\widehat{B}$ or to the pair $\widehat{A}$ and $\widehat{C}$, it is not possible to measure $\widehat{B}, \widehat{C}$ together.

3.3   Continuous Spectrum and Dirac Delta Functions

(a) Continuous Spectrum

So far we have considered the case when the eigenvalues of an operator $\widehat{A}$ are discrete. In case eigenvalues of an operator take on continuous values, the sum in the completeness relation (3.19) takes the form of an integral

$$\psi(x) = \int C(a)u_a(x)da \ , \tag{3.28}$$

where the label a corresponds to continuous set of eigenvalues and replaces the discrete label n in Eq. (3.19). For simplicity we first consider ψ to be a function of a single variable x only; the generalisation to three dimensions is straightforward.

Now using (3.28)

$$\int u_{a'}^*(x)\psi(x)dx \equiv (u_{a'}|\psi)$$

$$= \int da C(a)(u_{a'}|u_a) \ , \tag{3.29}$$

where a' lies in the domain of integration of a. Now the orthogonality condition (3.18) becomes

$$\int u_{a'}^*(x)u_a(x)dx \equiv (u_{a'}|u_a)$$

$$= 0 \quad \text{when } a \neq a' \ . \tag{3.30}$$

For $a = a'$, it does not have to vanish. In fact it must be infinitely large at $a = a'$ because if it is finite at $a = a'$, then the integral on right-hand side of Eq. (3.29) vanishes but $\psi(x)$ does not vanish in general. But this infinity must be such that

$$\int u_{a'}^*(x)\psi(x)dx = C(a') \int da(u_{a'}|u_a) \ , \tag{3.31}$$

so that in analogy with Eq. (3.20), we have

$$C(a') = \int u_{a'}^*(x)\psi(x)dx \tag{3.32}$$

with

$$\int da(u_{a'}|u_a) = 1 . \tag{3.33}$$

Thus $(u_{a'}|u_a)$ has the peculiar property: it is zero everywhere except at $a = a'$ [see Fig. 3.1] and at $a = a'$ it is infinitely large such that its integral is 1.

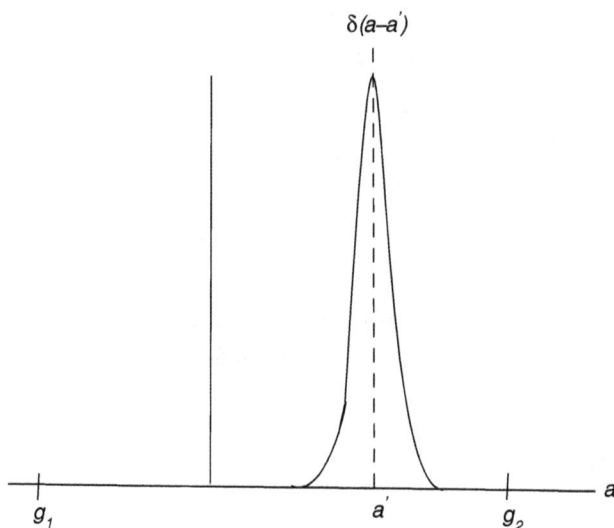

Fig. 3.1   The Dirac delta function.

Such a function is called Dirac δ-function and is written as

$$\delta(a - a') = 0 \qquad \text{for} \quad a \neq a'$$
$$= \infty \qquad \text{for} \quad a = a' \tag{3.34a}$$

such that

$$\int da\delta(a - a') = 1 . \tag{3.34b}$$

Thus we write

$$(u_{a'}|u_a) = \int u_{a'}^*(x)u_a(x)dx$$
$$= \delta(a - a') , \tag{3.35}$$

corresponding to orthonormality relation (3.17) for discrete set. Substituting Eq. (3.38) into Eq. (3.29), we have

$$\int u_{a'}^*(x)\psi(x)dx = \int daC(a)\delta(a - a') .$$

Comparing it with Eq. (3.35), we have

$$\int daC(a)\delta(a - a') = C(a') . \tag{3.36}$$

This is the fundamental property of the δ-function which we require.

The generalisation to three dimensions is obvious

$$x \to \mathbf{r}, \ a \to \mathbf{a}, \ a' \to \mathbf{a}' .$$

Thus Eqs. (3.28), (3.32), (3.34), (3.35) and (3.36) respectively become

$$\psi(\mathbf{r}) = \int C(\mathbf{a})u_{\mathbf{a}}(\mathbf{r})d\mathbf{a} \tag{3.37}$$

$$C(\mathbf{a}') = \int u_{\mathbf{a}'}^*(\mathbf{r})\psi(\mathbf{r})d\mathbf{r} \tag{3.38}$$

$$\begin{cases} \delta(\mathbf{a} - \mathbf{a}') = 0 & \text{for } \mathbf{a} \neq \mathbf{a}' , \\ = \infty & \text{for } \mathbf{a} = \mathbf{a}'. \end{cases} \tag{3.39a}$$

$$\int d\mathbf{a}\delta(\mathbf{a} - \mathbf{a}') = 1 , \tag{3.39b}$$

$$(u_{\mathbf{a}'}|u_{\mathbf{a}}) = \delta(\mathbf{a} - \mathbf{a}') , \tag{3.40}$$

$$\int C(\mathbf{a})\delta(\mathbf{a} - \mathbf{a}')d\mathbf{a} = C(\mathbf{a}') . \tag{3.41}$$

Here

$$\delta(\mathbf{a} - \mathbf{a}') = \delta(a_x - a_x')\delta(a_y - a_y')\delta(a_z - a_z') . \tag{3.42}$$

(b) Closure Relations We have

$$\psi(\mathbf{r}) = \sum_n C_n u_n(\mathbf{r}) \quad \text{(for discrete set)} \tag{3.43a}$$

$$= \int C(\mathbf{a})u_{\mathbf{a}}(\mathbf{r})d\mathbf{a} \quad \text{(for continuous set)} . \tag{3.43b}$$

Then

$$C_n = \int u_n^*(\mathbf{r}')\psi(\mathbf{r}')d\mathbf{r}' , \tag{3.44a}$$

$$C(\mathbf{a}) = \int u_{\mathbf{a}}^*(\mathbf{r}')\psi(\mathbf{r}')d\mathbf{r}' . \tag{3.44b}$$

Substituting in Eq. (3.43a), we have for discrete set

$$\psi(\mathbf{r}) = \sum_n \left(\int u_n^*(\mathbf{r}')\psi(\mathbf{r}')dr' \right) u_n(\mathbf{r})$$
$$= \int \left(\sum_n u_n(\mathbf{r})u_n^*(\mathbf{r}') \right) \psi(\mathbf{r}')dr' \ .$$

Thus it follows that

$$\sum_n u_n(\mathbf{r})u_n^*(\mathbf{r}') = \delta(\mathbf{r} - \mathbf{r}') \ . \tag{3.45a}$$

The corresponding relation for continuous set is

$$\int u_\mathbf{a}(\mathbf{r})u_\mathbf{a}^*(\mathbf{r}')da = \delta(\mathbf{r} - \mathbf{r}') \ . \tag{3.45b}$$

These are known as Closure Relations. These are equivalent to completeness relations, since one can write (e.g. discrete set)

$$\psi(\mathbf{r}) = \int \psi(\mathbf{r}')\delta(\mathbf{r} - \mathbf{r}')dr$$
$$= \int \psi(\mathbf{r}') \sum_n u_n(\mathbf{r})u_n^*(\mathbf{r}')dr$$
$$= \sum_n \left(\int u_n^*(\mathbf{r}')\psi(\mathbf{r}')dr' \right) u_n(\mathbf{r})$$
$$= \sum_n C_n u_n(\mathbf{r}) \ . \tag{3.46}$$

(c) A Simple Representation of δ-function Consider the function

$$\int e^{ia(x-x')}da \ .$$

Now $e^{ia(x-x')}$ is an oscillating function and the above integral is not defined. It is a question of agreeing to give a value to this integral. The prescription is

$$\int_{-\infty}^{\infty} e^{ia(x-x')}da = \lim_{\varepsilon \to 0} \left(\int_{-\infty}^{0} e^{ia(x-x')+\varepsilon a}da + \int_{0}^{\infty} e^{ia(x-x')-\varepsilon a}da \right)$$
$$= \lim_{\varepsilon \to 0} \left(\frac{1}{i(x-x')+\varepsilon} - \frac{1}{i(x-x')-\varepsilon} \right)$$
$$= \lim_{\varepsilon \to 0} \left(\frac{2\varepsilon}{\varepsilon^2 + (x-x')^2} \right) \ .$$

If $x \neq x'$, $(x - x')^2$ is a fixed number, no matter how small it may be. Now when $\varepsilon \to 0, \varepsilon^2$ can be neglected in comparison with $(x - x')^2$ so that the limit $= 0$. If $x = x'$

$$\lim_{\varepsilon \to 0} \left(\frac{2\varepsilon}{\varepsilon^2 + (x - x')^2} \right) = \lim_{\varepsilon \to 0} \frac{2}{\varepsilon} = \infty .$$

The behaviour is that of a δ-function, but we have to verify that its integral is 1. Thus we calculate

$$\int_{-\infty}^{\infty} \left(\int_{-\infty}^{\infty} e^{ia(x-x')} da \right) dx = \lim_{\varepsilon \to 0} \int_{-\infty}^{\infty} \frac{2\varepsilon}{\varepsilon^2 + (x - x')^2} dx$$

$$= \lim_{\varepsilon \to 0} \int_{-\infty}^{\infty} \frac{2\varepsilon}{\varepsilon^2 + \eta^2} d\eta, \quad \eta = (x - x')$$

$$= \lim_{\varepsilon \to 0} \left[2 \tan^{-1} \frac{\eta}{\varepsilon} \right]_{-\infty}^{\infty}$$

$$= 2 \lim_{\varepsilon \to 0} \left(\frac{\pi}{2} - \left(-\frac{\pi}{2} \right) \right)$$

$$= 2\pi .$$

Thus

$$\frac{1}{2\pi} \int_{-\infty}^{\infty} e^{ia(x-x')} da = \delta(x - x') . \tag{3.47a}$$

Its generalisation to three dimension is

$$\frac{1}{(2\pi)^3} \int e^{i\mathbf{a} \cdot (\mathbf{r} - \mathbf{r}')} da = \delta(\mathbf{r} - \mathbf{r}') . \tag{3.47b}$$

(d) Properties of δ-function

$$f(x)\delta(x - b) = f(b)\delta(x - b) ,$$

$$x\delta(x) = 0 ,$$

$$\delta(-x) = \delta(x) ,$$

$$\delta(bx) = \frac{1}{|b|} \delta(x) ,$$

$$\delta(x^2 - b^2) = \frac{1}{2b} (\delta(x - b) + \delta(x + b)) , \quad b > 0$$

$$\int \delta(a - x) dx \delta(x - b) = \delta(a - b) .$$

These equations have meaning only in the sense of integration; for example, the first one means

$$\int f(x)\delta(x - b) dx = f(b) .$$

$\delta(bx) = \frac{1}{|b|}\delta(x)$ means

$$\int \delta(bx)dx = \frac{1}{|b|} .$$

(e) Fourier Transform

The completeness relation, in terms of eigenfunctions $\frac{1}{\sqrt{2\pi}}e^{iax}$ becomes

$$f(x) = \frac{1}{\sqrt{2\pi}} \int_{-\infty}^{\infty} C(a)e^{iax}da , \qquad (3.48)$$

$$\frac{1}{\sqrt{2\pi}} \int_{-\infty}^{\infty} f(x)e^{-ia'x}dx = \frac{1}{(2\pi)} \int_{-\infty}^{\infty} daC(a) \int_{-\infty}^{\infty} e^{ix(a-a')}dx$$

$$= \int daC(a)\delta(a-a')$$

$$= C(a') .$$

Therefore

$$C(a) = \frac{1}{\sqrt{2\pi}} \int_{-\infty}^{\infty} f(x)e^{-iax}dx . \qquad (3.49)$$

$C(a)$ and $f(x)$ are called the Fourier transforms of each other.
The generalisation to three dimensions is

$$f(\mathbf{r}) = \frac{1}{(2\pi)^{3/2}} \int C(\mathbf{a})e^{i\mathbf{a}\cdot\mathbf{r}}da , \qquad (3.50a)$$

$$C(\mathbf{a}) = \frac{1}{(2\pi)^{3/2}} \int f(\mathbf{r})e^{-i\mathbf{a}\cdot\mathbf{r}}dr . \qquad (3.50b)$$

(f) Momentum Eigenfunctions (An Example of Continuous Spectrum of Eigenvalues)

The momentum operator in Schrödinger representation is

$$\hat{p} = -i\hbar\frac{\partial}{\partial x} .$$

Eigenvalue equation is

$$\hat{p}u_p(x) = pu_p(x) ,$$

$$-i\hbar\frac{\partial}{\partial x}u_p(x) = pu_p(x) . \qquad (3.51)$$

A solution of this equation is

$$u_p(x) = B\exp\left(\frac{i}{\hbar}px\right) . \qquad (3.52)$$

Here eigenvalue p is a continuous variable and takes on any value. Above we have taken the momentum in x direction. For three dimensions

$$\hat{\mathbf{p}} = -i\hbar\boldsymbol{\nabla}$$

and

$$\hat{\mathbf{p}}u_{\mathbf{p}}(\mathbf{r}) = \mathbf{p}u_{\mathbf{p}}(\mathbf{r}) ,$$
$$u_{\mathbf{p}}(\mathbf{r}) = B\exp\left(\frac{i}{\hbar}\mathbf{p}\cdot\mathbf{r}\right) . \tag{3.53}$$

Now

$$(u_{p'}|u_{\mathbf{p}}) = \int u_{p'}^{*}(\mathbf{r})u_{p}(\mathbf{r})d\mathbf{r}$$
$$= |B|^2\int e^{(i/\hbar)(\mathbf{p}-\mathbf{p}')\cdot\mathbf{r}}d\mathbf{r}$$
$$= |B|^2\hbar^3(2\pi)^3\delta(\mathbf{p}-\mathbf{p}') .$$

Thus if we select

$$B = \frac{1}{(2\pi\hbar)^{3/2}} ,$$

then

$$(u_{p'}|u_{\mathbf{p}}) = \delta(\mathbf{p}-\mathbf{p}') . \tag{3.54}$$

Thus normalised momentum eigenfunctions are

$$u_{\mathbf{p}}(\mathbf{r}) = \frac{1}{(2\pi\hbar)^{3/2}}e^{(i/\hbar)\mathbf{p}\cdot\mathbf{r}} . \tag{3.55}$$

Since these eigenfunctions form a complete set we can write

$$\psi(\mathbf{r}) = \frac{1}{(2\pi\hbar)^{3/2}}\int C(\mathbf{p})e^{(i/\hbar)\mathbf{p}\cdot\mathbf{r}}d\mathbf{p} , \tag{3.56a}$$

where

$$C(\mathbf{p}) = \frac{1}{(2\pi\hbar)^{3/2}}\int e^{-(i/\hbar)\mathbf{p}\cdot\mathbf{r}}\psi(\mathbf{r})d\mathbf{r} \tag{3.56b}$$

$\psi(\mathbf{r})$ and $C(\mathbf{p})$ are the Fourier transform of each other.

Now average value of the momentum operator is given by

$$\bar{\mathbf{p}} = \langle \mathbf{p} \rangle = \int \psi^*(r)\widehat{\mathbf{p}}\psi(\mathbf{r})d\mathbf{r}$$

$$= \frac{1}{(2\pi\hbar)^3} \iiint d\mathbf{p}' d\mathbf{p} C^*(\mathbf{p}')C(\mathbf{p})e^{-(i/\hbar)\mathbf{p}'\cdot\mathbf{r}}$$

$$\times(-i\hbar\boldsymbol{\nabla})e^{(i/\hbar)\mathbf{p}\cdot\mathbf{r}}d\mathbf{r}$$

$$= \frac{1}{(2\pi\hbar)^3} \iiint d\mathbf{p}' d\mathbf{p} C^*(\mathbf{p}')C(\mathbf{p})$$

$$\times \mathbf{p}e^{(i/\hbar)(\mathbf{p}-\mathbf{p}')\cdot\mathbf{r}}d\mathbf{r}$$

$$= \iint d\mathbf{p}' d\mathbf{p} C^*(\mathbf{p}')C(\mathbf{p})\mathbf{p}\delta(\mathbf{p} - \mathbf{p}')$$

$$= \int |C(\mathbf{p})|^2 \mathbf{p} d\mathbf{p} . \tag{3.57}$$

Thus $|C(\mathbf{p})|^2$ is the probability of momentum operator $\widehat{\mathbf{p}}$ having eigenvalue $\mathbf{p}$ when the system is in state $\psi(\mathbf{r})$. In other words, in a measurement of the momentum of a particle, the probability of finding the result $\mathbf{p}$ is $|C(\mathbf{p})|^2$. Thus $C(\mathbf{p})$ may be regarded as the wave function, in momentum space just as $\psi(\mathbf{r})$ is the wave function in $\mathbf{r}$ space. $C(\mathbf{p})$ is sometimes written as $\phi(\mathbf{p})$.

3.4 Uncertainty Principle and Non-Commutativity of Observables

In quantum mechanics, two observables $\widehat{A}$ and $\widehat{B}$ do not necessarily commute and obey a commutative law, i.e. in general

$$[\widehat{A}, \widehat{B}] \neq 0 . \tag{3.58}$$

This statement is essentially equivalent to the uncertainty principle which expresses the limitations on our knowledge imposed by mutual disturbances of observations. Eq. (3.58) implies that it is not possible to find simultaneous eigenfunctions of $\widehat{A}$ and $\widehat{B}$ i.e. we cannot have an exact knowledge of the result of measurement of $\widehat{A}$ and $\widehat{B}$ simultaneously.

We now show explicitly that Eq. (3.58) leads to the uncertainty principle. $\widehat{A}$ and $\widehat{B}$ being observables are hermitian. Let $\bar{a}$ and $\bar{b}$ denote the average values of large number of measurements of $\widehat{A}$ and $\widehat{B}$ respectively.

Define

$$\widehat{A} - \bar{a} = \delta\widehat{A} \equiv \alpha ,$$
$$\widehat{B} - \bar{b} = \delta\widehat{B} \equiv \beta . \tag{3.59}$$

It is clear from Eq. (3.59) that α and β are hermitian operators so that we have

$$\alpha^\dagger \alpha = \alpha^2 , \quad \beta^\dagger \beta = \beta^2 .$$

The mean square deviations of the measured values, of $\widehat{A}$ and $\widehat{B}$ about the mean are given by

$$(\overline{\Delta a})^2 = \langle (\widehat{A} - \overline{a})^2 \rangle = \langle \alpha^2 \rangle ,$$
$$(\overline{\Delta b})^2 = \langle (\widehat{B} - \overline{b})^2 \rangle = \langle \beta^2 \rangle . \qquad (3.60)$$

Now

$$(\overline{\Delta a})^2 \equiv \langle \alpha^2 \rangle = \int \psi^*(x) \alpha \alpha \psi(x) dx$$

$$= \int (\alpha^\dagger \psi)^* \alpha \psi dx$$

$$= \int (\alpha \psi)^* \alpha \psi dx$$

$$= \int |\alpha \psi|^2 dx .$$

Then

$$\langle \alpha^2 \rangle \langle \beta^2 \rangle = \left(\int |\alpha \psi|^2 dx \right) \left(\int |\beta \psi|^2 dx \right)$$

$$\geq \left| \int (\alpha \psi)^* \beta \psi dx \right|^2 . \qquad (3.61)$$

This follows from the Schwartz inequality

$$\left(\int |f|^2 dx \right) \left(\int |g|^2 dx \right) \geq \left| \int f^* g dx \right|^2 .$$

Hence

$$(\overline{\Delta a})^2 (\overline{\Delta b})^2 \geq \left| \int (\alpha \psi)^* \beta \psi dx \right|^2$$

$$= \left| \int (\alpha^\dagger \psi)^* \beta \psi dx \right|^2$$

$$= \left| \int \psi^* \alpha \beta \psi dx \right|^2$$

$$= \left| \int \psi^* \frac{\alpha\beta + \beta\alpha}{2} \psi dx + \int \psi^* \frac{\alpha\beta - \beta\alpha}{2} \psi dx \right|^2$$

$$= \left| P - iQ \right|^2 , \qquad (3.62)$$

where

$$P = \int \psi^* \frac{\alpha\beta + \beta\alpha}{2} \psi dx \ ,$$

$$Q = \int \psi^* i \frac{\alpha\beta - \beta\alpha}{2} \psi dx \ .$$

Since $\frac{\alpha\beta+\beta\alpha}{2}$ and $i\frac{\alpha\beta-\beta\alpha}{2}$ are hermitian operators and the average value of a large number of measurements of a hermitian operator is real, therefore P and Q are real numbers.

Thus Eq. (3.62) gives

$$(\overline{\Delta a})^2 (\overline{\Delta b})^2 \geq P^2 + Q^2 \geq Q^2$$

$$= \left| \int \psi^* i \frac{\alpha\beta - \beta\alpha}{2} \psi dx \right|^2$$

$$= \left| \int \psi^* i \frac{\hat{A}\hat{B} - \hat{B}\hat{A}}{2} \psi dx \right|^2 \ , \qquad (3.63)$$

since from Eq. (3.59) $\alpha\beta - \beta\alpha = \hat{A}\hat{B} - \hat{B}\hat{A}$.

In particular if $\hat{A} = \hat{p}$, $\hat{B} = \hat{x}$, $[\hat{p}, \hat{x}] = -i\hbar$

$$(\overline{\Delta a})^2 = (\overline{\Delta p})^2 \ ,$$

$$(\overline{\Delta a})^2 = (\overline{\Delta x})^2 \ ,$$

$$(\overline{\Delta p})^2 (\overline{\Delta x})^2 \geq \frac{\hbar^2}{4} |\int \psi^* \psi dx|^2 = \frac{\hbar^2}{4} \ .$$

The root mean square deviation is often called the uncertainty (standard deviation), i.e.

$$\Delta p = \sqrt{(\overline{\Delta p})^2} \ , \quad \Delta x = \sqrt{(\overline{\Delta x})^2} \ . \qquad (3.64)$$

Hence

$$\Delta p \Delta x \geq \hbar/2 \ , \qquad (3.65)$$

where Δp and Δx denote the uncertainties in the measured values of p and x. If

$$[\hat{A}, \hat{B}] = 0$$

there is no mutual disturbance, and the result of simultaneous measurements of observables $\hat{A}$ and $\hat{B}$ can be known exactly.

Example

Momentum and energy operators for a free particle are given by

$$\hat{A} = \hat{p} \rightarrow -i\hbar \frac{\partial}{\partial x}$$

$$\hat{B} = H = \frac{\hat{p}^2}{2m} \rightarrow -\frac{\hbar^2}{2m}\frac{\partial^2}{\partial x^2} \qquad (3.66)$$

so that

$$[\hat{p}, H]\psi(x) = \left[-i\hbar\frac{\partial}{\partial x} - \frac{\hbar^2}{2m}\frac{\partial^2}{\partial x^2} \right]\psi(x)$$
$$= 0$$

which is true for an arbitrary function $\psi(x)$. Thus

$$[\hat{p}, H] = 0 \ .$$

The energy and momentum of a free particle can be known exactly, simultaneously. In other words, it is possible to find a wave function which is a simultaneous eigenfunction of both momentum and energy.

3.5   Problems

3.1 If $\langle \hat{A} \rangle$ denotes the average value for a large number of measurements of an operator $\hat{A}$ for an arbitrary state function ψ, show that $\langle A \rangle$ is real if $\hat{A}$ is hermitian.

3.2 The state function for a free particle moving in x-direction is given by

$$\psi(x) = Ne^{(-x^2/2\delta^2 + ip_0x/\hbar)} \ .$$

Normalise this wave function. Find the state function $\phi(p)$ in momentum space.

(i) Show that for the state $\psi(x)$ given above

$$\langle x \rangle = 0 \ ,$$
$$\langle p \rangle = p_0 \ ,$$
$$\langle x^2 \rangle = \frac{\delta^2}{2} \ ,$$
$$\langle p^2 \rangle = p_0^2 + \frac{1}{2}\frac{\hbar^2}{\delta^2}.$$

Hence show that

$$(\overline{\Delta x})^2 = \langle (x - \langle x \rangle)^2 \rangle$$
$$= \frac{\delta^2}{2}$$

$$(\overline{\Delta p})^2 = \langle (p - \langle p \rangle)^2 \rangle$$
$$= \frac{1}{2}\frac{\hbar^2}{\delta^2}$$

so that

$$\Delta x \Delta p = \frac{1}{2}\hbar.$$

(ii) Using the relations

$$\langle p \rangle = \int p |\phi(p)|^2 dp$$

$$\langle p^2 \rangle = \int p^2 |\phi(p)|^2 dp$$

show that

$$\langle p \rangle = p_0 \ ,$$
$$\langle p^2 \rangle = p_0^2 + \frac{1}{2}\frac{\hbar^2}{\delta^2}.$$

$$\left(\text{Useful integral} \int_{-\infty}^{+\infty} x^{2n} e^{-\alpha x^2} dx = \frac{1}{\alpha^{n+\frac{1}{2}}} \frac{\sqrt{\pi}(2n)!}{2^{2n} n!} \right)$$

3.3 Show that for a particle of mass m, moving in a potential $V(\mathbf{r})$,

$$[H, \hat{\mathbf{r}}] = -i\hbar \frac{\hat{\mathbf{p}}}{m} \ ,$$
$$[\hat{\mathbf{p}}, H] = [\hat{\mathbf{p}}, V(\hat{\mathbf{r}})] = -i\hbar \nabla V \ .$$

Using the above results and the fact that H is hermitian, show that

$$m \frac{d}{dt} \langle \mathbf{r} \rangle = \langle \mathbf{p} \rangle \ ,$$

$$\frac{d}{dt} \langle \mathbf{p} \rangle = -\langle \nabla V \rangle \ .$$

(Note that Newton's law is valid for expectation values.)

3.4 Using the result

$$[H, x] = -i\hbar \frac{\hat{p}}{m} \ ,$$

for a particle of mass m moving in x-direction in a potential $V(x)$, show that the average value of its momentum in a stationary state with discrete energy is zero.

3.5 A particle is in a state

$$\psi(x) = \frac{1}{\sqrt{a}} \sin\left(\frac{5\pi x}{a}\right), \quad |x| \le a$$
$$= 0 \ \text{elsewhere} \, ,$$

show that the probability for the particle to be found with momentum p is given by

$$\frac{100\pi a}{2\hbar} \frac{\sin^2(pa/\hbar)}{(p^2 a^2/\hbar^2 - 25\pi^2)} \, .$$

3.6 A particle of mass m is confined by an infinite square well potential $V(x) = 0, |x| \le a; V(x) = \infty, |x| > a$. If the particle is in the state

$$\psi(x) = x, \quad |x| \le a \, ,$$
$$\psi(x) = 0, \quad |x| > a \, ,$$

find the probability that a measurement of energy will give the result

$$E_n = \frac{\hbar^2}{2m}\left(\frac{\pi^2}{4a^2}\right) n^2 \, .$$

3.7 For a free particle, find a wave funcion which is a simultaneous eigenfunction of both momentum and energy. This is not so for a particle moving in a potential since then $|\hat{p}, H| \ne 0$ [cf. Problem 3.3].

3.8 a) If a particle is in a state

$$\psi(x) = \left(\frac{2}{\pi\delta^2}\right)^{\frac{1}{4}} e^{-\frac{x^2}{\delta^2}}$$

find the probability of finding it in the momentum eigenstate

$$u_p(x) = \sqrt{\frac{1}{2\pi\hbar}} e^{-\frac{i}{\hbar}px} \, .$$

b) If it is in a state

$$u_p(x) = \sqrt{\frac{2}{\pi}} \sin nx, \qquad 0 \le x \le \pi$$

show that the probability of finding it with momentum p is given by

$$|\phi(p)|^2 = \frac{1}{\pi^2\hbar} \frac{\sin^4\left(\left(\frac{p}{\hbar} - n\right)\frac{\pi}{2}\right)}{\left(\frac{\frac{p}{\hbar}-n}{2}\right)^2} \, .$$

Chapter 4

Solution of Problems in Quantum Mechanics

We are now in a position to undertake the solution of physical problems in quantum mechanics. A typical problem in quantum mechanics is to find the allowed energy levels for a given system, which involves the following three steps: (i) What is the potential (physics)?, (ii) Write the Schrödinger equation and solve it (mathematics), (iii) Results and their interpretation (physics).

First we shall consider a potential which is constant, i.e. $V(x) = V$ (constant). It may be that V is a different constant in different regions of space.

4.1 One Dimensional Potential Step

The potential step is represented by the function

$$V(x) = V \quad \text{for} \quad x > 0,$$
$$V(x) = 0 \quad \text{for} \quad x < 0. \tag{4.1}$$

Let E denotes the total energy of the particle. Then

$$E = T(x) + V(x) = \text{constant}. \tag{4.2}$$

First we discuss the motion classically. As shown in Fig. 4.1, there are two cases: (i) $E - V(x) > 0$. This means that a particle with kinetic energy $T = E = p^2/2m$ approaches the potential barrier from the left. As the particle moves across the barrier, it enters the region of potential $V(x) = V$. The kinetic energy T_1 of the particle in this region is given by

$$T_1 = E - V = \frac{p_1^2}{2m}.$$

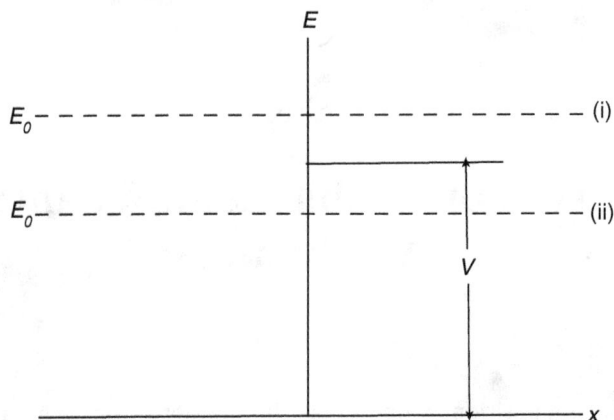

Fig. 4.1   The energy diagram of a potential step

The particle will, therefore, move with reduced momentum. Further, since the particles have sufficient energy to go over the barrier, there is total transmission. (ii) $E - V(x) < 0$, for $x > 0$. In this case the particles hitting the potential barrier from the left are unable to go across it and are completely reflected back. There is, therefore, no transmission through it.

Having discussed the classical aspect of the problem, we now take up the quantum mechanical treatment.

The quantum mechanical motion is determined by the eigenvalue equation

$$Hu = Eu \qquad (4.3a)$$

or

$$\left(-\frac{\hbar^2}{2m}\frac{\partial^2}{\partial x^2} + V(x)\right)u(x) = Eu(x). \qquad (4.3b)$$

Now

$$V(x) = 0 \quad x < 0,$$
$$= V \quad x > 0.$$

Here again we consider two cases: (i) $(E - V) > 0$. We define constants k_0 and k_1 by the equations

$$k_0^2 = \frac{2mE}{\hbar^2}, \quad k_1^2 = \frac{2m(E - V)}{\hbar^2}. \qquad (4.4)$$

From Eq. (4.3) we have

$$(\frac{\partial^2}{\partial x^2} + k_0^2)u_0(x) = 0 \quad x < 0, \tag{4.5a}$$

$$(\frac{\partial^2}{\partial x^2} + k_1^2)u_1(x) = 0 \quad x > 0. \tag{4.5b}$$

The general solutions of Eq. (4.5) are

$$u_0(x) = Ae^{ik_0x} + Be^{-ik_0x}, \tag{4.6a}$$

$$u_1(x) = De^{ik_1x} + Ce^{-ik_1x}. \tag{4.6b}$$

The first and second terms on the right-hand side of Eq. (4.6) represent plane waves moving in positive and negative directions of x-axis respectively.

We are interested in the situation where the particle approaches from the left and may be either transmitted or reflected. Thus,

$$u_0(x) = \underbrace{Ae^{ik_0x}}_{\text{(incident)}} + \underbrace{Be^{-ik_0x}}_{\text{(reflected)}}, \tag{4.7a}$$

$$u_1(x) = \underbrace{De^{ik_1x}}_{\text{(transmitted)}}. \tag{4.7b}$$

The constant C must be zero as there is no wave in negative direction of x-axis in the region $x > 0$.

The continuity conditions at $x = 0$ require

$$u_0(0) = u_1(0); \quad u_0'(0) = u_1'(0)$$

which gives us

$$A + B = D \tag{4.8a}$$

and

$$k_0(A - B) = k_1 D \tag{4.8b}$$

i.e. we have two equations for three constants. One constant can be determined from the normalisation condition

$$\int_{-\infty}^{\infty} |u(x)|^2 dx = 1. \tag{4.9}$$

From Eq. (4.8) we can find B and D in terms of A for any value E (i.e. no restriction on energy values). These are given by

$$B = \frac{(k_0 - k_1)A}{(k_0 + k_1)}; \quad D = \frac{2k_0 A}{(k_0 + k_1)}. \tag{4.10}$$

A can be determined from the normalisation condition but it is not very interesting. In order to find reflectivity and transmitivity of the potential step, we have to find the incident, transmitted and reflected probability currents.

Incident Current: (Subscript i denotes incident)

$$S_i(x) = \frac{\hbar}{2mi}\left(u_i^* \frac{\partial u_i}{\partial x} - u_i \frac{\partial u_i^*}{\partial x}\right) \tag{4.11}$$

$$= \frac{\hbar}{2mi}\left(e^{-ik_0 x} ik_0 e^{ik_0 x} - e^{ik_0 x}(-ik_0)e^{-ik_0 x}\right)|A|^2$$

$$= \frac{\hbar}{m}k|A|^2 \tag{4.12}$$

$$= \text{Incident flux } (\rho v_0,\, \rho = |A|^2 = \text{density of states}).$$

Transmitted Current:

$$S_t(x) = \frac{\hbar}{m}|D|^2 k_1$$

$$= \frac{\hbar}{m}\frac{4k_0^2}{(k+k_1)^2}k_1|A|^2. \tag{4.13}$$

Reflected Current:

$$S_r(x) = -\frac{\hbar}{m}|B|^2 k_0$$

$$= -\frac{\hbar}{m}\frac{(k_0 - k_1)^2}{(k_0 + k_1)^2}k_0|A|^2. \tag{4.14}$$

Transmitivity

$$T = \frac{|D|^2 k_1}{|A|^2 k_0}$$

$$= \frac{4k_0 k_1}{(k_0 + k_1)^2}. \tag{4.15}$$

Reflectivity

$$R = \frac{|B|^2}{|A|^2}$$

$$= \frac{(k - k_1)^2}{(k + k_1)^2}. \tag{4.16}$$

Therefore

$$R + T = \frac{(k_0 - k_1)^2}{(k_0 + k_1)^2} + \frac{4k_0 k_1}{(k_0 + k_1)^2} = 1.$$

Now k_0 is always greater than zero since $E > V$. k_1 however, can be zero and in that case $T = 0$. As E becomes smaller and smaller, the transmittivity goes down and when $E = V$, then $T = 0$, and $R = 1$ and everything is reflected.

In general R is not zero. It is in this respect that the quantum mechanical result differs from the classical theory. If $E \gg V$, then $k_0 \approx k_1$, $B = 0, D = A$, and $R \approx 0$ and everything is transmitted — which is, of course, the classical limit. In general, a part will be reflected and a part will be transmitted. In classical mechanics there is no reflection at the boundary if $E > V$, but in quantum mechanics there is reflection. The extreme quantum limit is for $E \ll |V|$ (i.e. when $k_1^2 = (2m/h^2)(E + |V|)$ and $k_1 \gg k_0$). In this case as is clear from Eq. (4.10) $B \approx -A$ and $D \approx 0$ and there is total reflection. Thus most interesting case is to take V large in magnitude, but negative (Fig. 4.2) so that we have a sudden large potential drop, through which classical particles pass with greatly increased momentum. However, in quantum mechanics, as seen above their is total reflection — the exact opposite of the classical prediction. This essentially quantum mechanical effect may be observed in nuclear physics when a low energy incident neutron, say, is reflected by the sudden onset of a highly attractive potential, as it approaches the surface of a nucleus.

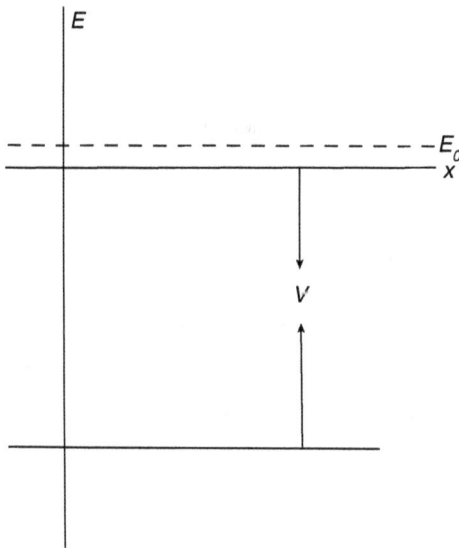

Fig. 4.2   The energy diagram of a potential step with negative V

(ii) $(E - V) < 0$, for $x > 0$. For $x < 0$, we have, as before

$$u_0(x) = A\,e^{ik_0x} + B\,e^{-ik_0x} \quad x < 0. \tag{4.17}$$

For $x > 0$, we put

$$K^2 = \frac{2m}{\hbar^2}(V - E)\,.$$

Then we have the Schrödinger equation as

$$\left(\frac{\partial^2}{\partial x^2} - K^2\right)u_1(x) = 0 \quad x > 0$$

and the solution is

$$u_1(x) = C\,e^{-Kx} + D\,e^{Kx}. \tag{4.18}$$

As $x \to \infty$, the second part of (4.18) is infinite. Since the wave function is to be finite, D must be zero. Therefore

$$u_1(x) = C\,e^{-Kx}. \tag{4.19}$$

Continuity conditions at $x = 0$ give

$$A + B = C\,, \tag{4.20a}$$

$$ik_0(A - B) = -KC. \tag{4.20b}$$

Therefore

$$B = \frac{(k_0 - iK)A}{(k_0 + iK)}\,; \quad C = \frac{2k_0A}{(k_0 + iK)} \tag{4.21}$$

for any value of E, so that again there is no restriction on possible energy values. Here

$$R = |B|^2/|A|^2 = 1\,,$$

i.e. we get total reflection as in the classical case.

There is, however, a difference from the classical result in that the probability of finding the particle in the classically forbidden region, i.e. $x > 0$ is not zero. This probability is given by

$$P(x) \propto |u_1(x)|^2 = |C|^2 e^{-2Kx} \quad x > 0\,.$$

Thus

$$P(x) \propto \frac{4k_0^2}{(k_0^2 + K^2)} e^{-2Kx}\,. \tag{4.22}$$

This is appreciable near the barrier edge and falls exponentially to negligible value at distances large compared with $1/K$. The wave function will thus look as shown in Fig. 4.3. The fact that reflectivity is 1 and still there is some probability of finding the particle in classically forbidden region may be reconciled by the uncertainty principle[1]: The uncertainty in position is $\Delta x = 1/K$. Now $\Delta x \Delta p \sim \hbar$, $\Delta p \sim \hbar/\Delta x = \hbar K$ and $\Delta T \sim \frac{(\Delta p)^2}{2m} = \frac{\hbar^2 K^2}{2m} = V - E$. Hence its final energy $E + \Delta T$ is sufficiently uncertain that we are no longer sure that total energy of particle is less than V.

To end this section we note that the complex number B/A given in Eq (4.21) can be rewritten in terms of the phase shift $\delta(E)$:

$$B/A = |B/A|e^{-2i\delta(E)} = e^{-2i\delta(E)} \tag{4.23}$$

where

$$\tan 2\delta(E) = \frac{2k_0 K}{K^2 - k_0^2}, \tag{4.24}$$

which gives

$$\tan \delta = k_0/K \quad \text{or} \quad k_0 \cot \delta = K. \tag{4.25}$$

In terms of the phase shift, Eqs (4.17) and (4.19) are given by

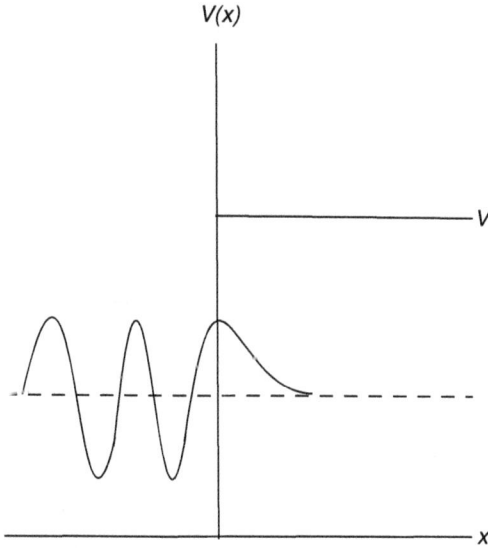

Fig. 4.3   The tunnelling of the wave function in classically forbidden region.

---

[1] G. Baym, Lectures on Quantum Mechanics. The Benjamin/Cummings Publishing Company, INC, (1981), p.93

$$u_0(x) = A[e^{ik_0x} + e^{-2i\delta(E)}e^{-ik_0x}]$$
$$= 2Ae^{-i\delta}\cos(k_0x + \delta),\tag{4.26}$$
$$u_1(x) = Ae^{-i(\frac{\pi}{2}-\delta)}e^{-Kx} = -Aie^{i\delta}e^{-Kx}.\tag{4.27}$$

In deriving Eq (4.27) we have used Eqs (4.21) and (4.25). Thus we conclude that the reflected wave is shifted in phase, at $x = 0$ from the incident wave at $x = 0$ by a factor $e^{-2i\delta(E)}$.

4.2 The Potential Barrier

We now discuss the quantum mechanical analysis of a particle going across the potential barrier as shown in Fig. 4.4.

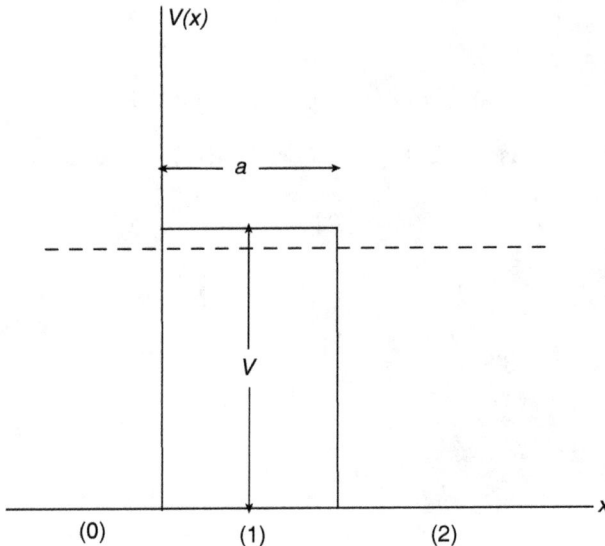

Fig. 4.4   The rectangular potential barrier.

The potential is given by
$$V(x) = 0 \quad x \le 0 \qquad \text{Region (0)},$$
$$= V \quad 0 < x < a \quad \text{Region (1)},$$
$$= 0 \quad x \ge a \qquad \text{Region (2)}.\tag{4.28}$$

The treatment for the case when $E > V$ is similar to that carried out for the potential step. We, therefore, discuss only the case when $E < V$.

Define

$$k^2 = \frac{2mE}{\hbar^2}, \quad K^2 = \frac{2m}{\hbar^2}(V - E) .$$

The Schrödinger equation in the three regions is

$$\left[\frac{\partial^2}{\partial x^2} + k^2\right] u_0(x) = 0, \quad x \le 0$$

$$\left[\frac{\partial^2}{\partial x^2} - K^2\right] u_1(x) = 0, \quad 0 < x < a$$

$$\left[\frac{\partial^2}{\partial x^2} + k^2\right] u_2(x) = 0. \quad x \ge a$$

The solutions of these equations are

$$u_0(x) = A e^{ikx} + B e^{-ikx} , \tag{4.29a}$$

$$u_1(x) = C e^{-Kx} + D e^{Kx} , \tag{4.29b}$$

$$u_2(x) = A' e^{ikx}. \tag{4.29c}$$

There is no reflected wave in region (2); we have only a transmitted wave as given in Eq. (4.29c). The continuity conditions give

$$u_0(0) = u_1(0), \quad u_1(a) = u_2(a),$$

$$u_0'(0) = u_1'(0), \quad u_1'(a) = u_2'(a).$$

These give

$$A + B = C + D \tag{4.30a}$$

$$ik(A - B) = -K(C - D) \tag{4.30b}$$

$$C e^{-Ka} + D e^{Ka} = A' e^{-ka} , \tag{4.31a}$$

$$-KC e^{-Ka} + KD e^{Ka} = ik_0 A' e^{ika}. \tag{4.31b}$$

From Eqs. (4.31), we get

$$D = \frac{A'}{2}\left(1 + \frac{ik}{K}\right) e^{(ik-K)a} \tag{4.32a}$$

and

$$C = \frac{A'}{2}\left(1 - \frac{ik}{K}\right) e^{(ik+K)a}. \tag{4.32b}$$

From Eqs. (4.30), we get

$$B = \frac{C}{2}\left(1 - \frac{iK}{k}\right) + \frac{D}{2}\left(1 + \frac{iK}{k}\right), \tag{4.33a}$$

$$A = \frac{C}{2}\left(1 + \frac{iK}{k}\right) + \frac{D}{2}\left(1 - \frac{iK}{k}\right). \tag{4.33b}$$

Let us put $K/k = \mu$; then from Eqs. (4.32) and (4.33)

$$A = \frac{1}{2} A' e^{ika} \frac{1}{\mu} (2\mu \cosh Ka - i(1 - \mu^2) \sinh Ka) , \qquad (4.34a)$$

$$B = \frac{1}{2} A' e^{ika} \frac{-i}{\mu} (1 + \mu^2) \sinh Ka . \qquad (4.34b)$$

Let us put

$$F = (2\mu \cosh Ka - i(1 - \mu^2) \sinh Ka) . \qquad (4.35)$$

Then from Eqs. (4.32) and (4.34), we have

$$A' = A \, 2\mu e^{-ika}/F , \qquad (4.36)$$

$$B = A(-i)(1 + \mu^2) \sinh Ka/F , \qquad (4.37)$$

$$C = A(-i)(1 + i\mu) e^{Ka}/F , \qquad (4.38a)$$

$$D = A(i)(1 - i\mu) e^{-Ka}/F. \qquad (4.38b)$$

Now reflectivity R and transmittivity T are given by

$$R = \frac{|B|^2}{|A|^2} = \frac{(1 + \mu^2)^2 \sinh^2 Ka}{[4\mu^2 \cosh^2 Ka + (1 - \mu^2)^2 \sinh^2 Ka]} , \qquad (4.39)$$

$$T = \frac{|F|^2}{|A|^2} = \frac{4\mu^2}{[4\mu^2 \cosh^2 Ka + (1 - \mu^2)^2 \sinh^2 Ka]} . \qquad (4.40)$$

First we note that when $E < V$, then the region (1), i.e. $0 < x < a$ is inaccessible classically and thus $T = 0$ and $R = 1$ classically. But here we see that in general $T \neq 0$ and there is a leakage of the particle through the barrier. The fact that something is transmitted at all is a purely quantum mechanical effect (this is called the tunnelling through the barrier).

If the barrier is broad, so that $Ka \gg 1$, then

$$\frac{1}{\sinh^2 Ka} = \frac{4}{e^{2Ka} + e^{-2Ka} - 2}$$
$$\rightarrow 4e^{-2Ka}$$

and

$$T = \frac{4\mu^2/\sinh^2 Ka}{(1 + \mu^2)^2 + 4\mu^2/\sinh^2 Ka}$$
$$\approx \left(4\mu/(1 + \mu^2)\right)^2 e^{-2Ka} . \qquad (4.41)$$

Using $\mu^2 = K^2/k^2 = (V - E)/E$, we obtain the transmitivity or the tunnelling probability when the barrier is broad as

$$T(E) = 16 \frac{E}{V} \left(1 - \frac{E}{V}\right) \exp\left(-\sqrt{\frac{8m}{\hbar^2}(V - E)} \, a\right). \qquad (4.42)$$

Thus we see that in quantum mechanics, it is possible for the particle to penetrate the barrier although its energy is less than the potential of the barrier. From Eq. (4.42), we see that the transmission probability is much less than one, and decreases exponentially with the increase of the width of barrier. This has important applications e.g. α-decay where α-particle having less energy than the potential barrier of the nucleus comes out of the nucleus. The shape of such a potential and the shape of the wave function is shown in Fig. 4.5.

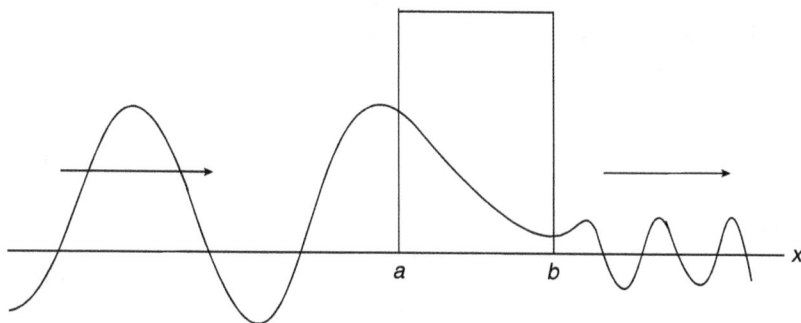

Fig. 4.5 The incident and transmitted waves for a rectangular potential barrier.

We expect this shape for the wave function because when $Ka \gg 1$ then, as is clear from Eq. (4.38b), $D \approx 0$; thus

$$u_0(x) = A \left(e^{ikx} + \frac{B}{A} e^{-ikx} \right), \tag{4.43a}$$

$$u_1(x) \approx C e^{-Kx}, \tag{4.43b}$$

$$u_2(x) \approx A' e^{ikx}. \tag{4.43c}$$

4.3 Parity

Suppose we have a potential such that

$$V(x) = V(-x) \tag{4.44}$$

that is, the potential is symmetric about the origin and then $H(x) = H(-x)$. Let $u_E(x)$ be the solution of eigenvalue equation

$$H\, u_E(x) = \left(-\frac{\hbar^2}{2m} \frac{\partial^2}{\partial x^2} + V(x) \right) u_E(x)$$
$$= E\, u_E(x). \tag{4.45}$$

Now changing $x \rightarrow -x$, we have

$$\left(-\frac{\hbar^2}{2m}\frac{\partial^2}{\partial x^2} + V(-x)\right)u_E(-x) = E\,u_E(-x) \tag{4.46}$$

so that using Eq. (4.44),

$$\left(-\frac{\hbar^2}{2m}\frac{\partial^2}{\partial x^2} + V(x)\right)u_E(-x) = E\,u_E(-x). \tag{4.47}$$

Thus we see that $u_E(-x)$ is also a solution of Eq. (4.45) for a given value of E. If we assume that there is only one linearly independent solution for a given eigenvalue E, then the two solutions can differ only by a constant, therefore

$$u_E(x) = \eta u_E(-x). \tag{4.48a}$$

Changing $x \rightarrow -x$

$$u_E(-x) = \eta u_E(x). \tag{4.48b}$$

Substituting Eq. (4.48b) in Eq. (4.48a), we have

$$u_E(x) = \eta^2 u_E(x). \tag{4.48c}$$

Therefore

$$\eta = \pm 1\ .$$

The constant η is called the parity of the state. For positive parity states, $(\eta = +1), u_E(x)$ is an even function of x; for negative parity state $(\eta = -1), u_E(x)$ is an odd function of x.

If there is more than one solution, and a particular solution does not have definite parity, then from $u_E(x)$ and $u_E(-x)$ one can always construct solutions

$$u_e(x) = \frac{1}{2}\left(u_E(x) + u_E(-x)\right) \tag{4.49a}$$

$$u_o(x) = \frac{1}{2}\left(u_E(x) - u_E(-x)\right) \tag{4.49b}$$

each of which has a definite parity. Thus it is always possible to construct solutions which are either symmetric or anti-symmetric about the origin provided that $V(-x) = V(x)$.

4.4 An Example of Bound State (Discrete Energy Levels)

In the previous examples we have considered potentials in which the particle could move off to infinity at least in one direction both classically and quantum mechanically. We now consider examples of a bound system in which this is not the case. Consider a potential of the form (see Fig. 4.6)

$$V(x) = 0, \quad |x| \leq a,$$
$$V(x) \to \infty, \quad |x| \geq a. \tag{4.50}$$

Classically the particle is confined to the region $|x| \leq a$, and whatever its energy, it bounces elastically off the potential walls. Since V becomes infinite for $|x| \geq a$, but other terms in the Schrödinger equation

$$\left(-\frac{\hbar^2}{2m} \frac{\partial^2}{\partial x^2} + V \right) u = Eu$$

for this region remain finite, we must impose the boundary condition

$$u(x) = 0 \quad |x| \geq a. \tag{4.51}$$

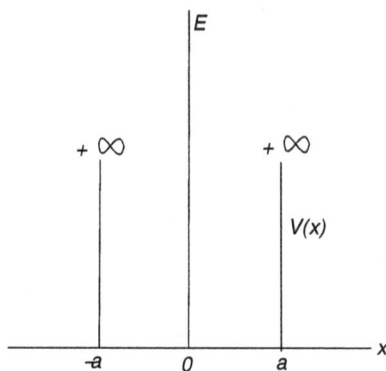

Fig. 4.6 The energy diagram of a one-dimensional infinite well where a particle is confined to the region

Thus the boundary condition at a surface at which there is an infinite potential step is that the wave function is zero.

For $|x| < a$, the Schrödinger equation is

$$-\frac{\hbar^2}{2m} \frac{d^2 u}{dx^2} = Eu, \quad k^2 = \frac{2mE}{\hbar^2} \tag{4.52a}$$

or

$$\left(\frac{d^2}{dx^2} + k^2 \right) u(x) = 0. \tag{4.52b}$$

The general solution of this equation is

$$u(x) = A \sin kx + B \cos kx.$$ (4.53)

Boundary condition at $x = \pm a$ gives

$$u(a) = 0 = u(-a),$$
$$A \sin ka + B \cos ka = 0,$$ (4.54a)
$$-A \sin ka + B \cos ka = 0.$$ (4.54b)

Therefore either

$$A \sin ka = 0$$

or

$$B \cos ka = 0.$$

We do not want both A and $B = 0$, since this would give a physically uninteresting solution $u(x) = 0$ everywhere. Also $\sin ka$ and $\cos ka$ cannot both be zero for a given value of k. Therefore there are two possible classes of solutions. For the first class

$$A = 0, \quad \cos ka = 0,$$

so that

$$ka = n\pi/2,$$

where n is an odd integer.

For the second class, $B = 0, \sin ka = 0$, so that $ka = (n\pi)/2$, where n is an even integer. Thus two classes of eigenfunctions and their energy eigenvalues are

$$u_n(x) = B \cos \frac{n\pi x}{2a}, \quad k_n = \frac{n\pi}{2a}, \quad E_n = \frac{\hbar^2}{2m}\left(\frac{\pi^2}{4a^2}\right)n^2; n \text{ odd}, \quad (4.55a)$$

$$u_n(x) = A \sin \frac{n\pi x}{2a}, \quad E_n = \frac{\hbar^2}{2m}\left(\frac{\pi^2}{4a^2}\right)n^2 ; n \text{ even.} \quad (4.55b)$$

Combining the above values of E_n we see that the possible energy levels are

$$E_n = \frac{\hbar^2}{2m}\left(\frac{\pi^2}{4a^2}\right)n^2, \quad n = 1, 2, \dots.$$ (4.56)

Thus a classically bound system when treated quantum mechanically has yielded discrete energy levels. This is a general feature as we shall see.

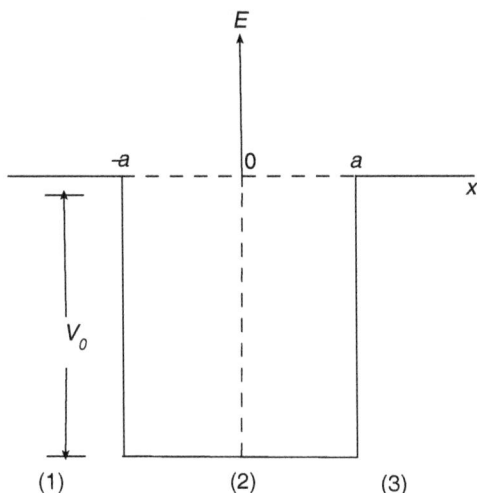

Fig. 4.7 The energy diagram of a square well potential.

4.5 Bound State (Square Well Potential)

The square well potential is an important concept in physics. The square well potential may be a reasonable approximation to the actual potential in some physical problems. Therefore a solution of this problem will give qualitative results for the actual problem.

Now

$$V(x) = -V \quad -a < x < a,$$
$$= 0 \quad \text{elsewhere.} \qquad (4.57)$$

Here we consider the case $E = T > 0$ for $|x| > a$ but $E < 0$ in $|x| < a$ so that the particle is confined in the potential well in the region $|x| \leq a$ i.e. the particle is in a bound state $(E - T - V < 0$ or $T = E + V < V$ for $|x| < a)$. [The other case when $E > 0$ in both the regions $|x| > a$ and $|x| < a$ is similar to the cases considered in Secs. 4.1 and 4.2 (see Problem 4.4).]

Let

$$\varepsilon = -E = |E| .$$

Define the wave numbers

$$K^2 = -\frac{2mE}{\hbar^2} = \frac{2m\varepsilon}{\hbar^2} \quad \text{(for regions (1) and (3))}, \qquad (4.58)$$

$$k^2 = \frac{2m(E + V)}{\hbar^2} = \frac{2m(V - \varepsilon)}{\hbar^2} \quad \text{(for region (2))}. \qquad (4.59)$$

The Schrödinger equation for eigenfunctions and eigenvalues is

$$\left(-\frac{\hbar^2}{2m}\frac{\partial^2}{\partial x^2} + V(x)\right)u(x) = E\,u(x). \tag{4.60}$$

This can be written for regions (1), (3) and (2) as

$$\left(\frac{d^2}{dx^2} - K^2\right)u_{1,3}(x) = 0, \tag{4.61a}$$

$$\left(\frac{d^2}{dx^2} + k^2\right)u_2(x) = 0. \tag{4.61b}$$

The following are solutions:

$$u_3(x) = A\,e^{-Kx}, \tag{4.62a}$$

(no term of the form e^{Kx} should appear since the wave function must not become infinite for large distances, i.e. $x \to \infty$)

$$u_2(x) = B\,e^{ikx} + C\,e^{-ikx}, \tag{4.62b}$$

$$u_1(x) = D\,e^{Kx}. \tag{4.62c}$$

(no terms of the form e^{-Kx} should appear since wave function must be finite for $x \to -\infty$). We can write $u_1(x)$ as

$$u_1(x) = D\,e^{-K|x|}. \tag{4.62d}$$

The potential (4.57) satisfies

$$V(x) = V(-x),$$

i.e. it is symmetric or even and we may confine ourselves to eigenfunctions of definite parity. Above solutions do not have definite parity, i.e. they are not of the form

$$u(x) = \pm u(-x).$$

Consider first $u_2(x)$. The solutions of definite parity, i.e. of even and odd parity are respectively

$$u_{2e}(x) = \frac{1}{2}(u_2(x) + u_2(-x))$$
$$= (B + C)\cos kx = B'\cos kx, \tag{4.63a}$$

$$u_{2o}(x) = \frac{1}{2}(u_2(x) - u_2(-x))$$
$$= i(B - C)\sin kx = C'\sin kx. \tag{4.63b}$$

The requirement of parity for u_1 and u_3 is

$$u_3(-x) = \pm u_1(x) \tag{4.64}$$

i.e.

$$D = \pm A .$$

Thus we have

$$u_3(x) = A e^{-Kx} , \tag{4.65a}$$
$$u_1(x) = \pm A e^{-K|x|}. \tag{4.65b}$$

Thus there are two possible types of solutions of definite parity, the ones with even parity

$$u_3(x) = A e^{-Kx} , \quad u_1(x) = A e^{-K|x|}$$
$$u_2(x) = B' \cos kx , \tag{4.66}$$

and the ones with odd parity

$$u_3(x) = A e^{-Kx} , \quad u_1(x) = -A e^{-K|x|}$$
$$u_2(x) = C' \sin kx. \tag{4.67}$$

For each solution we have two continuity conditions

$$u_2(a) = u_3(a) ,$$
$$u_2'(a) = u_3'(a). \tag{4.68}$$

Since an overall constant factor remains arbitrary unless determined by the normalisation condition, this means that there are two conditions to determine one constant. Thus the energy E must be regarded as an adjustable parameter. The continuity conditions can only be satisfied for certain discrete values, E_n, of the energy. The conditions (4.68) give for even parity solution

$$A e^{-Ka} = B' \cos ka , \tag{4.69a}$$
$$A K e^{-Ka} = B' k \sin ka. \tag{4.69b}$$

This can be satisfied only if

$$k \tan ka = K \quad \text{(even parity)}. \tag{4.70}$$

Similarly Eq. (4.68) give for odd parity solution

$$A e^{-Ka} = C' \sin ka \tag{4.71a}$$
$$-AK e^{-Ka} = C' k \cos ka \tag{4.71b}$$

which can be satisfied if

$$k \cot ka = -K \quad \text{(odd parity)}. \tag{4.72}$$

If we substitute Eqs. (4.58) and (4.59) for K and k, Eqs. (4.70) and (4.72) respectively determine the energy eigenvalues E_n corresponding to even and odd parity solutions. The equations can be solved graphically for E_n. In fact we notice from Eq (4.58) and (4.59), that Ka and ka can be combined to give

$$K^2a^2 + k^2a^2 = \frac{2mVa^2}{\hbar} \qquad (4.73)$$

which is the equation of a circle in the (Ka, ka) plane with a radius $\frac{\sqrt{2mVa^2}}{\hbar}$. Thus K and k of the energy eigenfunctions or eigenvalues are determined by the intersections of the plots of Eq (4.70)[for even modes] and Eq (4.72)[for odd modes] with the circles given in Eq. (4.73). Hence classically bound states when treated quantum mechanically yield discrete energy levels.

4.6 Problems

4.1 A particle of mass m, moves in a potential

$$V(x) = \infty \quad x \leq 0 \,,$$
$$V(x) = 0 \quad 0 < x < a \,,$$
$$V(x) = V \quad x \geq a \,.$$

For $E < V$, find the condition for allowed energy values.

4.2 A particle of mass m approaches a potential barrier

$$V(x) = 0 \quad x < 0$$
$$= V \quad 0 \leq x \leq a$$
$$= 0 \quad x > a$$

from $x = -\infty$. For $E < V$, determine reflection and transmission coefficients for $Ka \ll 1$, where

$$K^2 = \frac{2m}{\hbar^2}(V - E) \,.$$

4.3 Consider a particle of mass m in a potential

$$V = \quad 0 \quad x < 0 \,,$$
$$= -V \quad 0 \leq x \leq a \,,$$
$$= \quad 0 \quad x > a \,.$$

If $V \to \infty$ and $a \to 0$, such that $Va = \mu$, find the energy for the bound state $(E < 0)$.

4.4 Consider a rectangular potential barrier 4 eV high and 10^{-9} meters wide. Calculate a rough value for the probability for an electron of kinetic energy 3 eV to penetrate the barrier.

4.5 A particle of mass m, moves in a potential (with energy $E < V$)

$$V(x) = \infty \quad x \le 0$$
$$V(x) = 0 \quad 0 < x < a$$
$$V(x) = V \quad a \le x \le b$$
$$V(x) = 0 \quad x > b$$

the energy has a value as obtained in Problem 4.1, find the relative intensity at $x = b$ and $x = a$. For the region $x > b$, show that there is equal intensity in the beams travelling to left and right.

4.6 Find the transmitivity and reflectivity for a particle of energy $E(> 0)$ scattered by the square well potential of Fig. 4.7. Show that (i) for $K \ne k$, $T = 1$ and $R = 0$ for energies E given by

$$E = -V + \frac{n^2 \pi^2 \hbar^2}{8ma^2}$$

(ii) T has a minimum value for energies

$$E = -V + \frac{(2n + 1)^2 \pi^2 \hbar^2}{32ma^2}$$

(iii) as $E \to \infty$, $K \to k$, $T \to 1$.

Chapter 5

Simple Harmonic Oscillator

"The career of a young theoretical physicist consists of treating the harmonic oscillator in ever-increasing levels of abstraction." – Sidney Coleman.

5.1 Introduction

In classical mechanics, a harmonic oscillator consists of a particle moving under the action of a restoring force

$$F = -\frac{\partial V}{\partial x} = -m\omega^2 x \qquad (5.1)$$

so that

$$V(x) = \frac{1}{2}m\omega^2 x^2 . \qquad (5.2)$$

The equation of motion is

$$m\frac{d^2x}{dt^2} = -m\omega^2 x \qquad (5.3)$$

with the solution

$$x = a\cos(\omega t + \delta) \qquad (5.4)$$

representing an oscillator motion of angular frequency ω, time period $T = \frac{2\pi}{\omega}$ and amplitude a. The energy E of the oscillation is the potential energy at the extreme position $x = \pm a$ so that

$$E = \frac{1}{2}m\omega^2 a^2 . \qquad (5.5)$$

The simple harmonic potential is important in reality for the following reason:

Suppose a potential $V(x)$ has a minimum somewhere, say at $x = x_0$. We can then write the potential $V(x)$ as a power series around x_0

$$V(x) = V(x_0) + \frac{\partial V(x)}{\partial x}\Big|_{x=x_0}(x - x_0) + \frac{\partial^2 V(x)}{\partial x^2}\Big|_{x=x_0}\frac{(x - x_0)^2}{2}... \quad (5.6)$$

Now $V(x_0)$ is a constant and a constant does not matter in a potential and can be put equal to zero. But $\frac{\partial V(x)}{\partial x}\big|_{x=x_0} = 0$ because $V(x)$ has a minimum at $x = x_0$, therefore

$$V(x) = K + C(x - x_0)^2 + C'(x - x_0)^3... \quad (5.7)$$

If we are interested in the neighbourhood of x_0, higher terms will be small and hence

$$V(x) \approx C(x - x_0)^2. \quad (5.8)$$

Thus in this case it is a good approximation to approximate $V(x)$ by $C(x - x_0)^2$ which is a harmonic oscillator potential.

5.2   Quantum Theory of Simple Harmonic Oscillator

The hamiltonian for simple harmonic oscillator is

$$H = \frac{\hat{p}^2}{2m} + \frac{1}{2}m\omega^2\hat{x}^2 \quad (5.9)$$

where in the classical Hamiltonian, the dynamical variables p and x have been replaced by operators (correspondence principle). In Schrodinger picture:

$$\hat{x} = x, \quad \hat{p} = -i\hbar\frac{\partial}{\partial x}. \quad (5.10)$$

It is convenient to change the variable x to y:

$$y = \sqrt{\frac{m\omega}{\hbar}}x. \quad (5.11)$$

Using this variable:

$$H = \frac{\hbar\omega}{2}(-\frac{\partial^2}{\partial y^2} + y^2). \quad (5.12)$$

Define, the operators:

$$a^\dagger = \sqrt{\frac{1}{2m\hbar\omega}}(-i\hat{p} + m\omega\hat{x}) = \frac{1}{\sqrt{2}}(-\frac{\partial}{\partial y} + y), \quad (5.13)$$

$$a = \sqrt{\frac{1}{2m\hbar\omega}}(i\hat{p} + m\omega\hat{x}) = \frac{1}{\sqrt{2}}(\frac{\partial}{\partial y} + y). \quad (5.14)$$

Now

$$[a, a^\dagger]\Psi(y) = \frac{1}{2}[\frac{\partial}{\partial y} + y, -\frac{\partial}{\partial y} + y]\Psi(y)$$

$$= (\frac{\partial}{\partial y}y - y\frac{\partial}{\partial y})\Psi(y) = \Psi(y). \qquad (5.15)$$

Since $\Psi(y)$ is arbitrary;

$$[a, a^\dagger] = 1. \qquad (5.16)$$

In terms of the operators a and $a^\dagger$:

$$H = \frac{\hbar\omega}{2}(aa^\dagger + a^\dagger a) = \hbar\omega(aa^\dagger - \frac{1}{2})$$

$$= \hbar\omega(a^\dagger a + \frac{1}{2}). \qquad (5.17)$$

From Eqs. (5.16) and (5.17):

$$[H, a] = \hbar\omega[a^\dagger a, a] = \hbar\omega\{a^\dagger[a, a] + [a^\dagger, a]a\}$$

$$= -\hbar\omega a. \qquad (5.18a)$$

Similarly

$$[H, a^\dagger] = \hbar\omega a^\dagger. \qquad (5.18b)$$

Equations (5.18a) and (5.18b) are useful to determine eigenvalues and eigenstates of H:

$$Hu_n(y) = E_n u_n(y). \qquad (5.19)$$

Suppressing y, write the state

$$\chi = a u_n. \qquad (5.20)$$

Using the notation introduced in (ch. 3):

$$(\chi|\chi) = (au_n|\chi) = (u_n|a^\dagger\chi)$$

$$= (u_n|a^\dagger a u_n)$$

$$= \frac{1}{\hbar\omega}(u_n|(H - \frac{1}{2}\hbar\omega)u_n)$$

$$= \frac{1}{\hbar\omega}(E_n - \frac{1}{2}\hbar\omega)(u_n|u_n). \qquad (5.21)$$

The eigenstates of H, viz $u_n(y)$, form a complete orthonormal set:

$$(u_n|u_m) = \int u_n^*(y)u_m(y) = \delta_{mn}. \qquad (5.22)$$

Hence the state $\chi(y)$ can be expressed in terms of them:

$$\chi(y) = \sum_n C_n u_n(y) \tag{5.23}$$

and

$$(\chi|\chi) = \sum_n \sum_m C_n^* C_m (u_n|u_m)$$
$$= \sum_n |C_n|^2 \geq 0. \tag{5.24}$$

Hence from Eq. (5.21):

$$(E_n - \frac{1}{2}\hbar\omega) \geq 0$$

$$E_n \geq \frac{1}{2}\hbar\omega. \tag{5.25}$$

Thus E_n has a minimum value $E_0 = \frac{1}{2}\hbar\omega$ and let the corresponding eigen-function be $u_0(y)$. $E_0 = \frac{1}{2}\hbar\omega$ is called the zero point energy. Classically it is zero. It is important to note that the vacuum (ground state) energy not being zero has far reaching consequences and is known as the quantum fluctuation of the vacuum.

Now

$$H u_0(y) = E_0 u_0(y). \tag{5.26}$$

On using Eq. (5.18a)

$$H a u_0(y) = (-\hbar\omega a + a H) u_0(y)$$
$$= (E_0 - \hbar\omega) a u_0(y). \tag{5.27}$$

If $a u_0(y)$ is not zero, then $a u_0(y)$ is an eigenstate of H with eigenvalue $E_0 - \hbar\omega$ which contradicts that E_0 is the minimum value. Hence

$$a u_0(y) = 0. \tag{5.28}$$

Eqs. (5.27) and (5.28) show that a acts as an operator which annihilates energy in the system by $\hbar\omega$ and is called the annihilation operator. However, using Eq. (5.18b)

$$H a^\dagger u_0(y) = (\hbar\omega a^\dagger + a^\dagger H) u_0(y) = (E_0 + \hbar\omega) u_0(y)$$
$$= (\frac{3}{2}\hbar\omega) a^\dagger u_0(y). \tag{5.29}$$

Hence if $a^\dagger u_0(y) \neq 0$, it is an eigenstate of H with eigenvalue $E_0 + \hbar\omega = \frac{3}{2}\hbar\omega$. Thus $a^\dagger$ acts as an operator which creates energy in the system by

$\hbar\omega$ and is called the creation operator. Continuing this process n-times, we get

$$H(a^\dagger)^n u_0(y) = (n + \frac{1}{2})\hbar\omega(a^\dagger)^n u_0(y)) \qquad (5.30)$$

i.e. $(a^\dagger)^n u_0(y)$ is an eigenstate of H with eigenvalue $(n + \frac{1}{2})\hbar\omega$. Hence

$$u_n(y) = N_n(a^\dagger)^n u_0(y) \qquad (5.31)$$

where constant N_n is to be determined by normalization such that $(u_n|u_n) = 1$. Hence

$$Hu_n(y) = E_n u_n(y)$$
$$E_n = (n + \frac{1}{2})\hbar\omega \qquad , \qquad n = 0, 1, 2... \qquad (5.32)$$

i.e. simple harmonic oscillator has discrete eigenvalues in units of $\hbar\omega$. First we note (on using Eq. (5.18b)):

$$Hau_n(y) = (-\hbar\omega_a + aH)u_n(y)$$
$$= (E_n - \hbar\omega)au_n(y). \qquad (5.33)$$

Hence if $au_n(y) \neq 0$, it is an eigenstate of H, with eigenvalue

$$(E_n - \hbar\omega) = E_{n-1} = [(n-1) + \frac{1}{2}]\hbar\omega. \qquad (5.34)$$

Thus

$$\chi \equiv au_n(y) = C_n u_{n-1}(y) \qquad (5.35)$$

and

$$(\chi|\chi) = |C_n|^2(u_{n-1}|u_{n-1})$$
$$= |C_n|^2. \qquad (5.36)$$

Hence from Eqs. (5.21) and (5.36)

$$|C_n|^2 = n. \qquad (5.37)$$

Selecting the phase, so that C_n is real

$$C_n = \sqrt{n}. \qquad (5.38)$$

Hence

$$au_n(y) = \sqrt{n}u_{n-1}(y). \qquad (5.39)$$

Similarly, we get

$$a^\dagger u_n(y) = \sqrt{n+1}u_{n+1}(y). \qquad (5.40)$$

Now (cf Eq. (5.31))

$$
\begin{aligned}
u_n(y) &= N_n(a^\dagger)^n u_0(y) \\
&= N_n(a^\dagger)^{n-1} a^\dagger u_0(y) \\
&= N_n(a^\dagger)^{n-1} u_1(y) \\
&= N_n(a^\dagger)^{n-2} a^\dagger u_1(y) \\
&= N_n\sqrt{1.2}(a^\dagger)^{n-2} u_2(y) \\
&= N_n\sqrt{1.2.3}(a^\dagger)^{n-3} u_3(y).
\end{aligned}
$$

Continuing this process, we get

$$
\begin{aligned}
u_n(y) &= N_n\sqrt{1.2.3...n}\, u_n(y) \\
&= N_n\sqrt{n!}\, u_n(y).
\end{aligned}
\tag{5.41}
$$

Note in deriving Eq. (5.41), we have repeatedly used Eq. (5.40). From the normalization:

$$
(u_n|u_n) = 1
$$

we get

$$
N_n = \frac{1}{\sqrt{n!}}.
\tag{5.42}
$$

Hence

$$
u_n(y) = \frac{1}{\sqrt{n!}}(a^\dagger)^n u_0(y).
\tag{5.43}
$$

In order to obtain the explicit form of $u_n(y)$, we use following procedure. We note that the eigenvalue equation

$$
H u_n(x) = E_n u_n(x)
\tag{5.44}
$$

gives, on using the Eqs. (5.12) and (5.19),

$$
\frac{d^2 u_n}{dy^2} + (\epsilon_n - y^2)u_n = 0
\tag{5.45}
$$

where $E_n = \frac{\hbar\omega}{2}\epsilon_n$, so that

$$
\epsilon_n = (2n + 1).
\tag{5.46}
$$

For large values of y,

$$
\frac{d^2 u_n}{dy^2} - y^2 u_n = 0
\tag{5.47}
$$

which has solution $e^{\frac{-y^2}{2}}$. Therefore we write

$$
u_n(y) = e^{-\frac{1}{2}y^2} H_n(y).
\tag{5.48}
$$

Substituting back in Eq. (5.45), we obtain

$$H_n'' - 2yH_n'(y) + 2nH_n(y) = 0. \tag{5.49}$$

This is known as Hermite equation and its solutions are known as Hermite polynomials.

For $n = 0$,

$$H_0'' - 2yH_0' = 0.$$

Thus either $H_0(y)$ is a constant or

$$\frac{H_0''}{H_0'} = 2y$$

implying $ln\, H_0'(y) = y^2$ or

$$H_0'(y) = e^{y^2}.$$

But this solution is not possible as it blows up for $y \longrightarrow \pm\infty$, thus $H_0(y) = constant$ which we take as A_0. Thus

$$u_0(y) = A_0 e^{-\frac{1}{2}y^2}. \tag{5.50}$$

From Eq. (5.40):

$$
\begin{aligned}
u_1(y) &= a^\dagger u_0(y) \\
&= \frac{1}{\sqrt{2}}(-\frac{\partial}{\partial y} + y)u_0(y) = \sqrt{2}A_0 y e^{-\frac{1}{2}y^2},
\end{aligned} \tag{5.51}
$$

$$
\begin{aligned}
u_2(y) &= \frac{1}{\sqrt{2}}(a^\dagger)^2 u_0(y) \\
&= \frac{1}{\sqrt{2}}\frac{1}{2}(-\frac{\partial}{\partial y} + y)(-\frac{\partial}{\partial y} + y)A_0 e^{-\frac{1}{2}y^2} \\
&= \frac{A_0}{\sqrt{2}}(-1 + 2y^2)e^{-\frac{1}{2}y^2},
\end{aligned} \tag{5.52}
$$

$$
\begin{aligned}
u_3(y) &= \frac{1}{\sqrt{6}}(a^\dagger)^3 u_0(y) \\
&= \frac{1}{\sqrt{6}}\frac{1}{\sqrt{8}}(-\frac{\partial}{\partial y} + y)(-\frac{\partial}{\partial y} + y)(-\frac{\partial}{\partial y} + y)A_0 e^{-\frac{1}{2}y^2} \\
&= \frac{A_0}{\sqrt{6}}y(-3 + 2y^2)e^{-\frac{1}{2}y^2}.
\end{aligned} \tag{5.53}
$$

Continuing this process, we get

$$u_n(y) = AH_n(y)e^{-\frac{1}{2}y^2} \tag{5.54}$$

where $H_n(y)$ is a polynomial of degree n. From the above equations, we note that

$$H_n(-y) = -H_n(y) \ , \quad n : \text{odd integer}$$
$$H_n(-y) = H_n(y) \quad , \quad n : \text{even integer}. \tag{5.55}$$

To conclude: for a harmonic oscillator, the energy is quantized; the energy eigenvalues are given by

$$E_n = (n + \frac{1}{2})\hbar\omega \quad , \quad n = 0, 1, 2... \tag{5.56}$$

The lowest energy state $n = 0$, or ground state is not zero and has the value $\frac{1}{2}\hbar\omega$; it is called zero point energy. After this energy, the energy levels are equally spaced. The normalized energy eigenfunctions are given by

$$u_n(x) = (\frac{1}{2^n \sqrt{\pi} n!})^{\frac{1}{2}} (\frac{m\omega}{\hbar})^{\frac{1}{4}} e^{-(\frac{m\omega}{2\hbar})x^2} H_n(\sqrt{\frac{m\omega}{\hbar}} x). \tag{5.57}$$

The eigenvalues and the eigenfunctions for n=0,1,2 are shown in Fig. 5.1. Eq. (5.57) follows from Eq. (5.54) by putting $y = \sqrt{\frac{m\omega}{\hbar}} x$ and the relation

$$\int_{-\infty}^{\infty} e^{-y^2} H_n(y) H_m(y) dy = (\sqrt{\pi} 2^n n!) \delta_{nm} \tag{5.58}$$

i.e. $e^{-\frac{y^2}{2}} H_n$ and $e^{-\frac{y^2}{2}} H_m$ are orthogonal. This is not surprising because $e^{-\frac{y^2}{2}} H_n$ being energy eigenfunctions and eigenfunctions of an observable form a complete orthonormal set. The normalization constant for $e^{-\frac{y^2}{2}} H_n(y)$ is $(\sqrt{\pi} 2^n n!)^{\frac{-1}{2}}$.

5.3   Properties of Hermite Polynomial

i)   Hermite polynomial has a generating function

$$S(y, s) = e^{y^2 - (s-y)^2} = e^{-s^2 + 2sy}$$
$$= \sum_{n=1}^{\infty} \frac{H_n(y)}{n!} s^n. \tag{5.59}$$

Taylor expansion of $S(y, s)$ gives:

$$S(y, s) = S(y, 0) + \frac{\partial}{\partial s} S(y, s)|_{s=0} s + ... + \frac{\partial^n}{\partial s^n} S'(y, s)|_{s=0} \frac{s^n}{n!} ... \quad .$$

Thus

$$H_n(y) = \frac{\partial^n}{\partial s^n} S(y,s)|_{s=0}$$

$$= \frac{\partial^n}{\partial s^n} e^{(y^2 - (s-y)^2)}|_{s=0}$$

$$= (-1)^n e^{y^2} \frac{d^n}{dy^n} e^{-y^2}. \qquad (5.60)$$

From Eq. (5.60):

$$H_0(y) = 1$$
$$H_1(y) = 2y$$
$$H_2(y) = 2(-1 + 2y^2)$$

$$\cdot$$
$$\cdot$$
$$\cdot$$

ii)   Recursion relation

$$H_{n+1} - 2yH_n + 2nH_{n-1} = 0.$$

Differentiating Eq. (5.59) w.r.t s:

$$(-2s + 2y)e^{-s^2 + 2sy} = \sum_{n=1}^{\infty} \frac{H_n(y)}{n!} n s^{n-1}$$

which gives

$$-2\sum_n \frac{H_n(y)}{n!} s^{n+1} + 2y\sum_n \frac{H_n(y)}{n!} s^n - \sum_n n\frac{H_n(y)}{n!} s^{n-1} = 0.$$

Comparing s^n term:

$$H_{n+1} - 2yH_n + 2nH_{n-1} = 0. \qquad (5.61)$$

We now derive Eq. (5.58), by using the generating function, from Eq. (5.59).

$$\int_{-\infty}^{\infty} S(y,s)S(y,t)e^{-y^2} dy = \sum_n \sum_k \frac{s^n t^k}{n!k!} \int_{-\infty}^{\infty} H_n(y)H_k(y)e^{-y^2} dy. \qquad (5.62)$$

Now left-hand side

$$\int_{-\infty}^{\infty} S(y,s)S(y,t)e^{-y^2} dy = e^{2st} \int_{-\infty}^{\infty} e^{-(s+t-y)^2} dy$$

$$= e^{2st}\sqrt{\pi}$$

$$= \sqrt{\pi}\sum_n \frac{(2st)^n}{n!}. \qquad (5.63)$$

From Eqs. (5.62) and (5.63), we have

$$\int_{-\infty}^{\infty} H_n(y)H_k(y)e^{-y^2}\,dy = \delta_{nk}\sqrt{\pi}\frac{2^n}{n!}. \tag{5.64}$$

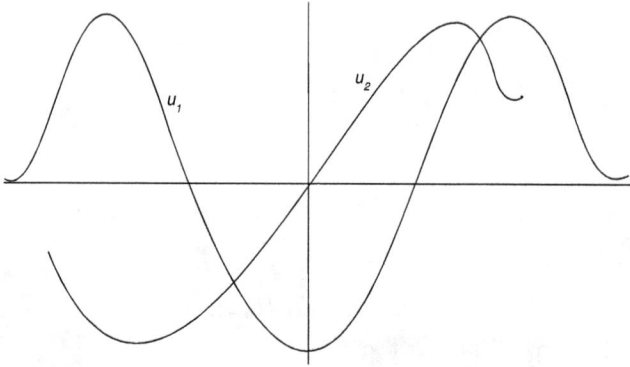

Fig. 5.1   Wave functions of stationary states of the harmonic oscillator for $n = 1$ and $n = 2$.

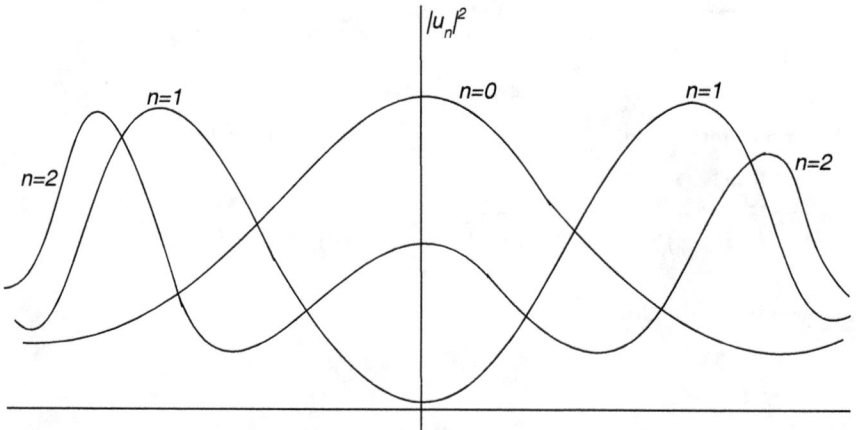

Fig. 5.2   Position probability density in the states $n = 0, 1, 2$.

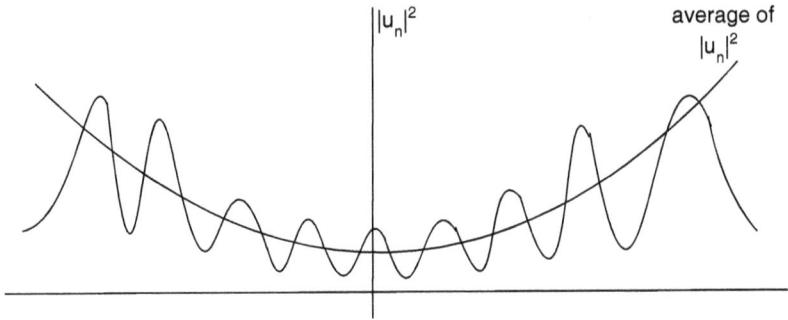

Fig. 5.3   Position probability density in a state with large n.

5.4   The Average Values

First, we note from Eqs. (5.13) and (5.14).

$$x = \sqrt{\frac{\hbar}{m\omega}}y = \sqrt{\frac{\hbar}{2m\omega}}(a + a^\dagger), \tag{5.65}$$

$$\hat{p} = -i\hbar\frac{\partial}{\partial x} = -i\sqrt{m\hbar\omega}\frac{\partial}{\partial y} = -i\sqrt{\frac{m\hbar\omega}{2}}(a - a^\dagger). \tag{5.66}$$

The energy eigenfunctions $u_n(x)$ of a simple harmonic oscillator form a complete orthonormal set, so that any arbitrary function $\Psi(x,t)$ can be expanded in terms of them

$$\Psi(x,0) = \Psi(x) = \sum_n C_n u_n(x), \tag{5.67}$$

$$\Psi(x,t) = \sum_n C_n u_n(x)e^{\frac{-iE_n t}{\hbar}}. \tag{5.68}$$

If $\Psi(x)$ is normalized:

$$\int \Psi^*(x)\Psi(x)dx = 1.$$

Thus

$$\int \sum_n \sum_k C_n^* C_k u_n^*(x) u_k(x) dx = 1$$

or

$$\sum_n \sum_k \delta_{nk} C_n^* C_k = 1$$

or

$$\sum_n |C_n|^2 = 1.$$

The average value of energy in the state $\Psi(x,t)$ is given by

$$\langle E \rangle \equiv \bar{E} = \int \Psi^*(x,t) H \Psi(x,t) dx$$

$$= \sum_n \sum_k C_n^* C_k e^{\frac{i(E_n - E_k)t}{\hbar}} \int u_n^*(x) H u_k(x) dx$$

$$= \sum_n \sum_k C_n^* C_k e^{\frac{i(E_n - E_k)t}{\hbar}} E_k \delta_{nk}$$

$$= \sum_n |C_n|^2 E_n.$$

Thus if a system is in state $\Psi(x,t)$, the probability of obtaining the result E_n is equal to $|C_n|^2$. The average value of x in the state $\Psi(x,t)$:

$$\langle x \rangle \equiv \bar{x} = \int \Psi^*(x,t) x \Psi(x,t) dx$$

$$= \sum_n \sum_k C_n^* C_k e^{i(n-k)\omega t} x_{nk}$$

where

$$x_{nk} = \int u_n^*(x) x u_k(x) dx = (u_n | x u_k)$$

$$= \sqrt{\frac{\hbar}{2m\omega}} (u_n |(a + a^\dagger) u_k).$$

On using Eq. (5.65). From Eqs. (5.39) and (5.40).

$$a u_k = \sqrt{k} u_{k-1},$$
$$a^\dagger u_k = \sqrt{k+1} u_{k+1}.$$

Hence, we have

$$x_{nk} = \sqrt{\frac{\hbar}{2m\omega}} [\sqrt{k} \delta_{n,k-1} + \sqrt{k+1} \delta_{n,k+1}]$$

$$= \sqrt{\frac{\hbar}{2m\omega}} [\sqrt{n+1} \delta_{k,n+1} + \sqrt{n} \delta_{k,n-1}]$$

and

$$\langle x \rangle \equiv \bar{x} = \sum_n \sum_k C_n^* C_k e^{i(n-k)\omega t} [\sqrt{\frac{\hbar}{2m\omega}} (\sqrt{n+1}\delta_{k,n+1} + \sqrt{n}\delta_{k,n-1})]$$

$$= \sqrt{\frac{\hbar}{2m\omega}} (\sum_n C_n^* C_{n+1} e^{-i\omega t}\sqrt{n+1} + \sum_n C_n^* C_{n-1} e^{i\omega t}\sqrt{n})$$

$$= \sqrt{\frac{\hbar}{2m\omega}} (\sum_{n=1}^{\infty} \sqrt{n}(C_n^* C_{n-1} e^{i\omega t} + C_{n-1}^*)C_n e^{-i\omega t})$$

$$= \sqrt{\frac{\hbar}{2m\omega}} (\sum_{n=1}^{\infty} \sqrt{n}(|C_n||C_{n-1}|e^{i(\omega t + \delta_{n-1} - \delta_n)}$$

$$+ |C_{n-1}||C_n|e^{-i(\omega t + \delta_{n-1} - \delta_n)}))$$

$$= \sqrt{\frac{2}{m\omega^2}} \sum_n \sqrt{n\hbar\omega} |C_n||C_{n-1}| \cos(\omega t + \delta_{n-1} - \delta_n)$$

where we have written $C_n = |C_n|e^{i\delta_n}$.

We conclude that x_{nk} can be expressed as a matrix, but $\bar{x}$ is a sum of functions, each of which represents a classical motion for a simple harmonic oscillator.

5.5   Problems

5.1 Show that

$$\langle x \rangle_n = \int u_n^*(x) x u_n(x) dx = (u_n | x u_n) = 0 ,$$

$$\langle p \rangle_n - \int u_n^*(x)(-i\hbar \frac{\partial}{\partial x}) u_n(x) dx - (u_n | \hat{p} u_n) - 0 .$$

5.2 Show that

$$(u_n | a^2 u_n) = 0,$$
$$(u_n | a^{\dagger^2} u_n) = 0.$$

5.3 Show that for a simple harmonic oscillator

$$\langle V \rangle_n = \frac{1}{2}(n + \frac{1}{2})\hbar\omega$$

$$\langle T \rangle_n = \frac{1}{2}(n + \frac{1}{2})\hbar\omega$$

$$\langle T \rangle_n = \langle V \rangle_n \quad : \quad Virial \ Theorem.$$

Hint: First show that

$$(u_n|(a + a^\dagger)^2 u_n) = (2n + 1),$$
$$(u_n|(a - a^\dagger)^2 u_n) = (2n + 1).$$

5.4 Show that for a simple harmonic oscillator

$$\Delta x \Delta p = \hbar(n + \frac{1}{2}),$$

$$\Delta x = [\langle (x - \langle x \rangle)^2 \rangle]^{\frac{1}{2}},$$
$$\Delta p = [\langle (p - \langle p \rangle)^2 \rangle]^{\frac{1}{2}}.$$

5.5 Show that

$$p_{nk} = -i\sqrt{m\omega\hbar}[\sqrt{n+1}\delta_{k,n+1} - \sqrt{n}\delta_{k,n-1}],$$

$$\langle p \rangle = -\sqrt{2m\omega t} \sum_n \sqrt{n}|C_n||C_{n-1}| \sin(\omega t + \delta_{n-1} - \delta_n).$$

Further show that

$$\langle x \rangle_t = \langle x \rangle_0 + \frac{\langle p_0 \rangle}{m\omega} \sin \omega t$$

$$\langle p \rangle_t = \langle p \rangle_0 \cos \omega t - m\omega\langle x \rangle_0 \sin \omega t.$$

in complete correspondence with classical equations.

5.6 Verify that

$$\Psi(x,t) = \sqrt{\frac{2}{3}}[(\frac{\beta}{\sqrt{\pi}})^{\frac{1}{2}}e^{\frac{-i\omega t}{2}} + (\frac{2\beta^3}{\sqrt{\pi}})^{\frac{1}{2}}e^{\frac{-i3\omega t}{2}}]e^{\frac{-\beta^2 x^2}{2}}$$

is a solution of time dependent Schrödinger equation for a particle of mass m in a simple harmonic oscillator potential. Calculate expectation values of x and p and show that

$$\frac{d\bar{x}}{dt} = \frac{\bar{p}}{m}.$$

Hint, note that 1st and 2nd term in $\Psi(x,0)$ are normalized eigenfunctions of the time independent Schrödinger equation $H\Psi = E\Psi$ with eigenvalues $\frac{\hbar\omega}{2}$ and $\frac{3\hbar\omega}{2}$ respectively.

5.7 A particle of mass m moves in a potential

$$V(x) = \frac{1}{2}m\omega^2 x^2 \quad x > 0,$$
$$= \infty \quad x < 0.$$

find the energy eigenvalues.

5.8 Show that the Schrödinger eqution for a simple harmonic oscillator in momentum space is given by:

$$\frac{\partial^2}{\partial p^2}\phi(p) + \frac{2}{m\hbar^2\omega^2}(E - \frac{p^2}{2m})\phi(p) = 0.$$

5.9 Obtain the ground state and first excited state wave functions $\phi_0(p)$ and $\phi_1(p)$ in momentum space.

Answer:

$$\phi_0(p) = \left(\frac{1}{\pi m\hbar\omega}\right)^{\frac{1}{4}} e^{-\frac{p^2}{2m\hbar\omega}},$$

$$\phi_1(p) = \left(\frac{4}{\pi}\frac{1}{(m\hbar\omega)^3}\right)^{\frac{1}{4}} e^{-\frac{i\pi}{2}} e^{-\frac{p^2}{2m\hbar\omega}}.$$

The phase factor can be omitted.

Chapter 6

Angular Momentum

6.1 Introduction

Classically we define angular momentum by the equation

$$\mathbf{L} = \mathbf{r} \times \mathbf{p}. \tag{6.1}$$

In quantum mechanics, by the correspondence principle $\mathbf{L}$ is an operator

$$\hat{\mathbf{L}} = \hat{\mathbf{r}} \times \hat{\mathbf{p}}. \tag{6.2}$$

In the Schrödinger representation $\hat{\mathbf{r}} = \mathbf{r}$ and

$$\hat{\mathbf{p}} = -i\hbar\nabla, \tag{6.3}$$

where we know that $\hat{p}_i$ and $\hat{x}_j$ satisfy the commutation relation

$$[\hat{p}_i, \hat{x}_j] = -i\hbar\delta_{ij}. \tag{6.4}$$

From Eqs. (6.2) and (6.3), it follows that the components of angular momentum can be written as

$$L_x = -i\hbar\left(y\frac{\partial}{\partial z} - z\frac{\partial}{\partial y}\right), \tag{6.5a}$$

$$L_y = -i\hbar\left(z\frac{\partial}{\partial x} - x\frac{\partial}{\partial z}\right), \tag{6.5b}$$

$$L_z = -i\hbar\left(x\frac{\partial}{\partial y} - y\frac{\partial}{\partial x}\right). \tag{6.5c}$$

6.2 Properties of Angular Momentum

(1) We now proceed to show that

$$[L_x, L_y] = i\hbar L_z.$$

87

Now

$$[L_x, L_y]\psi = (L_x L_y - L_y L_x)\psi$$

$$= -\hbar^2 \Big((y\frac{\partial}{\partial z} - z\frac{\partial}{\partial y})(z\frac{\partial}{\partial x} - x\frac{\partial}{\partial z}) - (z\frac{\partial}{\partial x} - x\frac{\partial}{\partial z})$$

$$\times (y\frac{\partial}{\partial z} - z\frac{\partial}{\partial y}) \Big)\psi$$

$$= -\hbar^2 \Big(y\frac{\partial\psi}{\partial x} + yz\frac{\partial^2\psi}{\partial z\partial x} - z^2\frac{\partial^2\psi}{\partial y\partial x} - yx\frac{\partial^2\psi}{\partial z^2} + zx\frac{\partial^2\psi}{\partial y\partial z}$$

$$- zy\frac{\partial^2\psi}{\partial x\partial z} + xy\frac{\partial^2\psi}{\partial z^2} + z^2\frac{\partial^2\psi}{\partial x\partial y} - x\frac{\partial\psi}{\partial y} - xz\frac{\partial^2\psi}{\partial z\partial y} \Big)$$

$$= -(i\hbar)^2 \Big(x\frac{\partial}{\partial y} - y\frac{\partial}{\partial x} \Big)\psi$$

$$= i\hbar L_z\psi.$$

Hence (since ψ is arbitrary)

$$[L_x, L_y] = i\hbar L_z. \tag{6.6a}$$

Similarly

$$[L_y, L_z] = i\hbar L_x, \tag{6.6b}$$

$$[L_z, L_x] = i\hbar L_y. \tag{6.6c}$$

The relations (6.6) can be written symbolically

$$\mathbf{L} \times \mathbf{L} = i\hbar\mathbf{L}. \tag{6.7a}$$

Relations (6.6) can also be written as

$$[L_i, L_j] = i\hbar\varepsilon_{ijk}L_k, \quad i, j, k = 1, 2, 3. \tag{6.7b}$$

Since L_x, L_y, L_z do not commute with each other, we cannot determine them simultaneously. In other words it is not possible to assign a fixed direction to angular momentum.

Proof. Suppose there exists a simultaneous eigenfunction ψ of L_x, L_y, L_z with eigenvalues $\lambda_x, \lambda_y, \lambda_z$. Then

$$L_x\psi = \lambda_x\psi$$
$$L_y\psi = \lambda_y\psi$$
$$L_z\psi = \lambda_z\psi$$

are satisfied simultaneously. We have

$$L_x(L_y\psi) = \lambda_y(L_x\psi)$$
$$= \lambda_y\lambda_x\psi.$$

Likewise

$$L_y(L_x\psi) = \lambda_x(L_y\psi)$$
$$= \lambda_x\lambda_y\psi,$$

so that

$$(L_xL_y - L_yL_x)\psi = 0.$$

But

$$(L_xL_y - L_yL_x) = i\hbar L_z,$$

therefore

$$(L_xL_y - L_yL_x)\psi = i\hbar L_z\psi$$
$$= i\hbar\lambda_z\psi.$$

Hence

$$\lambda_z = 0.$$

Similarly

$$\lambda_x = \lambda_y = 0.$$

Hence we cannot determine simultaneously the eigenvalues λ_x, λ_y and λ_z, the exception being the eigenvalue zero for each component of angular momentum.

(2) We now show that the magnitude of the angular momentum $L^2 = L_x^2 + L_y^2 + L_z^2$ commutes with L_x, L_y, L_z.

Proof. $[L^2, L_z] = \lfloor L_x^2, L_z\rfloor + (x \to y)$
$$= L_x[L_x, L_z] + [L_x, L_z]L_x + L_y[L_y, L_z] + [L_y, L_z]L_y$$
$$= L_x(-i\hbar L_y) + (-i\hbar L_y)L_x + L_y(i\hbar L_x) + (i\hbar L_x)L_y$$
$$= 0.$$

Similarly

$$[L^2, L_x] = 0, \quad [L^2, L_y] = 0. \tag{6.8}$$

This means we can determine L^2 and one of the components of L, say L_z simultaneously, that is to say we can simultaneously determine the eigenvalues of L^2 and L_z.

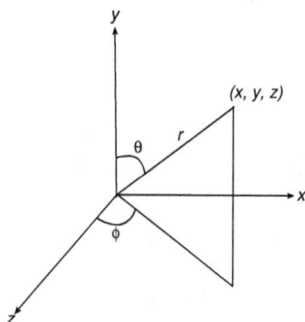

Fig. 6.1   The spherical polar coordinates of a point (x,y,z).

(3) It is convenient to express angular momentum operators in terms of spherical polar coordinates (these are natural coordinates for angular momentum),

$$x = r \sin \theta \cos \phi,$$
$$y = r \sin \theta \sin \phi,$$
$$z = r \cos \theta .$$

Then we have

$$L_z = -i\hbar \frac{\partial}{\partial \phi}, \tag{6.9a}$$

$$L_x = i\hbar \left(\sin \phi \frac{\partial}{\partial \theta} + \cot \theta \cos \phi \frac{\partial}{\partial \phi} \right), \tag{6.9b}$$

$$L_y = -i\hbar \left(\cos \phi \frac{\partial}{\partial \phi} - \cot \theta \sin \phi \frac{\partial}{\partial \phi} \right), \tag{6.9c}$$

$$L^2 = -\hbar^2 \left(\frac{1}{\sin \theta} \frac{\partial}{\partial \theta} \left(\sin \theta \frac{\partial}{\partial \theta} \right) + \frac{1}{\sin^2 \theta} \frac{\partial^2}{\partial \phi^2} \right) \tag{6.9d}$$

$$= -\hbar^2 \Omega ,$$

where

$$\Omega = \left(\frac{1}{\sin \theta} \frac{\partial}{\partial \theta} (\sin \theta \frac{\partial}{\partial \theta}) + \frac{1}{\sin^2 \theta} \frac{\partial^2}{\partial \phi^2} \right) \tag{6.9e}$$

is the angular part of the Laplacian ∇^2 in spherical polar coordinates.

Proof.

$$\mathbf{r} = r\hat{\mathbf{r}}_1$$

so that

$\hat{\mathbf{r}}_1 = \sin\theta\cos\phi\mathbf{i} + \sin\theta\sin\phi\mathbf{j} + \cos\theta\mathbf{k},$

$d\mathbf{r} = dr\hat{\mathbf{r}}_1 + r\sin\theta(-\sin\phi\mathbf{i} + \cos\phi\mathbf{j})d\phi + r\cos\theta d\theta(\cos\phi\mathbf{i} + \sin\phi\mathbf{j}) + r(-\sin\theta)d\theta\mathbf{k}.$

Compare it with

$$d\mathbf{r} = dr\hat{\mathbf{r}}_1 + rd\theta\hat{\boldsymbol{\theta}}_1 + r\sin\theta d\phi\hat{\boldsymbol{\phi}}_1.$$

Thus

$$\hat{\boldsymbol{\theta}}_1 = \cos\theta(\cos\phi\mathbf{i} + \sin\phi\mathbf{j}) - \sin\theta\mathbf{k},$$

$$\hat{\boldsymbol{\phi}}_1 = -\sin\theta\mathbf{i} + \cos\theta\mathbf{j}.$$

In polar coordinates

$$\boldsymbol{\nabla} = \hat{\mathbf{r}}_1\frac{\partial}{\partial r} + \hat{\boldsymbol{\theta}}_1\frac{1}{r}\frac{\partial}{\partial\theta} + \hat{\boldsymbol{\phi}}_1\frac{1}{r\sin\theta}\frac{\partial}{\partial\phi},$$

$$\nabla^2 = \frac{1}{r^2}\frac{\partial}{\partial r}(r^2\frac{\partial}{\partial r}) + \frac{1}{r^2}\Omega(\theta,\phi), \tag{6.10}$$

therefore $[\mathbf{r} = r\hat{\mathbf{r}}_1, \quad \hat{\mathbf{r}}_1\times\hat{\boldsymbol{\theta}}_1 = \hat{\boldsymbol{\phi}}_1, \quad \hat{\mathbf{r}}_1\times\hat{\boldsymbol{\phi}}_1 = -\hat{\boldsymbol{\theta}}_1]$

$$\mathbf{L} = -i\hbar\mathbf{r}\times\boldsymbol{\nabla}$$

$$= -i\hbar\left(\hat{\boldsymbol{\phi}}\frac{\partial}{\partial\theta} - \hat{\boldsymbol{\theta}}\frac{1}{\sin\theta}\frac{\partial}{\partial\phi}\right)$$

$$= -i\hbar\Big((-\sin\phi\mathbf{i} + \cos\phi\mathbf{j})\frac{\partial}{\partial\theta} - (\cos\theta\cos\phi\mathbf{i}$$

$$+ \cos\theta\sin\phi\mathbf{j} - \sin\theta\mathbf{k})\frac{1}{\sin\theta}\frac{\partial}{\partial\phi}\Big)$$

$$= -i\hbar\Big(\mathbf{i}(-\sin\phi\frac{\partial}{\partial\theta} - \cot\theta\cos\phi\frac{\partial}{\partial\phi}) + \mathbf{j}(\cos\phi\frac{\partial}{\partial\theta} - \cot\theta$$

$$\times\sin\phi\frac{\partial}{\partial\phi}) + \mathbf{k}\frac{\partial}{\partial\phi}\Big).$$

Hence we have Eqs. (6.9a), (6.9b), (6.9c).

It is convenient to introduce

$$L_+ \equiv L_x + iL_y = \hbar e^{i\phi}\left(\frac{\partial}{\partial\theta} + i\cot\theta\frac{\partial}{\partial\phi}\right), \tag{6.11a}$$

$$L_- \equiv L_x - iL_y = \hbar e^{-i\phi}\left(-\frac{\partial}{\partial\theta} + i\cot\theta\frac{\partial}{\partial\phi}\right). \tag{6.11b}$$

Then

$$L_+L_- = L_x^2 + L_y^2 - i[L_x, L_y] = L_x^2 + L_y^2 + L_z^2 - L_z^2 + \hbar L_z$$

$$L^2 = L_+L_- - L_z^2 - \hbar L_z$$

$$= \hbar^2 e^{i\phi}\left(\frac{\partial}{\partial\theta} + i\cot\theta\frac{\partial}{\partial\phi}\right)e^{-i\phi}\left(-\frac{\partial}{\partial\theta} + i\cot\theta\frac{\partial}{\partial\phi}\right) - \hbar^2\frac{\partial^2}{\partial\phi^2} + i\hbar^2\frac{\partial}{\partial\phi}$$

$$= -\hbar^2[\frac{\partial^2}{\partial\theta^2} + \cot\theta\frac{\partial}{\partial\theta} + \frac{1}{\sin^2\theta}\frac{\partial^2}{\partial\phi^2}]$$

giving

$$L^2 = -\hbar^2 \left(\frac{1}{\sin\theta} \frac{\partial}{\partial\theta} (\sin\theta \frac{\partial}{\partial\theta}) + \frac{1}{\sin^2\theta} \frac{\partial^2}{\partial\phi^2} \right).$$

So Eq. (6.9d) is proved.

6.3   Eigenfunctions and Eigenvalues of Angular Momentum

We now determine the simultaneous eigenfunctions of L^2 and L_z and their eigenvalues.

Since L^2 and L_z commute, it is possible to find simultaneous eigenfunctions $\psi_{\lambda m}$ satisfying the eigenvalue equations

$$L_z \psi_{\lambda m} = m\hbar \psi_{\lambda m} \tag{6.12a}$$
$$L^2 \psi_{\lambda m} = \lambda \hbar^2 \psi_{\lambda m}, \tag{6.12b}$$

where $m\hbar$ and $\lambda\hbar^2$ are eigenvalues of L_z and L^2 respectively corresponding to eigenfunctions $\psi_{\lambda m}$. We have extracted out factors $\hbar$ and $\hbar^2$ for convenience. Using Eq. (6.9a), we have from Eq. (6.12a)

$$-i\hbar \frac{\partial}{\partial\phi} \psi_{\lambda m} = \hbar m \psi_{\lambda m}. \tag{6.13}$$

The solution of this equation is of the form

$$\psi_{\lambda m} = f e^{im\phi}, \tag{6.14}$$

where f is independent of ϕ but is in general a function of r and θ. Since the same physical position is denoted by $\phi + 2n\pi$, n being an integer, we must impose the boundary condition that $\psi_{\lambda m}$ is periodic in ϕ with a period 2π that is

$$f e^{im\phi} = f e^{im(\phi+2n\pi)}$$

or

$$e^{2imn\pi} = 1.$$

This is possible only if

$$m = 0, \pm 1, \pm 2, \ldots$$

Thus eigenvalues of L_z are

$$m\hbar, \quad m = 0, \pm 1, \pm 2, \ldots$$

The factor depending on ϕ which characterises the eigenfunction of L_z is denoted by

$$\Phi_m(\phi) = \frac{1}{\sqrt{2\pi}} e^{im\phi}. \tag{6.15}$$

This function is normalised so that

$$\int_0^{2\pi} \Phi_m^*(\phi)\Phi_{m'}(\phi)d\phi = \delta_{mm'}. \tag{6.16}$$

We write

$$\psi_{\lambda m} = R(r)Y_{\lambda m}(\theta, \phi), \tag{6.17}$$

where we have separated the dependence on r, since neither L_z nor L^2 depend on r. We can then write the eigenvalue equations (6.12) as

$$L_z Y_{\lambda m}(\theta, \phi) = m\hbar Y_{\lambda m}(\theta, \phi), \tag{6.18a}$$

$$L^2 Y_{\lambda m}(\theta, \phi) = \lambda\hbar^2 Y_{\lambda m}(\theta, \phi). \tag{6.18b}$$

Since L_z depends only on ϕ and its eigenfunctions are characterised by $\Phi_m(\phi)$ we can write

$$Y_{\lambda m}(\theta, \phi) = P_{\lambda m}(\theta)\Phi_m(\phi). \tag{6.19}$$

Substituting Eq. (6.19) in Eq. (6.18b) with L^2 given in Eq. (6.9d), we have

$$\left(\frac{1}{\sin\theta}\frac{\partial}{\partial\theta}(\sin\theta\frac{\partial}{\partial\theta}) + \frac{1}{\sin^2\theta}\frac{\partial^2}{\partial\phi^2}\right) P_{\lambda m}(\theta)e^{im\phi} = -\lambda P_{\lambda m}(\theta)e^{im\phi}$$

or

$$\left(\frac{1}{\sin\theta}\frac{d}{d\theta}(\sin\frac{d}{d\theta}) - \frac{m^2}{\sin^2\theta}\right) P_{\lambda m}(\theta) = -\lambda P_{\lambda m}(\theta). \tag{6.20}$$

The dependence on ϕ has been eliminated. This is an eigenvalue equation for λ, which determines the possible eigenvalues of L^2. Since eigenfunctions are properly normalised

$$\int |Y_{\lambda m}|^2 d\Omega = 1. \tag{6.21a}$$

Now

$$d\Omega = \sin\theta d\theta d\phi.$$

This leads to (using Eq. (6.15))

$$\int_0^\pi |P_{\lambda m}(\theta)|^2 \sin\theta d\theta = 1. \tag{6.21b}$$

To find the eigenvalues λ of L^2, we proceed as follows. We have already introduced in Eq. (6.11)

$$L_\pm = L_x \pm iL_y$$

$$L_+L_- = L^2 - L_z^2 + \hbar L_z, \quad L_-L_+ = L^2 - L_z^2 - \hbar L_z \qquad (6.22)$$

where $L_\pm$ satisfy the commutation relation,

$$
\begin{aligned}
[L_z, L_\pm] &= [L_z, L_x] \pm i[L_z, L_y] \\
&= i\hbar L_y \pm i(-i\hbar L_x) \\
&= \pm \hbar L_\pm
\end{aligned}
\qquad (6.23a)
$$

$$
\begin{aligned}
[L_+, L_-] &= [L_x, L_x] - i[L_x, L_y] + i[L_y, L_x] + [L_y, L_y] \\
&= -i(i\hbar L_z) + i(-i\hbar L_z) \\
&= 2\hbar L_z
\end{aligned}
\qquad (6.23b)
$$

$$[L^2, L_\pm] = [L^2, L_x] \pm i[L^2, L_y] = 0. \qquad (6.23c)$$

Theorem: If $(L_x \pm iL_y)Y_{\lambda m} \equiv L_\pm Y_{\lambda m}$ are not zero then they are simultaneous eigenfunctions of L^2 and L_z belonging to eigenvalues $\lambda \hbar^2$, $(m \pm 1)\hbar$ respectively.

Proof.

$$
\begin{aligned}
L^2(L_\pm Y_{\lambda m}) &= L_\pm(L^2 Y_{\lambda m}) \\
&= \lambda \hbar^2 (L_\pm Y_{\lambda m}), \\
L_z(L_\pm Y_{\lambda m}) &= \{[L_z, L_\pm] + L_\pm L_z\}Y_{\lambda m} \\
&= \pm \hbar L_\pm Y_{\lambda m} + m\hbar L_\pm Y_{\lambda m} \\
&= (m \pm 1)\hbar (L_\pm Y_{\lambda m}). \qquad (6.24)
\end{aligned}
$$

L_+ is called the raising operator and L_- is called the lowering operator. We can write

$$L_\pm Y_{\lambda m} = C_{\lambda m}^\pm Y_{\lambda m \pm 1}, \qquad (6.25)$$

where $C_{\lambda m}^\pm$ are constants and can be fixed from the normalisation condition which gives

$$\int Y_{\lambda m \pm 1}^* Y_{\lambda m \pm 1} d\Omega = 1$$

or

$$
\begin{aligned}
|C_{\lambda m}^\pm|^2 &= \int Y_{\lambda m}^* (L_\mp)(L_\pm) Y_{\lambda m} d\Omega \\
&= \int Y_{\lambda m}^* (L^2 - L_z^2 \mp \hbar L_z) Y_{\lambda m} d\Omega \\
&= (\lambda - m^2 \mp m)\hbar^2. \qquad (6.26a)
\end{aligned}
$$

Hence

$$(\lambda - m^2 \mp m) \geq 0. \qquad (6.26b)$$

First taking the positive sign, we have

$$\lambda - m^2 + m \geq 0$$

or

$$\lambda + \frac{1}{4} \geq (m - 1/2)^2. \qquad (6.27a)$$

Hence we have the inequality

$$-\sqrt{\lambda + 1/4} + \frac{1}{2} \leq m < \sqrt{\lambda + 1/4} + \frac{1}{2}. \qquad (6.27b)$$

Thus m is bounded on both sides. Therefore m has a minimum and a maximum value which we respectively denote by m_2 and m_1. Then we must have

$$L_- Y_{\lambda m_2} = 0, \qquad (6.28a)$$

$$L_+ Y_{\lambda m_1} = 0, \qquad (6.28b)$$

since otherwise Eq. (6.24) shows $L_- Y_{\lambda m_2}$ is eigenfunction of L_z with eigenvalue $(m_2 - 1)\hbar$ which contradicts that $m_2\hbar$ is the minimum eigenvalue. Similar argument holds for $L_+ Y_{\lambda m_1}$. Thus from Eq. (6.25)

$$C^+_{\lambda m_1} = 0, \qquad (6.29a)$$

$$C^-_{\lambda m_2} = 0. \qquad (6.29b)$$

Hence [c.f. Eq. (6.26a)]

$$\lambda - m_1^2 - m_1 = 0, \quad \lambda - m_2^2 + m_2 = 0.$$

Therefore

$$\lambda = m_1(m_1 + 1) = m_2(m_2 - 1) \qquad (6.30)$$

so that

$$(m_1 + m_2)[m_1 - m_2 + 1] = 0.$$

Therefore

$$m_1 = -m_2 \qquad \text{[The other solution } m_1 = m_2 - 1$$
$$\text{is not possible as } m_1 > m_2]$$
$$= \sqrt{\lambda + 1/4} - \frac{1}{2}. \qquad (6.31)$$

It is customary to designate $m_1\hbar$ the maximum eigenvalue of L_z by $l\hbar$, where l must be an integer including zero as the eigenvalues of L_z are integral multiples of $\hbar$. Also from the fact that $m_1 = l = -m_2$,

$$-l \le m \le l.$$

Thus l must be a positive integer or zero. Then from Eq. (6.30)

$$\lambda = l(l+1), \qquad l = 0, 1, 2, \cdots$$

so that eigenvalues of L^2 are $l(l+1)\hbar$ and we write eigenvalue Eqs. (6.18)

$$L_z Y_{lm}(\theta, \phi) = m\hbar Y_{lm}(\theta, \phi), \tag{6.32a}$$

$$L^2 Y_{lm}(\theta, \phi) = l(l+1)\hbar^2 Y_{lm}(\theta, \phi), \tag{6.32b}$$

where

$$m = 0, \pm 1, \pm 2, \cdots, \pm l$$
$$= -l, -l+1, \cdots, 0, 1, 2, \cdots, (l-1), l.$$

That is $(2l+1)$ different values in all . From Eq. (6.26a) choosing the phases of $C_{lm}^{\pm}$ to be zero, we get

$$C_{lm}^{\pm} = \sqrt{l(l+1) - m(m \pm 1)}\ \hbar,$$
$$= \sqrt{(l \mp m)(l \pm m + 1)}\ \hbar. \tag{6.33}$$

Now Eq. (6.25) becomes

$$L_{\pm} Y_{lm}(\theta, \phi) = \sqrt{(l \mp m)(l \pm m + 1)}\hbar Y_{l,m\pm 1}(\theta, \phi)\ . \tag{6.34}$$

With $\lambda = l(l+1)$ and writing $P_{\lambda m}$ as P_l^m, we see from Eq. (6.20) that P_l^m satisfies

$$\frac{1}{\sin\theta}\frac{d}{d\theta}\left(\sin\theta\frac{dP_l^m(\theta)}{d\theta}\right) + \left(l(l+1) - \frac{m^2}{\sin^2\theta}\right)P_l^m(\theta) = 0 \tag{6.35}$$

which is known as Legendre's equation and $P_l^m(\theta)$ are called associated Legendre's polynomials.

Hence from Eqs. (6.15) and (6.19) normalised simultaneous eigenfunction of L^2 and L_z are given by

$$Y_{lm}(\theta, \phi) = N_{lm} P_l^m(\theta)\Phi_m(\phi)$$
$$= (-1)^m \left(\frac{(2l+1)(l-m)!}{4\pi(l+m)!}\right)^{1/2} P_l^m(\cos\theta)e^{im\phi}. \tag{6.36}$$

where N_{lm} is the normalizing constant determined from Eq. (6.21a). $Y_{lm}(\theta, \phi)$ are called normalised spherical harmonics. In Eq. (6.36) P_l^m are given by [see problem 6.7]

$$P_l^m(\cos\theta) = \frac{(-1)^l}{2^l l!}\sin^m\theta\left(\frac{d}{d(\cos\theta)}\right)^{l+m}(\sin\theta)^{2l}$$

or

$$P_l^m(\omega) = \frac{(1-\omega^2)^{m/2}}{2^l l!} \frac{d^{l+m}}{d\omega^{l+m}} (\omega^2 - 1)^l, \qquad (6.37)$$

where $\omega = \cos\theta$, $P_l^m(\omega)$ are known as Associated Legendre Polynomials.

It may be noted that for $m = 0$, we get

$$P_l^0(\omega) = \frac{1}{2^l l!} \frac{d^l}{d\omega^l} (\omega^2 - 1) = P_l(\omega), \qquad (6.38)$$

which are called Legendre Polynomials.

6.4 Properties of Associated Legendre Polynomials

The following properties of Associated Legendre Polynomials are given without proof

(1) $P_l^m(\omega)$ are orthogonal, that is

$$\int_{-1}^{+1} P_l^m(\omega) P_{l'}^m(\omega) d\omega = \delta_{ll'} \frac{2}{(2l+1)} \frac{(l+m)!}{(l-m)!}, \qquad (6.39)$$

(2) $\quad P_l^{-m}(\omega) = \dfrac{(-1)^m (l-m)!}{(l+m)!} P_l^m(\omega), \qquad (6.40)$

(3) $\quad P_l^m(-\omega) = (-1)^{l+m} P_l^m(\omega), \qquad (6.41)$

(4) $\quad Y_{l-m}(\theta, \phi) = (-1)^m \left[(2l+1) \dfrac{(l+m)!}{4\pi(l-m)!} \right]^{1/2} P_l^{-m}(\omega) e^{-im\phi}$

$$= (-1)^m Y_{lm}^*(P, \phi), \qquad (6.42)$$

(5) $\quad Y_{l0}(P, \phi) = \left(\dfrac{2l+1}{4\pi} \right)^{1/2} P_l(\cos\theta). \qquad (6.43)$

Equation (6.41) follows from Eq. (6.37) and Eq. (6.40) follows from the Leibnitz Theorem:

$$P_l^m(\omega) = \frac{(-1)^m}{2^l l!} \frac{(l+m)!}{(l-m)!} (1-\omega^2)^{-m/2} \frac{d^{l-m}}{d\omega^{l-m}} (\omega^2 - 1).$$

For $l = 0$, $m = 0$, $Y_{00} =$ constant. In other words the wave function of the state of a particle with zero angular momentum depends only on r, that is it has complete spherical symmetry.

The first four spherical harmonics corresponding to $l = 0$ and $l = 1$ are given by

l	$m = 1$	$m = 0$	$m = -1$
0		$Y_{00} = \frac{1}{(4\pi)^{\frac{1}{2}}}$	
1	$Y_{11} = -(\frac{3}{8\pi})^{\frac{1}{2}} \sin\theta e^{i\phi}$,	$Y_{10} = (\frac{3}{4\pi})^{\frac{1}{2}} \cos\theta$,	$Y_{1-1} = (\frac{3}{8\pi})^{\frac{1}{2}} \sin\theta e^{-i\phi}$

One final remark: Classically angular momentum vector may have any orientation. In quantum mechanics it is restricted to values for which its z-component is an integral multiple of $\hbar$. This fact is sometimes illustrated on a vector diagram. Consider, for example, $l = 2$ for which $l(l+1)\hbar^2 = 6\hbar^2$. Possible values of m are $-2, -1, 0, 1, 2$ and possible orientations of L are then illustrated as in the figure below (the radius of the semi-circle is $\sqrt{6}\hbar$)

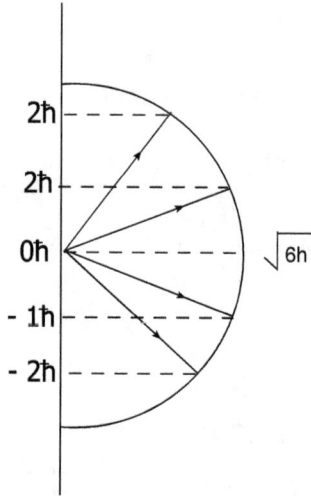

Fig. 6.2

6.5   Parity of a State

Previously we introduced the concept of parity in Sec. 4.3. We saw that in one dimensional problems if the potential is symmetric

$$V(x) = V(-x)$$

about the origin, then $\psi(-x)$ is also a solution of the Schrödinger equation

$$\left(-\frac{\hbar^2}{2m}\frac{\partial^2}{\partial x^2} + V(x)\right)\psi(x) = E\psi(x)$$

and the parity of $\psi(x)$ is defined by η

$$\psi(x) = \eta\psi(-x)$$

with $\eta = \pm 1$. The transformation $x \to -x$ corresponds in three dimensions to

$$\mathbf{r} \to \mathbf{r}' = -\mathbf{r}.$$

Such a transformation is called the reflection of coordinates. Now in spherical polar coordinates

$$\begin{cases} x = r \sin \theta \cos \phi, \\ y = r \sin \theta \sin \phi, \\ z = r \cos \theta. \end{cases} \tag{6.44}$$

If the axes are reflected to $\mathbf{r}' = (x', y', z')$ and if r', θ' and ϕ' are defined in the usual way, but with respect to x', y', z', then it is easy to see that $\mathbf{r} \to \mathbf{r}' = -\mathbf{r}$ corresponds to

$$\begin{cases} r' = r, \\ \theta' = \pi - \theta, \\ \phi' = \pi + \phi. \end{cases} \tag{6.45}$$

Consider now a particle in an angular momentum state $Y_{lm}(\theta, \phi) =$ constant $\times P_l^m(\cos\theta)e^{im\phi}$ which specifies the dependence of the particle's state on angles θ and ϕ. When $\phi \to \pi + \phi$, $e^{im\phi} \to e^{im\pi} e^{im\phi} = (-1)^m e^{im\phi}$. When $\theta \to \pi - \theta$, $P_l^m(\cos\theta) \to P_l^m(-\cos\theta) = (-1)^{l+m} \times P_l^m(\cos\theta)$. Thus under reflection of coordinates,

$$Y_{lm}(\theta, \phi) \to Y_{lm}(\theta', \phi') = \text{constant} \times (-1)^{l+m}(-1)^m P_l^m(\cos\theta)e^{im\phi}$$
$$= (-1)^l Y_{lm}(\theta, \phi). \tag{6.46}$$

Since, as is clear from Eq. (6.45) the dependence of the wave function on r is unchanged by reflection, the parity of any state specified by angular momentum l, m is equal to $(-1)^l$ and is determined by the l value only. $(-1)^l$ is often called the orbital parity of a state with an angular momentum specified by l and m values.

6.6 Problems

6.1 Consider an operator V_+ which satisfies the commutation relations

$$[L_+, V_+] = 0,$$
$$[L_z, V_+] = \hbar V_+.$$

Using these relations, show that

$$L_z(V_+ Y_{ll}) = (l+1)\hbar(V_+ Y_{ll}),$$
$$L^2(V_+ Y_{ll}) = (l+1)(l+2)\hbar^2(V_+ Y_{ll}).$$

6.2 Show that when the system is described by a state function $\psi(\mathbf{r})$,

$$\Delta L_x \Delta L_y \geq \frac{\hbar}{2}|\langle L_z \rangle|$$

where

$$(\Delta L_x)^2 = \langle (L_x - \langle L_x \rangle)^2 \rangle$$
$$(\Delta L_y)^2 = \langle (L_y - \langle L_y \rangle)^2 \rangle$$

and $\langle L_x \rangle$ and $\langle L_y \rangle$ denote the average values of L_x and L_y in the state $\psi(\mathbf{r})$. If $\psi(\mathbf{r})$ is normalised eigenfunction of L_z, with eigenvalue $m\hbar$, what is $\Delta L_x \Delta L_y$?

6.3 Find the commutator

$$[L^2, x_i], \quad i = 1, 2, 3$$

by using the result

$$[x_i, \hat{p}_j] = i\hbar \delta_{ij}$$

6.4 Show that for a state ψ_{lm}, such that

$$L_z \psi_{lm} = m\hbar \psi_{lm},$$
$$L^2 \psi_{lm} = l(l+1)\hbar^2 \psi_{lm},$$

the average values of L_x^2, L_y^2 are given by

$$\langle L_x^2 \rangle = \langle L_y^2 \rangle = \frac{l(l+1)\hbar^2 - m^2\hbar^2}{2} \quad .$$

6.5 Show that for an eigenstate ψ_{lm} of L_z, the average value of L_x is zero.

6.6 Show that

$$[L^2, [L^2, \hat{x}]] = 2\hbar^2(\hat{x}L^2 + L^2\hat{x}).$$

Hint: Use the commutation relation

$$[\hat{x}_i, \hat{p}_j] = ih\delta_{ij}$$

and the definition

$$L_i = \varepsilon_{ijk}\hat{x}_j\hat{p}_k$$

to show that

$$[L_i, \hat{x}_j] = +i\hbar\varepsilon_{ijt}\hat{x}_t,$$

$$[\mathbf{L}^2, \hat{x}_l] = (-ih)\sum_{ij}\varepsilon_{ijl}[L_i\hat{x}_j + \hat{x}_j L_i] \quad .$$

Using these relations, derive the final result.

6.7 Using the relation derived in the text

$$L_{\pm}Y_{lm}(\theta, \phi) = \sqrt{(l \mp m)(l \pm m + 1)}\hbar Y_{l,m\pm 1}(\theta, \phi)$$

and the expressions for $L_{\pm}$ in spherical polar coordinates, show that

$$P_l^{m\pm 1}(\theta) = [(l \mp m)(l \pm m + 1)]^{-1/2}(\pm\frac{\partial}{\partial\theta} - m\cot\theta)P_l^m(\theta)$$

$$= [(l \mp m)(l \pm m + 1)]^{-1/2}(\pm)\sin^{\pm m}\theta\frac{d}{d\theta}$$

$$\times \left(\sin^{\mp m}\theta P_{lm}(\theta)\right).$$

Chapter 7

Motion in Centrally Symmetric Field

7.1 Introduction

Consider a spinless particle of mass μ in a potential $V(r)$. The Hamiltonian is given by

$$H = \frac{\hat{p}^2}{2\mu} + V(r), \quad \hat{\mathbf{p}} = -i\hbar\nabla. \tag{7.1}$$

If the potential is spherically symmetric, $V(\mathbf{r}) = V(r)$

$$H = -\frac{\hbar^2}{2\mu}\nabla^2 + V(r). \tag{7.2}$$

The Schrödinger equation for such a system is

$$i\hbar\frac{\partial\psi}{\partial t} = H\psi$$

$$= \left(-\frac{\hbar^2}{2\mu}\nabla^2 + V(r)\right)\psi. \tag{7.3}$$

Such a system represents a hydrogen like atom in which an electron of mass μ is in a potential of the nucleus (which is regarded as fixed, because it is very heavy as compared with the electron), so that $V(r) = -\frac{Ze^2}{r}$ where Ze is the charge of nucleus. We are interested in the stationary states of the system, that is we want to find the energy eigenvalues of the system. For a stationary state

$$\psi(\mathbf{r}, t) = u(\mathbf{r})e^{-i(E/\hbar)t}. \tag{7.4}$$

Substituting in Eq. (7.3), we have

$$\left(-\frac{\hbar^2}{2\mu}\nabla^2 + V(r)\right)u(\mathbf{r}) = Eu(\mathbf{r}). \tag{7.5}$$

103

From a mathematical point of view, it is important to find a coordinate system in which the equation is simplified. Here V is a function of $r = |\mathbf{r}|$ alone, so it is convenient to use spherical polar coordinates r, θ, ϕ. Then

$$\nabla^2 u = \left(\frac{1}{r^2} \frac{\partial}{\partial r} \left(r^2 \frac{\partial}{\partial r} \right) + \frac{1}{r^2} \Omega(\theta, \phi) \right) u$$

$$= \frac{1}{r} \frac{\partial^2}{\partial r^2} (ru) + \frac{\Omega(\theta, \phi)}{r^2} u, \qquad (7.6a)$$

where

$$\Omega(\theta, \phi) = \frac{1}{\sin \theta} \frac{\partial}{\partial \theta} \left(\sin \theta \frac{\partial}{\partial \theta} \right) + \frac{1}{\sin^2 \theta} \frac{\partial^2}{\partial \phi^2}$$

$$= -\frac{1}{\hbar^2} L^2 \qquad (7.6b)$$

L is the magnitude of the orbital angular momentum of the particle. Substituting Eq. (7.6) in Eq. (7.5), we have

$$\left(-\frac{\hbar^2}{2\mu} \frac{1}{r^2} \frac{\partial}{\partial r} \left(r^2 \frac{\partial}{\partial r} \right) + \frac{L^2(\theta, \phi)}{2\mu r^2} + V(r) \right) u = Eu. \qquad (7.7)$$

Thus for a central field of force, the Hamiltonian operator can be written in the form

$$H = -\frac{\hbar^2}{2\mu} \frac{1}{r^2} \frac{\partial}{\partial r} \left(r^2 \frac{\partial}{\partial r} \right) + \frac{L^2(\theta, \phi)}{2\mu r^2} + V(r). \qquad (7.8)$$

Since L^2 and L_z do not depend on r and since

$$[L^2, L^2] = [L^2, L_z] = 0$$

it follows that

$$[H, L^2] = 0, \quad [H, L_z] = 0.$$

Thus it is possible to find simultaneous eigenfunctions of H, L^2 and L_z. But we know that

$$L^2 Y_{lm}(\theta, \phi) = l(l+1)\hbar^2 Y_{lm}(\theta, \phi),$$
$$L_z Y_{lm}(\theta, \phi) = m\hbar Y_{lm}(\theta, \phi).$$

Therefore we can write the solution of Eq. (7.7) in the form

$$u(r, \theta, \phi) = R(r) Y_{lm}(\theta, \phi), \qquad (7.9)$$

Hence using Eq. (7.9), we get from Eq. (7.7) the differential equation for the radial part

$$-\frac{\hbar^2}{2\mu} \frac{1}{r^2} \frac{\partial}{\partial r} \left(r^2 \frac{\partial R}{\partial r} \right) + \frac{l(l+1)}{2\mu r^2} \hbar^2 R + V(r) R = E R(r). \qquad (7.10)$$

We note that in this equation $L_z = m\hbar$ does not enter — this is in accordance with the $(2l + 1)$ fold degeneracy of the levels with which we are already familiar. We transform this equation into a more convenient form by making the substitution

$$R(r) = \frac{\chi(r)}{r}. \tag{7.11}$$

Thus Eq. (7.10) becomes

$$\left(-\frac{\hbar^2}{2\mu}\frac{\partial^2}{\partial r^2} + \frac{\hbar^2}{2\mu}\frac{l(l+1)}{r^2} + V(r)\right)\chi(r) = E\chi(r). \tag{7.12}$$

In this form in $H = T + V$

$$T = -\frac{\hbar^2}{2\mu}\frac{\partial^2}{\partial r^2} + \frac{l(l+1)\hbar^2}{2\mu r^2}. \tag{7.13}$$

Thus one has a simple physical interpretation

$$-\frac{\hbar^2}{2\mu}\frac{\partial^2}{\partial r^2} \longrightarrow \frac{\hat{p}_r^2}{2\mu}, \tag{7.14a}$$

where p_r is the radial momentum and then the transverse momentum p_t is given by

$$\frac{p_t^2}{2\mu} = \frac{l(l+1)\hbar^2}{2\mu r^2}, \tag{7.14b}$$

so that

$$T = \frac{1}{2\mu}(\hat{p}_r^2 + \hat{p}_t^2). \tag{7.14c}$$

Remarks:

(1) Element of volume in polar coordinates is $r^2 dr d\Omega$ with $d\Omega = \sin\theta d\theta d\phi$. But $Y_{lm}(\theta, \phi)$ are normalised over $d\Omega$. Hence normalisation condition on $R(r)$ is determined by

$$\int_0^\infty |R(r)|^2 r^2 dr = 1 \tag{7.15a}$$

and that for $\chi(r)$ by

$$\int |\chi(r)|^2 dr = 1. \tag{7.15b}$$

(2) The condition on the wave function $R(r)$ is that it should be finite everywhere. Since $R(r) = \frac{\chi(r)}{r}$, we require that $\chi(r) \to 0$ at least as fast as $r \to 0$. In particular $\chi(0) = 0$.

(3) Equation (7.12) is formally identical with Schrödinger's equation for one dimensional motion in a field of potential

$$V_l(r) = V(r) + \frac{\hbar^2}{2\mu} \frac{l(l+1)}{r^2}. \tag{7.15c}$$

The term $\frac{\hbar^2}{2\mu} \frac{l(l+1)}{r^2}$ is called the centrifugal potential or barrier; this gives a repulsive potential. For small r, it is likely to be large. It prevents a particle with nonzero angular momentum from getting too close to the origin, hence the name centrifugal barrier. We assume $r^2 V(r) \to 0$ as $r \to 0$ which is satisfied by the Coulomb potential.

7.2   Coulomb Potential

We can go as far as Eq. (7.12) without specifying $V(r)$. From now onwards we consider the specific problem of hydrogen like atoms for which

$$V(r) = -\frac{Ze^2}{r} \text{ (Coulomb potential)}$$

where Ze is the charge of nucleus. This is an important problem in atomic physics. From Eq. (7.12) for $\chi(r)$, we have for $V(r) = -\frac{Ze^2}{r}$

$$\frac{d^2\chi}{dr^2} + \frac{2\mu}{\hbar^2}\left(E + \frac{Ze^2}{r} - \frac{\hbar^2}{2\mu r^2}l(l+1)\right)\chi = 0. \tag{7.16}$$

We want to solve this equation under the boundary condition $\chi(r) \to 0$, as $r \to 0$. Let us put

$$x = \frac{\sqrt{2\mu}}{\hbar}r, \quad K = \frac{Ze^2}{\hbar}\sqrt{2\mu}, \quad E = -\varepsilon^2. \tag{7.17}$$

Equation (7.16) now becomes

$$\frac{d^2\chi}{dx^2} + \left(\frac{K}{x} - \frac{l(l+1)}{x^2} - \varepsilon^2\right)\chi = 0. \tag{7.18}$$

First we examine the asymptotic behaviour. As $x \to \infty$

$$\frac{K}{x} \text{ and } \frac{l(l+1)}{x^2} \longrightarrow 0$$

and then

$$\frac{d^2\chi}{dx^2} - \varepsilon^2\chi = 0. \tag{7.19}$$

Therefore $\chi \to e^{\pm\varepsilon x}$ for large x. We must distinguish between two cases $E > 0$ and $E < 0$. We consider here only the case $E < 0$. The solution

$\chi = e^{+\varepsilon x} \to \infty$ as $x \to \infty$ and therefore, is rejected. Thus the only possible solution is $\chi = e^{-\varepsilon x}$. We now examine the behaviour of Eq. (7.18) for small values of x. Multiply Eq. (7.18) by x^2 throughout. We have

$$x^2 \frac{d^2\chi}{dx^2} - l(l+1)\chi = 0. \tag{7.20}$$

We seek a solution of the form

$$\chi(x) = \text{constant} \times x^s. \tag{7.21a}$$

Substituting this in Eq. (7.20) gives

$$s(s-1) - l(l+1) = 0,$$
$$s = l+1 \quad \text{or} \quad s = -l. \tag{7.21b}$$

Now $l \geq 0$, therefore $s = -l$ is not a permissible solution as it contradicts the boundary condition $\chi(x) \to 0$ as $x \to 0$. Thus the boundary condition requires that

$$s = (l+1).$$

Therefore $\chi(x)$ can be written in the form

$$\chi(x) = e^{-\varepsilon x} x^{l+1} g(x), \tag{7.22}$$

where the expression (7.22) for $\chi(x)$ takes account of the behaviour of the function for large and small x and $g(x)$ is an unknown function. Substituting this expression for $\chi(x)$ in Eq. (7.18), we obtain

$$x \frac{d^2g}{dx^2} + 2(l+1-\varepsilon x)\frac{dg}{dx} + (K - 2\varepsilon l - 2\varepsilon)g = 0. \tag{7.23}$$

We now look for a solution of this equation in the form

$$g(x) = \sum_s C_s x^s = C_0 + C_1 x + C_2 x^2 + \cdots . \tag{7.24}$$

Then substituting Eq. (7.24) into Eq. (7.23), we get

$$\sum_s \left(C_s s(s-1)x^{s-1} + 2(l+1-\varepsilon x)C_s s x^{s-1} + (K - 2\varepsilon l - 2\varepsilon)C_s x^s \right) = 0, \tag{7.25}$$

which gives on equating the coefficient of x^{s-1} to zero, the following relations between successive coefficients C_s

$$C_s \left(s(s-1) + 2(l+1)s \right) = C_{s-1} \left(2\varepsilon s + 2\varepsilon l - K \right). \tag{7.26}$$

Equation (7.24) represents an infinite series. For large values of s, the relationship between the coefficients of successive powers of x is

$$\frac{C_s}{C_{s-1}} = \frac{2\varepsilon s + 2\varepsilon l - K}{s^2 + (2l+1)s} \longrightarrow \frac{2\varepsilon}{s}.$$

Let us compare with the known series

$$\sum_s \frac{1}{s!}(2\varepsilon x)^s \tag{7.27}$$

whose ratio of the coefficients of successive powers of x is

$$\frac{\frac{1}{s!}(2\varepsilon)^s}{\frac{1}{s-1!}(2\varepsilon)^{s-1}} = \frac{2\varepsilon}{s}$$

which is the same as the behaviour of C_s/C_{s-1} for large s. Series (7.27) converges to $e^{2\varepsilon x}$. Thus the series for $g(x)$ converges and it asymptotically behaves as $e^{2\varepsilon x}$ that is

$$g(x) \sim e^{2\varepsilon x}$$

for large x. Thus

$$R(x) = \frac{\sqrt{2\mu}}{\hbar} \frac{x^{l+1} e^{-\varepsilon x}}{x} g(x) \sim x^l e^{\varepsilon x}$$

for large x. Thus as $x \to \infty$, $R(x) \to \infty$, as ε is real. It will thus not represent a physically permissible solution. In order to avoid this, the series (7.24) has to be terminated at some finite value of s. Then $R(x)$ will be of the form

$$R(x) = e^{-\varepsilon x} P(x), \tag{7.28}$$

where $P(x)$ is a polynomial in x. This solution will satisfy all the requirements of the physical situation. The condition for termination of the series can be found as follows: Let the highest power of x in the series for $g(x)$ be $x^{n\prime}$ ($n\prime$ being an integer ≥ 0). Then the coefficients $C_{n\prime+1}$, $C_{n\prime+2}$ etc. must be zero so that from Eq. (7.26), putting $s = n\prime + 1$, we get $C_{n\prime+1} = 0$, if

$$2\varepsilon(n\prime + 1) + 2\varepsilon l - K = 0$$

or

$$\varepsilon = \frac{K}{2(n\prime + l + 1)} = \frac{K}{2n}, \tag{7.29}$$

where $n = n\prime + l + 1$ is a positive integer and for a given l is $\geq (l+1)$ as $n\prime \geq 0$. With the help of Eq. (7.17), Eq. (7.29) becomes

$$E = -\varepsilon^2 = -\frac{K^2}{4n^2}$$

$$= -\left(\frac{\mu e^4 Z^2}{2\hbar^2 n^2}\right). \tag{7.30}$$

First we note that energy is quantized and that energy levels do not depend on l and m. n is called principal quantum number ($n = 1, 2, 3, \cdots$). l and m are called azimuthal and magnetic quantum numbers ($l = 0, 1, 2, \cdots, (n-1)$) and ($m = -l, \cdots, l$) i.e. can take $(2l + 1)$ eigenvalues. n' is called the radial quantum number. That energy levels do not depend on m is a consequence of rotational symmetry of the potential. That they do not depend on l is peculiar to $1/r$ potential like Coulomb potential. This means that energy levels are degenerate. Since corresponding to each value of l, m can take $(2l+1)$ values, therefore degree of degeneracy is given by $\sum_{l=0}^{n-1} = n^2$. In other words for one energy value E, there are n^2 different eigenfunctions. Usually a degeneracy is associated with a symmetry which in this case is an external symmetry $[\mathbf{R}, H] = 0$, with

$$\mathbf{R} = \frac{1}{2m}(\mathbf{p} \times \mathbf{L} - \mathbf{L} \times \mathbf{p}) - \frac{e^2}{r}\mathbf{r}.$$

The operator $\mathbf{R}$ given above follows from the Lenz's vector in the classical Kepler problem by the correspondence principle.

Now $n = 1, 2, \ldots$ and for a given l, $n \geq (l+1)$. Thus for $E < 0$, that is for the bound states, formula (7.30) gives discrete set of energy levels. These are in agreement with experiment. We now introduce the spectroscopic notation. The states with different values of l are denoted by

$$l = 0, \ 1, \ 2, \ 3, \ 4, \ 5, \ \cdots s, \ p, \ d, \ f, \ g, \ h, \ \cdots \ .$$

The principal quantum number is written on the left of the above letters. Thus for $n = 1, l = 0$ we have $1s$ level; for $n = 2, l = 0, 1$, we have $2s$ and $2p$ levels which are degenerate. Since for $l = 0, m = 0$, the $1s$ and $2s$ levels are not further degenerate, but $2p$ level is three fold degenerate corresponding to $m = -1, 0, 1$. Thus for $n = 2$, we have four fold degeneracy.

We end this section, with following remarks.

1. The energy levels given in Eq. (7.30), can be written as

$$E_n == -\frac{Z^2 \alpha^2 (\mu c^2)}{2n^2}, \quad n = 1, 2, \cdots$$

$\alpha = \frac{e^2}{\hbar c}$ is called fine structure constant, $\alpha \approx 1/137$, α is dimensionless.

In particular for hydrogen atom, $Z = 1$, and since the nucleus is very heavy as compared to electron, $\frac{1}{\mu} = \frac{1}{m_e} + \frac{1}{m_N} \approx \frac{1}{m_e}$ ($\mu \approx m_e$). Hence for hydrogen atom, μ can be replaced by m_e. Note $m_e c^2 \approx 0.511 MeV$, the rest energy of electron.

2. Define the radius a_0, called the Bohr radius

$$\frac{1}{a_0} = \frac{\mu e^2}{\hbar^2} = \frac{(\mu c^2)\alpha}{c\hbar}; \quad E_n = -\frac{Z^2\alpha}{2n^2}\left(\frac{c\hbar}{a_0}\right)$$

$$c\hbar = 1.97 \times 10^{-13} MeVm$$

$$a_0 = \frac{c\hbar}{\alpha(m_e c^2)} = 0.528 \times 10^{-10} m$$

3. Define the Rydberg constant R:

$$R = \frac{(m_e c^2)\alpha^2}{2\hbar c}; \quad E_n = -Z\frac{(\hbar c)R}{n^2}$$

$$R \approx 1.0978 \times 10^5 cm^{-1}, \hbar cR \approx 13.61eV$$

7.3   Wave Functions of Hydrogen-like Atom

The wave functions of stationary states and the corresponding eigenvalues
are given by

$$u = R_{nl}Y_{lm}(\theta, \phi), \quad E_n = -\frac{\mu e^4 Z^2}{2\hbar^2 n^2}, \tag{7.31}$$

R_{nl} being defined (except for a normalisation constant) by $x e^l e^{-\varepsilon x} g(x)$ with

$$g(x) = \sum_s C_s x^s,$$

the series being terminated at $s = n' = n - l - 1$ and recurrence relations
for C_s being given in (7.26). Thus we can write

$$R_{nl} = B_{nl} x^l e^{-\varepsilon x} g_{nl}(x), \tag{7.32}$$

where $g_{nl}(x)$ is the polynomial

$$g_{nl}(x) = C_0 \sum_{s=0}^{n-l-1} (C_s/C_0)x^s \tag{7.33}$$

and constant $B_{nl}C_0$ is determined from the normalisation condition

$$\int R_{nl}^2 r^2 dr = 1. \tag{7.34}$$

Now from (7.26)

$$C_s s(2l + 1 + s) = C_{s-1}(2(s + l)\varepsilon - K), \tag{7.35}$$

with $\varepsilon = K/2n$, so that

$$
\begin{aligned}
C_s/C_0 &= \frac{K^s}{n^s}\frac{(s+l-n)(s+l-n-1)\cdots(2+l-n)(1+l-n)}{s!(s+2l+1)(s+2l+1-1)\cdots(2l+2)} \\
&= \frac{K^s(-1)s}{n^s}\frac{((n-l-1)(n-l-2)\cdots(n-l-(s-1))(n-l-s))(2l+1)!}{s!(s+2l+1)!} \\
&= \frac{K^s}{n^s}(-1)^s\frac{(n-l-1)!(2l+1)!}{s!(2l+1+s)!(n-l-s-1)!}.
\end{aligned}
$$

Thus

$$
g_{nl}(x) = C_0 \sum_{s=0}^{n-l-1} (-1)^s \frac{(n-l-1)!(2l+1)!(Kx/n)^s}{s!(2l+1+s)!(n-l-s-1)!}. \tag{7.36}
$$

We now introduce what are called associated Laguerre polynomials $L_{n+l}^{2l+1}(y)$ defined as

$$
L_{n+l}^{2l+1}(y) = \sum_{s=0}^{n-l-1} (-1)^{s+1} \frac{[(n+l)!]y^s}{(n-l-1-s)!(2l+1+s)!s!}. \tag{7.37}
$$

From Eqs. (7.33), (7.34), (7.37) and (7.38), it is clear that we can write

$$
R_{nl} = N_{nl}\left(\frac{Kx}{n}\right)^l e^{-\varepsilon x} L_{n+l}^{2l+1}\left(\frac{Kx}{n}\right), \tag{7.38}
$$

where

$$
N_{nl} = B_{nl}C_0(-1)\frac{(n/K)^l}{[(n+l)!]^2}(n-l-1)!(2l+1)! \tag{7.39}
$$

is the normalisation constant and can be determined from normalisation condition (7.35). It can be shown that the associated Laguerre polynomial $L_q^p(y)$ satisfies

$$
yL_q''^p(y) + (p+1-y)L_q'^p(y) + (q-p)_q^p = 0, \tag{7.40}
$$

this is the same equation as the differential equation for $y(x)$ with $y = \frac{Kx}{n}$, $q = n+l$, $p = 2l+1$ and $\varepsilon = \frac{K}{2n}$ as discussed previously. This is as it should be. The Laguerre polynomial $L_q(y)$ satisfies

$$
yL_q'' + (1-y)L_q' + qL_q = 0 \tag{7.41}
$$

and the associated Laguerre polynomial is

$$
L_q^p(y) = \frac{d^p}{dy^p}L_q(y), \tag{7.42}
$$

where $L_q(y)$ can be written as

$$
L_q(y) = e^y \frac{d^q}{dy^q}(y^q e^{-y}). \tag{7.43}
$$

A further property of $L_{n+l}^{2l+1}(y)$ is that

$$\int_0^\infty e^{-y} y^{2l} \left(L_{n+l}^{2l+1}(y) \right)^2 y^2 dy = \frac{2n[(n+l)!]^3}{(n-l-1)!}. \tag{7.44}$$

Using this relation, the normalisation constant N_{nl} can now be found and is given by

$$N_{nl} = -\left(\left(\frac{2Z}{na_0} \right)^3 \frac{(n-l-1)!}{2n[(n+l)!]^3} \right)^{1/2}, \tag{7.45}$$

where

$$a_0 = \frac{\hbar^2}{m_e e^2} \tag{7.46}$$

and is called the Bohr radius. (Here we have put $\mu = m_e$, the mass of electron as we are taking the nucleus to be at rest.) Since

$$Kx = 2m_e \frac{Ze^2}{\hbar^2} r = \frac{2Zr}{a_0}, \quad \varepsilon = \frac{K}{2n}$$

we can write the normalised energy eigenfunctions for hydrogen-like atom as

$$u_{nlm}(r, \theta, \phi) = R_{nl}(r) Y_{lm}(\theta, \phi),$$

$$R_{nl} = -\left(\left(\frac{2Z}{na_0} \right)^3 \frac{(n-l-1)!}{2n[(n+l)!]^3} \right)^{1/2} e^{-(Z/na_0)r} \left(\frac{2Zr}{na_0} \right)^l L_{n+l}^{2l+1} \left(\frac{2Zr}{na_0} \right). \tag{7.47}$$

The energy levels are given by

$$E_n = -\frac{Z^2 e^2}{2a_0 n^2}. \tag{7.48}$$

Radial functions for $n = 1$ and $n = 2$ are given below

$$n = 1: \quad l = 0$$

$$R_{10} = (Z/a_0)^{3/2} 2 e^{-Zr/a_0},$$

$$Y_{00} = \frac{1}{\sqrt{4\pi}}.$$

1s level: $u_{100} = (z/a_0)^{3/2} \frac{1}{\pi^{1/2}} e^{-Zr/a_0}$ corresponds to

$$E_1 = -\frac{Z^2 e^2}{2a_0},$$

$$n = 2: \quad l = 0, 1$$

$$R_{20} = (Z/2a_0)^{3/2} (2 - \frac{Zr}{a_0}) e^{-Zr/2a_0},$$

$$R_{21} = (Z/2a_0)^{3/2} \frac{Zr}{a_0 \sqrt{3}} e^{-Zr/2a_0},$$

$$l = 0, \quad m = 0,$$

$$l = 1, \quad m = -1, 0, -1.$$

$2s$ level:

$$u_{200} = \frac{Z^{3/2}}{4\sqrt{2\pi}a_0^{3/2}}(2 - Zr/a_0)e^{-Zr/2a_0}$$

corresponds to

$$E_2 = \frac{1}{4}E_1 .$$

Similarly using expressions for

$$Y_{1-1}, \quad Y_{10}, \quad Y_{11}$$

we can write the expressions for

$$u_{21-1}, \quad u_{210}, \quad u_{211}$$

corresponding to $2p$ levels. All these wave functions correspond to the same energy $E_2 = \frac{1}{4}E_1$. Thus for $n = 2$, we have in all 4 wave functions, one for $2s$ and three for $2p$ level, all corresponding to the same energy, i.e. degeneracy is $2^2 = 4$ fold as it should be.

It is useful to introduce a quantity, called radial probability density. Now the probability of finding the electron in the volume element $\mathbf{dr} = \mathbf{r}^2 dr \sin\theta d\theta d\phi$ is

$$u_{nlm}^*(\mathbf{r})u_{nlm}(\mathbf{r})\mathbf{dr} = R_{nl}^*(r)Y_{lm}^*(\theta,\phi)R_{nl}(r)Y_{lm}(\theta,\phi)r^2 dr \sin\theta d\theta d\phi.$$

Then

$$\int_0^\infty \int_0^\pi \int_0^{2\pi} u_{nlm}^*(r,\theta,\phi)u_{nlm}(r,\theta,\phi)r^2 dr \sin\theta d\theta d\phi = 1 .$$

Also eigenfunctions are orthogonal

$$\int_0^\infty \int_0^\pi \int_0^{2\pi} u_{n'l'm'}^*(r,\theta,\phi)u_{nlm}(r,\theta,\phi)r^2 dr \sin\theta d\theta d\phi = 0$$

unless $n' = n, l' = l, m' = m$.

Physically this means that if the particle is in eigenstate (n, l, m), the probability of finding it in another eigenstate (n', l', m') when n', l', m' are all different from n, l, m is zero.

Radial probability density $P_{nl}(r)$ is defined by

$$P_{nl}(r)dr = \int_0^\pi \int_0^{2\pi} u_{nlm}^* u_{nlm} r^2 dr \sin\theta d\theta d\phi$$

$$= r^2 R_{nl}^*(r)R_{nl}(r)dr$$

$$\times \int_0^\pi \int_0^{2\pi} Y_{lm}^*(\theta,\phi)Y_{lm}(\theta,\phi)\sin\theta d\theta d\phi$$

$$= R_{nl}^*(r)R_{nl}(r)r^2 dr,$$

which is the probability of finding the electron in a spherical shell of radius r and thickness dr. Let us now calculate the expectation value

$$\langle r_{nl} \rangle = \int_0^\infty r P_{nl}(r) dr = \int_0^\infty |R_{nl}(r)|^2 r^3 dr$$

$$= \frac{n^2 a_0}{Z}\{1 + \frac{1}{2}\{1 - \frac{l(l+1)}{n^2}\}\}.$$

This characterises the radius of the shell. This depends primarily on n while dependence on l is weak because of the factor $\frac{1}{2n^2}$ in l dependent term. This may be compared with the radii of circular orbits of Bohr $(r_n)_{Bohr} = \frac{n^2 a_0}{Z}$. All electrons in eigenstates with common n values have approximately the same $\langle r_{nl} \rangle$, independent of values of l or m. Such electrons are said to be in the same shell. Each shell also has the property that the associated eigenvalues have exactly the same value E_n. Note that for $r < a_0/Z$, $u_{nlm} \sim r^l, r \to 0$, so that the probability density $u_{nlm}^* u_{nlm} \sim r^{2l}, r \to 0$. Thus the value of probability density in a volume element near $r = 0$ is relatively large only for $l = 0$ and decreases rapidly with increasing l. Other useful expectation values are

$$\langle r^2 \rangle_{nl} = \frac{a_0^2 n^4}{2Z^2}[5 + \frac{1 - 3l(l+1)}{n^2}],$$

$$\langle \frac{1}{r} \rangle_{nl} = \frac{Z}{a_0 n^2},$$

$$\langle \frac{1}{r^2} \rangle_{nl} = \frac{Z^2}{a_0^2 n^3(l + \frac{1}{2})}.$$

7.4   Problems

7.1 In spectroscopic notation, a state is specified as $^{2s+1}l_j$, where s is the spin and j is the total angular momentum. In particular for hydrogen atom, $s = \frac{1}{2}$ i.e. the spin of the electron. Thus $j = l - \frac{1}{2}, l + \frac{1}{2}$. List all the possible states for $n = 1, n = 2, n = 3$. Find $(\frac{1}{\lambda_3} - \frac{1}{\lambda_2}), (\frac{1}{\lambda_3} - \frac{1}{\lambda_1}), (\frac{1}{\lambda_2} - \frac{1}{\lambda_1})$.
Show that for the transitions from the level

$$n = 3 \quad \text{to} \quad n = 2,$$
$$n = 3 \quad \text{to} \quad n = 1,$$
$$n = 2 \quad \text{to} \quad n = 1,$$

the photons of wavelengths $6559 A°$, $1025 A°$ and $1027 A°$ are emitted respectively.

7.2 A particle of mass m moves in a square well potential
$$V(\mathbf{r}) = -V_0 \quad r < a,$$
$$= 0 \quad r > a.$$
If bound s-state exists, show that energy eigenvalues are given by
$$-\gamma = k \cot \, ka$$
where
$$\gamma = \sqrt{2mW/\hbar^2}, \quad E = -W,$$
$$k = \sqrt{2m(V_0 - W)/\hbar^2} .$$
If there is only one such state and it is very weakly bound, show that
$$V_0 a^2 \sim \frac{\pi^2 \hbar^2}{8m} .$$

7.3 Assume that interaction between the neutron and proton that make up a deutron can be represented by a square well potential with $a \sim 2 \times 10^{-13}$ cms. If $l = 0$ energy level of the system (binding energy) is 2 MeV, calculate V_0 in MeV.

7.4 The radial wave function $\chi(\mathbf{r})$ satisfies the equation
$$\frac{d^2\chi}{dr^2} + \left(\frac{2m}{\hbar^2}(E - V) - \frac{l(l+1)}{r^2} \right) \chi = 0 .$$
Take $V(\mathbf{r}) = (1/2)m\omega^2 r^2$ and by expanding $\chi \exp(1/2\rho^2)$ in powers of $\rho = r/b$ with $b^2 = \hbar/m\omega$, show that energy levels are given by
$$E_n = (1/2)\hbar\omega(4n + 2l + 3)$$
where n is an integer ≥ 0.

7.5 A particle moves in a spherically symmetrical potential $V = \alpha/r^2 - \beta/r$, where α and β are constants. Find bound state energy levels.

7.6 Find the expectation values $\langle r \rangle$ and $\langle \frac{1}{r} \rangle$ for the ground state of a hydrogen-like atom.

7.7 Find the energy eigenvalues and eigenfunctions for a three-dimensional harmonic oscillator. The potential for such an oscillator is given by
$$V(x, y, z) = \frac{m}{2}(\omega_1^2 x^2 + \omega_2^2 y^2 + \omega_3^2 z^2).$$
Discuss the degeneracy of energy levels. Hint: Write the wave function as
$$u(x, y, z) = u_1(x)u_2(y)u_3(z)$$
and show that each of the functions $u_1(x)$, $u_2(y)$ and $u_3(z)$ satisfy the one dimensional harmonic oscillator equation.

7.8 For

$$H = \frac{\hat{p}^2}{2m} + V(\mathbf{r}) = \hat{T} - \frac{Ze^2}{\mathbf{r}}$$

show that

$$\langle \hat{T} \rangle_{nl} = -\frac{1}{2} \langle V \rangle_{nl}.$$

Chapter 8

Collision Theory

8.1 Two Body Problem: (Centre of Mass Motion)

Thus far, we have been considering the Schrödinger equation for the motion of a single particle in an external field of force. In many problems, such as the hydrogen atom, we have to consider the problem of the motion of two interacting particles (nucleus and electron in case of hydrogen atom). Previously we considered the nucleus as a fixed centre about which the electron moves. We shall discard this assumption now. We now show that the problem of the motion of two interacting particles can be reduced to that of one particle. Consider two particles with coordinates and masses $\mathbf{r}_1 = (x_1, y_1, z_1)$, m_1, and $\mathbf{r}_2 = (x_2, y_2, z_2)$, m_2 moving in a potential $V(\mathbf{r}_1, \mathbf{r}_2)$. The wave function of the system is denoted by $\psi(\mathbf{r}_1, \mathbf{r}_1, t)$. The Hamiltonian is

$$H = \frac{1}{2m_1}\hat{p}_1^2 + \frac{1}{2m_2}\hat{p}_2^2 + V(\hat{\mathbf{r}}_1, \hat{\mathbf{r}}_2).$$ (8.1)

The potential $V(\hat{\mathbf{r}}_1, \hat{\mathbf{r}}_2)$ includes the interaction energy between the two particles as well as other sources of potential energy.

In the Schrödinger representation

$$H = -\frac{\hbar^2}{2m_1}\nabla_1^2 - \frac{\hbar^2}{2m_2}\nabla_2^2 + V(\mathbf{r}_1, \mathbf{r}_2)$$ (8.2)

where

$$\nabla_1^2 = \frac{\partial^2}{\partial x_1^2} + \frac{\partial^2}{\partial y_1^2} + \frac{\partial^2}{\partial z_1^2}$$

and $\psi(\mathbf{r}_1, \mathbf{r}_2, t)$ now satisfies the Schrödinger equation

$$i\hbar\frac{\partial\psi}{\partial t} = H\psi$$

$$= \left(-\frac{\hbar^2}{2m_1}\nabla_1^2 - \frac{\hbar^2}{2m_2}\nabla_2^2 + V\right)\psi.$$ (8.3)

The interpretation of ψ is that $|\psi|^2 d\tau_1 d\tau_2$ ($d\tau_1 = d^3 r_1$, $d\tau_2 = d^3 r_2$) gives the probability that the first particle is at $\mathbf{r}_1$ in volume element $d\tau_1$ while at the same time the second particle is in volume element $d\tau_2$ at $\mathbf{r}_2$.

We now assume that $V(\mathbf{r}_1, \mathbf{r}_2)$ depends only on $\mathbf{r} = \mathbf{r}_1 - \mathbf{r}_2$, so that $V = V(\mathbf{r}_1 - \mathbf{r}_2)$. In many physical problems, the potential is of this form. Introduce the relative coordinates

$$\mathbf{r} = (\mathbf{r}_1 - \mathbf{r}_2), \qquad \mathbf{r} = (x, y, z) \tag{8.4a}$$

and the coordinates of the centre of mass

$$\mathbf{R} = \frac{m_1 \mathbf{r}_1 + m_2 \mathbf{r}_2}{m_1 + m_2}, \qquad \mathbf{R} = (X, Y, Z). \tag{8.4b}$$

Then using relations like

$$\begin{aligned}
\frac{\partial}{\partial x_1} &= \frac{\partial x}{\partial x_1}\frac{\partial}{\partial x} + \frac{\partial X}{\partial x_1}\frac{\partial}{\partial X} \\
&= \frac{\partial}{\partial x} + \frac{m_1}{m_1 + m_2}\frac{\partial}{\partial X},
\end{aligned} \tag{8.5}$$

we have

$$\nabla_1 = \nabla + \frac{m_1}{m_1 + m_2}\nabla_R. \tag{8.6a}$$

Similarly we have

$$\nabla_2 = -\nabla + \frac{m_2}{m_1 + m_2}\nabla_R. \tag{8.6b}$$

Hence we can write

$$-\frac{\hbar^2}{2m_1}\nabla_1^2 - \frac{\hbar^2}{2m_2}\nabla_2^2 = -\frac{\hbar^2}{2\mu}\nabla^2 - \frac{\hbar^2}{2M}\nabla_R^2$$

$$H = -\frac{\hbar^2}{2M}\nabla_R^2 - \frac{\hbar^2}{2\mu}\nabla^2 + V(\mathbf{r}) \tag{8.7}$$

where $M = m_1 + m_2$ is the total mass of the system and μ is the reduced mass

$$\frac{1}{\mu} = \frac{1}{m_1} + \frac{1}{m_2}.$$

Equation (8.3) can now be written as

$$i\hbar\frac{\partial \psi}{\partial t} = \left(\frac{-\hbar^2}{2M}\nabla_R^2 - \frac{\hbar^2}{2\mu}\nabla^2 + V(\mathbf{r})\right)\psi \tag{8.8}$$

with

$$\psi = \psi(\mathbf{r}, \mathbf{R}, t).$$

For a stationary state belonging to energy eigenvalue E_T of H, the time dependence will be of the form $e^{-iE_Tt/\hbar}$ so that

$$\psi(\mathbf{r}, \mathbf{R}, t) = \psi_0(\mathbf{r}, \mathbf{R})e^{-iE_Tt/\hbar}.$$

Substituting in Eq. (8.8), we have

$$\left(-\frac{\hbar^2}{2M}\nabla_R^2 - \frac{\hbar^2}{2\mu}\nabla^2 + V(\mathbf{r})\right)\psi_0(\mathbf{r}, \mathbf{R}) = E_T\psi_0(\mathbf{r}, \mathbf{R}). \qquad (8.9)$$

We note that no term in the parentheses depends on both $\mathbf{r}$ and $\mathbf{R}$. Hence we can write a solution of the form

$$\psi_0(\mathbf{r}, \mathbf{R}) = u(\mathbf{r})U(\mathbf{R}).$$

Substituting in Eq.(8.9) and dividing by $u(\mathbf{r})U(\mathbf{R})$, we get

$$\frac{\hbar^2}{2M}\frac{1}{U(\mathbf{R})}\nabla_R^2 U(\mathbf{R}) + \left(-\frac{\hbar^2}{2\mu}\frac{1}{u(\mathbf{r})}\nabla^2 u(\mathbf{r}) + V(\mathbf{r})\right) = E_T. \qquad (8.10)$$

This can have a solution for arbitrary $\mathbf{r}$ and $\mathbf{R}$ only if the part involving $\mathbf{R}$ and the part involving $\mathbf{r}$ are separately and identically constant. Thus we obtain

$$-\frac{\hbar^2}{2M}\frac{1}{U(\mathbf{R})}\nabla_R^2 U(\mathbf{R}) = \acute{E} = \text{constant}$$

or

$$-\frac{\hbar^2}{2M}\nabla_R^2 U(\mathbf{R}) = \acute{E}U(\mathbf{R}) \qquad (8.11)$$

and

$$-\frac{\hbar^2}{2\mu}\frac{1}{u(\mathbf{r})}\nabla^2 u(\mathbf{r}) + V(\mathbf{r}) = E_T - \acute{E}. \qquad (8.12)$$

Equation (8.11) is the eigenvalue equation for the motion of a free particle of mass M. This means that centre of mass of the system moves like a free particle with kinetic energy $\acute{E}$ and mass M. Let us denote the difference between the total energy E_T and the energy associated with the centre of mass $\acute{E}$ by the symbol E (relative energy) so that $E = E_T - \acute{E}$. Thus Eq. (8.12) for $u(\mathbf{r})$ becomes

$$-\frac{\hbar^2}{2\mu}\nabla^2 u(\mathbf{r}) + V(\mathbf{r})u(\mathbf{r}) = Eu(\mathbf{r}). \qquad (8.13)$$

This equation is the eigenvalue equation for the energy of the relative motion of the two particles and is the same as the equation for a particle that has reduced mass μ in an external field V.

Usually, in solving problems like that of the hydrogen atoms, we are not interested in the energy of the motion of the atom as a whole, but merely the energy resulting from the relative motion of the electron and nucleus, i.e. we are interested in the energy levels E associated with the relative motion. In practice we shall, therefore, usually solve only the equation for $u(\mathbf{r})$ and obtain the possible values of E.

For hydrogen-like atoms, in our previous result of energy levels if we allow for the motion of nucleus, we have to replace m_e by the reduced mass

$$m_e \longrightarrow \mu = \frac{m_e m_p}{m_e + m_p},$$

where m_p is the proton mass. Since $m_e \ll m_p$, this introduces a small correction which however is within experimental sensitivity.

The analysis into centre of mass and relative motion is essential for the collision of two particles which is our next topic.

Remarks: (1) μ = mass of one of the particles, say m_1 when the other particle mass $m_2 \to \infty$. (2) Fixed source approximation is a good one, whenever one particle (the target particle in two-particle collision) is very massive compared to the other (the projectile).

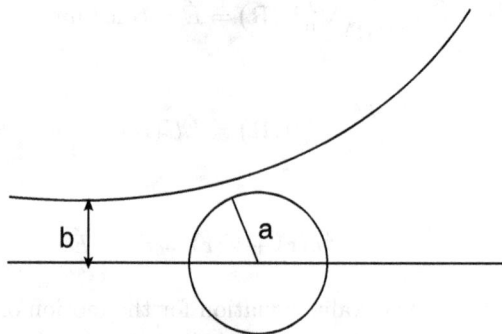

Fig. 8.1   Impact parameter b.

8.2   Collision Theory

We have considered two types of stationary states which involve (i) discrete spectrum of energy eigenvalues and correspond to bound states in which the particle is restrained by the potential to a particular region of space, and (ii) continuous spectrum of energy eigenvalues. The latter situation arises in collision problems in which case a particle arrives from infinity (large distance), is scattered by a field of forces and goes off to infinity (large distance). For these states the wave function may remain finite at infinity.

For bound states, for example hydrogen-like atoms, the boundary conditions at great distances were used to determine the discrete energy levels of the particle. In a collision problem, the energy is specified in advance, and the behaviour of the wave function at great distances is found in terms of it. This asymptotic behaviour can then be related to the probability that, as a result of the collision, the particle will deviate (or as, we say, will be scattered) through a given angle. The purpose of the theory is to calculate this probability, i.e. to say the probability that the bombarding particle will interact in a certain way with a target particle. Thus scattering processes are very important for studying mutual interactions of particles. For example to investigate nuclear potentials, one studies the mutual interaction of nucleons (proton or neutron) by scattering one off the other.

We first consider the kinematices of a scattering process.

8.3   Kinematices of a Scattering Process

Consider a collision between two particles with coordinates and masses given by

$$\mathbf{r}_1 = (x_1, y_1, z_1), m_1; \mathbf{r}_2 = (x_2, y_2, z_2), m_2.$$

The wave function ψ satisfies the Schrödinger equation

$$i\hbar \frac{\partial \psi}{\partial t} = \left(-\frac{\hbar^2}{2m_1}\nabla_1^2 - \frac{\hbar^2}{2m_2}\nabla_2^2 + V(\mathbf{r}_1 - \mathbf{r}_2)\right)\psi.$$

By introducing the variables

$$M = m_1 + m_2$$
$$\frac{1}{\mu} = \frac{1}{m_1} + \frac{1}{m_2}$$
$$\mathbf{r} = \mathbf{r}_1 - \mathbf{r}_2$$
$$\mathbf{R} = \frac{m_1\mathbf{r}_1 + m_2\mathbf{r}_2}{M}$$

(as shown previously), the Schrödinger equation is separable and we can write the wave function

$$\psi(\mathbf{r}, \mathbf{R}, t) = u(\mathbf{r})U(\mathbf{R})e^{-i(E+\acute{E})t/\hbar}. \tag{8.14}$$

$u(\mathbf{r})$ and $U(\mathbf{R})$ then satisfy the equations

$$-\frac{\hbar^2}{2\mu}\nabla^2 u + Vu = Eu, \tag{8.15}$$

$$-\frac{\hbar^2}{2M}\nabla_R^2 U = \acute{E}U. \tag{8.16}$$

Equation (8.16) implies that centre of mass moves like a free particle. Equation (8.15) describes relative motion of two particles. The results obtained from Eq. (8.15) therefore describe the collision with reference to a system of coordinates moving with the centre of mass (called centre of mass system) and must be transformed appropriately to give the motion of particles in system of coordinates fixed in the laboratory in which the observations are made. The laboratory frame is defined in which one of the particles (called the target particle) is initially at rest. Take m_2 to be target particle and m_1 to be the bombarding particle.

We define the appropriate quantities in the two frames as follows: Let $\mathbf{q}_1$ and $\acute{\mathbf{q}}_1$ be initial and final momenta of the bombarding particle in the laboratory system. Let $\acute{\mathbf{q}}_2$ be recoil momentum of the target particle in this frame. Conservation of momentum gives

$$\mathbf{q}_1 = \acute{\mathbf{q}}_1 + \acute{\mathbf{q}}_2. \tag{8.17}$$

Velocity of centre of mass in this frame is given by

$$\mathbf{V} = \dot{\mathbf{R}} = \frac{1}{M}m_1\mathbf{r}_1$$

$$= \frac{\mathbf{q}_1}{M} = \frac{\acute{\mathbf{q}}_1 + \acute{\mathbf{q}}_2}{M}. \tag{8.18}$$

In the centre of mass frame, let initial momenta of particles 1 and 2 be $\mathbf{p}_1$ and $\mathbf{p}_2$ and let final momenta be $\acute{\mathbf{p}}_1$ and $\acute{\mathbf{p}}_2$. In this frame, centre of mass is at rest

$$M\dot{\mathbf{R}} = 0, \tag{8.19}$$

this gives

$$\mathbf{p}_1 + \mathbf{p}_2 = 0 = \acute{\mathbf{p}}_1 + \acute{\mathbf{p}}_2$$

or

$$\mathbf{p}_1 = -\mathbf{p}_2 = \mathbf{p}, \quad \acute{\mathbf{p}}_1 = -\acute{\mathbf{p}}_2 = \mathbf{p}'. \tag{8.20a}$$

Relative velocity $\mathbf{v}$ is given by

$$\mathbf{v} = \frac{\mathbf{p}_1}{m_1} - \frac{\mathbf{p}_2}{m_2} = \frac{\mathbf{p}}{\mu}. \tag{8.20b}$$

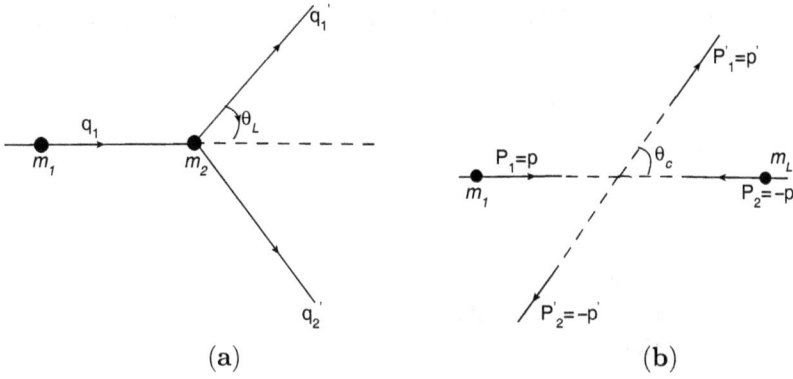

Fig. 8.2 (a)- Scattering of two particles in the laboratory system. (b)- Scattering of two particles in the centre of mass system.

The collision process in two frames is shown in Fig. 8.1.
We have

$$\mathbf{r}_1 = \dot{\mathbf{R}} + \mathbf{u}, \tag{8.20c}$$

where $\mathbf{u}$ is the initial velocity of the particle 1 in the centre of mass frame. Equation (8.20c) can be written as

$$\frac{\mathbf{q}_1}{m_1} = \frac{\mathbf{q}_1}{M} + \frac{\mathbf{p}}{m_1} \tag{8.20d}$$

or

$$\mathbf{p} = \frac{m_2}{M}\mathbf{q}_1. \tag{8.21}$$

Similarly (after collision $\mathbf{r}_1 = \frac{\mathbf{q}_1'}{m_1}$, $\dot{\mathbf{R}} = \frac{1}{M}(\mathbf{q}_1' + \mathbf{q}_2') = \frac{1}{M}\mathbf{q}_1$, $\mathbf{u}' = \frac{\mathbf{p}'}{m}$)

$$\frac{\mathbf{q}_1'}{m_1} = \frac{\mathbf{q}_1}{M} + \frac{\mathbf{p}'}{m_1}$$

or

$$\mathbf{q}_1' = \frac{m_1}{M}\mathbf{q}_1 + \mathbf{p}'. \tag{8.22}$$

Also

$$\frac{\mathbf{q}_2'}{m_2} = \frac{\mathbf{q}_1}{M} - \frac{\mathbf{p}'}{m_2}$$

or

$$\mathbf{q}_2' = \frac{m_2}{M}\mathbf{q}_1 - \mathbf{p}'$$
$$= \mathbf{p} - \mathbf{p}'. \tag{8.23}$$

Take $\mathbf{q}_1$ along z-axis. Then from Eq. (8.21) we see that $\mathbf{p}$ is also along z-axis

$$\mathbf{q}_1 \equiv (0, 0, q), \mathbf{p} \equiv (0, 0, p).$$

From Fig. 8.1a, b and Eq. (8.22) we have

$$q_1' \cos\theta_L = \frac{m_1}{M} q + p' \cos\theta_C,$$

$$q_1' \sin\theta_L = p' \sin\theta_C.$$

We now confine ourselves to elastic scattering $|\mathbf{p}| = |\mathbf{p}'| = p$. Then from the above equations, we have

$$\tan\theta_L = \frac{\sin\theta_C}{\cos\theta_C + \frac{m_1}{M}\frac{q}{p}}$$

$$= \frac{\sin\theta_C}{\cos\theta_C + \frac{m_1}{m_2}}. \qquad (8.24a)$$

The azimuthal angle ϕ about the bombarding particle is the same in both frames

$$\phi_L = \phi_C. \qquad (8.24b)$$

Now kinetic energy T_C in centre of mass frame is given by

$$T_C = p^2/2\mu = \frac{1}{2\mu}\frac{m_2^2}{M^2} q^2$$

$$= \frac{m_2}{m_1 + m_2} T_L, \quad \left(T_L = \frac{q^2}{2m_1}\right). \qquad (8.25)$$

Equations (8.24) and (8.25) give the relations between the two frames for the elastic scattering.

Remarks: (1) In the limit $m_2 \to \infty$, the laboratory frame is equivalent to the centre of mass frame and neglecting the recoil of the target is a very reasonable approximation if $m_1/m_2 \ll 1$. (2) When $m_1 = m_2$, as in the case of proton–neutron scattering, the distinction between the two frames is essential. Then

$$T_C = \frac{1}{2} T_L$$

$$\tan\theta_L = \frac{\sin\theta_C}{\cos\theta_C + 1} = \tan(\theta_C/2) \qquad (8.26)$$

that is

$$\theta_L = \theta_C/2.$$

8.4 Scattering Cross Section

In the centre of mass frame, the Schrödinger equation for stationary states:

$$[-\frac{\hbar^2}{2\mu}\nabla^2 + V(\mathbf{r})]u(r,\theta,\phi) = Eu(r,\theta,\phi). \qquad (8.27)$$

In a scattering process, an incoming beam of particles of energy $T = \frac{p^2}{2\mu}$ is moving in z-direction. As it enters the scattering region, it is deviated from its original path (scattered) and is detected in a detector (Fig. (8.2)). The incoming beam is represented by a plane wave e^{ikz}, $p = \hbar k$. The outgoing scattered wave is represented by an outgoing spherical wave $\frac{e^{ikr}}{r}$. Therefore in the region $V = 0$:

$$u(r,\theta,\phi) \xrightarrow[r\to\infty]{} A[\underbrace{e^{ikz}}_{\text{incident wave}} + \underbrace{f(\theta,\phi)\frac{1}{r}e^{ikr}}_{\text{outgoing scattered wave}}]. \qquad (8.28)$$

The number of particles detected by the detector gives the probability of scattering at angle θ. Thus the probability which is related to the scattering cross section is proportional to the number of particles detected at an angle between θ and $\theta + d\theta$ divided by the incoming flux.

Now incoming flux:

$$\rho v = |Ae^{ikz}|^2 v = |A|^2 v. \qquad (8.29)$$

Outgoing probability density:

$$P = |Af(\theta,\phi)\frac{1}{r}e^{ikr}|^2. \qquad (8.30)$$

Therefore the number of particles scattered in volume element between r and $r + dr$ in the solid angle $d\Omega$ is given by

$$P d\tau = \Gamma r^2 dr d\Omega$$
$$= |A|^2 |f(\theta,\phi)|^2 d\Omega dr. \qquad (8.31)$$

Flux received in volume element $d\tau$:

$$P d\tau \frac{1}{dt} = |A|^2 |f(\theta,\phi)|^2 d\Omega \frac{dr}{dt}$$
$$= |A|^2 |f(\theta,\phi)|^2 d\Omega \, v. \qquad (8.32)$$

Therefore, the number of particles scattered into $d\Omega$ per unit time per unit incident flux:

$$d\sigma = |f(\theta,\phi)|^2 d\Omega. \qquad (8.33)$$

The above quantity has dimension of area; $f(\theta, \phi)$ has dimension of length called scattering amplitude.

Now

$$\frac{d\sigma}{d\Omega} = |f(\theta, \phi)|^2 \quad : \text{differential cross section,} \qquad (8.34)$$

$$\sigma = \int (\frac{d\sigma}{d\Omega}) d\Omega = \int |f(\theta, \phi)|^2 d\Omega \quad : \text{total cross section.} \qquad (8.35)$$

Scattering is an important tool to probe the structure of matter. Ever

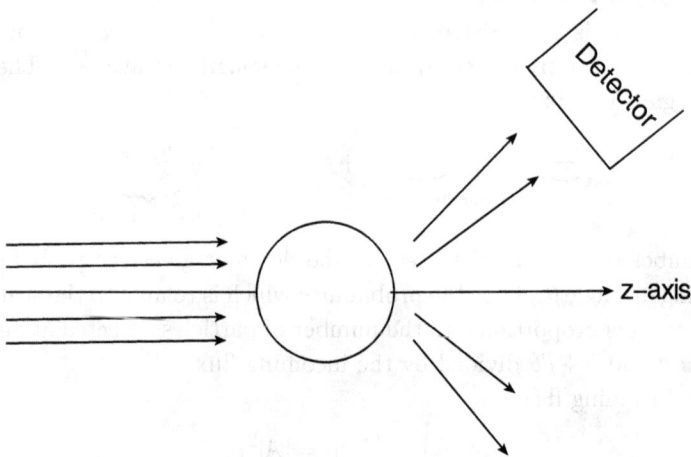

Fig. 8.3   The incident plane wave and the scattered wave moving radially outwards.

since Lord Rutherford discovered the atomic nucleus by scattering of α particles on atoms; the structure of matter has been probed using scattering of electrons, protons and photons on various targets. Major discoveries have been made in atomic, condensed matter, nuclear and particle physics. The scattering cross section gives information about the interaction which caused the deviation of incident particle to the detector.

The relation between the differential cross section in the centre of mass frame and that in the laboratory frame follows from the fact that the effective cross section in the two frames must be same:

$$\left(\frac{d\sigma}{d\Omega}\right)_C d\Omega(\theta_C, \phi_C) = \left(\frac{d\sigma}{d\Omega}\right)_L d\Omega(\theta_L, \phi_L). \qquad (8.36)$$

From Eq. (8.24b)

$$\phi_c = \phi_L$$

$$\sin \theta_L d\theta_L = \frac{(1 + \frac{m_1}{m_2} + \cos \theta_C) \sin \theta_C d\theta_C}{(\frac{m_1}{m_2} + \cos \theta_C)^2 [1 + 2\frac{m_1}{m_2} \cos \theta_C + (\frac{m_1}{m_2})^2]^{\frac{1}{2}}}. \tag{8.37}$$

Hence from Eqs. (8.36) and (8.37)

$$\left(\frac{d\sigma}{d\Omega}\right)_L = \frac{(\frac{m_1}{m_2} + \cos \theta_C)^2 [1 + 2\frac{m_1}{m_2} \cos \theta_C + (\frac{m_1}{m_2})^2]^{\frac{1}{2}}}{(1 + \frac{m_1}{m_2} \cos \theta_C)} \left(\frac{d\sigma}{d\Omega}\right)_C. \tag{8.38}$$

8.5 Scattering by a Spherical Symmetry Potential

Most of the potentials in physics are spherically symmetric viz. $V(\mathbf{r}) = V(r)$. For a spherical potential $f(\theta, \phi) = f(\theta)$, $d\Omega = 2\pi \sin \theta d\theta$. Our aim is to calculate $f(\theta)$ in terms of phase shifts. This is particularly convenient at low energies, where only a few partial waves are important. It is also useful, when in a scattering process, a resonance with a definite angular momentum is excited.

We start the section by semiclassical argument: The angular momentum of a particle traveling with momentum p in the incident beam at a distance b closest to the scattering centre is pb, where b is called the impact parameter (Fig. 8.3). Thus

$$pb = \ell\hbar.$$

If R is the radius of interaction, the scattering takes place if

$$b \leq R$$
$$\frac{\ell\hbar}{p} \leq R$$

or

$$\ell \leq \frac{pR}{\hbar} = kR.$$

If the energy of the beam is sufficiently small

$$kR < 1, \quad \ell < 1.$$

Hence, at low energies, $\ell = 0$ (s-wave) is relevant.

The above argument shows that it is convenient to analyse the scattering amplitude $f(\theta)$ in terms of its angular momentum components.

The incident wave

$$e^{ikz} = e^{ikr\cos\theta} \tag{8.39}$$

includes all the components of angular momentum about the origin.

$$e^{ikr\cos\theta} = \sum_\ell j_\ell(kr)P_\ell\cos(\theta)$$

$$= \sqrt{4\pi}\sum_\ell \frac{1}{\sqrt{2\ell+1}}a_\ell j_\ell(kr)Y_{\ell 0}(\theta,\phi) \qquad (8.40)$$

where $a_\ell = (2\ell+1)i^\ell$ and $j_\ell(kr)$ is the spherical Bessel function which is regular at $r = 0$ and has asymptotic behavior $r \to \infty$: $\frac{1}{kr}\sin(kr - \frac{\ell\pi}{2}) = \frac{e^{-i\frac{\ell\pi}{2}}(e^{ikr}-(-1)^\ell e^{-ikr})}{2ikr}$. Further we have replaced Legendre polynomial $P_\ell(\cos\theta)$ by the spherical harmonic:

$$Y_{\ell 0} = \sqrt{\frac{2\ell+1}{4\pi}}P_\ell(\cos\theta). \qquad (8.41)$$

For s-wave; project out $\ell = 0$ component in Eq. (8.40)

$$\int e^{ikr\cos\theta}Y_{00}^* d\Omega = \sqrt{4\pi}\sum_\ell \frac{1}{\sqrt{2\ell+1}}a_\ell j_\ell(kr)\int Y_{00}^* Y_{\ell 0}d\Omega$$

$$= \sqrt{4\pi}\sum_\ell \frac{1}{\sqrt{2\ell+1}}a_\ell j_\ell(kr)\delta_{\ell 0}$$

$$= \sqrt{4\pi}a_0 j_0(kr). \qquad (8.42)$$

Now

$$\int e^{ikr\cos\theta}Y_{00}^* d\Omega = \frac{1}{\sqrt{4\pi}}\int e^{ikr\cos\theta}\sin\theta d\theta d\phi$$

$$= \frac{1}{\sqrt{4\pi}}2\pi\frac{1}{ikr}[e^{ikr} - e^{-ikr}]. \qquad (8.43)$$

Therefore

$$a_0 j_0(kr) = \frac{1}{2ikr}[e^{ikr} - e^{-ikr}]. \qquad (8.44)$$

Hence for s-wave

$$e^{ikr\cos\theta} = \frac{1}{2ikr}[e^{ikr} - e^{-ikr}]. \qquad (8.45)$$

The same result can also be obtained by the following simple argument. Since s-wave scattering is isotropic, the $\ell = 0$ component in the incident wave is

$$\frac{\int e^{ikr\cos\theta}d\Omega}{\int d\Omega} = \frac{2\pi\int e^{ikr\cos\theta}\sin\theta d\theta}{4\pi}$$

$$= \frac{1}{2ikr}[e^{ikr} - e^{-ikr}]. \qquad (8.46)$$

The effect of scattering potential can only be to alter the outgoing wave e^{ikr}. Hence for s-wave:

$$u(r) \sim \frac{S e^{ikr} - e^{-ikr}}{2ikr}. \tag{8.47}$$

The flux of incoming wave is proportional to $|e^{-ikr}|^2 = 1$.

The flux of outgoing wave is proportional to $|S e^{ikr}|^2 = |S|^2$.

For elastic scattering, the flux of incoming wave must be equal to flux of outgoing wave:

$$|S|^2 = 1, \quad S = e^{2i\delta}. \tag{8.48}$$

where δ is real. δ is called the phase shift.

Reexpressing Eq. (8.47) in the form

$$
\begin{aligned}
u(r) &\sim \frac{e^{2i\delta} e^{ikr} - e^{ikr}}{2ikr} \\
&= \frac{1}{2ikr}[(e^{2i\delta} - 1)e^{ikr} + (e^{ikr} - e^{-ikr})] \\
&= \underbrace{\frac{e^{ikr} - e^{-ikr}}{2ikr}}_{\text{incident wave}} + \frac{e^{2i\delta} - 1}{2ik} \frac{e^{ikr}}{r}
\end{aligned}
\tag{8.49}
$$

and comparing it with Eq. (8.28):

$$f = \frac{e^{2i\delta} - 1}{2ik} = \frac{e^{i\delta}(e^{i\delta} - e^{-i\delta})}{2ik} = \frac{e^{i\delta} \sin \delta}{k}. \tag{8.50}$$

Thus for s-wave scattering

$$\frac{d\sigma}{d\Omega} = |f(\theta)|^2 = \frac{\sin^2 \delta}{k^2}, \tag{8.51}$$

$$\sigma = 4\pi \frac{\sin^2 \delta}{k^2} = \pi \left(\frac{2 \sin \delta}{k} \right)^2. \tag{8.52}$$

Hence $\left(\frac{2 \sin \delta}{k} \right)$ is effective radius of target. Since $\sin \delta \leq 1$,

$$\sigma \leq \frac{4\pi}{k^2} \tag{8.53}$$

i.e. the upper limit for s-wave scattering at an energy corresponding to k, is purely geometric and is independent of the interaction potential.

Finally we note from Eqs. (8.48) and (8.50):

$$S = 1 + 2ikf. \tag{8.54}$$

Then from the unitarity of S-matrix viz.

$$SS^* = 1 \tag{8.55}$$

we have

$$1 + 2ik(f - f^*) + 4k^2|f|^2 = 1$$

$$\text{Im} f = k|f|^2 = \frac{k}{4\pi}\sigma_{\text{total}} \tag{8.56}$$

called the optical theorem.

The generalization to any partial wave is straightforward. For large r i.e. $r \to \infty$, using Eq. (8.40) and the asymptotic form of $j_\ell(kr)$ given just after Eq. (8.40), the incident wave behaves as

$$e^{ikr\cos\theta} \xrightarrow[r\to\infty]{} \sum_\ell \frac{2\ell+1}{2ikr}[e^{ikr} - (-1)^\ell e^{-ikr}]P_\ell(\cos\theta). \tag{8.57}$$

After scattering, one expects that only outgoing spherical wave would change its phase. Thus

$$u(r,\theta) \xrightarrow[r\to\infty]{} \sum_\ell \frac{2\ell+1}{2ikr}[e^{2i\delta_\ell}e^{ikr} - (-1)^\ell e^{-ikr}]P_\ell(\cos\theta)$$

$$= \sum_\ell \frac{2\ell+1}{2ikr}[(e^{2i\delta_\ell} - 1)e^{ikr} + (e^{ikr} - (-1)^\ell e^{-ikr})]P_\ell(\cos\theta)$$

$$= e^{ikr\cos\theta} + \sum_\ell \frac{2\ell+1}{2ik}(e^{2i\delta_\ell} - 1)\frac{1}{r}e^{ikr}P_\ell(\cos\theta). \tag{8.58}$$

Hence comparing above equation with Eq. (8.28), we get

$$f(\theta) = \frac{1}{2ik}\sum_\ell(2\ell+1)(e^{2i\delta_\ell} - 1)P_\ell(\cos\theta)$$

$$= \sum_\ell(2\ell+1)f_\ell(k)P_\ell(\cos\theta). \tag{8.59}$$

where $f_\ell(k)$ is the scattering amplitude for the ℓth partial wave:

$$f_\ell(k) = \frac{1}{2ik}(e^{2i\delta_\ell(k)} - 1)$$

$$= \frac{1}{k}e^{i\delta_\ell}\sin\delta_\ell, \tag{8.60}$$

$$\text{Im} f_\ell(k) = \frac{1}{k}\sin^2\delta_\ell(k). \tag{8.61}$$

Now

$$f(0) = \sum_\ell(2\ell+1)f_\ell(k)P_\ell(1) = \sum_\ell(2\ell+1)f_\ell(k).$$

Thus

$$\text{Im} f(0) = \sum_\ell(2\ell+1)\frac{1}{k}\sin^2\delta_\ell. \tag{8.62}$$

The total cross section:

$$\sigma = \int |f(\theta)|^2 d\Omega = \int |f(\theta)|^2 2\pi \sin\theta d\theta$$

$$= \sum_\ell \sum_{\ell'} (2\ell+1)(2\acute{\ell}+1) f_\ell f_{\ell'}^* \int P_\ell(\cos\theta) P_{\ell'}(\cos\theta) d\Omega$$

$$= \sum_\ell \sum_{\ell'} (2\ell+1)(2\acute{\ell}+1) f_\ell f_{\ell'}^* \sqrt{\frac{4\pi}{2\ell+1}} \sqrt{\frac{4\pi}{2\ell'+1}} Y_{\ell 0} Y_{\ell' 0} d\Omega$$

$$= \sum_\ell \sum_{\ell'} 4\pi \sqrt{(2\ell+1)(2\acute{\ell}+1)} f_\ell f_{\ell'}^* \delta_{\ell\acute{\ell}} = 4\pi \sum_\ell (2\ell+1)|f_\ell|^2$$

$$= 4\pi \sum_\ell (2\ell+1) \sin^2 \delta_\ell. \tag{8.63}$$

Hence from Eqs. (8.62) and (8.63), we obtain the optical theorem:

$$\mathrm{Im} f(0) = \frac{k}{4\pi}\sigma. \tag{8.64}$$

Each term in Eq. (8.59) corresponds to a definite value of the angular momentum ℓ. Eq. (8.63) shows that each angular momentum contributes independently to the total cross section. Since $\sin^2 \delta_\ell \leq 1$, each ℓ contributes to the sum in Eq. (8.63) at most a partial cross section,

$$(\sigma_\ell)_{\mathrm{max}} = \frac{4\pi}{k^2}(2\ell+1), \tag{8.65}$$

an important result. Eq. (8.59) for the differential scattering cross section shows that the the contributions of different angular momenta are not additive in each direction but may interfere with the each other.

We note that the dimension of cross section is that of area. The cross section is measured in barn;

$$1 \mathrm{barn} = 10^{-24} \mathrm{cm}^2 = 10^{-28} \mathrm{meter}^2. \tag{8.66}$$

8.6 Phase Shifts

By determining the phase shifts at various energies empirically from the experimental data using Eqs. (8.59) and (8.64), one can get information about V or the interaction between the bombarding and target particles. The method of partial wave for expressing the scattering amplitude is quite general and does not depend on the detailed form of $V(r)$ provided that it falls faster than $\frac{1}{r}$ at large r.

For a given $V(r)$ and given energy, the asymptotic form of $u(r,\theta)$ in Eq. (8.58) determines δ_ℓ in much the same way that the boundary conditions determine the energy levels of a bound system.

Finally, we derive an expression for δ_ℓ in terms of the potential. For a spherically symmetric potential, the simultaneous eigenfunctions of Hamiltonian and angular momentum L^2 and L_z are $R_\ell(r)Y_{\ell m}(\theta,\phi)$. Hence the general solution of Eq. (8.27) can be written in the form

$$u(r,\theta) = \sum_{\ell=0}^{\infty} A_\ell(k)R_\ell(r)P_\ell(\cos\theta).$$

Now the wave function $u(r,\theta)$ for each partial wave is

$$R_\ell(r)P_\ell(\cos\theta) = \frac{G_\ell}{r}P_\ell(\cos\theta)$$

where the radial wave functions $R_\ell(r)$ and $G_\ell(r)$ satisfy the differential equations.

$$\frac{1}{k^2 r^2}\frac{d}{dr}\left(r^2\frac{dR_\ell}{dr}\right) + \left(1 - \frac{2\mu}{\hbar^2 k^2}V(r) - \frac{\ell(\ell+1)}{k^2 r^2}\right)R_\ell = 0, \quad (8.67)$$

$$\frac{1}{k^2}\frac{d^2 G_\ell}{dr^2} + \left(1 - \frac{2\mu}{\hbar^2 k^2}V(r) - \frac{\ell(\ell+1)}{k^2 r^2}\right)G_\ell = 0. \quad (8.68)$$

For the free particle, the differential equation for R_ℓ^0 and G_ℓ^0 are:

$$\frac{1}{x^2}\frac{d}{dx}\left(x^2\frac{dR_\ell^0}{dx}\right) + \left(1 - \frac{\ell(\ell+1)}{x^2}\right)R_\ell^0 = 0 \quad (8.69)$$

$$\frac{1}{k^2}\frac{d^2 G_\ell^0}{dr^2} + \left(1 - \frac{\ell(\ell+1)}{k^2 r^2}\right)G_\ell^0 = 0 \quad (8.70)$$

where $x = kr$. First, we note that Eq. (8.69) is the Bessel equation, whose solution, which is regular at $r = 0$, is the spherical Bessel function $j_\ell(x) = j_\ell(kr)$:

$$R_\ell^0(r) = j_\ell(kr) \xrightarrow[kr \to 0]{} \frac{(kr)^\ell}{1.3.5.....(2\ell+1)}. \quad (8.71)$$

Now from Eqs. (8.68) and (8.70)

$$\left[G_\ell\frac{d^2 G_\ell^0}{dr^2} - G_\ell^0\frac{d^2 G_\ell}{dr^2}\right] = -G_\ell^0\frac{2\mu V(r)}{\hbar^2}G_\ell. \quad (8.72)$$

Integrating Eq. (8.72), from 0 to r:

$$G_\ell\frac{dG_\ell^0}{dr} - G_\ell^0\frac{dG_\ell^0}{dr}\Big|_0^r = -\int_0^r G_\ell^0(\acute{r})\frac{2\mu V(\acute{r})}{\hbar^2}G_\ell(\acute{r})d\acute{r}. \quad (8.73)$$

Now $G_\ell(r)$ and $G_\ell^0(r)$ vanish at $r = 0$. Thus we have,

$$G_\ell \frac{dG_\ell^0}{dr} - G_\ell^0 \frac{dG_\ell}{dr} = -\frac{2\mu}{\hbar^2} \int_0^r G_\ell^0(\acute{r})V(\acute{r})G_\ell(\acute{r})d\acute{r}. \qquad (8.74)$$

Now from Eq. (8.58), we obtain

$$G_\ell(r) \xrightarrow[r\to\infty]{} \frac{1}{2ik} \left[e^{2i\delta_\ell} e^{ikr} - (-1)^\ell e^{-ikr} \right]$$

$$= \frac{1}{2ik} e^{i\delta_\ell} \left[e^{i\delta_\ell} e^{ikr} - e^{i\ell\pi} e^{-i\delta_\ell} e^{-ikr} \right]$$

$$= \frac{1}{2ik} e^{i(\delta_\ell + \frac{\ell\pi}{2})} \left[e^{i(kr + \delta_\ell - \frac{\ell\pi}{2})} - e^{i(kr + \delta_\ell - \frac{\ell\pi}{2})} \right]. \qquad (8.75)$$

Hence, for $r \to \infty$,

$$G_\ell(r) \longrightarrow \frac{1}{k} \sin(kr + \delta_\ell - \frac{\ell\pi}{2}) \qquad (8.76)$$

$$G_\ell^0(r) \longrightarrow \frac{1}{k} \sin(kr - \frac{\ell\pi}{2}) \qquad (8.77)$$

up to and overall phase factors $e^{i(\delta_\ell + \frac{\ell\pi}{2})}$ and $e^{\frac{i\ell\pi}{2}}$, which cancel out, on both sides of Eq. (8.74). Using Eqs. (8.76) and (8.77), we get from Eq. (8.74)

$$\sin\delta_\ell = -\frac{2\mu}{\hbar^2} k \int_0^\infty G_\ell(r)V(r)G_\ell^0(r)dr$$

$$= -\frac{2\mu}{\hbar^2} k \int_0^\infty R_\ell(r)V(r)j_\ell(kr)r^2 dr. \qquad (8.78)$$

Eq. (8.78) is not very useful, since $R_\ell(r)$ is not known. However if the potential $V(r)$ is weak and regarded as perturbation, then to first order in perturbation, $R_\ell(r)$ can be replaced by the free particle solution $j_\ell(kr)$. For this case

$$\sin\delta_\ell \approx -\frac{2\mu k}{\hbar^2} \int V(r)j_\ell^2(kr)r^2 dr. \qquad (8.79)$$

This is called Born approximation.

Now $j_\ell(kr)$, for $kr \ll 1$, is given in Eq. (8.71). Thus it follows that if $ka \leq 1$, where a is the radius so that $V(r)$ is very small for $a \geq r$, then in the region $r \leq a \ll \frac{1}{k}$, that is the region where $V(r)$ is appreciable, the phase shift δ_ℓ for the ℓth wave, $\ell > ka$, will be very small. It follows from Eq. (8.59):

$$f(\theta) = \sum_{\ell=0}^{\ell_{max}} (2\ell + 1)f_\ell(k)P_\ell(\cos\theta) \qquad (8.80)$$

where $\ell_{\max} \sim ka$. Only the terms for ℓ: $0 \le \ell < ka$ will contribute. Hence the method is easier to apply for small ka.

Now the solution of the Bessel Eq. (8.69), which is regular at $r = 0$, is

$$j_\ell(kr) = (-1)^\ell \frac{1}{k^{\ell+1}} r^\ell \left(\frac{1}{r}\frac{d}{dr}\right)^\ell \frac{\sin kr}{r}. \tag{8.81}$$

The second solution of the Bessel equation which is irregular at $r = 0$ is

$$n_\ell(kr) = (-1)^{\ell-1} \frac{r^\ell}{k^{\ell+1}} \left(\frac{1}{r}\frac{d}{dr}\right)^\ell \frac{\cos kr}{r}. \tag{8.82}$$

For s-wave scattering in the Born approximation, we have from Eqs. (8.79) and (8.81) for $\ell = 0$,

$$\sin \delta_0 = -\frac{2\mu}{\hbar^2} k \int V(r) \frac{\sin^2 kr}{k^2 r^2} r^2 dr.$$

For very slow velocities $\frac{\sin kr}{kr} \to 1$, so that

$$\delta_0 \approx \sin \delta_0 \approx -\frac{2\mu}{\hbar^2} k \int V(r) r^2 dr$$

$-\frac{\delta_0}{k}$ is called the s-wave scattering length:

$$a_s \to \lim_{k \to 0} -\frac{\delta_0}{k} \tag{8.83}$$

where

$$a_s = \frac{2\mu}{\hbar^2} \int V(r) r^2 dr$$

$a_s < 0$, if V is attractive and > 0 if V is repulsive.
The general solution of Eq. (8.69):

$$R_\ell^0 = (A_\ell j_\ell(kr) + B_\ell n_\ell(kr)). \tag{8.84}$$

The following limits for the spherical Bessel functions are useful:

$$j_\ell(kr) \xrightarrow[kr \to 0]{} \frac{(kr)^\ell}{(2\ell+1)!!},$$

$$n_\ell(kr) \xrightarrow[kr \to 0]{} -\frac{(2\ell-1)!!}{k^{\ell+2} r^{\ell+1}},$$

$$j_\ell(kr) \xrightarrow[kr \to \infty]{} \frac{1}{kr} \sin\left(kr - \frac{\ell\pi}{2}\right),$$

$$n_\ell(kr) \xrightarrow[kr \to \infty]{} -\frac{1}{kr} \cos\left(kr - \frac{\ell\pi}{2}\right). \tag{8.85}$$

As an application of the above limits, we consider the behavior of the phase shift δ_ℓ near $k = 0$. First we note that $V(r) \approx 0$, for $r > a$. We consider the radial equation (8.68) in these regions:

i) $0 \leq r \leq a$: k^2 is negligible:

$$\frac{d^2 G_\ell}{dr^2} - \left[\frac{2\mu}{\hbar^2} V(r) + \frac{\ell(\ell+1)}{r^2} \right] G_\ell = 0$$

ii) $a \leq r \ll k^{-1} (ka \leq kr \ll 1)$:

$$\frac{d^2 G_\ell}{dr^2} - \frac{\ell(\ell+1)}{r^2} G_\ell = 0$$

which has solution

$$G_\ell(r) = C_1 r^{\ell+1} + C_2 r^{-\ell}$$

where C_1 and C_2 are independent of k.

iii) $r \sim k^{-1}$ up to ∞

$$\frac{d^2 G_\ell}{dr^2} + \left[k^2 - \frac{\ell(\ell+1)}{r^2} \right] G_\ell = 0$$

which has general solution given in Eq. (8.84)

$$G_\ell(r) = r R_\ell(r) = r[A_\ell j_\ell(kr) + B_\ell n_\ell(kr)]$$
$$\longrightarrow r \left[A_\ell \frac{k^r r^\ell}{(2\ell+1)!!} - B_\ell \frac{(2\ell_1)!}{k^{\ell+1}} r^{-\ell-1} \right].$$

for $kr \to 0$. Then connecting the solutions in regions (ii) and (iii) for $kr \ll 1$, we obtain

$$\frac{A_\ell k^\ell}{(2\ell+1)!!} = C_1, \quad -\frac{B_\ell(2\ell+1)!!}{k^{\ell+1}} r^{-\ell-1} = C_2.$$

Thus using the asymptotic limits (8.85)

$$G_\ell \sim \frac{A_\ell}{k} \sin(kr - \frac{1}{2}\ell\pi) - \frac{B_\ell}{k} \cos(kr - \frac{1}{2}\ell\pi).$$

Putting $A_\ell = C'_\ell \cos \delta_\ell$, $B_\ell = -C'_\ell \sin \delta_\ell$, so that the last equation takes the correct asymptotic form:

$$G_\ell \sim \frac{C'_\ell}{k} \sin(kr - \frac{\ell\pi}{2} + \delta_\ell)$$

where

$$\tan \delta_\ell = \frac{B_\ell}{A_\ell} = \frac{C_2}{C_1} \frac{k^{2\ell+1}}{(2\ell+1)!!(2\ell_1)!!} \,. \tag{8.86}$$

Note that C_2 and C_1 are independent of k. Thus as $k \to 0$, $k^{2\ell+1} \cot \delta_\ell \to$ constant which depends on ℓ. In particular $k \cot \delta \to -\frac{1}{a_s}$, where a_s is called the s-wave scattering length. (Note that in contrast to (8.83), this result does not depend on the Born approximation). Since $\delta_\ell \sim k^{2\ell+1}$, for small k, all phase shifts with $\ell \neq 0$ are small in comparison with δ_0 i.e. at low k, the s-wave dominates.

8.7   Optical Theorem

We now discuss the optical theorem:

$$\operatorname{Im} f(0) = \frac{1}{k} \sum_{\ell=0}^{\infty} (2\ell + 1) \sin^2 \delta_\ell$$

$$= \frac{k}{4\pi} \sigma_{\text{tot}}. \tag{8.87}$$

We have so far discussed elastic scattering for which

$$S_\ell = e^{2i\delta_\ell} \tag{8.88}$$

$$|S_\ell|^2 = 1 \tag{8.89}$$

and phase shifts δ_ℓ are real. In many scattering experiments, there can be inelastic scattering viz. the target may also get excited or change its state, or another particle may emerge. This means that there is an absorption of the incident beam in such experiments and if we describe the scattering process as a one channel picture, then δ is complex. This is because

$$(\text{Flux})_{\text{in}} - (\text{Flux})_{\text{out}} = \text{Absorption},$$

$$(\text{Flux})_{\text{in}} \neq (\text{Flux})_{\text{out}}. \tag{8.90}$$

In order to satisfy the above condition, we write

$$\delta_\ell \longrightarrow \delta_\ell + i\beta_\ell \quad (\delta_\ell \text{ real}),$$

$$S_\ell(k) = e^{-2\beta_\ell(k)} e^{2i\delta_\ell(k)}$$

$$= \eta_\ell(k) e^{2i\delta_\ell(k)}, \tag{8.91}$$

$$|S_\ell(k)|^2 = |\eta_\ell(k)|^2, \quad \eta_\ell(k) = e^{-2\beta_\ell(k)}. \tag{8.92}$$

Thus, since for absorption β is positive,

$$|S_\ell|^2 \neq 1$$

$$|S_\ell|^2 \leq 1 \tag{8.93}$$

or

$$0 \leq \eta_l \leq 1. \tag{8.94}$$

The partial wave scattering amplitude $f_\ell(k)$ is now given by

$$f_\ell(k) = \frac{S_\ell(k) - 1}{2ik}$$

$$= \frac{\eta_\ell e^{2i\delta_\ell} - 1}{2ik}$$

$$= \frac{\eta_\ell \sin 2\delta_\ell}{2k} + i \frac{1 - \eta_\ell \cos 2\delta_\ell}{2k}. \tag{8.95}$$

Thus the total elastic cross section is

$$\sigma_{el} = 4\pi \sum_\ell (2\ell + 1)|f_\ell(k)|^2$$

$$= \frac{4\pi}{k^2} \frac{1}{4} \sum_\ell (2\ell + 1)|1 - S_\ell|^2$$

$$= 4\pi \sum_l (2\ell + 1)\frac{1 + \eta_\ell^2 - 2\eta_\ell \cos 2\delta_\ell}{4k^2}. \tag{8.96}$$

There is also a cross section for inelastic processes. Now

$$\frac{1}{(\text{Flux})_{in}}|(\text{Flux})_{in} - (\text{Flux})_{out}| = 1 - |S_\ell|^2$$

and thus

$$\sigma_{abs} = \frac{4\pi}{k^2} \frac{1}{4} \sum_\ell (2\ell + 1)(1 - |S_\ell|^2)$$

$$= \frac{\pi}{k^2} \sum_\ell (2\ell + 1)(1 - \eta_\ell^2). \tag{8.97}$$

Thus the total cross section is

$$\sigma_{tot} = \sigma_{el} + \sigma_{abs}$$

$$= \frac{\pi}{k^2} \sum_\ell (2\ell + 1)[2 - 2\eta_\ell \cos 2\delta_\ell]. \tag{8.98}$$

But

$$\text{Im } f_\ell(k) = \frac{1 - \eta_\ell \cos 2\delta_\ell}{2k},$$

$$\text{Im } f(0) = \sum_\ell (2\ell + 1)\text{Im } f_l(k)$$

$$= \frac{1}{2k} \sum_\ell (2\ell + 1)(1 \quad \eta_\ell \cos 2\delta_\ell)$$

$$= \frac{k}{4\pi}\sigma_{tot}. \tag{8.99}$$

Thus the optical theorem is indeed satisfied and is a consequence of conservation of flux or probability.

Now $\sigma_{el,l}$ is maximum, when

$$S_\ell = -1, \quad \text{since} \quad |S_\ell|^2 \leq 1 \tag{8.100a}$$

$$(\sigma_{el,\ell})_{max} = \frac{4\pi}{k^2}(2\ell + 1) \tag{8.100b}$$

$$\sigma_{abs,\ell} = 0. \tag{8.100c}$$

Maximum absorption occurs when $S_\ell = 0$, then

$$\sigma_{el,\ell} = \frac{\pi}{k^2}(2\ell + 1), \tag{8.101a}$$

$$\sigma_{abs,\ell} = \frac{\pi}{k^2}(2\ell + 1). \tag{8.101b}$$

The incoming and outgoing waves are coherent in elastic scattering (because S_ℓ enters in σ_{el} as opposed to $|S_\ell|^2$ in σ_{abs}) and therefore can interfere constructively or destructively.

We see that when $S_\ell = 0$ or $\eta_\ell = 0$, we have total absorption. Nevertheless there is still elastic scattering in that partial wave. This becomes clear in scattering by a black disc [Fig. 8.4]. Let a be the radius of the disc. Now only those partial waves are important for which

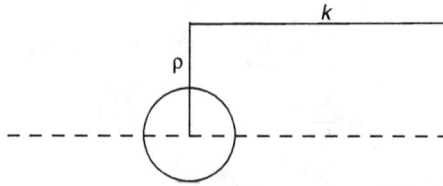

Fig. 8.4   Scattering by a black disc of radius a.

$$l \leq L = ka$$

and we shall take here $ka \gg 1$. Write $\ell = \rho k$, so that $\rho \leq a$. Thus for the black disc limit $\ell < L$ ($\rho < a$), there is complete absorption:

$$\eta_\ell = 0, \ e^{-2\beta_\ell} = 0,$$
$$S_\ell = 0.$$

$\rho > a, \ell > L$, no elastic scattering and no absorption:

$$\eta_\ell = 1, \ e^{-2\beta_\ell} = 1, \ S_\ell = 1.$$

Hence we have from Eq. (8.91), Eq. (8.96) and Eq. (8.97) [using $L = ka \gg 1$]

$$\sigma_{el} = \frac{\pi}{k^2} \sum_{\ell=0}^{L}(2\ell + 1) \approx \frac{\pi}{k^2}L^2 = \pi a^2,$$

$$\sigma_{abs} = \frac{\pi}{k^2} \sum_{\ell=0}^{L}(2\ell + 1) \approx \pi a^2,$$

$$\sigma_{tot} = \sigma_{el} + \sigma_{abs} = 2\pi a^2.$$

On classical grounds we expect that the cross section cannot exceed the area presented by the disc. The above result can be understood as follows:

The absorptive disc takes flux proportional to πa^2 out of the incident beam and this leads to a shadow behind the disc. The incident beam is diffracted from the edge of the disc and that gives rise to elastic scattering. The elastic scattering that accompanies absorption is called shadow or diffraction scattering. It is strongly peaked forward. This can be seen as follows:

Since elastic scattering is coherent,

$$\frac{d\sigma^{el}}{d\Omega} = \frac{1}{4k^2}\Big|\sum_\ell (2\ell+1)(1-S_\ell(k))P_\ell(\cos\theta)\Big|^2. \tag{8.102}$$

For a black disc

$$S_\ell = 0 \quad \ell < L, \quad \rho \le a,$$
$$S_\ell = 1 \quad \ell > L, \quad \rho > a, \tag{8.103}$$

hence

$$\frac{d\sigma^{el}}{d\Omega} = \frac{1}{4k^2}\Big|\sum_{\ell=0}^{L} (2\ell+1)P_\ell(\cos\theta)\Big|^2. \tag{8.104}$$

Now for high energy $ka \gg 1$, $L = ka$ and we replace the summation by the integration

$$x = k\rho \longrightarrow l+\frac{1}{2}$$

$$\sum_\ell = \int dx$$

$$\frac{d\sigma^{el}}{d\Omega} = \frac{1}{k^2}\Big|\int_0^{ka} x\,dx\,P_x(\cos\theta)\Big|^2. \tag{8.105}$$

Now we use the formula [valid for large n and small θ]

$$P_n(\cos\theta) \equiv J_0(n\sin\theta)$$
$$= J_0(n\theta), \tag{8.106}$$

$$\frac{d\sigma^{el}}{d\Omega} = \frac{1}{k^2}\Big|\int_0^{ka} x\,dx\,J_0(x\theta)\Big|^2, \tag{8.107}$$

where J_0 is the Bessel function of zeroth order. Using the formulae

$$z J_0(z) = \frac{d}{dz} z J_1(z), \tag{8.108a}$$

$$J_0(x\theta) = \frac{1}{x\theta} \frac{d}{d(x\theta)} (x\theta J_1(x\theta)), \tag{8.108b}$$

we get,

$$\frac{d\sigma^{el}}{d\Omega} = \frac{1}{k^2} \left| \int_0^{ka\theta} \frac{x\theta d(x\theta) J_0(x\theta)}{\theta^2} \right|^2$$

$$= \frac{1}{k^2 \theta^4} |ka\theta J_1(ka\theta)|^2$$

$$= k^2 a^4 |J_1(ka\theta)/ka\theta|^2. \tag{8.109}$$

A plot of $\frac{d\sigma^{el}}{d\Omega}$ as a function of θ is shown in Fig. 8.5. Now $\frac{d\sigma^{el}}{d\Omega}$ is maximum

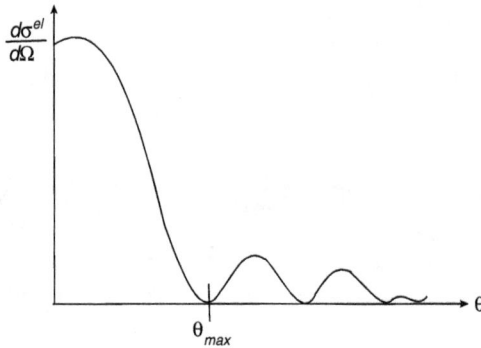

Fig. 8.5   The incident plane wave and the scattered wave moving radially outwards.

$(= a^4 k^2)$ when $\theta = 0$; it is zero when

$$ka\theta = 1$$

or

$$\theta = 1/ka.$$

Thus $\frac{d\sigma^{el}}{d\Omega}$ has the approximate value $a^4 k^2$ in the forward direction and decreases rapidly in the angular region

$$0 < \theta \le 1/ka.$$

Thus

$$\sigma^{el} = a^4 k^2 2\pi \int_0^{1/ka} \sin\theta d\theta$$

$$= \pi a^2.$$

These features are observed in nuclear scattering and in particle physics scattering at high energies.

8.8 Resonances, Dispersion Formula

A resonance is characterized by its mass (energy E_0) and width Γ.

A decaying particle produced at time t is described by the wave function,

$$\psi(t) = 0, \quad t < 0$$
$$\psi(t) = \psi(0)e^{(-iE_0t-\frac{\Gamma}{2}t)/\hbar}, \quad t \geq 0 \tag{8.110}$$

where

$$w = \frac{\Gamma}{\hbar} \quad \text{is the decay probability,}$$

$$\tau = \frac{\hbar}{\Gamma} = \frac{t_{1/2}}{\ln 2} = 1.44t_{1/2}, \tag{8.111}$$

τ is the mean life time, $t_{1/2}$, the half life of resonance. Now

$$|\psi(t)|^2 = |\psi(0)|e^{-\frac{\Gamma t}{\hbar}} = |\psi(0)|^2 e^{-\frac{t}{\tau}}. \tag{8.112}$$

The Fourier transform of $\psi(t)$:

$$\psi(E) = \int_0^\infty e^{\frac{iEt}{\hbar}} \psi(t)dt$$

$$= \psi(0) \int_0^\infty e^{i(E-E_0+i\frac{\Gamma}{2})\frac{t}{\hbar}} dt$$

$$= \frac{i\hbar\psi(0)}{(E-E_0)+i\frac{\Gamma}{2}},$$

$$|\psi(E)|^2 = \frac{|\psi(0)|^2\hbar^2}{(E-E_0)^2+\frac{\Gamma^2}{4}}. \tag{8.113}$$

The probability density for finding the particle with energy E:

$$P(E) = |\psi(E)|^2 = \frac{|\psi(0)|^2\hbar^2}{(E-E_0)^2+\frac{\Gamma^2}{4}}. \tag{8.114}$$

The unknown constant $|\psi(0)|^2$ is fixed by the normalization:

$$\int_{-\infty}^\infty P(E)dE = 1. \tag{8.115}$$

Using the formula

$$\pi\delta(E - E_0) = \lim_{\epsilon \to 0} \frac{\epsilon}{(E - E_0)^2 + \epsilon^2}, \tag{8.116}$$

$$\int_{-\infty}^{\infty} P(E)dE = |\psi(0)|^2\hbar^2 \int_{-\infty}^{\infty} \frac{2}{\Gamma} \left[\frac{\frac{\Gamma}{2}}{(E - E_0)^2 + \frac{\Gamma^2}{4}} \right] dE$$

$$= |\psi(0)|^2\hbar^2 \frac{2}{\Gamma} \int_{-\infty}^{\infty} \pi\delta(E - E_0)dE$$

$$= (2\pi)|\psi(0)|^2\frac{\hbar^2}{\Gamma}. \tag{8.117}$$

From Eq. (8.115):

$$|\psi(0)|^2 = \frac{\Gamma}{2\pi\hbar^2}. \tag{8.118}$$

Hence

$$P(E) = |\psi(E)|^2 = \frac{\Gamma}{2\pi} \frac{1}{(E - E_0)^2 + \frac{\Gamma^2}{4}}. \tag{8.119}$$

Eq. (8.119) gives the natural line shape of decaying particle. Now $P(E)$ has its maximum value $\frac{2}{\pi}\Gamma$ at $E = E_0$. At $E = E_0 \pm \frac{\Gamma}{2}$, $P(E) = \frac{1}{\pi}\Gamma$. $P(E)$ as a function of E given in Eq. (8.119) is shown in the Fig. 8.6. Γ is the width of resonance.

A striking feature of a scattering process is the appearance of high, narrow peaks in the scattering cross section expressed as a function of energy. The peaks are called resonances.

Partial wave scattering amplitude is

$$f_\ell(k) = \frac{1}{k}e^{i\delta_\ell} \sin \delta_\ell = \frac{1}{k \cot \delta_\ell - ik}. \tag{8.120}$$

At resonance E_0, $f_\ell(k)$ should be maximum i.e. at $E = E_0$, $\cot \delta_\ell = 0$. Let us expand $\cot \delta_\ell$, about $E - E_0$:

$$\cot \delta_\ell(E) = \cot \delta_\ell(E_0) + \cot' \delta_\ell(E)\Big|_{(E=E_0)} (E - E_0). \tag{8.121}$$

Put

$$\cot' \delta_\ell(E)\Big|_{(E=E_0)} = -\frac{2}{\Gamma}$$

so that near resonance

$$\cot \delta_\ell(E) = -\frac{2}{\Gamma}(E - E_0), \tag{8.122}$$

and

$$f_\ell(k) = \frac{1}{k[-\frac{2}{\Gamma}(E - E_0) - i]}$$

$$= \frac{-\frac{\Gamma}{2}}{E - E_0 + i\frac{\Gamma}{2}}.$$
(8.123)

At resonance

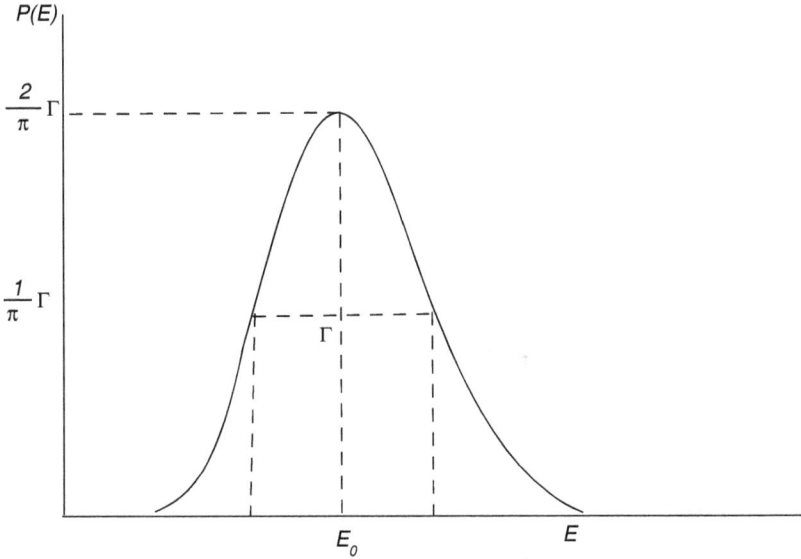

Fig. 8.6 Natural line shape of a decaying state. Γ is the full width at half maximum.

$$E \to E_0, \quad \cot \delta_\ell \to 0, \quad \delta_\ell \to \frac{\pi}{2}.$$
(8.124)

If the resonance occurs in partial wave ℓ (i.e. angular momentum of resonance is ℓ), then near resonance only the ℓth partial wave in expansion of scattering amplitude given in Eq. (8.59) is important, so that

$$f(\theta) = (2\ell + 1)f_\ell(k)P_\ell(\cos \theta).$$
(8.125)

The partial wave amplitude $f_\ell(k)$, near resonance has the form in Eq. (8.123). Hence near the resonance

$$f(\theta) \approx (2\ell + 1)\frac{-\frac{\Gamma}{2}}{k[(E - E_0) + i\frac{\Gamma}{2}]}P_\ell(\cos \theta)$$
(8.126)

and then

$$f(0) = \frac{(2\ell + 1)(-\frac{\Gamma}{2})}{k[(E - E_0) + i\frac{\Gamma}{2}]} = \frac{(2\ell + 1)(-\frac{\Gamma}{2})[(E - E_0) - i\frac{\Gamma}{2}]}{k[(E - E_0)^2 + \frac{\Gamma^2}{4}]} \quad (8.127)$$

and

$$\text{Im} f(0) = \frac{(2\ell + 1)(\frac{\Gamma}{2})^2}{k[(E - E_0)^2 + \frac{\Gamma^2}{4}]}. \quad (8.128)$$

Using the optical theorem, near resonance, σ_{tot}:

$$\sigma_{\text{tot}} = \frac{4\pi}{k} \text{Im} f(0) = \frac{4\pi}{k^2} \frac{(2\ell + 1)(\frac{\Gamma}{2})^2}{(E - E_0)^2 + \frac{\Gamma^2}{4}}. \quad (8.129)$$

In general if resonance of spin J occurs in a scattering process

$$\sigma_{\text{tot}} = \frac{4\pi}{k^2} \frac{(2J + 1)(\frac{\Gamma}{2})^2}{(E - E_0)^2 + \frac{\Gamma^2}{4}}. \quad (8.130)$$

This is called Breit–Wigner formula. If both the projectile and the target has spin J_1 and J_2 respectively, then

$$\sigma_{\text{tot}} = \frac{1}{(2J_1 + 1)(2J_2 + 1)} \frac{4\pi}{k^2} \frac{(2J + 1)(\frac{\Gamma}{2})^2}{(E - E_0)^2 + \frac{\Gamma^2}{4}}. \quad (8.131)$$

8.9 Bound States and Resonances: Square Well Potential

We illustrate the concept of bound states and resonances, for scattering by a square well potential.

$$E > 0, \quad V = -|V|, \quad k^2 = \frac{2mE}{\hbar^2}, \quad K^2 = \frac{2m(E + |V|)}{\hbar^2}. \quad (8.132)$$

The Schrödinger equation for a stationary state is

$$\left[\frac{d^2}{dx^2} + k^2\right] u_0(x) = 0 \quad x < -a, \quad (8.133)$$

$$\left[\frac{d^2}{dx^2} + K^2\right] u_1(x) = 0 \quad -a < x < a, \quad (8.134)$$

$$\left[\frac{d^2}{dx^2} + k^2\right] u_2(x) = 0 \quad x > 0. \quad (8.135)$$

The solutions are

$$u_0(x) = Ae^{ikx} + Be^{-ikx}, \quad (8.136)$$

$$u_1(x) = Ce^{iKx} + De^{-iKx}, \quad (8.137)$$

$$u_2(x) = Fe^{ikx}. \quad (8.138)$$

Boundary condition gives: (see Ch. 4)

$$\frac{F}{A} = \frac{2e^{-2ika}}{[2\cos 2ka - i(\frac{K}{k} + \frac{k}{K})\sin 2ka]}$$
$$\equiv e^{-2ika} S(E) \qquad (8.139)$$

where $S(E)$ is called the S-matrix:

$$S(E) = \frac{2}{[2\cos 2ka - i(\frac{K}{k} + \frac{k}{K})\sin 2ka]}. \qquad (8.140)$$

Now the transmissivity is given by

$$T(E) = |\frac{F}{A}|^2 = |S(E)|^2$$
$$= \frac{4}{4\cos^2 2ka + (\frac{K}{k} + \frac{k}{K})^2 \sin^2 2ka}. \qquad (8.141)$$

The S-matrix $S(E)$, regarded as an analytic function of E, has poles when

$$2\cos 2ka - i(\frac{K}{k} + \frac{k}{K})\sin 2ka = 0. \qquad (8.142)$$

Clearly k must be imaginary, otherwise no solution exists. Since $k = \frac{\sqrt{2mE}}{\hbar}$ i.e. for poles, E must be negative. Now

$$\frac{\cos 2ka}{\sin 2ka} = \frac{1}{2}(\cot Ka - \tan ka)$$
$$= \frac{1}{2}\frac{\cot^2 Ka - 1}{\cot ka}. \qquad (8.143)$$

Hence from Eq. (8.142)

$$\cot^2 Ka - i(\frac{K}{k} + \frac{k}{K})\cot Ka - 1 = 0. \qquad (8.144)$$

The two solutions of the quadratic Eq. (8.144):

$$\text{i)} \quad k\cot Ka = iK,$$
$$\text{ii)} \quad K\cot Ka = ik. \qquad (8.145)$$

These solutions can be put in the form

$$\text{i)} \quad K\tan Ka = -ik,$$
$$\text{ii)} \quad K\cot Ka = ik. \qquad (8.146)$$

The solutions (i) and (ii) give the bound states of even and odd parity respectively on negative E-axis.

In the neighborhood of a resonance $E_r > 0$, the S-matrix:

$$S(E) = \frac{1}{\cos 2ka[1 - \frac{i}{2}(\frac{K}{k} + \frac{k}{K})\tan 2ka]} \Rightarrow \pm 1 \qquad (8.147)$$

so that

$$\tan 2ka \to 0$$
$$\cos 2ka \to \pm 1 \qquad (8.148)$$

as $E \to E_r$. By Taylor expansion near resonance a function

$$f(E) = f(E_r) + (E - E_r)\acute{f}(E)\Big|_{E=E_r}. \qquad (8.149)$$

In our case

$$f(E) = (\frac{K}{k} + \frac{k}{K})\tan 2ka,$$
$$f(E_r) = 0. \qquad (8.150)$$

Put

$$\acute{f}(E)\Big|_{E=E_r} = \frac{4}{\Gamma} \qquad (8.151)$$

so that near resonance

$$S(E) = \frac{\pm 1}{[1 - \frac{i}{2}\frac{4}{\Gamma}(E - E_r)]}$$
$$= \frac{\pm\frac{\Gamma}{2}}{\frac{\Gamma}{2} - i(E - E_r)} = \frac{\pm i\frac{\Gamma}{2}}{(E - E_r) + \frac{i\Gamma}{2}}. \qquad (8.152)$$

Thus $S(E)$ appears to have a pole at

$$E = E_r - i\frac{\Gamma}{2} \qquad (8.153)$$

as shown in the Fig. 8.6. From Eq. (8.147)

$$S(E) = \frac{\cos 2Ka + \frac{i}{2}(\frac{K}{k} + \frac{k}{K})\sin 2K}{[\cos^2 2Ka + \frac{1}{4}(\frac{K}{k} + \frac{k}{K})\sin^2 2Ka]}$$
$$= |S(E)|e^{i\delta(E)} \qquad (8.154)$$

where

$$\tan \delta(E) = \frac{1}{2}(\frac{K}{k} + \frac{k}{K})\tan 2Ka. \qquad (8.155)$$

Now near resonance

$$\tan \delta(E) = \frac{2}{\Gamma}(E - E_r)$$

$$\frac{d}{dE}[\tan \delta(E)] = \frac{2}{\Gamma}. \tag{8.156}$$

Thus

$$\frac{d\delta(E)}{dE} = \frac{2\Gamma}{[\Gamma^2 + 4(E - E_r)^2]}$$

$$\left.\frac{d\delta(E)}{dE}\right|_{E=E_r} = \frac{2}{\Gamma}. \tag{8.157}$$

8.10   δ-function Potential

$$V(x) = V_0 \delta(x)$$

Schrödinger Equation is

$$\left[-\frac{\hbar^2}{2m}\frac{d^2}{dx^2} + V(x)\right]u(x) = Eu(x) \tag{8.158}$$

or

$$\left[\frac{d^2}{dx^2} + k^2\right]u(x) = \frac{2m}{\hbar}V(x)u(x)$$

$$= \frac{2m}{\hbar}V_0\delta(x)u(x) \tag{8.159}$$

where

$$k^2 = \frac{2mE}{\hbar^2}, \quad E > 0.$$

First we determine the boundary conditions for the δ-function potential. Integrate Eq. (8.159) between the limits $0 - \epsilon$ and $0 + \epsilon$:

$$\int_{0-\epsilon}^{0+\epsilon} \frac{d^2}{dx^2}u(x)dx + \int_{0-\epsilon}^{0+\epsilon} k^2 u(x)dx = \frac{2mV_0}{\hbar}\int_{0-\epsilon}^{0+\epsilon} \delta(x)u(x)dx, \tag{8.160}$$

which gives

$$\left.\frac{du}{dx}\right|_{0-\epsilon}^{0+\epsilon} + \left.k^2 u(x)\right|_{0-\epsilon}^{0+\epsilon} = \frac{2mV_0}{\hbar}u(0) \tag{8.161}$$

or in the limit $\epsilon \to 0$:

$$\left(\frac{du}{dx}\right)_{0+\epsilon} - \left(\frac{du}{dx}\right)_{0-\epsilon} = \frac{2mV_0}{\hbar}u(0). \tag{8.162}$$

Hence, we have the boundary conditions: $u(x)$ is continuous, but $\frac{du}{dx}$ is not as given in Eq. (8.162).

In the regions $x < 0$ and $x > 0$, $V(x) = 0$:

$$u(x) = Ae^{ikx} + Be^{-ikx}, \qquad x < 0 \qquad (8.163)$$

$$u(x) = Ce^{ikx}, \qquad x > 0. \qquad (8.164)$$

The boundary conditions given above give

$$u(0) = A + B = C, \qquad (8.165)$$

$$C(ik) - ik(A - B) = \frac{2mV_0}{\hbar^2}u(0)$$

$$= \frac{2mV_0}{\hbar^2}C. \qquad (8.166)$$

From Eq. (8.165) and (8.166):

$$\frac{C}{A} = \frac{ik\hbar^2}{ik\hbar^2 - mV_0}, \qquad (8.167)$$

$$\frac{B}{A} = \frac{C}{A} - 1 = \frac{mV_0}{ik\hbar^2 - mV_0}. \qquad (8.168)$$

Hence from Eqs. (8.163) and (8.164):

$$u(x) = A[e^{ikx} + \frac{B}{A}e^{-ikx}]$$

$$\stackrel{!}{=} A[e^{ikx} + \frac{mV_0}{ik\hbar^2 - mV_0}e^{-ikx}], \quad x < 0 \qquad (8.169)$$

$$u(x) = A\frac{ik\hbar^2}{ik\hbar^2 - mV_0}, \quad x > 0. \qquad (8.170)$$

The transmitivity and reflectivity are given by

$$T = |\frac{C}{A}|^2 = \frac{E}{E + \frac{mV_0^2}{2\hbar^2}}, \qquad (8.171)$$

$$R = |\frac{B}{A}|^2 = \frac{\frac{mV_0^2}{2\hbar^2}}{E + \frac{mV_0^2}{2\hbar^2}}, \qquad (8.172)$$

$$R + T = 1.$$

The S-matrix

$$S(E) = \frac{C}{A} = \frac{\text{Coefficient of transmitted wave}}{\text{Coefficient of incident wave}}$$

$$= \frac{ik\hbar^2}{ik\hbar^2 - mV_0}. \qquad (8.173)$$

$S(E)$ regarded as an analytic function of k has a pole at

$$ik\hbar^2 - mV_0 = 0$$

or

$$E = -\frac{mV_0^2}{2\hbar^2}. \tag{8.174}$$

Pole is at negative value of E. A pole at negative E corresponds to a bound state. This can be seen as follows:
For

$$E < 0 , \quad E = -|E|,$$
$$k = iK , \quad K = \sqrt{\frac{2m|E|}{\hbar^2}}. \tag{8.175}$$

Then the solutions in Eqs. (8.163) and (8.164) are replaced by

$$u(x) = Ae^{-Kx}, \qquad x > 0,$$
$$u(x) = Be^{-K|x|}, \qquad x < 0. \tag{8.176}$$

Boundary conditions, namely u continuous and the one given in Eq. (8.162) give

$$u(0) = A = B \tag{8.177}$$

$$-KA - KA = \frac{2m}{\hbar^2} V_0 A \tag{8.178}$$

so that

$$A \neq 0, \quad K = -\frac{mV_0}{\hbar^2}. \tag{8.179}$$

Hence

$$E = -\frac{mV_0}{2\hbar^2}. \tag{8.180}$$

i.e. δ-function potential has only one bound state.

8.11   The Born Approximation

Suppose we have a particle in momentum state $|p_i\rangle$ at some early time $t_i(- \to \infty)$ i.e. it is in an eigenstate of $H_0 = \frac{p^2}{2m}$ with momentum $\mathbf{p}_i$, and we turn on a potential $V(r)$ for $t > t_i$ and then we observe the particle at a later $t_f \to \infty$ when it is again in an eigenstate of H_0 with momentum $\mathbf{p}_f$. This is what happens in a scattering process. Thus for $t_i < t < t_f$, the system is in a time dependent state $\psi(t)$ which satisfies the Schrödinger equation

$$i\hbar \frac{\partial \psi(t)}{\partial t} = (H_0 + V)\psi(t) \tag{8.181}$$

with the boundary conditions

$$\psi(t) = u_{\mathbf{p}_i}(\boldsymbol{r})e^{\frac{-i}{\hbar}E_{\mathbf{p}_i}t} \quad t < t_i$$
$$= u_{\mathbf{p}_f}(\boldsymbol{r})e^{\frac{-i}{\hbar}E_{\mathbf{p}_f}t} \quad t > t_f \tag{8.182}$$

where $E_p = \dfrac{\boldsymbol{p}^2}{2m}$. It is convenient to write the solution of Eq. (8.181) as

$$\psi_i(\boldsymbol{r}, t) = \sum_{\boldsymbol{p}} a_{\mathbf{p}\mathbf{p}_i}(t)u_{\mathbf{p}}(\mathbf{r})e^{\frac{-iE_p}{\hbar}t}. \tag{8.183}$$

The subscript on ψ indicates that we are using boundary condition for $t < t_i$. Substituting Eq. (8.183) in Eq. (8.181) and using that $H_0 u_{\mathbf{p}}(\mathbf{r}) = E_p u_{\mathbf{p}}(r)$, we obtain

$$\sum_{\boldsymbol{p}} i\hbar(\dot{a})_{\boldsymbol{p}\boldsymbol{p}_i}(t)u_{\mathbf{p}}(r)e^{\frac{-iE_p}{\hbar}t} = \sum_{\boldsymbol{p}} a_{\boldsymbol{p}\boldsymbol{p}_i}(t)Vu_{\mathbf{p}}(r)e^{\frac{-iE_p}{\hbar}t}.$$

By multiplying on the right by $u_{\mathbf{p}'}^*(r)$ and integrating over d^3r and using the orthogonality relation

$$\int u_{\boldsymbol{p}'}^*(r)u_{\boldsymbol{p}}(r)d^3r = \delta_{\boldsymbol{p}'\boldsymbol{p}}, \tag{8.184}$$

we obtain

$$i\hbar \dot{a}_{\boldsymbol{p}\boldsymbol{p}_i}(t) = \sum_{\boldsymbol{p}} a'_{\boldsymbol{p}\boldsymbol{p}_i}(t)e^{\frac{-i}{\hbar}(E_p - E_{\tilde{p}})t}\langle p'|V| \to p\rangle \tag{8.185a}$$

where

$$\langle \boldsymbol{\acute{p}}|V|\boldsymbol{p}\rangle = \int d^3r u_{\boldsymbol{\acute{p}}}(r)Vu_{\boldsymbol{p}}(r). \tag{8.185b}$$

Now integration over t, with the boundary condition

$$a'_{\boldsymbol{p}\boldsymbol{p}}(t) = \delta_{\boldsymbol{\acute{p}}\boldsymbol{p}}, \quad t < t_i$$

gives

$$a'_{p'p_i}(t) = \delta_{\acute{p}p_i} + \frac{i}{\hbar} \int\limits_{t_i}^{t} dt \sum_{p} \langle \acute{p}|V|p\rangle e^{\frac{-i}{\hbar}(E_p - E_{\acute{p}})\frac{t}{\hbar}} a_{pp_i}(t). \quad (8.186)$$

For $t < t_i$, $V = 0$ and for $t = t_i$, the integral vanishes and we recover the boundary condition. This is an exact result. We now assume that V is small perturbation. Then to the zeroth order

$$a_{\acute{p}p_i}(t) = \delta_{\acute{p}p_i}$$

and to the first order [known as Born approximation]

$$a_{\acute{p}p_i}(t) \approx \delta_{\acute{p}p_i} + \frac{1}{i\hbar} \int_{t_i}^{t} dt \sum_{p} \langle \acute{p}|V|p\rangle e^{\frac{-i}{\hbar}(E_p - E_{\acute{p}})t} \delta_{pp_i}$$

$$= \delta_{\acute{p}p_i} + \frac{1}{i\hbar} \int_{t_i}^{t} dt \langle \acute{p}|V|p_i\rangle e^{\frac{-i}{\hbar}(E_{p_i} - E_{\acute{p}})\frac{t}{\hbar}}. \quad (8.187)$$

Taking $t \to \infty$, $\acute{p} \to p_f$, we have [since V vanishes for $t < t_i$]

$$a_{p_f p_i}(t) = \delta_{p_f p_i} + \frac{1}{i\hbar} \langle p_f|V|p_i\rangle \int_{\infty}^{\infty} e^{\frac{-i}{\hbar}(E_{p_i} - E_{p_f})t}$$

$$= \delta_{p_f p_i} + \frac{1}{\hbar} \langle p_f|V|p_i\rangle 2\pi\delta(\frac{E_{p_i} - E_{p_f}}{\hbar})$$

$$= \delta_{p_f p_i} - 2\pi i\delta(E_{p_i} - E_{p_f})\langle p_f|V|p_i\rangle. \quad (8.188)$$

Now the probability that the interaction V causes a transition from the state $|p_i\rangle$ to a state $|p'\rangle$ is given by

$$P_{\acute{p}p_i} = \left| \int d^3r u^*_{p'} \psi_i(r, t) \right|^2$$

which on using Eq. (8.183) and the orthogonality relation (8.184) gives

$$P_{\acute{p}p_i} = |a'_{pp_i}|^2. \quad (8.189)$$

Thus for $\acute{p} = p_f \neq p_i$, we get from Eq. (8.188) transition probability amplitude:

$$a_{p_f p_i} = -2\pi i\delta(E_{p_i} - E_{p_f})\langle p_f|V|p_i\rangle$$

$$= 2\pi i\delta(E_{p_i} - E_{p_f})T_{fi} \quad (8.190)$$

so that the scattering matrix in Born approximation is

$$T_B = -\langle p_f|V|p_i\rangle$$

$$= -\int u^*_{p_f}(r)V(r)u_{p_i}(r)d^3r \quad (8.191)$$

where momentum eigenfunctions are given by

$$u_{\mathbf{p}_i}(\mathbf{r}) = \frac{1}{(2\pi\hbar)^{\frac{3}{2}}} e^{\frac{i\mathbf{p}_i \cdot \mathbf{r}}{\hbar}},$$

$$u_{\mathbf{p}_f}(\mathbf{r}) = \frac{1}{(2\pi\hbar)^{\frac{3}{2}}} e^{\frac{i\mathbf{p}_f \cdot \mathbf{r}}{\hbar}}. \tag{8.192}$$

Hence from Eqs.(8.191) and (8.192):

$$T_B = -\frac{1}{(2\pi\hbar)^3} \int e^{-\frac{i\mathbf{p}_f \cdot \mathbf{r}}{\hbar}} V(\mathbf{r}) e^{\frac{i\mathbf{p}_i \cdot \mathbf{r}}{\hbar}} d^3r$$

$$= -\frac{1}{(2\pi\hbar)^3} \int e^{\frac{i\mathbf{q} \cdot \mathbf{r}}{\hbar}} V(r) d^3r \tag{8.193}$$

where

$$\mathbf{q} = (\mathbf{p}_i - \mathbf{p}_f) = \hbar(\mathbf{k}_i - \mathbf{k}_f) = \hbar\mathbf{k} \tag{8.194}$$

is the momentum transfer.

Now T_B has dimension $\frac{L^3}{E^2 t^3}$, where as scattering amplitude $f(\theta, \phi)$ has dimension of length viz. L. Hence $f(\theta, \phi)$ in the Born approximation:

$$f_B(\theta, \phi) = -\frac{\mu}{\hbar^2} \frac{1}{2\pi} \int e^{\frac{i\mathbf{q} \cdot \mathbf{r}}{\hbar}} V(\mathbf{r}) d^3r \tag{8.195}$$

$$= -\frac{\mu}{2\pi\hbar^2} \int e^{i\mathbf{k} \cdot \mathbf{r}} V(\mathbf{r}) d^3r.$$

Now

$$q^2 = p_i^2 + p_f^2 - 2\mathbf{p}_i \cdot \mathbf{p}_f = 2p^2(1 - \cos\theta)$$

$$= 4p^2 \sin^2 \frac{\theta}{2} \tag{8.196}$$

where θ is the angle between $\mathbf{p}_i$ and $\mathbf{p}_f$, i.e. the scattering angle. For elastic scattering

$$\mathbf{p}_i^2 = \mathbf{p}_f^2 = \mathbf{p}^2 = p^2. \tag{8.197}$$

For a spherically symmetric potential:

$$V(\mathbf{r}) = V(r), \quad f_B(\theta, \phi) = f_B(\theta). \tag{8.198}$$

Now $\mathbf{r}$ in polar co-ordinates: $\acute{\theta}$, $\acute{\phi}$

$$\mathbf{r} = r(\sin\acute{\phi}\sin\acute{\theta}, \cos\acute{\phi}\sin\acute{\theta}, \cos\acute{\theta}).$$

Take $\mathbf{q}$ along $\acute{\theta} = 0$ axis, $\mathbf{r} \cdot \mathbf{q} = rq\cos\acute{\theta}$

$$d^3r = r^2 dr \sin\acute{\theta} d\acute{\theta} d\acute{\phi} = 2\pi r^2 dr \sin\acute{\theta} d\acute{\theta}.$$

Thus, we have

$$f_B(\theta) = -\frac{\mu}{\hbar^2} \int V(r)r^2 \left[\int_0^\pi e^{ikr\cos\theta} \sin\theta d\theta \right] dr$$

$$= -\frac{\mu}{\hbar^2} \int \frac{[e^{ikr} - e^{-ikr}]}{ikr} r^2 V(r) dr$$

$$= -\frac{2\mu}{\hbar^2 k} \int_0^\infty \sin(kr) r V(r) dr. \tag{8.199}$$

For the limiting case of small velocities, $\frac{\sin kr}{kr} \sim 1$

$$f_B(\theta) = -\frac{2\mu}{\hbar^2} \int_0^\infty V(r)r^2 dr. \tag{8.200}$$

On the other hand for large velocities or high energies, from Eqs. (8.194) and (8.196), $k = \frac{2p}{\hbar \sin\frac{\theta}{2}}$, and in Eq. (8.199) for $f_B(\theta)$ it appears in the denominator. Thus the scattering is mainly through the small angles: $\sin\frac{\theta}{2} \simeq \frac{\theta}{2}$ and

$$f_B = -\frac{2\mu}{\hbar^2} \int_0^\infty V(r) \frac{\sin(\frac{2p}{\hbar}r\theta)}{(\frac{2p}{\hbar})\theta r} r^2 dr$$

$$= \Phi(p\,\theta), \text{ say.} \tag{8.201}$$

Thus the scattering cross-section

$$\sigma = 2\pi \int_0^\pi \sin\theta d\theta |f(\theta)|^2$$

$$\simeq 2\pi \int_0^\pi \theta d\theta |\Phi(p\,\theta)|^2. \tag{8.202}$$

Note that for small θ, $\sin\theta \sim \theta$ and for large θ, $|\sin\theta| < 1$ and $\Phi(p\,\theta) \sim \frac{1}{\theta}$. Thus the integral over θ converges so rapidly that the integration can be extended to ∞ without any great error. Thus putting $\frac{p}{\hbar}\theta = x$,

$$\sigma = \frac{2\pi\hbar^2}{p^2} \int_0^\infty x dx |\Phi(x)|^2$$

i.e.

$$\sigma \propto \frac{\hbar^2}{p^2} = \frac{\hbar^2}{2\mu E}$$

or

$$\lim_{E\to\infty} (\sigma E) = \text{constant}, \tag{8.203}$$

which gives the high energy limit of the cross-section.

As a first application of the Born approximation, we apply it to scattering by the Yukawa potential:

$$V(r) = g^2 e^{(-r/r_0)} \frac{1}{r}. \tag{8.204}$$

For this potential, from Eq. (8.199):

$$f_B(\theta) = -\frac{\mu}{\hbar^2} \frac{1}{ik} g^2 \int_0^\infty \left[e^{(ik - \frac{1}{r_0})r} - e^{(-ik - \frac{1}{r_0})r} \right] dr$$

$$= -\frac{\mu}{\hbar^2} \frac{(-1)}{ik} g^2 \left[\frac{1}{ik - \frac{1}{r_0}} + \frac{1}{ik + \frac{1}{r_0}} \right]$$

$$= -\frac{2\mu g^2}{\hbar^2} \frac{1}{k^2 + (\frac{1}{r_0})^2}$$

$$= -\frac{2\mu g^2}{\hbar^2} \frac{\hbar^2}{q^2 + (\frac{\hbar}{r_0})^2} = -2g^2 \frac{(\mu c^2)}{(c\hbar)^2} \frac{\hbar^2}{q^2 + (\frac{\hbar}{r_0})^2}. \tag{8.205}$$

For Rutherford scattering i.e. scattering of a charged particle with charge $Z_1 e$ on a target of charge $Z_2 e$,

$$V(r) = \frac{Z_1 Z_2 e^2}{r}. \tag{8.206}$$

For the Coulomb potential given in Eq. (8.206) take $r_0 \to \infty$, i.e. $\frac{1}{r_0} \to 0$ and $g^2 \to Z_1 Z_2 e^2$ in Eq. (8.205) and hence from Eq. (8.206) we obtain for the Rutherford scattering

$$f_B(\theta) = -\frac{2\mu Z_1 Z_2 e^2}{\hbar^2} \frac{\hbar^2}{q^2} = -2\mu \frac{Z_1 Z_2 e^2}{q^2}. \tag{8.207}$$

Note the remarkable fact, that for the Rutherford scattering, the dependance on Planck's constant is canceled. Hence Rutherford scattering cross section is given by

$$\frac{d\sigma}{d\Omega} = \frac{4(\mu c^2)^2 Z_1^2 Z_2^2 e^4}{(cq)^4} = \mu^2 \frac{Z_1^2 Z_2^2 e^4}{4p^4 \sin^4 \frac{\theta}{2}}. \tag{8.208}$$

This formula was first derived by Lord Rutherford in classical mechanics. This is probably the only formula which is not modified in quantum mechanics. It is important to remark here that the formula (8.208) also agrees with the result obtained from the exact treatment of the Coulomb scattering in quantum mechanics. However in the exact treatment, the scattering amplitude is different from (8.207):

$$f(\theta) = \frac{n'}{p(1 - \cos\theta)} e^{2i\delta_0'} \exp[in' \ln(1 - \cos\theta)] \tag{8.209a}$$

$$n' = -\frac{Z_1 Z_2 e^2 \mu}{\hbar^2 k}, \qquad e^{2i\delta_0} = \frac{\Gamma(1 - in')}{\Gamma(1 + in')}. \tag{8.209b}$$

$\frac{d\sigma}{d\Omega}$ gives the scattering cross section of a spinless particle of charge $Z_1 e$ with a point particle of charge $Z_2 e$.

For α-particle scattering of heavy nucleus of mass M and charge Ze:

$$\frac{1}{\mu} = \frac{1}{m_\alpha} + \frac{1}{M} \approx \frac{1}{m_\alpha},$$

$$\frac{d\sigma}{d\Omega} = m_\alpha \frac{Z^2 e^2}{p^4 \sin^4 \frac{\theta}{2}}. \tag{8.210}$$

For electron scattering (ignoring its spin) with a point nucleus of charge Ze:

$$\left(\frac{d\sigma}{d\Omega}\right)_R = 4\frac{(m_e c^2)^2 Z^2 e^4}{(cq)^4} = \frac{(m_e c^2)^2 Z^2 e^4}{4(cp)^4 \sin^4 \frac{\theta}{2}}. \tag{8.211}$$

Taking into account the spin of electron, the cross section was calculated by Mott. The Mott scattering cross section for scattering of electron of spin $1/2$, on a spinless point target of charge Ze is given by:

$$\left(\frac{d\sigma}{d\Omega}\right)_{\text{Mott}} = 4(Ze^2)^2 \frac{E^2}{(cq)^4} \left(1 - \frac{v^2}{c^2} \sin^2 \frac{\theta}{2}\right). \tag{8.212}$$

For $\frac{v}{c} \ll 1$, $E \to m_e c^2$, it reduces to $\left(\frac{d\sigma}{d\Omega}\right)_R$. For high energy electron scattering i.e. $\frac{v}{c} \to 1$: $E \to cp$

$$\left(\frac{d\sigma}{d\Omega}\right)_{\text{Mott}} = 4(Ze^2)^2 \frac{E^2 \cos^2 \frac{\theta}{2}}{4E^4 \sin^4 \frac{\theta}{2}} = (Ze^2)^2 \frac{\cos^2 \frac{\theta}{2}}{4E^4 \sin^4 \frac{\theta}{2}}. \tag{8.213}$$

We now consider the scattering of an electron from a charge distribution of a spinless nucleus or proton, having structure. First ignoring the spin of electron, from Eq. (8.193)

$$T = T_B = -\frac{1}{(2\pi\hbar)^3} \int e^{\frac{i\mathbf{q}\cdot\mathbf{r}}{\hbar}} V(\mathbf{r}) d^3 r, \tag{8.214}$$

$$\nabla^2 V(r) = 4\pi Z e^2 \rho(\mathbf{r}). \tag{8.215}$$

From Green's theorem:

$$\int \left(\phi \nabla^2 \psi - \psi \nabla^2 \phi\right) d^3 r = \int [\phi \nabla \psi - \psi \nabla \phi] \cdot d\mathbf{S}. \tag{8.216}$$

If ψ and $\nabla \psi$ vanish on large surface:

$$\int [\phi \nabla \psi - \psi \nabla \phi] \cdot d\mathbf{S} = 0. \tag{8.217}$$

For

$$\phi = e^{\frac{i\mathbf{q}\cdot\mathbf{r}}{\hbar}}, \quad \text{and} \quad \psi = V(\mathbf{r}),$$

we have from Eqs. (8.216) and (8.217):

$$
\begin{aligned}
\int e^{\frac{i\mathbf{q}\cdot\mathbf{r}}{\hbar}}\nabla^2 V(\mathbf{r})d^3r &= \int V(\mathbf{r})\nabla^2 e^{\frac{i\mathbf{q}\cdot\mathbf{r}}{\hbar}}d^3r \\
&= (\frac{i\mathbf{q}}{\hbar})^2 \int V(\mathbf{r})e^{\frac{i\mathbf{q}\cdot\mathbf{r}}{\hbar}}d^3r \\
&= -\frac{q^2}{\hbar^2}\int e^{\frac{i\mathbf{q}\cdot\mathbf{r}}{\hbar}}V(\mathbf{r})d^3r. \quad (8.218)
\end{aligned}
$$

Hence, on using Eq. (8.215):

$$
\begin{aligned}
T &= -\frac{1}{(2\pi\hbar)^3}\int e^{\frac{i\mathbf{q}\cdot\mathbf{r}}{\hbar}}V(\mathbf{r})d^3r \\
&= -\frac{1}{(2\pi\hbar)^3}(-\frac{\hbar^2}{q^2})\int e^{\frac{i\mathbf{q}\cdot\mathbf{r}}{\hbar}}\nabla^2 V(\mathbf{r})d^3r \\
&= \frac{Ze^2}{2\pi^2\hbar}\frac{1}{q^2}\int e^{\frac{i\mathbf{q}\cdot\mathbf{r}}{\hbar}}\rho(\mathbf{r})d^3r, \quad (8.219)
\end{aligned}
$$

$$
\begin{aligned}
f(\theta) &= 4\pi^2\frac{m_e}{\hbar^2}\hbar^3 T \\
&= \frac{2m_e}{q^2}Ze^2\int e^{\frac{i\mathbf{q}\cdot\mathbf{r}}{\hbar}}\rho(\mathbf{r})d^3r \quad (8.220)
\end{aligned}
$$

and

$$
\begin{aligned}
\frac{d\sigma}{d\Omega} &= \frac{4m_e^2(Ze^2)^2}{q^4}|F(q^2)|^2 \\
&= (\frac{d\sigma}{d\Omega})_R|F(q^2)|^2 \quad (8.221)
\end{aligned}
$$

where

$$F(q^2) = \int e^{\frac{i\mathbf{q}\cdot\mathbf{r}}{\hbar}}\rho(\mathbf{r})d^3r \quad (8.222)$$

$$\rho(\mathbf{r}) = (\frac{1}{2\pi})^3 \int e^{-\frac{i\mathbf{q}\cdot\mathbf{r}}{\hbar}}F(q^2)d^3q. \quad (8.223)$$

The form factor $F(q^2)$ denotes the structure of the target. $\rho(\mathbf{r})$ gives the charge of the target. For low momentum transfer (small q^2), we can expand

$$F(q^2) = F(0) + \frac{\partial F(q^2)}{\partial q^2}\Big|_{q^2=0}\frac{q^2}{\hbar^2} + O(q^4). \quad (8.224)$$

For $qR \ll 1$ (R nuclear radius):

$$e^{\frac{i\mathbf{q}\cdot\mathbf{r}}{\hbar}} = \left(1 + i\frac{\mathbf{q}\cdot\mathbf{r}}{\hbar} + \frac{(i\mathbf{q}\cdot\mathbf{r})^2}{2!\hbar^2} + \cdots \right),$$

$$F(q^2) \approx \int \left(1 + i\frac{\mathbf{q}\cdot\mathbf{r}}{\hbar} + \frac{(i\mathbf{q}\cdot\mathbf{r})^2}{2!\hbar^2}\right) \rho(\mathbf{r}) r^2 dr d\Omega. \qquad (8.225)$$

Now

$$\int_0^\pi \mathbf{q}\cdot\mathbf{r} \, d\Omega = 0.$$

Thus, we have

$$F(q^2) = 4\pi \int \rho(r) r^2 dr - \frac{1}{2}\int \frac{(\mathbf{q}\cdot\mathbf{r})^2}{\hbar^2}\rho(r)r^2 d\Omega dr$$

$$= 4\pi \int \rho(r) r^2 dr - \frac{1}{2}\frac{q^2}{\hbar^2}\left(\frac{4\pi}{3}\right)\int r^4 \rho(r) dr. \qquad (8.226)$$

Hence we have

$$F(0) = 4\pi \int \rho(r) r^2 dr = \int \rho(r) d^3 r,$$

$$\left.\frac{\partial F(q^2)}{\partial q^2}\right|_{q^2=0} = -\frac{2\pi}{3}\int \rho(r) r^4 dr$$

$$= -\frac{1}{6}\int \rho(r) r^2 d^3 r = -\frac{1}{6}\langle r^2 \rangle. \qquad (8.227)$$

$\langle r^2 \rangle$ is called the mean square radius. Electron spin can be taken into account, by replacing $\left(\frac{d\sigma}{d\Omega}\right)_R$ in Eq. (8.221) by $\left(\frac{d\sigma}{d\Omega}\right)_{\text{Mott}}$.

8.12 Quasi-Classical Approximation

For most of the potentials, no exact solution is available. As such one has to resort to approximation techniques. The Wentzel–Kramers–Brillioum or WKB approximation is specially useful when one is dealing with slowly varying potentials; what we mean by such potentials will become clear as we develop the WKB approximation.

The Schrödinger equation for stationary states for a particle of mass m moving in a potential $V(r)$ is

$$\left(-\frac{\hbar^2}{2m}\nabla^2 + V(\mathbf{r})\right)u(\mathbf{r},t) = Eu(\mathbf{r},t). \qquad (8.228)$$

Let us write the solution of Eq. (20.183) as

$$u(\mathbf{r}) = \omega(\mathbf{r})e^{iS(\mathbf{r})/\hbar}, \tag{8.229}$$

where $\omega(\mathbf{r})$ and $S(\mathbf{r})$ are real functions. Substituting Eq. (20.184) in Eq. (20.183), we have

$$-\frac{\hbar^2}{2m}\left(-\frac{1}{\hbar^2}(\boldsymbol{\nabla}S)^2\omega + \frac{i}{\hbar}(2\boldsymbol{\nabla}\omega\cdot\boldsymbol{\nabla}S + \omega\boldsymbol{\nabla}^2 S) + \boldsymbol{\nabla}^2\omega\right) + V(\mathbf{r})\omega = E\omega. \tag{8.230}$$

Separating out the real and imaginary parts, we get

$$2\boldsymbol{\nabla}\omega\cdot\boldsymbol{\nabla}S + \omega\boldsymbol{\nabla}^2 S = 0, \tag{8.231}$$

$$\boldsymbol{\nabla}^2\omega - \frac{\omega}{\hbar^2}(\boldsymbol{\nabla}S)^2 + \omega\big(k(\mathbf{r})\big)^2 = 0, \tag{8.232}$$

where

$$k(\mathbf{r}) = \sqrt{(2m/\hbar^2)(E - V(\mathbf{r}))}. \tag{8.233}$$

Let us suppose that both $\omega(\mathbf{r})$ and $S(\mathbf{r})$ remain finite as $\hbar \longrightarrow 0$. We obtain the classical result if we put $\hbar = 0$, viz.

$$\frac{1}{2m}(\boldsymbol{\nabla}S)^2 = (E - V(\mathbf{r})),$$

so that if $\mathbf{p} = \boldsymbol{\nabla}S$, we get

$$E = \frac{p2}{2m} + V(\mathbf{r}).$$

Thus, the semi-classical approximation is valid if

$$\hbar^2\boldsymbol{\nabla}^2\omega \ll \omega(\boldsymbol{\nabla}S)^2. \tag{8.234}$$

For one dimensional motion in the x-direction Eqs. (20.186), (20.187) and (20.190) give,

$$\frac{d^2 S/dx^2}{dS/dx} = -2\frac{d\omega}{dx}, \tag{8.235}$$

$$\frac{d^2\omega}{dx^2} - \frac{\omega}{\hbar^2}\left(\frac{dS}{dx}\right)^2 + \omega[k(x)]^2 = 0, \tag{8.236}$$

and

$$\hbar^2\frac{d^2\omega/dx^2}{\omega} \ll \left(\frac{dS}{dx}\right)^2. \tag{8.237}$$

The solution of Eq. (20.190) is

$$\ln\frac{dS}{dx} = -2\ln\omega + \ln A$$

i.e.

$$\frac{dS}{dx} = \frac{A}{\omega^2} \tag{8.238}$$

and using the condition (20.192), Eq. (20.191) then gives

$$\frac{A}{\omega^2} = \frac{dS}{dx} = \pm \hbar k(x), \tag{8.239}$$

i.e.

$$S(x) = \pm \hbar \int_{x_0}^{x} k(x')dx' \tag{8.240}$$

and

$$\omega(x) = \left(\frac{A}{\hbar k(x)}\right)^{1/2}, \tag{8.241}$$

so that a general solution of Eq. (8.228) for the one dimensional problem is

$$u(x) = \left(\frac{A}{\hbar k(x)}\right)^{1/2} \left\{ C_1 \exp\left(i \int_{x_0}^{x} k(x')dx'\right) \right.$$

$$+ C_2 \exp\left(-i \int_{x_0}^{x} k(x')dx'\right) \right\}$$

$$= \frac{C}{\sqrt{k(x)}} \sin\left\{ \int_{x_0}^{x} k(x') + \alpha \right\}, \quad E > V(x). \tag{8.242}$$

The region where $E < V(x)$ is called the classically unattainable region. In that region $k(x)$ is an imaginary function and we have

$$k(x) = i|k(x)|$$

and requiring that the wave function be finite everywhere, we get the solution in this region

$$u(x) = \frac{C'}{\sqrt{|k(x)|}} e^{-\int_{x_0}^{x} |k(x')|dx'}, \quad E < V(x). \tag{8.243}$$

The values of x_i for which $E = V(x_i)$ are called classical turning points. They correspond to those points in space where the classical particle comes to a halt $[k(x_i) = 0]$ and turns back. The wave function given in Eq. (20.196) or (20.197) become infinite at these points and the WKB approximation breaks down. The problem then is to connect the solutions (20.196) and (20.197) on two sides of a turning point. Such connection formulae can be derived as follows.

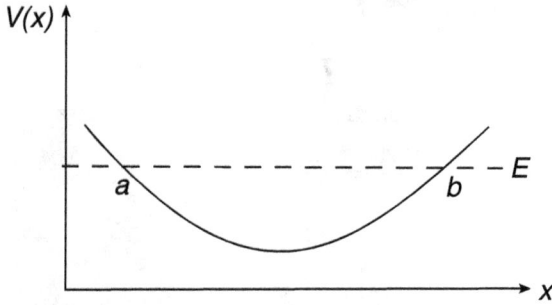

Fig. 8.7 The classical turning points a and b. The particle is confined in the region $a \leq x \leq b$.

Suppose the potential $V(x)$ has the form shown in Fig 8.7. There are two turning points a and b. Away from the turning points, the solution is given by the WKB approximation. Since the particle is confined between a and b, the wave function is real and in the WKB approximation is given in this region by [c.f. Eq. (20.196)]

$$u(x) = \frac{C}{\sqrt{k}} \sin(\eta(x) + \alpha), \qquad (8.244a)$$

where

$$\eta(x) = \int_a^x k(x)dx, \qquad (8.244b)$$

and α is an arbitrary phase factor.

In the region $x > b$, the wave function is given by Eq. (20.197)

$$u(x) = \frac{C'}{\sqrt{|k|}} \exp\left(-\int_b^x |k(x)|dx\right), \qquad (8.245)$$

whereas in the region $x < a$, it is given by

$$u(x) = \frac{C''}{\sqrt{|k|}} \exp\left(-\int_x^a |k(x)|dx\right). \qquad (8.246)$$

In order to find the solution near the turning points, we solve the Schrödinger equation exactly by using the following approximation for the potential $V(x)$. Near the turning point $x = a$, we expand $V(x)$ in powers of $(x - a)$ and retain only the first power,

$$V(x) = V(a) + \frac{\partial V}{\partial x}\Big|_{x=a}(x - a)$$
$$= V(a) - F(x - a), \qquad (8.247)$$

where $F = -\frac{\partial V}{\partial x}\big|_{x=a}$ is a positive number. Thus near $x = a$, we can write [absorbing $\frac{1}{\hbar^2}$ in F]

$$k^2(x) = 2mF(x - a)$$
$$= 2mFy, \tag{8.248a}$$

$$\eta(x) = \sqrt{2mF}\frac{2}{3}y^{3/2}, \tag{8.248b}$$

where $y = x - a$.

Using the above approximation, the Schrödinger equation near $x = a$ can be written as

$$\frac{d^2u}{dy^2} + 2mFyu = 0. \tag{8.249}$$

By writing

$$u(y) = y^{1/2}\phi(\sqrt{2mF}\frac{2}{3}y^{3/2}) = y^{1/2}\phi(\eta), \tag{8.250}$$

one can easily see that $\phi(\eta)$ satisfies Bessel's equation

$$\frac{d^2\phi}{d\eta^2} + \frac{1}{\eta}\frac{d\phi}{d\eta} + (1 - \frac{1}{9\eta^2})\phi = 0. \tag{8.251}$$

The general solution of Eq. (20.205) is given by

$$\phi(\eta) = A_1 J_{1/3}(\eta) + B_1 J_{-1/3}(\eta). \tag{8.252}$$

This is the solution of the Schrödinger equation near the turning point $x = a$ to the right of a. To obtain the solution near $x = a$ to the left of a, we simply change $\eta \to i|\eta|$ so that the solution is now given by

$$\phi(|\eta|) = A_2 i^{-1/3} J_{1/3}(i|\eta|) + B_2 i^{1/3} J_{-1/3}(i|\eta|). \tag{8.253}$$

For large η, the behaviour of the Bessel function is given by

$$J_{\pm 1/3}(\eta) \to (1/2\pi\eta)^{-1/2}\cos(\eta \mp \frac{\pi}{6} - \frac{\pi}{4}), \tag{8.254}$$

$$i^{\mp 1/3}J_{\pm 1/3}(i|\eta|)$$
$$\to i^{-1/3}(\frac{1}{2}i\pi|\eta|)^{-1/2}(e^{-|\eta|}e^{(\mp\pi/6-\pi/4)i} + e^{|\eta|}e^{-(\mp\pi/6-\pi/4)i})\frac{1}{2}$$
$$= (2\pi|\eta|)^{-\frac{1}{2}}(e^{|\eta|} + e^{-|\eta|}e^{-(1/2\pm 1/3)\pi i}). \tag{8.255}$$

Using Eqs. (20.208) and (20.209), we see that for large η, the solution (20.206) takes the form of Eq. (8.244a, 8.244b), i.e.

$$u = y^{1/2}\phi(\eta) \rightarrow 3\frac{A_1}{\sqrt{\pi}}(2mFy)^{-1/4}\cos(\eta - \frac{\pi}{4})$$

$$= 3\frac{A_1}{\sqrt{\pi k}}\sin(\eta + \pi/4) \tag{8.256}$$

provided $A_1 = B_1$, whereas the solution (20.207) takes the form of Eq. (8.246), i.e.

$$u = |y|^{1/2}\phi(|\eta|) \longrightarrow -\frac{3}{2}\frac{A_2}{\sqrt{\pi}}(2mF|y|)^{-1/4}e^{-|\eta|}$$

$$= -\frac{3}{2}\frac{A_2}{\sqrt{\pi|k|}}e^{-|\eta|} \tag{8.257}$$

provided $A_2 = -B_2$. Now for large η, the $u = y^{1/2}\phi(\eta)$ and $u = |y|^{1/2}\phi(|\eta|)$ should go over to the WKB solution given in Eqs. (8.244a, 8.244b) and (8.246) respectively. We see that this is so if

$$3\frac{A_1}{\sqrt{\pi}} = C, \alpha = \pi/4, C'' = -\frac{3}{2}\frac{A_2}{\sqrt{\pi}}.$$

Now the wave function should be continuous at $y = 0$. This condition gives $-\frac{1}{\sqrt{2}}A_2 = A_1$. Hence we have the final result that the WKB solution to the right of the turning point a is given by

$$u(x) = \frac{C}{\sqrt{k}}\sin\left(\int_a^x k(x)dx + \frac{\pi}{4}\right), \tag{8.258}$$

whereas to the left of a it is given by [c.f. Eq. (8.246)]

$$u(x) = \frac{C}{\sqrt{2|k|}}\exp\left(-\int_x^a |k(x)|dx\right). \tag{8.259}$$

In the region $a \leq x \leq b$, the particle is confined in a potential and the energy spectrum is discrete. The WKB solution to the left of the turning point b is given by

$$u(x) = \frac{C'}{\sqrt{k}}\sin\left(\int_x^b k(x)dx + \frac{\pi}{4}\right). \tag{8.260}$$

Now the wave functions given in Eqs. (8.258) and (8.260) must be the same in the whole region a to b. This requires that

$$\int_a^x k(x)dx + \frac{\pi}{4} + \int_x^b k(x)dx + \frac{\pi}{4} = (n+1)\pi$$

with $C = (-1)^n C'$. Hence we have the quantum condition

$$\int_a^b k(x)dx = (n+1)\pi - \frac{\pi}{2}$$

or

$$\int_a^b p(x)dx = (n + \frac{1}{2})\pi\hbar$$

or

$$\oint p(x)dx = 2\pi\hbar(n + \frac{1}{2}), \tag{8.261}$$

where $\oint p(x)dx = 2\int_a^b p(x)dx$ is the integral taken over the whole period of quasi-classical motion of the particle.

We now consider the radial part of the Schrödinger equation

$$\frac{d^2\chi}{dr^2} + \frac{2m}{\hbar^2}\left(E - V(r) - \frac{\hbar}{2m}\frac{l(l+1)}{r^2}\right)\chi(r) = 0. \tag{8.262}$$

This equation is identical with the Schrödinger's equation for one dimensional motion in a field of effective potential

$$V_l(r) = V(r) + \frac{\hbar^2}{2m}\frac{l(l+1)}{r^2}. \tag{8.263}$$

Hence the solution of this equation in the WKB approximation can easily be written in the same way as discussed earlier.

We now apply the WKB approximation to the problem of tunneling through a barrier. Let us consider a particular example for which we take

$$V(r) = -V_0 \quad r < a$$
$$= 0 \quad r > a, \tag{8.264}$$

so that

$$V_l(r) = -V_0 + \frac{\hbar^2}{2m}\frac{l(l+1)}{r^2}, \quad r < a,$$
$$= \frac{\hbar^2}{2m}\frac{l(l+1)}{r^2}, \quad r > a. \tag{8.265}$$

Inside the potential well, we ignore the angular momentum. Now if we plot $V_l(r)$ versus r, we have a situation resembling that shown in Fig. 8.8. In the region $a < x < b$ (b is the second turning point at which $E - V_l(r) = 0$), the WKB wave function is given by

$$\exp\left(-\int |k(r)|dr\right),$$

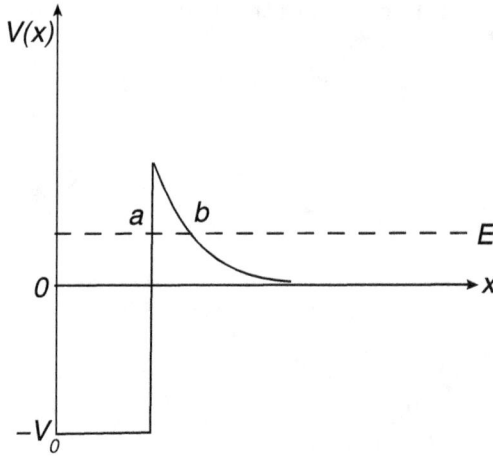

Fig. 8.8 The plot resembling the effective potential $V_l(r)$.

where

$$|k(r)| = \left(\frac{l(l+1)}{r^2} - \frac{2m}{\hbar^2}E\right)^{1/2}. \qquad (8.266)$$

The transmission coefficient T is given by

$$T = \frac{|\text{Flux}|_{\text{outside barrier}}}{|\text{Flux}|_{\text{inside the well}}} \times \exp\left(-2\int_a^b |k(r)|dr\right)$$

$$= \left(\frac{E}{E+V_0}\right)^{1/2} \exp\left(-2\int_a^b \left(\frac{l(l+1)}{r^2} - \frac{2m}{\hbar^2}E\right)^{1/2}dr\right).$$

$$(8.267)$$

Now each time the particle hits the wall of the well, the probability of transmission is given by Eq. (8.267). The number of times per second the particle hits the wall of the potential well is given by

$$\left(\frac{p}{m}\right)\frac{1}{2a} = \left(\frac{E+V_0}{2m}\right)^{1/2}\frac{1}{a}. \qquad (8.268)$$

Hence the total probability of transmission is given by

$$\frac{1}{\tau} = \left(\frac{E+V_0}{2ma^2}\right)^{1/2} T$$

$$= \left(\frac{E}{2ma^2}\right)^{1/2} \exp\left(-2\int_a^b \left(\frac{l(l+1)}{r^2} - \frac{2m}{\hbar^2}E\right)^{1/2}dr\right),$$

$$(8.269)$$

where τ is the lifetime of the bound system. We note that

$$b = \hbar\sqrt{\frac{l(l+1)}{2mE}}. \tag{8.270}$$

The integral occurring in Eq. (20.221) can be expressed as

$$\int_a^b \left(\frac{l(l+1)}{r^2} - \frac{2m}{\hbar^2}E\right) dr = \sqrt{l(l+1)}\int_\gamma^1 (\frac{1}{x^2} - 1)^{1/2}dx, \tag{8.271}$$

where

$$x = \sqrt{\frac{2mE}{l(l+1)\hbar^2}}\, r, \tag{8.272}$$

$$\gamma = \sqrt{\frac{2mE}{l(l+1)}}\,\frac{a}{\hbar}. \tag{8.273}$$

For a special case when $\gamma \ll 1$, the integral

$$\int_\gamma^1 \frac{1}{x}(1-x^2)^{1/2}dx = -\ln\gamma. \tag{8.274}$$

Hence

$$\tau = \left(\frac{2ma^2}{E}\right)^{1/2}\gamma^{-2\sqrt{l(l+1)}}$$
$$= \left(\frac{2ma^2}{E}\right)^{1/2}\left(\frac{l(l+1)\hbar^2}{2mEa^2}\right)^{\sqrt{l(l+1)}}. \tag{8.275}$$

We now apply the WKB approximation to s-wave bound states. For the s-wave, the radial Schrödinger equation is

$$\frac{d^2\chi}{dr^2} + \left[\frac{2\mu}{\hbar^2}(E - V(r))\right]\chi = 0.$$

In the WKB approximation, the wave function $\chi(r)$ for s-wave bound states is given by

$$\chi(r) = \left[N/\sqrt{k(r)}\right]\sin\left[\int_r^{r_1} k(r')dr' + \pi/4\right], \tag{8.276}$$

where r_1 is the first turning point ($k(r_1) = 0$). The above solution is to the left of the turning point. Now

$$k(r) = \left[\frac{2\mu}{\hbar^2}(E - V(r))\right]^{1/2}.$$

The boundary condition $\chi(0) = 0$ gives the quantisation condition.

$$\int_0^{r_1} k(r)dr + \pi/4 = n\pi \tag{8.277a}$$

$$\text{or} \quad \int_0^{r_1} [2\mu(E - V(r)]^{1/2}\,dr = (n - \frac{1}{4})\pi\hbar, \tag{8.277b}$$

where $n = 1, 2, 3, \ldots$ is the principle quantum number.

We now apply the above results for a linear potential viz. $V(r) = \beta r$. This potential is relevant for the resonances J/ψ (3100 MeV) and ψ' (3684 MeV). The conventional interpretation of the resonances is that they are regarded as s-wave bound states of charmed quark and antiquark ($c\bar{c}$ called charmonium). Furthermore, since quarks have spin $\frac{1}{2}$, c and $\bar{c}$ spins may be combined to form a total spin S, which is either 0 or 1 (Sec. 11.6). Since resonances ψ and ψ' have quantum numbers of a photon viz. $J^P = 1^-$, they can be identified with 1^3S_1 and 2^3S_1 states of the charmonium. The mass of a charmed quark is high (1500 MeV – 2000 MeV), therefore it is a good approximation to treat the bound $c\bar{c}$ system non-relativistically. It is assumed that the potential $V(r) = \beta r$ keeps them bound together.

For the linear potential, the turning point r_1 is given by

$$r_1 = (1/\beta)E. \tag{8.278a}$$

Let us put

$$r = (1/\beta)Ex. \tag{8.278b}$$

Then we have

$$\int_0^{r_1} k(r)dr = \frac{\sqrt{2\mu}}{\hbar} \frac{(E)^{3/2}}{\beta} \int_0^1 (1-x)^{1/2}dx$$

$$= \frac{\sqrt{2\mu}}{\hbar} \frac{2}{3\beta}(E)^{3/2}.$$

Hence from Eq. (8.277a, 8.277b), we have the energy levels for s-states

$$E_n = (2\mu)^{-\frac{1}{3}} \left[\frac{3\pi}{2}(\hbar\beta)(n - \frac{1}{4})\right]^{2/3}. \tag{8.279}$$

Hence the mass spectrum for s-states of the charmonium is given by ($\mu = \frac{1}{2}m_c$).

$$m(n^3S_1)c^2 = 2m_cc^2 + E_n$$

$$= 2m_cc^2 + (2\mu)^{-1/3}\left[\frac{3}{2}(\pi\hbar\beta)(n - \frac{1}{4})\right]^{2/3}. \tag{8.280}$$

For $m_c = 1850$ MeV and $\beta = 1/(b^2 \hbar c)$ with $b = (1950$ MeV$)^{-1}$, one gets from Eq. (20.230)

$$m(2^3 S_1) - m(1^3 S_1) \simeq 590 \text{ MeV},$$

to be compared with the experimental value of 580 MeV.

We now determine the mean square radius of charmonium using the WKB approximation. For this purpose we first determine the normalisation constant N. Now

$$|N|^2 \int_0^{r_1} \frac{1}{k(r)} \left\{ \sin \left[\int_r^{r_1} k(r')dr' + \frac{\pi}{4} \right] \right\}^2 dr = 1. \qquad (8.281)$$

In order to evaluate the integral, we assume that the average value of the $\sin^2$ term is $\frac{1}{2}$.[1] We have

$$N^2 \frac{1}{2} \int_0^{r_1} \frac{1}{k(r)} dr = \frac{N^2}{2} \int_0^{r_1} \left[\frac{2\mu}{\hbar^2} (E - \beta r) \right]^{-1/2} dr$$
$$= 1. \qquad (8.282)$$

This gives

$$N^2 = \sqrt{2\mu} \frac{\beta}{\hbar} (E)^{-1/2}. \qquad (8.283)$$

The mean square radius is given by

$$\langle \psi_s | r^2 | \psi_s \rangle \equiv \int_0^{r_1} r^2 |\chi(r)|^2 dr$$
$$= \frac{N^2}{2} \int_0^{r_1} \frac{r^2 dr}{k(r)}$$
$$= \frac{N^2 E^3}{2\beta^3} \sqrt{\frac{\hbar^2}{2\mu E}} \int_0^1 \frac{x^2}{(1-x)^{1/2}} dx$$
$$= \frac{N^2 E^3}{2\beta^3} \sqrt{\frac{\hbar^2}{2\mu E}} \frac{\Gamma(1/2)\Gamma(3)}{\Gamma(7/2)}$$
$$= \frac{8}{15} \frac{E^2}{\beta^2}. \qquad (8.284)$$

Hence [with $2\mu = m_c$] and using Eq. (20.229),

$$\langle \psi_{ns} | r^2 | \psi_{ns} \rangle = \frac{8}{15} \frac{E_n^2}{\beta^2}$$
$$= \frac{8}{15} (\pi)^{4/3} \left(\frac{4}{9} \frac{m_c}{\hbar^2} \right)^{-2/3} (n - \frac{1}{4})^{4/3}. \qquad (8.285)$$

---

[1] It is usual practice to take the average value of the oscillatory part to be $\frac{1}{2}$. See for example equation 4.12 of C. Quigg (Fermilab)and Jonathan L. Rosner (Minnesota U.), Phys. Rept. 56 (1979) 167–235.

8.13   Problems

8.1 The wave function

$$u(r, \theta, \phi) \xrightarrow[r \to \infty]{} A\left(e^{ikz} + f(\theta, \phi)\frac{e^{ikr}}{r}\right)$$

represents the incident plane wave and the scattered wave radially moving outward. Let $\mathbf{S}$ denotes the probability current density. Show that $\mathbf{S}_{\text{out}} = \frac{v}{r^3}|A|^2|f(\theta, \phi)|^2\mathbf{r} +$ higher terms which can be neglected for large r. Calculate $\mathbf{S}_{\text{out}} \cdot d\mathbf{\Sigma}$ where $|d\mathbf{\Sigma}| = d\Sigma$ is the surface element $r^2 d\Omega$. Hence show that

$$\frac{\mathbf{S}_{\text{out}} \cdot d\mathbf{\Sigma}}{S_{\text{in}}} = |f(\theta, \phi)|^2 d\Omega = d\sigma.$$

8.2 Consider the scattering of a beam of particles by a hard sphere of radius a:

$$V = \infty \quad r < a,$$
$$ = 0 \quad\; r > a.$$

Show that the phase shift δ_ℓ is given by

$$\tan \delta_\ell = \frac{j_\ell(ka)}{n_\ell(ka)},$$

where j_ℓ and n_ℓ have the usual meaning and k is the wave number. Find the cross section when $a \ll 1/k$. For $a \gg 1/k$, show that

$$\delta_\ell \approx -ka + \ell\pi/2.$$

8.3 Consider the scattering of a particle of mass m by a hard sphere of radius a

$$V = \infty \quad r < a,$$
$$ = 0 \quad\; r > a.$$

Treat the case for which the particle moves sufficiently slowly so that $\ell \geq 2$ phase shifts are negligible. Show that

$$\tan \delta_0 = -\tan(ka),$$
$$\tan \delta_1 = \frac{ka - \tan ka}{1 + ka \tan ka},$$

$k^2 = 2mE/\hbar^2$. If ka is small and the phase shifts δ_0 and δ_1 are small, show that

$$\delta_0 \approx -ka, \qquad \delta_1 \approx -\frac{(ka)^3}{3}.$$

Show further that

$$\frac{d\sigma}{d\Omega} \sim a^2 \left(1 - \frac{(ka)^2}{3} + 2(ka)^2 \cos\theta + O(ka)^3 \right).$$

8.4 Consider the scattering of a particle at low energies (i.e. small k, so that one can confine to s-wave only) by an attractive potential $V(r)$, the range of the potential being $r = a$. Suppose there exists among a discrete spectrum of negative energy levels a bound state (with angular momentum $\ell = 0$) at energy $E = -\epsilon$ ($\epsilon > 0$) where ϵ is very small. Show that the s-wave phase shift is given by

$$\tan\delta_0 = -\sqrt{E/\epsilon}$$

and

$$\sigma(E) = \frac{2\pi\hbar^2}{\mu(E + \epsilon)}.$$

Sketch $\sigma(E)$.

8.5 Use Born approximation to find the angular distribution (that is behaviour of $\sigma(\theta)$ with respect to θ) of the electrons in the elastic scattering of electrons by an atom represented by a shielded Coulomb potential

$$V(r) = -\frac{Ze^2}{r}\exp(-r/a).$$

Find also the total cross section.

8.6 Consider the potential

$$V = V_0 \qquad \text{for} \quad r < a,$$
$$= 0 \qquad \text{for} \quad r > a.$$

Find the scattering amplitude $f(\theta)$ in the Born approximation.

8.7 A solution of the equation

$$(\nabla^2 + k^2)G_k(\mathbf{r}) = \delta^3(\mathbf{r})$$

is given by

$$G_k(\mathbf{r}) = -\left(C_+ e^{ikr} + C_- e^{-ikr} \right)\frac{1}{4\pi r},$$

where
$$C_+ + C_- = 1.$$

Verify that
$$\psi(\mathbf{r}) = e^{i\mathbf{k}\cdot\mathbf{r}} + \frac{2\mu}{\hbar^2}\int G_k(\mathbf{r} - \mathbf{r}')\psi(\mathbf{r}')V(\mathbf{r}')d^3r'$$

is solution of
$$\left(-\frac{\hbar^2}{2\mu}\nabla^2 + V(\mathbf{r})\right)\psi(\mathbf{r}) = E_k\psi(\mathbf{r}),$$

where
$$E_k = \frac{\hbar k^2}{2\mu^2}.$$

Show that in the limit of zero energy the asymptotic form of ψ is
$$\psi \underset{r\to\infty}{\sim} 1 - \frac{a}{r},$$

where
$$a = \frac{\mu}{2\pi\hbar^2}\int \psi(\mathbf{r})V(\mathbf{r})d^3r.$$

Hence show that
$$\lim_{k\to 0}\delta_0/k = -a.$$

8.8 The scattering amplitude in terms of phase shift is given by

$$f(\theta) = \frac{1}{2ik}\sum_{l=0}^{\infty}(2l+1)(e^{2i\delta_\ell} - 1)P_\ell(\cos\theta).$$

The Born formula for the scattering amplitude $f(\theta)$ is

$$f_B(\theta) = -\frac{2\mu}{\hbar^2}\int_0^{\infty}\frac{\sin qr}{qr}V(r)r^2dr.$$

For small phase shifts, $e^{2i\delta_\ell} - 1 \approx 2i\delta_\ell$. Using the formula
$$\frac{\sin qr}{qr} = \sum_{\ell}(2\ell+1)P_\ell(\cos\theta)\big(j_\ell(kr)\big)^2,$$

show that for small δ_ℓ:

$$\delta_\ell^B \approx \frac{2\mu}{\hbar^2}k\int_0^{\infty}V(r)j_\ell^2(kr)r^2dr.$$

Since for $\ell > \ell_{max}$, the phase shifts are small, we can use the Born amplitude as a device for summing up the partial wave series for $\ell > \ell_{max}$. Using the above statement show that

$$f(\theta) = \frac{1}{2ik}\sum_{\ell=0}^{\ell_{max}}(2\ell+1)\big((e^{2i\delta_\ell} - 1)$$

$$-(e^{2i\delta_\ell^B} - 1)\big)P_\ell(\cos\theta) + f_B(\theta).$$

8.9 Consider the scattering of a particle of mass m by the potential

$$V = V_0 \quad r < a,$$
$$= 0 \quad r > a.$$

Estimate the differential scattering cross section for $\hbar^2 k^2 \gg 2mV_0$ and discuss the validity of the approximation used.

In the limit $ka \ll 1$, show that the cross section takes on the form

$$\frac{d\sigma}{d\Omega} = A + B\cos\theta$$

with $B \ll A$.

8.10 Consider an electron confined inside a sphere of radius R. What is the pressure exerted on the surface of the sphere, if the electron is in (i) the lowest s state (ii) the lowest p-state?

Hint:
$$V = 0 \quad r < a,$$
$$= \infty \quad r > a.$$

and $P = -\frac{\partial E}{\partial V}$ [V here is volume].

8.11 Show that for s-wave bound states, the wave function at the origin is given by

$$|\psi_s(0)|^2 = \frac{2\mu}{4\pi}\left\langle \frac{dV}{dr}\right\rangle,$$

where V is the potential between two particles and μ is their reduced mass.

8.12 Using the WKB approximation, show that for the s-wave bound states for the logarithmic potential $V(r) = C\ln r/r_0$, the energy eigenvalues are given by

$$E_n = C\ln[(2n - \frac{1}{2})\sqrt{\pi}] + C\ln[(1/\sqrt{m_c C})1/r_0],$$

where m_c is the mass of charmed quark. Show that the mean square radius of the charmonium for the logarithmic potential is given by

$$\langle\psi_{ns}|r^2|\psi_{ns}\rangle = (1/Cm_c)(4\pi/\sqrt{3})(n - 1/4)^2.$$

Hint: The quantisation condition is

$$\int_0^{r_1} [m_c(E - C\ln(r/r_0))]^{1/2}dr = (n - 1/4)\pi\hbar,$$

where $r_1 = r_0 e^{E/C}$. In order to perform the integration, put $(E/C - y^2) = \ln(r/r_0)$.

Chapter 9

Operators

9.1 State Vectors

In developing quantum mechanics in the previous chapters we have introduced two concepts: (i) State of a system, which has no classical analogue, and (ii) Dynamical variables, which are represented by hermitian operators, called observables. Although they have a classical analogue they are treated here quite differently. For example, in the Schrödinger representation x is an algebraic variable, the dynamical variable $\hat{A}$ is an operator $\hat{A}$ $= \hat{A}\left(x, \frac{\partial}{\partial x}\right)$, (for example, the momentum operator $\hat{p} = -i\hbar\frac{\partial}{\partial x}$) and the state of a system is represented by the wave or state function $\psi(x)$.

We now want to evolve a more general formulism. In this formulism, it is convenient to represent states in quantum mechanics by vectors, called state vectors in a certain vector space, usually with an infinite number of dimensions. This is because vectors have simple transformation laws. The state function $\psi(x)$ is then a particular representation of the state vector.

Following Dirac, we shall denote the state vector by $|\rangle$, called the ket vector and we shall label a particular one by $|\psi\rangle$. We also introduce a bra vector corresponding to the ket vector $|\psi\rangle$ and denote it by $\langle\psi|$.

We now define the scalar product as

$$\langle\psi|\phi\rangle \,.$$

This being a scalar product is a number, in general complex. The complex conjugate of $\langle\phi|\psi\rangle$ is defined as

$$\langle\phi|\psi\rangle^* = \langle\psi|\phi\rangle \,. \tag{9.1}$$

Operators:

In quantum mechanics, dynamical variables are linear operators. A linear operator $\hat{A}$ is defined by the relation

$$|F\rangle = \hat{A}|\psi\rangle \qquad (9.2)$$

i.e. a linear operator operating on a ket produces a new ket. Its linearity is specified by the following relations

$$(\hat{A} + \hat{B})|\psi\rangle = \hat{A}|\psi\rangle + \hat{B}|\psi\rangle \qquad (9.3a)$$

$$\hat{A}(|\psi\rangle + |\phi\rangle) = \hat{A}|\psi\rangle + \hat{A}|\phi\rangle \qquad (9.3b)$$

$$\hat{A}(C|\psi\rangle) = C\hat{A}|\psi\rangle \qquad (9.3c)$$

where C is a number. The product $\hat{A}\hat{B}$ of two linear operators $\hat{A}$ and $\hat{B}$ is defined by

$$(\hat{A}\hat{B})|\psi\rangle = \hat{A}(\hat{B}|\psi\rangle). \qquad (9.4)$$

In general

$$(\hat{A}\hat{B})|\psi\rangle \neq (\hat{B}\hat{A})|\psi\rangle \qquad (9.5)$$

i.e.

$$\hat{A}\hat{B} \neq \hat{B}\hat{A}. \qquad (9.6)$$

In a special case when $\hat{A}\hat{B} = \hat{B}\hat{A}$, we say $\hat{A}$ and $\hat{B}$ commute.

Consider an operator $\hat{A}$, such that

$$|F\rangle = \hat{A}|\phi\rangle.$$

Perform the scalar product

$$\langle\psi|F\rangle = \langle\psi|\hat{A}|\phi\rangle. \qquad (9.7)$$

In particular if C is a number

$$\langle\psi|C|\phi\rangle = C\langle\psi|\phi\rangle. \qquad (9.8)$$

Quantities like

$$\langle\phi|\hat{A}|\psi\rangle$$

are called matrix elements.

The *Hermitian Conjugate* $\hat{A}^\dagger$ of $\hat{A}$ is defined by

$$\langle\psi|F\rangle^* = \langle\psi|\hat{A}|\phi\rangle^* \qquad (9.9a)$$

$$= \langle\phi|\hat{A}^\dagger|\psi\rangle. \qquad (9.9b)$$

But, by definition

$$\langle\psi|F\rangle^* = \langle F|\psi\rangle.$$

Therefore we can write

$$\langle F| = \langle\phi|\hat{A}^\dagger. \tag{9.10}$$

One can easily prove that

$$(\hat{A}^\dagger)^\dagger = \hat{A}$$
$$(\hat{A} + \hat{B})^\dagger = \hat{A}^\dagger + \hat{B}^\dagger$$
$$(\lambda\hat{A})^\dagger = \lambda^*\hat{A}^\dagger$$

where λ is a number, and

$$(\hat{A}\hat{B})^\dagger = \hat{B}^\dagger\hat{A}^\dagger.$$

If $\hat{A}^\dagger = \hat{A}$, $\hat{A}$ is called hermitian. $\hat{A}$ is called unitary if $\hat{A}^\dagger\hat{A} = \hat{A}\hat{A}^\dagger = 1$. Recall that an observable in quantum mechanics corresponds to a hermitian operator. Thus for a hermitian operator $\hat{A}$,

$$\langle\phi|\hat{A}|\psi\rangle = \langle\psi|\hat{A}|\phi\rangle^*. \tag{9.11}$$

Definitions:
Two state vectors $|\phi\rangle$ and $|\psi\rangle$ are said to be orthogonal if

$$\langle\phi|\psi\rangle = 0 \tag{9.12a}$$

or

$$\langle\psi|\phi\rangle = 0. \tag{9.12b}$$

A state vector is said to be normalised if

$$\langle\psi|\psi\rangle = 1. \tag{9.13}$$

Eigenvectors:

If there is an operator $\hat{A}$ such that operating upon some state $|a_n\rangle$, it gives the same state, i.e. to say

$$\hat{A}|a_n\rangle = a_n|a_n\rangle, \tag{9.14}$$

where a_n is a number, then $|a_n\rangle$ is called an eigenvector of $\hat{A}$ belonging to the eigenvalue a_n.

Theorem:
(i) The eigenvalues of a hermitian operator are real.
(ii) The eigenvectors of a hermitian operator are orthogonal.

Proof:
Eigenvalue equation is

$$\hat{A}|a_n\rangle = a_n|a_n\rangle$$

so that

$$\langle a_m|\hat{A}|a_n\rangle = \langle a_n|\hat{A}^\dagger|a_m\rangle^* = \langle a_n|\hat{A}|a_m\rangle^*$$
$$= a_m{}^*\langle a_n|a_m\rangle^*$$
$$= a_m{}^*\langle a_m|a_n\rangle$$

Therefore

$$a_n\langle a_m|a_n\rangle = a_m{}^*\langle a_m|a_n\rangle$$

or

$$(a_n - a_m{}^*)\langle a_m|a_n\rangle = 0.$$

Hence for

$$(i) \quad m = n, \quad a_n = a_n{}^*, \quad \langle a_n|a_n\rangle \neq 0$$
$$(ii) \quad m \neq n : \text{We must have} \quad \langle a_m|a_n\rangle = 0$$

i.e. they are orthogonal. If the eigenstates are normalised:

$$\langle a_n|a_m\rangle = 1$$

In this case

$$\langle a_m|a_n\rangle = \delta_{mn} = \begin{bmatrix} 1 & m = n, \\ 0 & m \neq 0. \end{bmatrix}$$

Completeness:

We assume that eigenvectors of an observable form a complete set so that a state vector $|\psi\rangle$ can be expressed as a linear expansion of the eigenstates $|a_n\rangle$ of an observable $\hat{A}$. Thus

$$|\psi\rangle = \sum_n C_n|a_n\rangle, \quad \text{for the discrete case} \qquad (9.15a)$$

$$= \int C(a)|a\rangle da, \quad \text{for continuous eigenvalues.} \qquad (9.15b)$$

Then for the discrete case

$$\langle a_m | \psi \rangle = \sum_n C_n \langle a_m | a_n \rangle$$

$$= \sum_n C_n \delta_{mn} = C_m. \tag{9.16}$$

Therefore

$$C_n = \langle a_n | \psi \rangle. \tag{9.17}$$

Substituting in Eq.(9.15a)

$$| \psi \rangle = \sum_n | a_n \rangle \langle a_n | \psi \rangle. \tag{9.18}$$

Thus formally one can put

$$\sum_n | a_n \rangle \langle a_n | = \hat{1}, \tag{9.19}$$

where $\hat{1}$ denotes the unit operator. Equation (9.19) is a formal statement of the completeness condition.

For the continuous case we have from Eq. (9.15b)

$$\langle a' | \psi \rangle = \int C(a) \langle a' | a \rangle da$$

$$= \int C(a)(a' - a) da$$

$$= C(a')$$

or

$$C(a) = \langle a | \psi \rangle. \tag{9.20}$$

Substituting in Eq. (9.15b)

$$| \psi \rangle = \int | a \rangle \langle a | \psi \rangle da. \tag{9.21}$$

Thus in this case, the formal statement of completeness is

$$\int | a \rangle \langle a | da = \hat{1}.$$

Thus for any set of eigenvectors $|a\rangle$, the completeness condition can be formally expressed as

$$S_a | a \rangle \langle a | = \hat{1}, \tag{9.22a}$$

where

$$S_a = \sum_{a_n} \quad \text{if } a = a_n, \text{ discrete}$$

$$= \int \cdots da \quad \text{if } a \text{ is a continuous variable,} \qquad (9.22b)$$

Eq. (9.17) or (9.20) provides a representation of the state $|\psi\rangle$ in the space defined by the basic vectors

$$|a_n\rangle \quad \text{or} \quad |a\rangle \qquad (9.23)$$

and completely determine the state.

Similarly in this basis, an operator $\hat{\alpha}$ is represented by the matrix elements α_{mn}:

$$\langle a_n|\hat{\alpha}|a_m\rangle = \alpha_{mn} \quad \text{for the discrete case.} \qquad (9.24)$$

Thus

$$\hat{\alpha}|a_m\rangle = \sum_{a_n} |a_n\rangle\langle a_n|\hat{\alpha}|a_m\rangle\langle a_m| = \sum_{a_n} \alpha_{nm}|a_n\rangle$$

and

$$\hat{\alpha} = \sum_n \sum_m |a_n\rangle\langle a_n|\hat{\alpha}|a_m\rangle\langle a_m| = \sum_n \sum_m |a_n\rangle\langle a_m|, \qquad (9.25)$$

and similarly for the continuous case. If $\hat{\alpha}$ and $\hat{\beta}$ are two linear operators then

$$(\alpha\beta)_{nm} = \langle n|\hat{\alpha}\hat{\beta}|m\rangle = \sum_k \langle n|\hat{\alpha}|k\rangle\langle k|\hat{\beta}|n\rangle$$

$$= \sum_k \alpha_{nk}\beta_{kn}.$$

Thus the matrix representation of product of two operators is the product of matrix representations of the two operators.

In terms of state vectors the average value of repeated measurements of an observable $\hat{A}$ when the system is in a normalised state $|\psi\rangle$ is given by

$$\bar{a}_\psi = \langle \psi|\hat{A}|\psi\rangle. \qquad (9.26a)$$

Using the completeness relation, we can write

$$\bar{a}_\psi = \sum_n \sum_m \langle\psi|a_n\rangle\langle a_n|\hat{A}|a_m\rangle\langle a_m|\psi\rangle \qquad (9.26b)$$

$$= \sum_n \sum_m \langle\psi|a_n\rangle a_m \delta_{mn}\langle a_m|\psi\rangle$$

$$= \sum_n |\langle a_n | \psi \rangle|^2 a_n. \tag{9.26c}$$

Thus the weighting factor, which appears in the calculation of the average, namely,

$$|\langle a_n | \psi \rangle|^2 \tag{9.27}$$

gives the probability of a result a_n in a measurement of $\hat{A}$ on $|\psi\rangle$.

By the use of the completeness relations, we can easily see that the quantum mechanics of the previous chapters follows as a particular case of the present formalism.

Finally for a Unitary operator $\hat{U}$ with eigenvector $|u\rangle$ and corresponding eigenvalue λ:

$$\hat{U}|u\rangle = \lambda |u\rangle$$

implies

$$\langle u | \hat{U}^\dagger = \lambda^* \langle u |$$

and since

$$\hat{U}^\dagger \hat{U} = 1$$
$$\langle u | u \rangle = \langle u | \hat{U}^\dagger \hat{U} | u \rangle$$
$$= \langle u | \lambda^* \lambda | u \rangle$$
$$= |\lambda|^2 \langle u | u \rangle$$

i.e.

$$|\lambda|^2 = 1.$$

Thus a unitary operator has eigenvalues which are of the type $e^{i\theta}$.

9.2 Schrödinger Representation

Let us take the basis vectors to be eigenvectors of the position operator $\hat{x}$:

$$\hat{x}|x\rangle = x|x\rangle, \tag{9.28a}$$
$$\langle x' | x \rangle = \delta(x' - x), \tag{9.28b}$$
$$\langle x' | \hat{x} | x \rangle = x \delta(x' - x). \tag{9.28c}$$

In this basis, the state $|\psi\rangle$ is determined by quantities

$$\langle x | \psi \rangle$$

which we denote as

$$\psi(x) = \langle x|\psi\rangle \tag{9.29}$$

and call this the state function in x or the Schrödinger representation. Then

$$\psi^*(x) = \langle \psi|x\rangle. \tag{9.30}$$

The expression $\langle x|\psi\rangle$ may be thought of as the component of a state vector $|\psi\rangle$; when x is a continuous variable, we have an infinity of components which run together to form the state function

$$\psi(x) = \langle x|\psi\rangle.$$

Note that the average value of repeated measurements of position is given by

$$\begin{aligned}
\bar{x}_\psi &= \langle \psi|\hat{x}|\psi\rangle \\
&= \int \int \langle \psi|x'\rangle\langle x'|\hat{x}|x\rangle\langle x|\psi\rangle dx dx' \\
&= \int \int \langle \psi|x'\rangle x\delta(x' - x)\langle x|\psi\rangle dx' dx \\
&= \int |\langle x|\psi\rangle|^2 x dx.
\end{aligned} \tag{9.31}$$

This shows that $|\psi(x)|^2 = |\langle x|\psi\rangle|^2$, determines the probability density $P(x)$ of the particle at x.

Let us now express $\bar{a}_\psi$ in the Schrödinger or x-representation

$$\begin{aligned}
\bar{a}_\psi &= \langle \psi|\hat{A}|\psi\rangle \\
&= \int \int \langle \psi|x\rangle\langle x|\hat{A}|x'\rangle\langle x'|\psi\rangle dx' dx.
\end{aligned} \tag{9.32a}$$

This reduces to our previous definition of $\bar{a}_\psi$ in the Schrödinger representation, namely

$$\bar{a}_\psi = \int \psi^*(x)\hat{A}\left(x, \frac{\partial}{\partial x}\right)\psi(x) dx, \tag{9.32b}$$

if we write

$$\langle x|\hat{A}|x'\rangle = \delta(x' - x)\hat{A}\left(x', \frac{\partial}{\partial x'}\right). \tag{9.33}$$

Example:

The momentum operator $\hat{p}$ in the Schrödinger Representation. We have the quantum condition

$$\hat{x}\hat{p} - \hat{p}\hat{x} = i\hbar. \tag{9.34}$$

Then
$$\langle x|(\hat{x}\hat{p} - \hat{p}\hat{x})|x'\rangle = i\hbar\langle x|x'\rangle. \tag{9.35}$$

Now
$$\langle x|\hat{p}\hat{x}|x'\rangle = x'\langle x|\hat{p}|x'\rangle, \tag{9.36}$$
$$\begin{aligned}\langle x|\hat{x}\hat{p}|x'\rangle &= \langle x'|(\hat{x}\hat{p})^\dagger|x\rangle^* \\ &= \langle x'|\hat{p}\hat{x}|x\rangle^* \\ &= x\langle x'|\hat{p}|x\rangle^* \\ &= x\langle x|\hat{p}|x'\rangle.\end{aligned} \tag{9.37}$$

Therefore Eq. (9.35) becomes
$$\begin{aligned}(x - x')\langle x|\hat{p}|x'\rangle &= i\hbar\langle x|x'\rangle \\ &= i\hbar\delta(x - x').\end{aligned} \tag{9.38}$$

Consider now
$$\begin{aligned}&\int_{-g}^{g} f(x)\frac{\partial}{\partial x}\delta(x - x')dx \qquad -g < x' < g \\ &= f(x)\delta(x - x')|_{-g}^{g} - \int_{-g}^{g} f'(x)\delta(x - x')dx.\end{aligned} \tag{9.39}$$

The first term on the second line gives zero, since $\delta(\pm g - x') = 0$. Thus
$$\begin{aligned}\int_{-g}^{g} f(x)\frac{\partial}{\partial x}\delta(x - x')dx &= -\int_{-g}^{g} f'(x)(x - x')dx \\ &= -f'(x').\end{aligned} \tag{9.40}$$

Therefore
$$f(x)\frac{\partial}{\partial x}\delta(x - x') = -f'(x)\delta(x - x'). \tag{9.41a}$$

Similarly
$$\begin{aligned}\int_{-g}^{g} f(x')\frac{\partial}{\partial x'}\big(\delta(x - x')\big)dx' &\qquad -g < x < g \\ &= -\int_{-g}^{g} \frac{\partial f(x')}{\partial x'}\delta(x - x')dx' \\ &= -\frac{\partial f(x')}{\partial x'}\Big|_{x'=x}.\end{aligned} \tag{9.41b}$$

Therefore
$$f(x')\frac{\partial}{\partial x'}\delta(x - x') = -\frac{\partial f(x')}{\partial x'}\delta(x - x'). \tag{9.41c}$$

Take $f(x) = x - x'$. Then from Eq.(9.41a)
$$(x - x')\frac{\partial}{\partial x}\delta(x - x') = -\delta(x - x'). \tag{9.41d}$$

Therefore we can write Eq.(9.38) as

$$(x - x')\langle x|\hat{p}|x'\rangle = (x - x')\left(-i\hbar\frac{\partial}{\partial x}\delta(x - x')\right).$$

Hence

$$\langle x|\hat{p}|x'\rangle = -i\hbar\frac{\partial}{\partial x}\delta(x - x')$$

$$= -i\hbar\frac{\partial}{\partial x}\langle x|x'\rangle. \tag{9.42}$$

The meaning of this equation is as follows [use $\frac{\partial}{\partial x}\delta(x - x') = -\frac{\partial}{\partial x'}\delta(x - x')$ and Eq. (9.41c)]:

$$\langle x|\hat{p}|x'\rangle\psi(x') = -i\hbar\left(\frac{\partial}{\partial x}\delta(x - x')\right)\psi(x')$$

$$= -i\hbar\delta(x - x')\frac{\partial}{\partial x'}\psi(x'). \tag{9.43a}$$

Therefore when $\langle x|\hat{p}|x'\rangle$ is multiplied with a function $\psi(x')$ on the right, its effect can be represented by [see Eq. (9.33)]

$$\langle x|\hat{p}|x'\rangle = (x - x')\left(-i\hbar\frac{\partial}{\partial x'}\right). \tag{9.43b}$$

Consider now more general matrix elements

$$\langle x|\hat{p}|\psi\rangle = \int dx'\langle x|\hat{p}|x'\rangle\langle x'|\psi\rangle$$

$$= \int dx'(x - x')\left(-i\hbar\frac{\partial}{\partial x'}\right)\langle x'|\psi\rangle$$

$$= -i\hbar\frac{\partial}{\partial x}\langle x|\psi\rangle. \tag{9.44}$$

It is clear from Eqs. (9.43a, 9.43b) and (9.44) that in the Schrödinger representation

$$\hat{p} = -i\hbar\frac{\partial}{\partial x}. \tag{9.45}$$

Eigenfunctions:

The eigenvalue equation

$$\hat{A}|a_n\rangle = a_n|a\rangle \tag{9.46}$$

becomes in the Schrödinger representation $\langle x|\hat{A}|a_n\rangle = a_n\langle x|a_n\rangle$. Using completeness conditions, we can write

$$\int dx'\langle x|\hat{A}|x'\rangle\langle x'|a_n\rangle = a_n\langle x|a_n\rangle \tag{9.47}$$

or

$$\int dx'(x - x')\hat{A}(x', \frac{\partial}{\partial x'})\langle x'|a_n\rangle = a_n\langle x|a_n\rangle, \tag{9.48}$$

i.e.

$$\hat{A}(x, \frac{\partial}{\partial x})\langle x|a_n\rangle = a_n\langle x|a_n\rangle. \tag{9.49}$$

We write

$$u_{a_n}(x) = \langle x|a_n\rangle. \tag{9.50}$$

The above equation shows that $u_{a_n}(x)$ is an eigenfunction of $\hat{A}(x, \frac{\partial}{\partial x})$ belonging to the eigenvalue a_n.

9.3 Relation between the Momentum and Schrödinger Representations

Let $|p\rangle$ denote eigenvectors of the momentum operator

$$\hat{p}|p\rangle = p|p\rangle \tag{9.51a}$$

$$\langle p'|p\rangle = (p' - p) \tag{9.51b}$$

$$\langle p'|\hat{p}|p\rangle = p(p' - p), \tag{9.51c}$$

so that in the momentum representation $\hat{p}$ is an algebraic variable p. Now

$$\langle x|\hat{p}|p\rangle = p\langle x|p\rangle. \tag{9.52}$$

Using Eq. (9.44)

$$\langle x|\hat{p}|p\rangle = -i\hbar\frac{\partial}{\partial x}\langle x|p\rangle, \tag{9.53}$$

therefore Eq. (9.53) gives

$$-i\hbar\frac{\partial}{\partial x}\langle x|p\rangle = p\langle x|p\rangle. \tag{9.54a}$$

The normalised solution of Eq. (9.54a) is

$$\langle x|p\rangle = \frac{1}{\sqrt{2\pi\hbar}}e^{(ip/\hbar)x}. \tag{9.54b}$$

This is an eigenfunction of the momentum operator in the Schrödinger representation. Now we can expand the state $|\psi\rangle$ in terms of the eigenvectors of the momentum operator:

$$|\psi\rangle = \int dp|p\rangle\langle p|\psi\rangle. \tag{9.55}$$

Then taking the Schrödinger representation

$$\langle x|\psi\rangle = \int dp\langle x|p\rangle\langle p|\psi\rangle$$
$$-\frac{1}{\sqrt{2\pi\hbar}}\int dp e^{(ip/\hbar)x}\langle p|\psi\rangle. \tag{9.56}$$

On the other hand, expanding $|\psi\rangle$ in terms of eigenvectors of the position operator

$$|\psi\rangle = \int dx|x\rangle\langle x|\psi\rangle. \tag{9.57}$$

Therefore taking the momentum representation

$$\langle p|\psi\rangle = \int dx\langle p|x\rangle\langle x|\psi\rangle$$
$$= \frac{1}{\sqrt{2\pi\hbar}}\int dx e^{-i(p/\hbar)x}\langle x|\psi\rangle. \tag{9.58}$$

Therefore $\langle x|\psi\rangle$ and $\langle p|\psi\rangle$ are Fourier transforms of each other. Previously we labelled $\langle p|\psi\rangle$ by $C(p)$ and $\langle x|\psi\rangle$ by $\psi(x)$. The probability of the system having momentum p when in state $|\psi\rangle$ is $|\langle p|\psi\rangle|^2$.

Examples:

(i) Simultaneous eigenfunctions of angular momentum L^2 and z-component of $\mathbf{L}$, L_z:

Using polar coordinates, select the basis vectors to be $|\theta, \phi\rangle$. Since L^2, L_z depend on θ and ϕ only, the simultaneous eigenfunctions of L^2 and L_z are given by

$$Y_{lm}(\theta, \phi) = \langle \theta, \phi|lm\rangle, \qquad (9.59)$$

where the simultaneous eigenvector $|lm\rangle$ of L^2 and L_z satisfy

$$\begin{aligned} L^2|lm\rangle &= \hbar^2 l(l+1)|lm\rangle \\ L_z|lm\rangle &= \hbar m|lm\rangle, \end{aligned} \qquad (9.60)$$

$l = 0, 1, 2, \cdots$, $m = 0, \pm1, \pm2, \cdots, \pm l$. In particular

$$\langle \phi|m\rangle = \frac{1}{\sqrt{2\pi}}e^{im\phi}. \qquad (9.61)$$

(ii) Discrete eigenfunctions of the hydrogen atom. Select the basis vectors to be $|r, \theta, \phi\rangle$, since the Hamiltonian depends on all three coordinates r, θ, ϕ. Then the eigenfunctions can be written as

$$u_{nlm}(r, \theta, \phi) = \langle r, \theta, \phi|nlm\rangle, \qquad (9.62)$$

where n specifies the energy eigenvalues

$$E_n = -\left(\frac{me^4 Z^2}{2\hbar^2 n^2}\right), \qquad (9.63)$$

$n = 1, 2, \cdots$; $n \geq (l+1)$ for a given l. $|nlm\rangle$ are simultaneous eigenvectors of H, L^2 and L_z.

Also we note that in the Schrödinger representation, the orthonormality relations and completeness condition

$$\langle a_m|a_n\rangle = \delta_{mn}$$
$$\sum_n |a_n\rangle\langle a_n| = \hat{1} \qquad (9.64)$$

become

$$\int \langle a_m|x\rangle\langle x|a_n\rangle dx = \delta_{mn}$$

or

$$\int u^*_{a_m}(x) u_{a_n}(x) dx = \delta_{mn}$$

$$\sum_n \langle x|a_n\rangle\langle a_n|x'\rangle = \langle x|x'\rangle \qquad (9.65)$$

$$= \delta(x - x') \qquad (9.66)$$

or

$$\sum_n u_{a_n}(x) u^*_{a_n}(x') = \delta(x - x'). \qquad (9.67)$$

9.4 Matrix Mechanics

An operator is diagonal in the space defined by its eigenvectors. Consider for example the Hamiltonian H. Introduce two sets of basis vectors. The eigenvectors $|E_n\rangle$ form one set of basis vectors. In this basis H is diagonal:

$$H|E_n\rangle = E_n|E_n\rangle, \qquad (9.68)$$

$$\langle E_m|E_n\rangle = \delta_{mn}, \qquad (9.69)$$

$$\langle E_m|H|E_n\rangle = E_n\delta_{nm}.$$

Using the completeness relation

$$\sum_n |E_n\rangle\langle E_n| = 1 \qquad (9.70)$$

$$H = \sum_n \sum_m |E_n\rangle\langle E_n|H|E_m\rangle\langle E_m|$$

$$= \sum_n \sum_m |E_n\rangle E_m \delta_{mn}\langle E_m|$$

$$= \sum_n E_n|E_n\rangle\langle E_n|. \qquad (9.71)$$

Consider another set of basis vectors

$$\langle a_i|a_j\rangle = \delta_{ij}, \qquad (9.72)$$

$$\sum_i |a_i\rangle\langle a_i| = 1. \qquad (9.73)$$

In this basis

$$H = \sum_i \sum_j |a_j\rangle\langle a_j|H|a_i\rangle\langle a_i|$$

$$= \sum_i \sum_j |a_j\rangle H_{ij}\langle a_i| \qquad (9.74)$$

and since basis vectors $|E_n\rangle$ and $|a_i\rangle$ form a complete set, one can write

$$|E_n\rangle = \sum_i |a_i\rangle\langle a_j|E_n\rangle$$

$$= \sum_i U_{ni}|a_i\rangle \tag{9.75}$$

where

$$U_{ni} = \langle a_i|E_n\rangle,$$
$$U_{ni}{}^* = \langle a_i|E_n\rangle^* = \langle E_n|a_i\rangle. \tag{9.76}$$

Now

$$|a_i\rangle = \sum_n |E_n\rangle\langle E_n|a_i\rangle$$

$$= \sum_n U_{ni}{}^*|E_n\rangle$$

$$= (U^\dagger)_{in}|E_n\rangle. \tag{9.77}$$

Thus on using Eqs. (9.75) and (9.76)

$$\langle E_m|E_n\rangle = \sum_i U_{ni}\langle E_m|a_i\rangle$$

$$= \sum_i U_{ni}U_{mi}{}^* = \sum_i U_{ni}(U^\dagger)_{im}. \tag{9.78}$$

Hence, we have

$$\sum_i U_{ni}(U^\dagger)_{im} = U_{ni}U^\dagger{}_{im} = \delta_{nm},$$

$$UU^\dagger = 1. \tag{9.79}$$

Thus one can go from one set of basis vectors to another set by a unitary transformation. The unitary matrix U_{ni} is overlap $\langle a_n|E_n\rangle$ between two set of basis vectors.
Now

$$\langle E_n| = \langle a_i|U_{ni}{}^*$$

and

$$\langle a_i| = \sum_n \langle E_n|U_{in}. \tag{9.80}$$

Therefore

$$\langle E_n|H|E_m\rangle = \sum_i \sum_j \langle E_m|a_j\rangle\langle a_j|H|a_i\rangle\langle a_i|E_m\rangle$$

$$= \sum_i \sum_j U_{nj}{}^* H_{ji} U_{mi} = \sum_i \sum_j U_{jn}{}^\dagger H_{ji} U_{mi}$$

$$= (U^\dagger H U)_{mn}. \tag{9.81}$$

Hence if the Hamiltonian (or any other observable) is given in a basis in which it is not diagonal, it can be diagonalised by a unitary transformation U, which is an overlap between the two basis vectors as given in Eqs.(9.76). We note that

$$\langle a_i|H|E_n\rangle = E_n\langle a_i|E_n\rangle$$

or

$$\sum_j \langle a_i|H|a_j\rangle\langle a_j|E_n\rangle = E_n\langle a_i|E_n\rangle. \tag{9.82}$$

Therefore

$$\sum_j (H_{ij} - E_n\delta_{ij})\langle a_j|E_n\rangle = 0. \tag{9.83}$$

Hence the eigenvalues of H are given by

$$\det |H - EI| = 0. \tag{9.84}$$

This is also true for any other observable $\hat{F}(H \to \hat{F}, E \to F)$.
Finally, since both sets of basis vectors viz. $|E_n\rangle$ and $|a_i\rangle$ form complete sets, any arbitrary ket $|\psi\rangle$ can be written in terms of them:

$$|\psi\rangle = \sum_n |E_n\rangle\langle E_n|\psi\rangle = \sum_n \beta_n|E_n\rangle \tag{9.85}$$

and

$$|\psi\rangle = \sum_i |a_i\rangle\langle a_i|\psi\rangle = \sum_i \alpha_i|a_i\rangle. \tag{9.86}$$

Now, using Eqs. (9.80) and (9.85)

$$\alpha_i = \langle a_i|\psi\rangle = \sum_n \sum_m \langle E_n|U_{in}\beta_m|E_m\rangle = \sum_n U_{in}\beta_n$$

$$= U_{in}\beta_n$$

or

$$\alpha = U\beta \tag{9.87}$$

where α and β are column matrices

$$\alpha = \begin{pmatrix} \alpha_1 \\ \alpha_2 \\ \cdot \\ \cdot \\ \cdot \end{pmatrix}, \quad \beta = \begin{pmatrix} \beta_1 \\ \beta_2 \\ \cdot \\ \cdot \\ \cdot \end{pmatrix}.$$

9.5   Simple Harmonic Oscillator

The Hamiltonian operator is given by

$$H = \frac{\hat{p}^2}{2m} + \frac{1}{2}m\omega^2\hat{x}^2. \tag{9.88}$$

We want to solve the eigenvalue problem, i.e. to find eigenvalues and eigenvectors of H without introducing any explicit representations for the operator.

Let us introduce

$$a^\dagger = \frac{1}{\sqrt{2m\hbar\omega}}(\hat{p} + im\omega\hat{x}),$$

$$a = \frac{1}{\sqrt{2m\hbar\omega}}(\hat{p} - im\omega\hat{x}). \tag{9.89}$$

Note that the operators a and $a^\dagger$ are not hermitian. Using the basic commutation relation

$$[\hat{x}, \hat{p}] = i\hbar, \tag{9.90}$$

we see that

$$[a, a^\dagger] = \frac{1}{2m\hbar\omega}\{im\omega[\hat{p}, \hat{x}] - im\omega[\hat{x}, \hat{p}]\}$$

$$= \frac{1}{2m\hbar\omega}2m\hbar\omega = 1. \tag{9.91}$$

It is easy to see that the Hamiltonian operator H is related to a and $a^\dagger$ as

$$\hbar\omega a a^\dagger = H + \frac{1}{2}\hbar\omega,$$

$$\hbar\omega a^\dagger a = H - \frac{1}{2}\hbar\omega. \tag{9.92}$$

From Eq. (9.91), we can prove by induction

$$a(a^\dagger)^n - (a^\dagger)^n a = n(a^\dagger)^{n-1}. \tag{9.93}$$

Now

$$[a, H] = aH - Ha$$

$$= \hbar\omega a. \tag{9.94a}$$

Therefore

$$aH = (H + \hbar\omega)a. \tag{9.94b}$$

Similarly we have from Eq. (9.94a), by taking its hermitian conjugate,

$$[a^\dagger, H] = -\hbar\omega a^\dagger. \tag{9.95a}$$

Therefore

$$Ha^\dagger = a^\dagger (H + \hbar\omega). \tag{9.95b}$$

Let $|n\rangle$ denote the normalized eigenstate of H belonging to eigenvalue E_n:

$$H|n\rangle = E_n|n\rangle. \tag{9.96a}$$

Let

$$|F\rangle = a|n\rangle$$

then

$$\langle F| = \langle n|a^\dagger. \tag{9.96b}$$

Now,

$$
\begin{aligned}
\langle F|F\rangle &= \langle n|a^\dagger a|n\rangle \\
&= \sum_m \langle n|a^\dagger|m\rangle\langle m|a|n\rangle \\
&= \sum_m |\langle m|a|n\rangle|^2 \geq 0.
\end{aligned}
\tag{9.97}
$$

Therefore

$$\langle n|a^\dagger a|n\rangle \geq 0. \tag{9.98a}$$

Similarly we can show that

$$\langle n|aa^\dagger|n\rangle \geq 0. \tag{9.98b}$$

Then from Eq.(9.98a), using Eq. (9.92), we have

$$\langle n|(H - \frac{1}{2}\hbar\omega)|n\rangle \geq 0$$

or

$$\langle n|H|n\rangle \geq \frac{1}{2}\hbar\omega\langle n|n\rangle$$

or

$$E_n\langle n|n\rangle \geq \frac{1}{2}\hbar\omega\langle n|n\rangle$$

i.e.

$$E_n \geq \frac{1}{2}\hbar\omega. \tag{9.99}$$

Hence the eigenvalues of H are positive and have a minimum given by

$$\lambda = \frac{1}{2}\hbar\omega. \tag{9.100}$$

Let $|0\rangle$ denote the eigenstate of H belonging to the least eigenvalue of H which we have denoted by $\lambda = \frac{1}{2}\hbar\omega$. Thus

$$H|0\rangle = \lambda|0\rangle. \tag{9.101}$$

Now

$$Ha|0\rangle = \lambda a|0\rangle. \tag{9.102}$$

But using Eq. (9.94a), we have from Eq. (9.102)

$$Ha|0\rangle = (aH - \hbar\omega a)|0\rangle = (\lambda - \hbar\omega)a|0\rangle, \tag{9.103}$$

showing that unless $a|0\rangle = 0$, $a|0\rangle$ is an eigenstate of H with eigenvalue $\lambda - \hbar\omega < \lambda$, which contradicts that λ is the least eigenvalue. Thus we must have

$$a|0\rangle = 0. \tag{9.104}$$

We now construct a system of vectors

$$|0\rangle, a^\dagger|0\rangle, a^\dagger a^\dagger|0\rangle, \cdots, (a^\dagger)^n|0\rangle. \tag{9.105}$$

From Eq. (9.95b), we have

$$
\begin{aligned}
Ha^\dagger|0\rangle &= a^\dagger H|0\rangle + \hbar\omega a^\dagger|0\rangle \\
&= (\lambda + \hbar\omega)a^\dagger|0\rangle \\
&= \frac{3}{2}\hbar\omega a^\dagger|0\rangle.
\end{aligned} \tag{9.106}
$$

Hence if $a^\dagger|0\rangle \neq 0$, $a^\dagger|0\rangle$ is an eigenstate of H belonging to eigenvalue $\frac{3}{2}\hbar\omega$.
Similarly,

$$
\begin{aligned}
Ha^\dagger a^\dagger|0\rangle &= a^\dagger Ha^\dagger|0\rangle + \hbar\omega a^\dagger a^\dagger|0\rangle \\
&= a^\dagger(l + \hbar\omega)a^\dagger|0\rangle + \hbar\omega a^\dagger a^\dagger|0\rangle \\
&= (\lambda + 2\hbar\omega)a^\dagger a^\dagger|0\rangle \\
&= \frac{5}{2}\hbar\omega a^\dagger a^\dagger|0\rangle.
\end{aligned} \tag{9.107}
$$

Continuing this process we get

$$H(a^\dagger)^n|0\rangle = (n + \frac{1}{2})\hbar\omega(a^\dagger)^n|0\rangle, \tag{9.108}$$

showing that $(a^\dagger)^n|0\rangle$ is an eigenvector of H belonging to eigenvalue

$$E_n = (n + \frac{1}{2})\hbar\omega. \tag{9.109}$$

Consider [use Eqs.(9.93) and (9.104)]

$$\langle 0|a^n(a^\dagger)^n|0\rangle = \langle 0|a^{n-1}a(a^\dagger)^n|0\rangle$$
$$= \langle 0|a^{n-1}(n(a^\dagger)^{n-1} + (a^\dagger)^n a)|0\rangle$$
$$= n\langle 0|a^{n-1}(a^\dagger)^{n-1}|0\rangle. \qquad (9.110)$$

Repeating this process we find

$$\langle 0|a^n(a^\dagger)^n|0\rangle = n(n-1)\cdots 1\langle 0|0\rangle$$
$$= n! \qquad (9.111)$$

Therefore the eigenstate $(a^\dagger)^n|0\rangle$ is not normalised, but $\frac{1}{\sqrt{n!}}(a^\dagger)^n|0\rangle$ is normalised. We denote this normalised eigenstate of H belonging to eigenvalue $E_n = (n + \frac{1}{2})\hbar\omega$ by

$$|n\rangle = \frac{1}{\sqrt{n!}}(a^\dagger)^n|0\rangle. \qquad (9.112)$$

Then

$$a^\dagger|n\rangle = \frac{1}{\sqrt{n!}}(a^\dagger)^{n+1}|0\rangle$$
$$= \frac{1}{\sqrt{n!}}\sqrt{(n+1)!}\frac{1}{\sqrt{(n+1)!}}(a^\dagger)^{n+1}|0\rangle$$
$$= \sqrt{n+1}|n+1\rangle, \qquad (9.113)$$

where $|n+1\rangle$ is a normalised eigenstate of H belonging to the eigenvalue

$$E_{n+1} = (n + 1 + \frac{1}{2}\hbar\omega) = E_n + \hbar\omega. \qquad (9.114a)$$

Similarly,

$$a|n\rangle = \frac{1}{\sqrt{n!}}a(a^\dagger)^n|0\rangle$$
$$= \frac{1}{\sqrt{n!}}\{n(a^\dagger)^{n-1} + (a^\dagger)^n a\}|0\rangle$$
$$= \frac{n}{\sqrt{n!}}(a^\dagger)^{n-1}|0\rangle$$
$$= \frac{n}{\sqrt{n!}}\sqrt{(n-1)}\frac{1}{\sqrt{(n-1)!}}(a^\dagger)^{n-1}|0\rangle$$
$$= \sqrt{n}|n-1\rangle, \qquad (9.114b)$$

where $|n-1\rangle$ is a normalised eigenstate H with eigenvalue

$$E_{n-1} = (n - 1 + \frac{1}{2})\hbar\omega = E_n - \hbar\omega. \qquad (9.115)$$

Thus, we see that a and $a^\dagger$ are the operators which respectively annihilate and create energy in the system in units of $\hbar\omega$.

We also see that

$$\langle m|a^\dagger|n\rangle = \sqrt{n+1}\langle m|n+1\rangle$$
$$= \sqrt{n+1}\delta_{m,n+1}, \qquad (9.116a)$$
$$\langle m|a|n\rangle = \sqrt{n}\langle m|n-1\rangle$$
$$= \sqrt{n}\delta_{m,n-1}. \qquad (9.116b)$$

The above expressions give the matrix elements of $a^\dagger$ and a in the basis defined by eigenvectors of the Hamiltonian of a simple harmonic oscillator. From Eqs. (9.116a, 9.116b) one can also determine the matrix elements of x and p by using Eq. (9.89).

9.6   Supersymmetric Oscillator

The simplest example of a supersymmetric system is provided by simple harmonic oscillator for which the Hamiltonian is [$\hbar = 1, m = 1, \omega = 1$]

$$H_B = \frac{1}{2}(p^2 + q^2)$$
$$= \frac{1}{\sqrt{2}}(q+ip)(q-ip) - \frac{1}{2}[q,p]$$
$$= a^\dagger a + \frac{1}{2}. \qquad (9.117)$$

where $a^\dagger$ and a are creation and annihilation operators which satisfy commutation relation $[a, a^\dagger] = 1$. The eigenstates of H_B are $|0\rangle_B, |1\rangle_B, |2\rangle_B, ...$

$$a|0\rangle_B = 0,$$
$$|n\rangle_B \propto (a^\dagger)^n|0\rangle_B,$$
$$H_B = (n+\frac{1}{2})|n\rangle_B. \qquad (9.118)$$

We now introduce the fermionic oscillator, which is defined by the operators d and $d^\dagger$ satisfying anti-commutation relations $\{d, d^\dagger\} = 1$ [see problem 9.6]. The Hamiltonian is $H_F = d^\dagger d - \frac{1}{2}$ and eigenstates of H_F are $|0\rangle_F$ and $d^\dagger|0\rangle_F$ so that

$$d|0\rangle_F = 0. \qquad (9.119)$$

For the combined system, the total Hamiltonian is

$$H = H_B + H_F$$
$$= a^\dagger a + d^\dagger d \qquad (9.120)$$

and eigenstates are

$$|0\rangle, \quad |1\rangle = a^\dagger |0\rangle, \quad |2\rangle = (a^\dagger)^2 |0\rangle ...$$
$$|1\rangle = d^\dagger |0\rangle, \quad |2\rangle = d^\dagger |1\rangle ... \tag{9.121}$$

Note the degeneracy of states in supersymmetric harmonic oscillator. The degeneracy in energy indicates that there is a symmetry.

Define the operators $Q = a^\dagger d, Q^\dagger = a d^\dagger$. Q is no longer a pure bosonic or fermionic object, and Q and $Q\dagger$ provide the simplest supersymmetric algebra

$$\{Q, Q\} = \{Q^\dagger, Q^\dagger\} = 1,$$
$$\{Q, Q^\dagger\} = H,$$
$$[Q, H] = [Q^\dagger, H] = 0. \tag{9.122}$$

These relations, which can be easily derived, explain degeneracy mentioned above. The above considerations show that any enlargement of the space-time group requires generators which anti-commute.

Finally since H_B is a diagonal matrix in the basis $|n\rangle_B$ with eigenvalues $n_B = 0, 1, 2, ...$, the statistical Boltzmann factor is given by

$$\text{Tr}_{H_B} e^{-\beta H_B} = 1 + e^{-\beta} + e^{-2\beta} + ...$$
$$= \frac{1}{1 - e^{-\beta}} \quad \text{(Bose–Einstein)}.$$

Similarly for H_F in the eigenvalues $n_F = 0, 1$,

$$\text{Tr}_{H_F} e^{-\beta H_F} = 1 + e^{-\beta} \quad \text{(Fermi–Dirac)}$$

and [taking into account degeneracy implied in Eq.(9.121)]

$$\text{Tr}_H e^{-\beta H} = 1 + 2e^{-\beta} + 2e^{-2\beta} + 2e^{-3\beta} + ...$$
$$= (1 + e^{-\beta})(1 - e^{-\beta})^{-1}$$
$$= \frac{1 + e^{-\beta}}{1 - e^{-\beta}} \quad \text{(Supersymmetric)}.$$

9.7   Problems

9.1 $\hat{A}$ and $\hat{B}$ are two arbitrary hermitian operators. Which of the following operators
 (i) $\hat{A}\hat{B}$, (ii) $\hat{A}^2$, (iii) $\hat{A}\hat{B} - \hat{B}\hat{A}$, (iv) $\hat{A}\hat{B} + \hat{B}\hat{A}$, (v) $\hat{A}\hat{B}\hat{A}$
(a) are hermitian,
(b) have real non-negative expectation values,

(c) have pure imaginary expectation values,

(d) are purely numerical operators?

9.2 If an operator $\hat{A}$ has the property that

$$\hat{A}^4 = 1,$$

what are its eigenvalues?

9.3 The Hamiltonian

$$H = \frac{\hat{p}^2}{2m} + V(\hat{x})$$

has a set of eigenstates $|n\rangle$ with energy eigenvalues E_n. The lowest eigenstate $|0\rangle$ has the energy E_0. Show that

$$\sum_n (E_n - E_0)|\langle n|x|0\rangle|^2 = \frac{\hbar^2}{2m}.$$

Hint:

$$[\hat{x}, \hat{p}] = i\hbar,$$
$$[\hat{x}, [H, \hat{x}]] = \frac{\hbar^2}{m}.$$

Verify the above sum rule for a simple harmonic oscillator.

9.4 In the momentum representation, show that the position operator $\hat{x}$ is represented by

$$\langle p|\hat{x}|p'\rangle = i\hbar \frac{\partial}{\partial p} \langle p|p'\rangle$$

$$\langle p|\hat{x}|\psi\rangle = i\hbar \frac{\partial}{\partial p} \langle p|\psi\rangle$$

that is

$$\hat{x} = i\hbar \frac{\partial}{\partial p}.$$

9.5 If $|n\rangle$ denotes a normalised eigenstate of a simple harmonic oscillator, belonging to the eigenvalue $E_n = (n + \frac{1}{2})\hbar\omega$, show that

$$\langle n|\hat{x}|m\rangle = \begin{cases} \sqrt{\hbar/m\omega}(\frac{n+1}{2})^{1/2}, & m = n+1 \\ \sqrt{\hbar/m\omega}(\frac{n}{2})^{1/2}, & m = n-1 \\ 0 & \text{otherwise.} \end{cases} \qquad (9.123)$$

9.6 If an operator $\hat{A}$ has the following properties

$$\hat{A}^2 = 0,$$

$$\hat{A}\hat{A}^\dagger + \hat{A}^\dagger\hat{A} = 1,$$

show that

(i) $N = \hat{A}^\dagger \hat{A}$ is hermitian.

(ii) $N^2 = N$. Hence show that N has eigenvalues 0 and 1.

(iii) $[N, \hat{A}^\dagger] = \hat{A}^\dagger$.

(iv) Let $|0\rangle$ and $|1\rangle$ be eigenvectors of N belonging to eigenvalues 0 and 1. Show that

$$\hat{A}^\dagger |0\rangle = 1,$$
$$\hat{A}^\dagger |1\rangle = 0.$$

9.7 Let $\hat{B}$ and $\hat{C}$ be two anticommutating operators

$$\hat{B}\hat{C} + \hat{C}\hat{B} = 0.$$

Let $|\psi\rangle$ be a simultaneous eigenstate of both $\hat{B}$ and $\hat{C}$. What can be said about the corresponding eigenvalues? If $\hat{C}^2 = 1$ holds, what does it imply for the eigenvalue of $\hat{B}$?

9.8 Show that for a simple harmonic oscillator

$$\langle |\hat{x}^2|0\rangle = \begin{cases} \hbar/2\sqrt{2}m\omega, & \text{for } n = 2 \\ 0, & \text{for } n \neq 2 \end{cases} \tag{9.124}$$

Further show that

$$\langle 0|\hat{x}^2|0\rangle = \frac{\hbar}{2m\omega}.$$

9.9 In the three dimensional vector space with an orthonormal set of basis vector $\{|1\rangle, |2\rangle, |3\rangle\}$,

(a) Find the matrix representation of the following operators

$$|1\rangle\langle 1| \ , \quad |2\rangle\langle 2|, \quad |1\rangle\langle 2| - |2\rangle\langle 1|, \quad |1\rangle\langle 3| - |2\rangle\langle 3| + |2\rangle\langle 1|$$
$$2|1\rangle\langle 1| - \frac{1}{\sqrt{2}}|3\rangle\langle 2| + i|2\rangle\langle 2|.$$

(b) Which of the above operators are hermitian?

(c) Let $\hat{A} = i|1\rangle\langle 2| - i|2\rangle\langle 1| + |3\rangle\langle 3|$. Show that $\hat{A}$ is hermitian and find the eingevalues and eigenstates of $\hat{A}$.

(d) Show that eigenstates of $\hat{A}$ are orthogonal.

9.10 For the Hamiltonian

$$H = \frac{\hat{p}^2}{2m} + V(r),$$

$$[\hat{p}_i, H] = -i\hbar \frac{\partial}{\partial x_i} V(r).$$

Using the above result, show that

$$[\hat{p}_j, [\hat{p}_i, H]] = (-i\hbar)^2 \frac{\partial}{\partial x_j} \frac{\partial}{\partial x_i} V(r).$$

From the relation:

$$\langle m | [\hat{p}_j, [\hat{p}_i, H]] | m \rangle = (-i\hbar)^2 \langle m | \left(\frac{\partial}{\partial x_j} \frac{\partial}{\partial x_i} V(r) \right) | m \rangle,$$

derive the sum rule

$$\sum_n (E_m - E_n) \langle m | \hat{p}_i | n \rangle \langle n | \hat{p}_i | m \rangle = -\frac{\hbar^2}{2} \langle m | \nabla^2 V | m \rangle,$$

then show that

$$\sum_n (E_n - E_m) |(\vec{p})_{mn}|^2 = \frac{\hbar^2}{2} \int |\psi_m(\vec{x})|^2 \nabla^2 V d^3 x$$

$$= \frac{\hbar^2}{2} \quad 4\pi e^2 Z |\psi_m(0)|^2$$

where for the Coulomb potential

$$\nabla^2 V = 4\pi e^2 Z \delta^3(\vec{x}).$$

9.11 Show that 3×3 matrices

$$(G_i)_{jk} = -i\hbar \epsilon_{ijk}$$

satisfy the relation

$$(G_i G_j - G_j G_i)_{mk} = -i\hbar \epsilon_{ijn} (G_n)_{mk}$$

i.e.

$$[G_i, G_j] = i\hbar \epsilon_{ijk} G_n,$$

showing that matrices G's satisfy the commutation relation of angular momentum and as such gives the adjoint representation of group O_3 and represent spin 1. Write these matrices explicitly:

$$G_1 = \hbar \begin{pmatrix} 0 & 0 & 0 \\ 0 & 0 & -i \\ 0 & -i & 0 \end{pmatrix}, \quad G_2 = \hbar \begin{pmatrix} 0 & 0 & i \\ 0 & 0 & 0 \\ -i & 0 & 0 \end{pmatrix},$$

$$G_3 = \hbar \begin{pmatrix} 0 & -i & 0 \\ i & 0 & 0 \\ 0 & 0 & 0 \end{pmatrix}.$$

Show that the eigenvalues of G_3 are $\hbar, 0, -\hbar$.

In the basis in which G_3 is diagonal the corresponding eigenvectors $|x\rangle$ are:

$$|1\rangle = \begin{pmatrix} 1 \\ 0 \\ 0 \end{pmatrix}, |0\rangle = \begin{pmatrix} 0 \\ 1 \\ 0 \end{pmatrix}, |-1\rangle = \begin{pmatrix} 0 \\ 0 \\ 1 \end{pmatrix}.$$

Show that the eigenvectors $|y\rangle$ in the basis in which G_3 is not diagonal are:

$$|G_3 = +1\rangle = \frac{1}{\sqrt{2}} \begin{pmatrix} -1 \\ -i \\ 0 \end{pmatrix}, |G_3 = 0\rangle = \begin{pmatrix} 0 \\ 0 \\ 1 \end{pmatrix},$$

$$|G_3 = -1\rangle = \frac{1}{\sqrt{2}} \begin{pmatrix} -1 \\ i \\ 0 \end{pmatrix}.$$

Now the matrix which connect the two bases is

$$U_{mi} = \langle y_m | x_i \rangle.$$

Show that

$$U = \frac{1}{\sqrt{2}} \begin{pmatrix} -1 & i & 0 \\ 0 & 0 & \sqrt{2} \\ -1 & -i & 0 \end{pmatrix}.$$

Chapter 10

Heisenberg Equation of Motion, Invariance Principle and Path Integral

10.1 Introduction

In quantum mechanics, dynamical variables are operators which do not in general commute e.g.,

$$[\hat{x}, \hat{p}] = i\hbar.$$

In the Schrödinger representation $\hat{x} \longrightarrow x$ (i.e. it is treated as an algebraic variable)

$$\hat{p} \to -i\hbar \frac{\partial}{\partial x}. \tag{10.1}$$

Now dimensionally

$$[x][p] = \left(\frac{\text{Action}}{\hbar} \right).$$

Also we see that dimensionally

$$[t][E] = [\hbar].$$

We would expect time t to be always treated as an algebraic variable, thus in analogy with Eq. (10.1) it is plausible to postulate

$$H \to i\hbar \frac{\partial}{\partial t}, \tag{10.2}$$

where t is treated as an algebraic variable and the sign chosen is conventional but turns out to be convenient. The above equation is an operator equation which operates on time dependent state vectors $|\Psi(t)\rangle$. Thus

$$i\hbar \frac{\partial}{\partial t} |\Psi(t)\rangle = H|\Psi(t)\rangle. \tag{10.3}$$

We now show that in the Schrödinger representation it leads to the ordinary Schrödinger equation for the state function

$$\Psi(x, t) = \langle x|\Psi(t)\rangle. \tag{10.4}$$

From Eq. (10.3), we have

$$i\hbar \frac{\partial}{\partial t}\langle x|\Psi(t)\rangle = \langle x|H|\Psi(t)\rangle$$

$$= \int d\acute{x}\langle x|H|\acute{x}\rangle\langle\acute{x}|\Psi(t)\rangle$$

where

$$\langle x|H|\acute{x}\rangle = \delta(x - \acute{x})H(\acute{x}, -i\hbar\frac{\partial}{\partial\acute{x}}) \qquad (10.5)$$

so that we have in the Schrödinger representation

$$i\hbar\frac{\partial}{\partial t}\langle x|\Psi(t)\rangle = H(x, -i\hbar\frac{\partial}{\partial x})\langle x|\Psi(t)\rangle. \qquad (10.6)$$

A particular solution to this equation is

$$\langle x|\Psi(t)\rangle_{E_n} = e^{-iE_n t/\hbar}u_n(x) \qquad (10.7)$$

where $u_n(x)$ is the energy eigenfunction:

$$H(x, -i\hbar\frac{\partial}{\partial x})u_n(x) = E_n u_n(x). \qquad (10.8)$$

We may write

$$u_n(x) = \langle x|E_n\rangle \qquad (10.9)$$

so that from Eq. (10.7)

$$\langle x|\Psi(t)\rangle_{E_n} = e^{-iE_n t/\hbar}\langle x|E_n\rangle. \qquad (10.10)$$

A general solution of Eq. (10.6) is the linear sum of the above solution

$$\langle x|\Psi(t)\rangle = \sum_n e^{-iE_n t/\hbar}\langle x|E_n\rangle a(E_n), \qquad (10.11a)$$

$$\langle x|\Psi(0)\rangle = \sum_n \langle x|E_n\rangle a(E_n). \qquad (10.11b)$$

Let us write $\Psi(0) = \psi$,

$$\langle x|\psi\rangle = \sum_n \langle x|E_n\rangle a(E_n)$$

$$= \sum_n \langle x|E_n\rangle\langle E_n|\psi\rangle \qquad (10.12)$$

so that

$$a(E_n) = \langle E_n|\psi\rangle. \qquad (10.13)$$

Thus the time dependent solution, which satisfies Eq. (10.6) is

$$\langle x|\Psi(t)\rangle = \sum_n e^{-iE_n t/\hbar}\langle x|E_n\rangle\langle E_n|\psi\rangle. \qquad (10.14)$$

10.2   Heisenberg Equation of Motion

Schrödinger's equation of motion is

$$i\hbar\frac{\partial}{\partial t}|\Psi(t)\rangle = H|\Psi(t)\rangle. \tag{10.15}$$

We see that all the time dependence is in the state vector. This is called the Schrödinger picture. In this picture, the operators representing the dynamical variables are regarded as being independent of time. The state vectors represent the observed systems and these change with time, as do the results of observations or measurements. Thus the average result of repeated measurements of an observable $\hat{A}$, made at time t, on a system in the state $|\Psi(t)\rangle$ is

$$\begin{aligned}
\bar{a}_{\Psi(t)} &= \langle\Psi(t)|\hat{A}|\Psi(t)\rangle \\
&= \int dx d\acute{x}\langle\Psi(t)|x\rangle\langle x|\hat{A}|\acute{x}\rangle\langle\acute{x}|\Psi(t)\rangle \\
&= \int dx\langle\Psi(t)|x\rangle\hat{A}(x,\frac{\partial}{\partial x})\langle x|\Psi(t)\rangle \\
&= \int dx\psi^*(x,t)\hat{A}(x,\frac{\partial}{\partial x})\psi(x,t). \tag{10.16}
\end{aligned}$$

This is of course a function of time.

Now in classical theory we have no concept of state, and we deal with dynamical variables which change with time. We now want to go to a different picture of motion, in which states correspond to state vectors which are independent of time and dynamical variables correspond to time dependent linear operators. We shall see that this new picture has a classical analogue and is more in the spirit of classical description.

The trick to go from the Schrödinger picture to the new picture is simple. We note that Eq (10.15) has the formal solution

$$\begin{aligned}
|\Psi(t)\rangle &= e^{-iHt/\hbar}|\Psi(0)\rangle \\
&= e^{-iHt/\hbar}|\psi\rangle. \tag{10.17}
\end{aligned}$$

Note that the Hamiltonian H is a hermitian operator. Thus

$$\begin{aligned}
\bar{a}_{\Psi(t)} &= \langle\Psi(t)|\hat{A}|\Psi(t)\rangle \\
&= \langle\psi|e^{iHt/\hbar}\hat{A}e^{-iHt/\hbar}|\psi\rangle. \tag{10.18}
\end{aligned}$$

Let us now define the time dependent linear operator

$$\hat{A}(t) = e^{iHt/\hbar}\hat{A}e^{-iHt/\hbar}. \tag{10.19}$$

Then

$$\bar{a}_{\Psi(t)} = \langle\psi|\hat{A}(t)|\psi\rangle = \bar{a}(t)_\psi. \tag{10.20}$$

The quantity on the right-hand side represents the average value of the time dependent operator, for a state $|\psi\rangle$ specified at $t = 0$. This picture of motion is quite different from the Schrödinger picture. Here the operator $A(t)$ represents the making of an observation at time t on a state which is specified at $t = 0$. This new picture is called the Heisenberg picture. This is much closer to the classical description, and one may expect the time dependent operators $A(t)$ to be rather closely related to the corresponding time-dependent classical dynamical variables.

Let us differentiate Eq. (10.19) with respect to t. For this purpose we rewrite Eq. (10.19) as

$$e^{-iHt/\hbar}\hat{A}(t) = \hat{A}e^{-iHt/\hbar}.$$

Thus

$$-\frac{iH}{\hbar}e^{-iHt/\hbar}\hat{A}(t) + e^{-iHt/\hbar}\frac{d\hat{A}(t)}{dt}$$

$$= \hat{A}\frac{-i}{\hbar}He^{-iHt/\hbar}.$$

Multiply both sides on the left by $i\hbar e^{iHt/\hbar}$

$$e^{iHt/\hbar}He^{-iHt/\hbar}\hat{A}(t) + i\hbar\frac{d\hat{A}(t)}{dt}$$

$$= e^{iHt/\hbar}\hat{A}He^{-iHt/\hbar}$$

$$= e^{iHt/\hbar}\hat{A}e^{-iHt/\hbar}e^{iHt/\hbar}He^{-iHt/\hbar}.$$

But

$$e^{iHt/\hbar}He^{-iHt/\hbar} = H(t),$$
$$e^{iHt/\hbar}\hat{A}e^{-iHt/\hbar} = \hat{A}(t). \tag{10.21}$$

Therefore

$$i\hbar\frac{d\hat{A}(t)}{dt} = \hat{A}(t)H(t) - H(t)\hat{A}(t)$$

$$= [\hat{A}(t), H(t)]. \tag{10.22}$$

In particular for $\hat{A}(t) = H(t)$. Then

$$i\hbar\frac{dH(t)}{dt} = [H(t), H(t)] = 0 \tag{10.23}$$

thus

$$H(t) = H(0) = H. \tag{10.24}$$

i.e. H is constant of motion.

Thus we can write

$$i\hbar\frac{d\hat{A}(t)}{dt} = \left[\hat{A}(t), H\right]. \tag{10.25}$$

This is known as the Heisenberg Equation of Motion.

Any observable which commutes with the Hamiltonian is a constant of motion, since then

$$i\hbar\frac{d\hat{F}(t)}{dt} = [\hat{F}, H] = 0.$$

This implies that

$$\frac{d}{dt}\langle\psi|\hat{F}(t)|\psi\rangle = \langle\psi|\frac{d\hat{F}}{dt}|\psi\rangle = 0$$

i.e. the average value of an observable which commutes with the Hamiltonian does not change with time. In particular if the system is in an eigenstate of $\hat{F}$ at $t = 0$, the state will be an eigenstate at any subsequent time.

To sum up

	S-picture	H-picture
States represented by vectors	Moving or time dependent	fixed or independent of time
	$i\hbar\frac{d}{dt}|\Psi(t)\rangle_S$	$i\hbar\frac{d}{dt}|\Psi\rangle_H = 0$
	$= H|\Psi(t)\rangle_S$	$|\psi\rangle_H = |\Psi(0)\rangle_S$
Dynamical	$i\hbar\frac{d\hat{A}}{dt} = 0$	$i\hbar\frac{d\hat{A}(t)}{dt}$
variables are	$\hat{A} = \hat{A}(0)$	$= \left[\hat{A}(t), H\right]$
represented by		
linear operators		

The Heisenberg equation of motion has a classical analogue; in particular it is analogous to classical equations of motion in the Hamiltonian form.

One way to make the transition from classical mechanics to quantum mechanics by correspondence principle is discussed in Ch. 2. An alternate

way to make this transition is as follows. Consider a dynamical variable
$F(p_i, q_i, t)$, which is a function of generalized coordinates q_i and conjugate
momenta p_i:

$$\frac{dF}{dt} = \frac{\partial F}{\partial t} + \sum_i \left(\frac{\partial F}{\partial q_i} \dot{q}_i + \frac{\partial F}{\partial p_i} \dot{p}_i \right). \tag{10.26}$$

The Hamiltonian:

$$H = T + V = \sum_i p_i \dot{q}_i - L, \quad L = T - V. \tag{10.27}$$

Hamilton's canonical equations of motion are

$$\frac{dq_i}{dt} \equiv \dot{q}_i = \frac{\partial H}{\partial p_i}, \quad \frac{dp_i}{dt} \equiv \dot{p}_i = -\frac{\partial H}{\partial q}. \tag{10.28}$$

Then (we will confine ourselves to the case when F does not depend explicitly on time)

$$\frac{dF}{dt} = \sum_i \left(\frac{\partial f}{\partial q_i} \frac{\partial H}{\partial p_i} - \frac{\partial f}{\partial p_i} \frac{\partial f}{\partial q_i} \right)$$

$$= (F, H)_{PB}. \tag{10.29}$$

We note that for

$$F = H, \quad \frac{dH}{dt} = 0$$

i.e. the energy is conserved. (F, H) in Eq. (10.29) is the classical Poisson Bracket (PB). In general we define the Poisson bracket between two dynamical variables A and B:

$$(A, B)_{PB} = \sum_i \left(\frac{\partial A}{\partial q_i} \frac{\partial B}{\partial p_i} - \frac{\partial A}{\partial p_i} \frac{\partial B}{\partial q_i} \right) \tag{10.30}$$

$$(A, B)_{PB} = -(B, A)_{PB}.$$

We note that the Eq. (20.32) is analogous to Heisenberg's equation of
motion (10.25) if the classical PB is replaced by the quantum PB defined
by (Dirac)

$$(\hat{A}(t), H) = \frac{1}{\iota \hbar} [\hat{A}(t), H] \tag{10.31}$$

and the dynamical variable $\hat{A}(t)$ and the Hamiltonian are regarded as operators.

Thus the rule to go from classical mechanics to quantum mechanics is to treat dynamical variables A and B as hermitian operators and replace the classical poisson bracket by the quantum commutator:

$$(A, B)_{PB} \rightarrow \frac{1}{i\hbar}[\hat{A}, \hat{B}]. \tag{10.32}$$

Let us apply this rule to canonical coordinates q_i, p_i. Then from Eq.(20.33):

$$(q_i(t), q_j(t))_{PB} = 0 = (p_i(t), p_j(t))_{PB},$$

$$(q_i(t), p_j(t))_{PB} = \sum_r \left(\frac{\partial q_i}{\partial q_r} \frac{\partial p_j}{\partial p_r} - \frac{\partial q_i}{\partial p_r} \frac{\partial p_j}{\partial q_r} \right) = \sum_r (\delta_{ir} \delta_{jr}) = \delta_{ij}. \tag{10.33}$$

Thus the rule (20.33) gives

$$[\hat{q}_i(t), \hat{q}_j(t)] = 0 = [\hat{p}_i(t), \hat{p}_j(t)],$$

$$[\hat{q}_i(t), \hat{p}_j(t)] = i\hbar \delta_{ij}. \tag{10.34}$$

These are known as canonical quantization conditions. Note the important fact that these hold for equal time [see prob. 10.1]

10.3   Free Particle Propagator

As an application of the Heisenberg picture, we calculate correlation function

$$K(t_2, t_1) = \langle q_2(t_2) | q_1(t_1) \rangle_H. \tag{10.35}$$

First we have to construct the states $|q, t\rangle$ and $|p, t\rangle$ which are "instantaneous" eigenstates of the Heisenberg picture operators $\hat{q}(t)$ and $\hat{p}(t)$:

$$\hat{q}|q\rangle = q|q\rangle$$
$$e^{\frac{i}{\hbar}Ht} \hat{q}|q\rangle = q e^{\frac{i}{\hbar}Ht}|q\rangle$$
$$e^{\frac{i}{\hbar}Ht} \hat{q} e^{-\frac{i}{\hbar}Ht} e^{\frac{i}{\hbar}Ht}|q\rangle = q e^{\frac{i}{\hbar}Ht}|q\rangle$$
$$\hat{q}(t)|q(t)\rangle = q|q(t)\rangle, \tag{10.36}$$

where $|q(t)\rangle = e^{\frac{i}{\hbar}Ht}|q\rangle$. Now using the completeness relation,

$$\int dq |q\rangle \langle q| = 1,$$

we get

$$\int dq e^{\frac{t}{\hbar} H t} |q\rangle \langle q| e^{-\frac{t}{\hbar} H t} = e^{\frac{t}{\hbar} H t} e^{-\frac{t}{\hbar} H t}$$
$$= 1$$

i.e.

$$\int dq |q(t)\rangle \langle q(t)| = 1. \tag{10.37}$$

Further

$$\langle \psi, -t | q, t \rangle = \langle \psi | e^{\frac{t}{\hbar} H(-t)} e^{\frac{t}{\hbar} H t} | q \rangle = \langle \psi | q \rangle = \langle q | \psi \rangle^*$$
$$= \psi^*(q)$$

while

$$\langle q, t | \psi \rangle = \langle q | e^{-\frac{t}{\hbar} H t} | \psi \rangle$$
$$= \langle q | \psi(t) \rangle = \psi(q, t). \tag{10.38}$$

Thus

$$K(t_2, t_1) = \langle q_2(t_2) | q_1(t_1) \rangle_H$$
$$= \langle q_2(t_2) | e^{\frac{t}{\hbar} H t_1} | q_1 \rangle.$$

Now for a free particle, $H = \dfrac{\hat{p}^2}{2m}$ and using $\hat{p}|p\rangle = p|p\rangle$, we get

$$e^{\frac{t}{\hbar} \frac{\hat{p}^2}{2m} t_1} |p\rangle = e^{\frac{t}{\hbar} \frac{p^2}{2m} t_1} |p\rangle.$$

Then

$$K(t_2, t_1) = \int dp \langle q_2(t_2) | e^{\frac{t}{\hbar} H t_1} | p \rangle \langle p | q_1 \rangle$$
$$= \int dp e^{\frac{t}{\hbar} \frac{p^2}{2m} t_1} \langle q_2(t_2) | p \rangle \langle p | q_1 \rangle.$$

Further using

$$\langle q_2(t_2) | p \rangle = \langle q_2 | e^{-\frac{t}{\hbar} H t_2} | p \rangle$$
$$= \langle q_2 | e^{-\frac{t}{\hbar} \frac{\hat{p}^2}{2m} t_2} | p \rangle = e^{-\frac{i}{\hbar} \frac{p^2}{2m} t_2} \langle q_2 | p \rangle,$$

we get

$$K(t_2, t_1) = \int dp\, e^{-\frac{i}{\hbar}\frac{p^2}{2m}(t_2-t_1)} \langle q_2|p\rangle\langle p|q_1\rangle$$

$$= \frac{1}{2\pi\hbar} \int dp\, e^{-\frac{i}{\hbar}\frac{p^2}{2m}t} e^{\frac{i}{\hbar}pq}, \tag{10.39}$$

where we have used $\langle q_2|p\rangle = \frac{1}{\sqrt{2\pi\hbar}} e^{\frac{i}{\hbar}pq_2}$, $\langle p|q_1\rangle = \frac{1}{\sqrt{2\pi\hbar}} e^{-\frac{i}{\hbar}pq_1}$, and $t = t_2 - t_1$, $q = q_2 - q_1$. Thus

$$K(t) = \frac{1}{2\pi\hbar} \int dp\, e^{-\frac{i}{\hbar}\frac{t}{2m}(p-\frac{m}{t}q)^2} e^{+\frac{i}{\hbar}\frac{t}{2m}\frac{m^2}{t^2}q^2}$$

$$= \frac{1}{2\pi\hbar}\left(\frac{2\pi m\hbar}{it}\right)^{\frac{1}{2}} e^{\frac{i}{\hbar}\frac{m}{2}\frac{q^2}{t}}$$

where we have used the Gaussian Integral

$$\int dx\, e^{-a\frac{x^2}{2}} = \sqrt{\frac{2\pi}{a}}.$$

Thus finally

$$K(t,q) = \left(\frac{m}{2\pi\hbar it}\right)^{\frac{1}{2}} e^{-\frac{m}{2i\hbar t}q^2} \tag{10.40}$$

which is the free particle propagator, Green's function for the Schrödinger operator $\left(-\frac{\hbar}{2m}\frac{\partial^2}{\partial q^2} - i\hbar\frac{\partial}{\partial t}\right)$.

10.4 Unitary Transformation

The transformation (10.17) or (10.19) viz.

$$|\Psi(t)\rangle = e^{\frac{-iHt}{\hbar}}|\psi\rangle$$

$$= U^\dagger|\psi\rangle \tag{10.41}$$

$$\hat{A}(t) = U\hat{A}U^\dagger \tag{10.42}$$

where

$$U = e^{\frac{iHt}{\hbar}} \tag{10.43a}$$

$$UU^\dagger = 1 = U^\dagger U \tag{10.43b}$$

is an example of unitary transformation. Here H is the generator of the unitary transformation:

$$U = e^{\frac{iHt}{\hbar}}.$$

Under the above unitary transformation, we go from one physical picture (Schrödinger) to another physically equivalent picture (Heisenberg).

In general a unitary transformation can be written as

$$U = e^{i\epsilon \hat{F}} \tag{10.44}$$

where ϵ is real and $\hat{F}$ is hermitian so that

$$U^\dagger = e^{-i\epsilon \hat{F}^\dagger}$$
$$= e^{-i\epsilon \hat{F}} = U^{-1}.$$

Consider a state represented by the state vector $|\psi\rangle$. We can form a new state by making a unitary transformation.

$$|\psi^u\rangle = U|\psi\rangle,$$
$$\langle \psi^u| = \langle \psi|U^\dagger. \tag{10.45}$$

If $|\psi\rangle$ is normalised so that

$$\langle \psi|\psi\rangle = 1$$

then

$$\langle \psi^u|\psi^u\rangle = \langle \psi|U^\dagger U \psi\rangle$$
$$= \langle \psi|\psi\rangle \tag{10.46}$$

so that transformed state $|\psi^u\rangle$ is also normalised.

If the $|a_n\rangle$'s form a complete set of states so that

$$|\psi\rangle = \sum_n C_n|a_n\rangle$$
$$= \sum_n |a_n\rangle\langle a_n|\psi\rangle, \tag{10.47a}$$

then

$$|\psi^u\rangle = \sum_n C_n U|a_n\rangle$$
$$= \sum_n C_n|a_n^u\rangle \tag{10.47b}$$

i.e.

$$|a_n^u\rangle = U|a_n\rangle \tag{10.48}$$

also form a complete set.

The unitary operators enable us to make transformations from one description of a system to another physically equivalent description. Under unitary transformation we define the transformation law on an operator as

$$A^u = UAU^\dagger. \tag{10.49}$$

Equations (10.45) and (10.49) define unitary transformations.

Theorem 1: under a unitary transformation eigenvalues of an operator remain unchanged.

Consider the eigenvalue equation

$$A|a_n\rangle = a_n|a_n\rangle. \tag{10.50}$$

Then

$$UAU^\dagger U|a_n\rangle = a_n U|a_n\rangle$$
$$A^u|a_n^u\rangle = a_n|a_n^u\rangle$$

proving the theorem.

Theorem 2: The average value of a large number of measurements of $\hat{A}$ on a system in state $|\psi\rangle$ remains unchanged under a unitary transformation:

$$\begin{aligned}
\bar{a}_\psi &= \langle\psi|\hat{A}|\psi\rangle \\
&= \langle\psi|U^\dagger U\hat{A}U^\dagger U|\psi\rangle \\
&= \langle\psi^u|\hat{A}^u|\psi^u\rangle \\
&= \bar{a}_{\psi^u}. \tag{10.51}
\end{aligned}$$

Consequences:

A unitary transformation does not change the physics; it is just a transformation from one description to another physically equivalent one.

10.5 Invariance Principles, Conservation Laws

In quantum mechanics, a transformation is associated with a unitary operator:

$$U = e^{i\epsilon\hat{F}}, \quad \hat{F}^\dagger = \hat{F}. \tag{10.52}$$

$\hat{F}$ is called the generator of transformation.

Consider the matrix element $\langle f|H|i\rangle$ or a transition from an initial state $|i\rangle$ to a final state $|f\rangle$ by an operator S viz. the matrix elements $\langle f|S|i\rangle$.

Now

$$\begin{aligned}
\langle f|S|i\rangle &= \langle f|U^\dagger USU^\dagger U|i\rangle \\
&= \langle f^\mu|USU^\dagger|i^\mu\rangle \tag{10.53}
\end{aligned}$$

which is an identity. However if

$$USU^\dagger = S \qquad (10.54)$$

then

$$\langle f|S|i \rangle = \langle f^\mu|S|i^\mu \rangle. \qquad (10.55)$$

Hence invariance of S, viz. Eq. (10.54) implies that the result of an experiment remains unchanged in a situation where the states are transformed. For example, for a closed system, due to homogeneity of space, there is no preferred origin and we have the invariance under unitary transformation corresponding to translation in time and space implying that the result of an experiment will be same now, in the past and future; here and elsewhere. Thus if we examine the light emitted by a quaser, we find that the atoms were behaving (i.e. obeying the same laws) billions of light years away as they are now. Similarly for a closed system there is no preferred direction due to isotropy of space and hence we have rotational invariance. As a result the generators of these transformations viz. energy, momentum and angular momentum (see next section) are conserved for a closed system. To see this, from Eq. (10.54):

$$USU^\dagger U = SU$$

or

$$US = SU; \quad [U,S] = 0 = [U,H] \qquad (10.56)$$

i.e. S-matrix and H commutes with U. Under infinitesimal transformation

$$U = 1 + i\epsilon\hat{F}, \qquad (10.57)$$

and Eq. (10.54) gives

$$S + i\epsilon[\hat{F},S] + O(\epsilon^2) = S$$

i.e.

$$[\hat{F},S] = 0. \qquad (10.58)$$

Hence the invariance under transformation U means, the generator of transformation $\hat{F}$ commutes with S-matrix or Hamiltonian.

$$\hat{F}|F_i \rangle = F_i|F_i \rangle$$
$$\hat{F}|F_f \rangle = F_f|F_f \rangle$$
$$\langle F_f|[S,\hat{F}]|F_i \rangle = \langle F_f|S\hat{F} - \hat{F}S|F_i \rangle$$
$$(F_i - F_f)\langle F_f|S|F_i \rangle = 0. \qquad (10.59)$$

Hence, from Eqs. (10.58) and (10.59)

$$F_i = F_f, \quad if \quad \langle F_f|S|F_i \rangle \neq 0. \tag{10.60}$$

i.e. eigenvalues of generator $\hat{F}$ are conserved.

Examples

1- Under translation in time

$$t \longrightarrow t + \tau$$

$$|\psi(t - \tau)\rangle = e^{-iH(t-\tau)}|\psi\rangle = e^{\frac{iH\tau}{\hbar}}|\psi(t)\rangle$$
$$|\psi(t + \tau)\rangle = e^{\frac{-iH\tau}{\hbar}}|\psi(t)\rangle \tag{10.61}$$

and

$$A(t + \tau) = e^{\frac{iH\tau}{\hbar}} \hat{A}(t) e^{\frac{-iH\tau}{\hbar}}. \tag{10.62}$$

Thus the unitary operator corresponding to the translation in time $t \to t + \tau$ is

$$U = e^{\frac{iH\tau}{\hbar}}.$$

Invariance under translation in time implies the generator of translation in time viz. H is conserved.

2- Translation in space: [for simplicity, consider one dimensional case; then generalize to 3-dimensions]:

$$x \longrightarrow x + a.$$

Associated with translation in space is a unitary operator U_T:

$$U_T|x\rangle = |x - a\rangle$$
$$U_T^\dagger|x\rangle = U_T^{-1}|x\rangle = |x + a\rangle, \tag{10.63}$$
$$U_T|\psi\rangle = |\psi^T\rangle$$
$$\psi^T(x) = \langle x|\psi^T\rangle, \tag{10.64}$$
$$\text{where } \langle x|U_T|\psi\rangle = \langle x + a|\psi\rangle$$
$$= \psi(x + a). \tag{10.65}$$

Now by Taylor expansion, for infinitesimal a:

$$\psi^T(x) = \psi(x + a)$$
$$= \psi(x) + a\frac{\partial}{\partial x}\psi(x) + \mathcal{O}(a^2)$$
$$= \left(1 + a\frac{\partial}{\partial x} + ...\right)\psi(x)$$

i.e.

$$\langle x|U_T|\psi\rangle = \langle x|\psi^T\rangle = \langle x|1 + a\frac{\partial}{\partial x} + ...|\psi\rangle. \qquad (10.66)$$

But $-i\hbar\frac{\partial}{\partial x} = \hat{p}$, the momentum operator. Thus

$$U_T = e^{\frac{ia\hat{p}}{\hbar}}$$

i.e. the momentum operator $\hat{p}$ is the generator of translation in space [For an alternate approach see Prob. 10.6].

For 3-dimensional case

$$\mathbf{x} \longrightarrow \mathbf{x} + \mathbf{a}; \quad U_T = e^{\frac{i\mathbf{a}\cdot\hat{\mathbf{p}}}{\hbar}}. \qquad (10.67)$$

Hence if

$$[\hat{\mathbf{p}}, H] \quad or \quad [\hat{\mathbf{p}}, S] = 0 \qquad (10.68)$$

then momentum is conserved i.e. translational invariance implies conservation of momentum.

Now [using prob. 10.3]

$$U_T\hat{x}U_T^\dagger = e^{\frac{ia\hat{p}}{\hbar}}\hat{x}e^{\frac{-ia\hat{p}}{\hbar}} \qquad (10.69)$$

$$= \hat{x} + \frac{ia}{\hbar}[\hat{p}, \hat{x}] + \frac{\hbar^2}{2!}[\hat{p}, [\hat{p}, \hat{x}]] + ...$$

$$= \hat{x} + \frac{ia}{\hbar}(-i\hbar) = \hat{x} + a. \qquad (10.70)$$

Hence in general

$$\hat{A}(\mathbf{x} + \mathbf{a}) = e^{\frac{i\mathbf{a}\cdot\hat{\mathbf{p}}}{\hbar}}\hat{A}(\mathbf{x})e^{-\frac{i\mathbf{a}\cdot\hat{\mathbf{p}}}{\hbar}}. \qquad (10.71)$$

We now consider several examples of a closed system (not acted upon by external forces). The simplest example is that of a free particle, for which the Hamiltonian

$$H = \frac{\hat{\mathbf{p}}^2}{2m}, \quad \hat{\mathbf{p}} = -i\hbar\nabla \qquad (10.72)$$

is obviously translationally invariant:

$$[\hat{\mathbf{p}}, H] = 0. \qquad (10.73)$$

Thus momentum is conserved and it corresponds to Newton's first law of motion (law of inertia) in classical mechanics.

The Hamiltonian for an isolated system of particles mutually interacting with one another:

$$H = \frac{\hat{\mathbf{p}}^2}{2m} + V(\mathbf{x}_i - \mathbf{x}_j) \qquad (10.74)$$

is also translationally invariant, $[\hat{p}, H] = 0$.

The third example is the periodic potential $V(x) = V(x + a)$, a being lattice constant. Let us write unitary transformation U_T as $T(a)$ for lattice translation:

$$T(a)|x\rangle = |x - a\rangle, \tag{10.75}$$

$$T^\dagger(a)|x\rangle = |x + a\rangle. \tag{10.76}$$

Obviously the Hamiltonian with periodic potential is invariant under lattice translation.

$$T(a)HT^\dagger(a) = H,$$

$$[T(a), H] = 0. \tag{10.77}$$

Thus it is possible to find simultaneous eigenstates of H and $T(a)$. Let $|n\rangle$ be an eigenstate of H with eigenvalue $|n\rangle$, n denotes the nth lattice site:

$$H|n\rangle = E_0|n\rangle, \tag{10.78}$$

$$HT(a)|n\rangle = T(a)H|n\rangle = E_0 T(a)|n\rangle. \tag{10.79}$$

Hence

$$T(a)|n\rangle \equiv |n - 1\rangle \tag{10.80}$$

is also an eigenstate of H with eigenvalue E_0. We have n-fold degeneracy. $T(a)$ is unitary but not hermitian, hence its eigenvalues are of type $e^{i\theta}$. The simultaneous eigenstates of H and $T(a)$ are constructed as follows. Define the state

$$|ka\rangle = \sum_{n=-\infty}^{\infty} e^{inka}|n\rangle. \tag{10.81}$$

For this state, using Eq. (10.78) and (10.79)

$$H|ka\rangle = \sum_n e^{inka} H|n\rangle = E_0|ka\rangle \tag{10.82}$$

while

$$T(a)|ka\rangle = \sum_{n=-\infty}^{\infty} e^{inka}|n - 1\rangle$$

$$= \sum_{n=-\infty}^{\infty} e^{i(n+1)ka}|n\rangle$$

$$= e^{ika}|ka\rangle. \tag{10.83}$$

Hence $|ka\rangle$ is simultaneously an eigenstate of H and $T(a)$ with eigenvalues E_0 and e^{ika} respectively. Now from Eq. (10.83):

$$\langle x|T(a)|ka\rangle = e^{ika}\langle x|ka\rangle. \qquad (10.84)$$

But on using Eq. (10.76)

$$\langle x|T(a)|ka\rangle = \langle x+a|ka\rangle. \qquad (10.85)$$

Hence

$$\langle x+a|ka\rangle = e^{ika}\langle x|ka\rangle \qquad (10.86)$$

or

$$\psi_k(x+a) = e^{ika}\psi_k(x). \qquad (10.87)$$

Eq. (10.87) implies that $\psi_k(x)$ can be written in terms of Bloch function, $u_k(x)$:

$$\psi_k(x) = e^{ikx}u_k(x). \qquad (10.88)$$

To see this, we note that Bloch function has the property

$$u_k(x+a) = u_k(x). \qquad (10.89)$$

Thus from Eq. (10.88)

$$\psi_k(x+a) = e^{ik(x+a)}u_k(x+a)$$
$$= e^{ikx}e^{ika}u_k(x) = e^{ika}\psi_k(x).$$

The translational invariance does not hold for a particle moving in a harmonic oscillator potential centered on some fixed point, since the potential defines a natural coordinate origin. For such a system momentum is not conserved.

3- Rotation operator:

Just as there is a relation between the momentum operator $\hat{p}$ and displacement operator $e^{\frac{i\hat{p}a}{\hbar}}$, there exists a relationship between the angular momentum operator and rotation operator. Consider a rotation of coordinates

$$x_i \longrightarrow \acute{x_i} = R_{ij}x_j. \qquad (10.90)$$

The length

$$x_i^2 = x_1^2 + x_2^2 + x_3^2 = x^2 + y^2 + z^2 \qquad (10.91)$$

is invariant under the rotation of coordinates:

$$
\begin{aligned}
\acute{x_i}^2 &= R_{ij}x_j R_{ik}x_k \\
&= (R)_{ij}(R^T)_{ki}x_j x_k \\
&= (R^T)_{ki}R_{ij}x_j x_k = x_i^2.
\end{aligned}
\tag{10.92}
$$

Hence

$$
\begin{aligned}
(R^T)_{ki}R_{ij} &= \delta_{kj}, \\
(R^T R)_{kj} &= \delta_{kj}, \\
R^T R &= 1 = RR^T.
\end{aligned}
\tag{10.93}
$$

For example, for rotation about z-axis, by an angle θ

$$
\begin{aligned}
\acute{x} &= x\cos\theta + y\sin\theta, \\
\acute{y} &= -x\sin\theta + y\cos\theta, \\
\acute{z} &= z.
\end{aligned}
\tag{10.94}
$$

For infinitesimal rotation:

$$
\acute{\mathbf{x}} = \mathbf{x} - \boldsymbol{\theta} \times \mathbf{x}, \quad \boldsymbol{\theta} = \theta\hat{e}_z.
\tag{10.95}
$$

For a general infinitesimal transformation

$$
\begin{aligned}
\acute{\mathbf{x}} &= \mathbf{x} - \boldsymbol{\omega} \times \mathbf{x}, \\
\acute{x_i} &= x_i - \epsilon_{ijk}\omega_j x_k.
\end{aligned}
\tag{10.96}
$$

Corresponding to rotation of coordinates, there is a unitary operator U_R:

$$
U_R|\psi\rangle = |\psi^R\rangle,
\tag{10.97}
$$

$$
\begin{aligned}
U_R|\mathbf{x}\rangle &= |\mathbf{x} + \boldsymbol{\omega} \times \mathbf{x}\rangle, \\
U_R^\dagger|\mathbf{x}\rangle &= |\mathbf{x} - \boldsymbol{\omega} \times \mathbf{x}\rangle.
\end{aligned}
\tag{10.98}
$$

$$
\begin{aligned}
\psi_R(\mathbf{x}) = \langle \mathbf{x}|U_R|\psi\rangle &= \langle \mathbf{x} - \boldsymbol{\omega} \times \mathbf{x}|\psi\rangle \\
&= \psi(\mathbf{x} - \boldsymbol{\omega} \times \mathbf{x})
\end{aligned}
\tag{10.99}
$$

By Taylor expansion

$$
\psi_R(\mathbf{x}) = \psi(\mathbf{x}) - \boldsymbol{\omega}.(\mathbf{x} \times \boldsymbol{\nabla})\psi(\mathbf{x}),
\tag{10.100}
$$

so that

$$
\begin{aligned}
U_R(\omega) &= 1 - \boldsymbol{\omega}.(\mathbf{x} \times \boldsymbol{\nabla}) \\
&= 1 - \frac{i}{\hbar}\boldsymbol{\omega}.\mathbf{L}.
\end{aligned}
$$

Exponentiation gives for finite rotation,

$$U_R(\omega) = e^{-\frac{i}{\hbar}\boldsymbol{\omega}\cdot\mathbf{L}}. \tag{10.101}$$

where $\boldsymbol{L} = -i\hbar\boldsymbol{x} \times \boldsymbol{\nabla}$ is the orbital angular momentum operator. It is easy to check that L_i satisfy the commutation relation

$$[L_i, L_j] = i\hbar\epsilon_{ijk}L_k.$$

Hence angular momentum is the generator of rotation. Intrinsic spin, a purely quantum mechanical concept with no classical analogue can be taken into account by replacing $\mathbf{L}$ by $\mathbf{J}$ with the same commutation relations as those for $\mathbf{L}$:

$$[J_i, J_j] = i\hbar\epsilon_{ijk}J_k. \tag{10.102}$$

(In quantum mechanics, the commutation relations are fundamental entitles). The commutation relations define the angular momentum J_i. Hence the generators of rotation group $SO(3)$ are J_i (see ch. 11)

$$U_R = e^{-\frac{i}{\hbar}\boldsymbol{\omega}\cdot\mathbf{J}}. \tag{10.103}$$

Rotational invariance implies

$$H = U_R H U_R^\dagger,$$

which gives

$$[\mathbf{J}, H] = 0 \tag{10.104}$$

i.e. angular momentum is conserved. This holds for an isolated system (i.e. a system not in an external field) or a system in a centrally symmetric potential for which:

$$H = \frac{\hat{\mathbf{p}}^2}{2m} + V(r) \tag{10.105}$$

which is invariant under rotation, so that $[\mathbf{L}, H] = 0$

10.6 Discrete Transformation

In section (10.5), we have discussed the continuous transformations. Space reflection and time reversal viz. $\mathbf{x} \longrightarrow -\mathbf{x}$, $t \longrightarrow -t$ are discrete transformations:

i- Space reflection

$$\mathbf{x} \longrightarrow -\mathbf{x}. \tag{10.106}$$

Associated with this transformation is a unitary operator $\hat{P}$, called the parity operator. Thus

$$\hat{P}|\mathbf{x}\rangle = |-\mathbf{x}\rangle,$$
$$\hat{P}\hat{P}|\mathbf{x}\rangle = \hat{P}|-\mathbf{x}\rangle = |\mathbf{x}\rangle. \tag{10.107}$$

Hence

$$\hat{P}^2 = 1 \tag{10.108}$$

i.e. $\hat{P}$ has two eigenvalues $+1$ and -1. $\hat{P}$ is both hermitian and unitary:

$$\hat{P}^\dagger = \hat{P} = \hat{P}^{-1}. \tag{10.109}$$

Now

$$\hat{P}\psi(\mathbf{x}) = \psi(-\mathbf{x}) = \langle -\mathbf{x}|\psi\rangle = \langle \mathbf{x}|\hat{P}|\psi\rangle. \tag{10.110}$$

Any ket $|\psi\rangle$ can be expressed as

$$|\psi\rangle = \frac{1}{2}[(1+\hat{P})|\psi\rangle + (1-\hat{P})|\psi\rangle]$$
$$= |\psi_+\rangle + |\psi_-\rangle \tag{10.111}$$

where

$$|\psi_+\rangle = \frac{(1+\hat{P})}{2}|\psi\rangle \tag{10.112}$$

$$|\psi_-\rangle = \frac{(1-\hat{P})}{2}|\psi\rangle \tag{10.113}$$

and have the property (using Eq. (10.108)):

$$\hat{P}|\psi_\pm\rangle = \pm|\psi_\pm\rangle.$$

Hence $|\psi_+\rangle$ and $|\psi_-\rangle$ are eigenstates of $\hat{P}$, with eigenvalues $+1$ and -1 respectively. Thus $|\psi_+\rangle$ and $|\psi_-\rangle$ form a complete set, so that any arbitrary state can be expressed in terms of them. From Eq. (10.110), it follows that

$$\hat{P}\psi_+(\mathbf{x}) = \psi_+(-\mathbf{x}) = \psi_+(\mathbf{x}), \tag{10.114}$$
$$\hat{P}\psi_-(\mathbf{x}) = \psi_-(-\mathbf{x}) = -\psi_-(\mathbf{x}). \tag{10.115}$$

If Hamiltonian H is invariant under space reflection, then

$$\hat{P}H\hat{P}^{-1} = H$$
$$\hat{P}H = H\hat{P}$$

i.e.

$$[\hat{P}, H] = 0 \tag{10.116}$$

so that

$$\frac{d\hat{P}}{dt} = 0$$

i.e. parity is conserved. This is a type of conservation law which plays as important role in quantum mechanics and has no classical analogue.

The law of conservation of parity is not universal. There is class of interactions called weak interactions (responsible for β-decay), for which this law is violated.

ii-Time reversal

$$t \longrightarrow -t. \tag{10.117}$$

Under time reversal

$$\mathbf{x} \longrightarrow \mathbf{x},$$
$$\mathbf{p} \longrightarrow -\mathbf{p},$$
$$\mathbf{J} \longrightarrow -\mathbf{J}. \tag{10.118}$$

Hence, under time reversal, the basic commutation relations of the quantum mechanics viz.

$$[x_i, p_j] = i\hbar \delta_{ij},$$

$$[J_i, J_j] = i\hbar \epsilon_{ijk} J_k.$$

are not invariant, in contrast to space reflection($\mathbf{x} \longrightarrow -\mathbf{x}$, $\mathbf{p} \longrightarrow -\mathbf{p}$, $\mathbf{J} \longrightarrow \mathbf{J}$), in which case they are invariant. Thus an operator corresponding to time reversal cannot be a unitary operator as it does not preserve the fundamental laws of quantum mechanics. A way out of this difficulty is to simultaneously change $i \longrightarrow -i$. Accordingly we will take the complex conjugate of all complex numbers to preserve the invariance of physical laws under time reversal. For time reversal, the corresponding operator T is not unitary i.e. $T^\dagger \neq T^{-1}$.

Under time reversal, for a ket vector $|\psi(t)\rangle$:

$$|\psi^{(t)}(t)\rangle = T|\psi(t)\rangle = \langle \psi(-t)|, \tag{10.119}$$
$$\langle \mathbf{x}|\psi^{(t)}(t)\rangle = \langle \psi(-t)|\mathbf{x}\rangle = \langle \mathbf{x}|\psi(-t)\rangle^*. \tag{10.120}$$

Thus

$$\psi^{(t)}(\mathbf{x}, t) = \psi^*(\mathbf{x}, -t).$$

or

$$\psi^{(t)}(\mathbf{x}, -t) = \psi^*(\mathbf{x}, t) \tag{10.121}$$

The time dependent Schrödinger equation is

$$i\hbar\frac{\partial}{\partial t}\psi(x,t) = H\psi(x,t).$$ (10.122)

Taking the complex conjugate (H is hermitian)

$$-i\hbar\frac{\partial}{\partial t}\psi^*(x,t) = H\psi^*(x,t).$$

Changing $t \to -t$

$$i\hbar\frac{\partial}{\partial t}\psi^*(x,-t) = H\psi^*(x,-t).$$

Then using Eq. (10.121)

$$i\hbar\frac{\partial}{\partial t}\psi^t(x,t) = H\psi^t(x,t)$$

i.e. the Schrödinger equation is invariant under time reversal.

10.7 The Path Integration Formulation of Quantum Mechanics[1]

10.7.1 *Introduction*

There are two ways to effect a transition between classical and quantum physics:

(1) Canonical quantization

$$[\hat{q}(t), \hat{p}(t)] = i\hbar$$ (10.123)

which we discussed in Sec. 10.2, and originated from Hamiltonian mechanics. As is clear from Eq. (10.34), which hold for equal time, time plays a central role.

(ii) Path integral The path integral approach was formulated by Dirac, Feynman and is essentially tied to Lagrangian formulation of mechanics based on Action functional.

$$\mathcal{S}([q], t_i, t_f) = \int_{t_i}^{t_f} dt L(q, \dot{q}).$$ (10.124)

All fundamental laws of classical physics can then be understood in terms of the action $\mathcal{S}$. Furthermore there is one to one correspondence between the symmetries of the action and existence of conserved quantities. Another important point in this approach is that time is treated on equal footing with the other coordinates.

---

[1]For Further reading and more details, see
(i) A. Zee. *Quantum Field Theory*, Princeton University Press 2003, Chapter 1.2.
(ii) P. Raymond, *Field Theory, A Modern Primer*, The Benjamin Cummings Publishing Company, 1981, Chap. II

10.7.2 *Path Integral Method*

Just as in classical mechanics one can go from Hamiltonian formulation to the Lagrangian, or vice versa, we start from the Schrödinger formalism, based on Hamiltonian, and make a transition to one based on the Lagrangian. An important quantity to calculate in quantum mechanics is the transition amplitude.

$$\mathcal{K}(\psi, \phi) = \langle \psi, t_F | \phi, t_I \rangle \qquad (10.125)$$

where states involved are in the Schrödinger picture at time t_I and t_F, and $|\psi(t)\rangle$ satisfies the Schrödinger equation.

$$i\hbar \frac{\partial}{\partial t} |\psi(t)\rangle_s = H |\psi(t)\rangle$$

which has a a formal solution

$$|\psi(t)\rangle_s = e^{-\frac{i}{\hbar}Ht} |\psi\rangle, \quad |\psi\rangle = |\psi(0)\rangle.$$

We need to construct the states $|q, t\rangle$ and $|p, t\rangle$ which are "instantaneous" eigenstates of the Heisenberg picture operators $\hat{q}(t)$ and $\hat{p}(t)$. This we have already done in Sec. 10.3:

$$\hat{q}(t) |q(t)\rangle = q |q(t)\rangle,$$

$$\int dq |q(t)\rangle \langle q(t)| = 1.$$

Then the transition amplitude from a point q_I to a point q_F in time $T = t_F - t_I$ is given by

$$\mathcal{K} = \langle q_F | e^{-\frac{i}{\hbar}HT} | q_I \rangle$$
$$\equiv \langle q_F, t_F | q_I, t_I \rangle \qquad (10.126)$$

where $q_I = q(0)$ and $q_F = q(T)$. Divide T into N segments, each lasting $\delta t = \frac{T}{q}$. Then

$$\mathcal{K} = \langle q_F | e^{-\frac{i}{\hbar}H\delta t} e^{-\frac{i}{\hbar}H\delta t} .... e^{-\frac{i}{\hbar}H\delta t} | q_I \rangle.$$

Now insert a complete set of one particle states $\int dq |q\rangle \langle q| = 1$, so that

$$\mathcal{K} = \prod_{i=1}^{N-1} \int dq_i \langle q_F | e^{-\frac{i}{\hbar}H\delta t} | q_{N-1} \rangle \langle q_{N-1} | e^{-\frac{i}{\hbar}H\delta t} | q_{N-2} \rangle .....$$

$$\langle q_2 | e^{-\frac{i}{\hbar}H\delta t} | q_1 \rangle \langle q_1 | e^{-\frac{i}{\hbar}H\delta t} | q_I \rangle. \qquad (10.127)$$

Consider now

$$I_j = \langle q_{j+1} | e^{-\frac{i}{\hbar} H \delta t} | q_j \rangle$$

with

$$H = \frac{\hat{p}^2}{2m} + V(\hat{q}).$$

Introduce a complete set of one particle states $\int dp |p\rangle \langle p| = 1$ so that

$$I_j = \int dp \langle q_{j+1} | p \rangle \langle p | e^{-\frac{i}{\hbar} \delta t [\frac{\hat{p}^2}{2m} + V(\hat{q})]} | q_j \rangle.$$

To proceed, a convention is selected that after the expansion, all p operators will be moved to the left of all q operators, the effect of the commutator $[\hat{p}^2, V(q)]$ is $O(\delta t^2)$ and is irrelevant for $\delta t \to 0$. Then using $\langle q_{j+1} | p \rangle = \frac{1}{\sqrt{2\pi\hbar}} e^{\frac{i}{\hbar} p q_{j+1}}$, $\langle p | q_j \rangle = \frac{1}{\sqrt{2\pi\hbar}} e^{-\frac{i}{\hbar} p q_j}$,

$$I_j = \frac{1}{2\pi\hbar} \int dp \, e^{\frac{i}{\hbar} p (q_{j+1} - q_j)} e^{-\frac{i}{\hbar} \delta t [\frac{p^2}{2m}]} e^{-\frac{i}{\hbar} \delta t V(q_j)}.$$

Then using the Gaussian integral

$$\int dx \, e^{-a\frac{x^2}{2} + bx} = \sqrt{\frac{2\pi}{a}} e^{\frac{b^2}{2a}} \tag{10.128}$$

we get

$$I_j = \frac{1}{2\pi\hbar} \left(\frac{2\pi}{i\delta t/m\hbar} \right)^{\frac{1}{2}} e^{\frac{i}{\hbar} \delta t [\frac{m}{2} (\frac{q_{j+1} - q_j}{\delta t}) - V(q_j)]}.$$

Thus

$$\mathcal{K} = \left(\frac{-im}{2\pi\hbar\delta t} \right)^{N/2} \prod_{i=1}^{N-1} \int dq_i e^{\frac{i}{\hbar} \delta t [\sum_{j=0}^{N-1} \frac{m}{2} (\frac{q_{j+1} - q_j}{\delta t})^2 - V(q_j)]}. \tag{10.129}$$

Now take the continuum limit $\delta t \to 0$, $N \to \infty$

$$\frac{q_{j+1} - q_j}{\delta t} \to \dot{q}$$

$$\delta t \sum_{j=1}^{N} \to \int_T^t dt \tag{10.130}$$

where $q(T) = q_I$ and $q(t) = q_F$. Then

$$\mathcal{K} = \langle q_F | e^{i\frac{H}{\hbar}(t-T)} | q_I \rangle$$

$$= \int \mathcal{D}q(t) e^{\int_T^t \frac{i}{\hbar} [\frac{1}{2} m \dot{q}^2 - V(q)]}$$

$$= \int \mathcal{D}q(t) e^{\frac{i}{\hbar} \int_T^t L(q, \dot{q})} \tag{10.131}$$

where

$$\int Dq(t) = \lim_{N\to\infty} \left(\frac{-im}{2\pi\delta t}\right)^{N/2} \prod_{i=0}^{N-1} \int dq_i \qquad (10.132)$$

and

$$L(q,\dot{q}) = \frac{1}{2}m\dot{q} - V(q) \qquad (10.133)$$

is the Lagrangian. In terms of the action

$$\mathcal{S}([q],t,T) = \int_T^t dt\, L(\dot{q},q) \qquad (10.134)$$

we obtain

$$\mathcal{K} = \langle q_F, t_F | q_I, t_I \rangle$$
$$= \int Dq(t) e^{\frac{i}{\hbar}\mathcal{S}([q],t,T)}. \qquad (10.135)$$

This implies that to obtain $\mathcal{K}$, we simply have to integrate over all possible paths $q(t)$ such that $q(T) = q_I$ and $q(t) = q_F$, weighted by the exponential of $\frac{i}{\hbar}$ times the action evaluated for the particular path.

In the classical limit $\hbar \to 0$, the integral is determined by the minimum of $\mathcal{S}$ i.e. by the classical path $q_c(t)$ that extremizes the action and must obey the Euler–Lagrange equations:

$$\frac{d}{dt}\left(\frac{\partial L}{\partial \dot{q}}\right) - \frac{\partial L}{\partial q} = 0.$$

"Thus the path integral formalism clearly brings out the distinction between the classical and quantum mechanics. In the former the particle takes one path to go from q_I to q_F i.e. $q_c(t)$, while all paths contribute to the latter. [P.Raymond]"

Finally the transition amplitude from some initial state $|\phi\rangle$ at t_I to a final state $|\psi\rangle$ at $t = t_F$ is given by

$$\mathcal{K}(\psi,\phi) \equiv \langle \psi, t_F | \phi, t_I \rangle$$
$$= \int dq_F dq_I \langle \psi_F | q_F(t_F)\rangle \langle q_F(t_F) | q_I(t_I)\rangle \langle q_I(t_i)|\phi_I\rangle$$
$$= \int dq_F dq_I \psi_F^*(q_F)\mathcal{K}(q_F,t_F,q_I,t_I)\phi_I(q_I) \qquad (10.136)$$

where

$$\mathcal{K}(q_F,t_F,q_I,t_I) = \int Dq(t) e^{\frac{i}{\hbar}\mathcal{S}([q],t_I,t_F)} \qquad (10.137)$$

and is called the propagator, since it contains all the information regarding the development of the system.

10.7.3   Free Particle Propagator from Path Integral Formulism

To illustrate how one applies path integral formalism, we calculate the free particle propagator which we have already calculated in Sec. (10.3). We now calculate it by the path integral method where $L = \frac{1}{2} m \dot{q}^2$ and from Eqs. (10.129) and (10.132)

$$
\mathcal{K}(q_F, t_F, q_I, t_I) = \int \left(\frac{-im}{2\pi \hbar \delta t} \right)^{\frac{N}{2}} dq_1 .... dq_{N-1} e^{\frac{i}{\hbar} \sum \frac{m}{2\delta t}(q_{j+1}...q_j)^2}.
$$

Consider

$$
\int dq_j e^{\frac{im}{2\hbar \delta t}[(q_{j+1}-q_j)^2 + (q_j - q_{j-1})^2]} = \int dq_j e^{\frac{im}{2\hbar \delta t}[q_j - \frac{1}{2}(q_{j-1}-q_{j+1})]^2} . e^{\frac{im}{4\hbar \delta t}(q_{j+1}-q_{j-1})^2}
$$

$$
= \left(\frac{2\pi i \hbar \delta t}{2m} \right)^{\frac{1}{2}} e^{\frac{1}{4}\frac{im}{\delta t \hbar}(q_{j+1}-q_{j-1})^2}
$$

where we have used the Gaussian integral (10.128). Thus the first integration $[j = 1, \ q_0 = q_I]$ gives

$$
\sqrt{\frac{2\pi i \delta t \hbar}{2m}} e^{\frac{im}{4\delta t \hbar}(q_2 - q_I)^2}.
$$

The second integration $[j = 2]$ gives

$$
\int dq_2 e^{\frac{m}{4\delta t \hbar}(q_2-q_I)^2} e^{\frac{im}{2\delta t \hbar}(q_3-q_2)^2} = \int dq_2 e^{\frac{im}{4\delta t \hbar}[\frac{3}{2}(q_2 - \frac{2}{3}q_3 - \frac{1}{3}q_3)^2 . e^{\frac{im}{4\delta t \hbar}\frac{1}{3}(q_3 - q_I)^2}]}
$$

$$
= \left(\frac{4\pi i \delta t \hbar}{3m} \right)^{\frac{1}{2}} e^{\frac{i\delta m}{2.3\delta t \hbar}(q_3 - q_I)^2}
$$

where we have again the Gaussian integral (10.128) with $b = 0$. Therefore after $(N - 1)$ integrations, we get

$$
\mathcal{K}(q_F, t_F, q_I, t_I) = \left(\frac{m}{2\pi i \delta t \hbar} \right)^{\frac{N}{2}} \prod_{j=1}^{N-1} \left(\frac{2\pi i j \delta t \hbar}{(j+1)} \right)^{\frac{1}{2}} e^{\frac{im}{2N\delta t}(q_F - q_I)^2}.
$$

Now $N\delta t = t_F - t_I = T$, $\quad q = q_F - q_I$ while

$$
\prod_{j=1}^{N-1} a \frac{j}{j+1} = a^{N-1} \frac{1.....(N-1)}{2....N} = \frac{a^{N-1}}{N}
$$

where $a = 2\pi i \frac{\delta t \hbar}{m}$. Thus

$$
\mathcal{K}(q_F, t_F, q_I, t_I) = \left(\frac{m}{2\pi i N \delta t \hbar} \right)^{1/2} e^{-\frac{m}{2i\hbar T}q^2}
$$

$$
= \left(\frac{m}{2\pi i \hbar T} \right)^{1/2} e^{-\frac{m}{2i\hbar T}q^2} \tag{10.138}
$$

which agrees with the one found in Sec. (10.3)

This example also brings out the role of classical action. The classical equation of motion for a free particle is

$$\ddot{q} = 0$$

which with boundary conditions $q(0) = q_I$ and $q(T) = q_F$ has the solution $[T = t_F - t_I]$

$$q(t) = \frac{q_F - q_I}{T} t + q_I.$$

Thus the classical action is

$$S_c = \int_0^T \frac{1}{2} m \dot{q}_c^2 \, dt$$

$$= \int_0^T \frac{1}{2} m \frac{(q_F - q_I)^2}{T^2} \, dt$$

$$= \frac{m}{2} \frac{(q_f - q_I)^2}{T}$$

so that

$$e^{\frac{i}{\hbar} S_c} = e^{\frac{i}{\hbar} \frac{m(q_F - q_I)^2}{2T}}. \tag{10.139}$$

It is apparent that the argument of the exponential appearing in the free particle propagator Eq. (10.138), is given precisely by the $\frac{i}{\hbar}$ times the value of the action along the classical trajectory $q_c(t)$. It can be shown that this is general feature of all quadratic actions [see problem 10.8].

10.7.4   *Motion with a Source: Generating Functional for Time-ordered Products of Heisenberg Operators* $\hat{q}_i(t)$

We now discuss the path integral describing a particle moving in the potential $V(q)$ and undergoing external time dependent force $J(t)$ for which

$$L(\dot{q}, q) = \frac{1}{2} m \dot{q}^2 - V(q) + J(t) \tag{10.140}$$

and

$$H = \frac{\hat{p}^2}{2m} + V(\hat{q}) - J(t)\hat{q}. \tag{10.141}$$

Following the procedure in Sec. (10.7.2),

$$Z_{FI}(J) = \left(\frac{-im}{2\pi\hbar\delta t} \right)^{N/2} \prod_{i=1}^{N-1} \int dq_i e^{i\frac{\delta t}{\hbar} [\sum_{j=0}^{N-1} \frac{m}{2} \left(\frac{q_{j+1} - q_j}{\delta t} \right)^2 - V(q_j) + J(k_j)q_j]}$$

$$= \int Dq(t) e^{\frac{i}{\hbar} \int_{t_I}^{t_F} dt[\frac{1}{2}m\dot{q}^2 - V(q) + J(q)q]}. \tag{10.142}$$

We put $V(q) = 0$ and then

$$Z_{FI}(J) = \left(\frac{m}{2\pi i \hbar \delta t}\right)^{N/2} \int dq_1 .... dq_{N-1} e^{\frac{i}{\hbar} \sum_{j=0}^{N-1} [\frac{m}{2\delta t}(q_{J+1}-q_j)^2 + q_j J(t_j)\delta t]}.$$

(10.143)

We first look at $e^{\frac{i}{\hbar} \sum\limits_{j=0}^{N-1} q_j J(t_j)\delta t}$. The exponent multiplying $\frac{i}{\hbar}\delta t[q_0 - q_I]$ is

$q_0 J_0(t_0) + q_1 J(t_1)..... q_{N-1} J(t_{N-1})$

$$= q_0 [J_0(t_0) + q_1 J(t_1).... + q_{N-1} J(t_{N-1})]$$

$$+ (q_1 - q_0)J(t_1) + (q_2 - q_0)J(t_2)......(q_{N-1} - q_0)J(t_{N-1})$$

$$= q_I \sum_{j=0}^{N-1} J(t_j) + \{(q_1 - q_I)(J(t_1) + J(t_2) + ...J(t_{N-1})$$

$$+ (q_2 - q_1)(J(t_2) + J(t_3)....J(t_{N-1}))$$

$$\vdots$$

$$+ (q_{N-2} - q_{N-3})(J(t_{N-2}) + J(t_{N-1}))$$

$$+ (q_{N-1} - q_{N-2})(J(t_{N-1}))\}$$

$$= q_I \sum_{j=0}^{N-1} J(t_j) + \sum_{j=0}^{N-1} (q_{j+1} - q_j) \sum_{k=j+1}^{N-1} J(t_k).$$

(10.144)

Thus the full exponent in Eq. (10.143) is

$$\frac{i}{\hbar} \sum_{j=0}^{N-1} \left\{ q_I \sum_{j=0}^{N-1} J(t_j)\delta t + \frac{m}{2\delta t}(q_{j+1} - q_j)^2 + (q_{j+1} - q_j) \sum_{k=j+1}^{N-1} J(t_k)\delta t \right\}$$

$$= \frac{\iota}{\hbar} \left\{ \sum_{j=0}^{N-1} \left\{ q_I J(t_j)\delta t + \frac{m}{2\delta t} \sum_{j=0}^{N-1} \left[(q_{j+1} - q_j) + \frac{\delta t}{m} \sum_{k=j+1}^{N-1} J(t_k)\delta t \right]^2 \right. \right.$$

$$\left. \left. - \frac{m}{2\delta t} \sum_{j=0}^{t_{N-1}} \left(\frac{\delta\iota}{m}\right)^2 \left[\sum_{k=j+1}^{N-1} J(t_k)\delta t \right]^2 \right\} \right\}.$$

(10.145)

We now take the continuum limit

$$I(t_{j+1}) = \sum_{k=j+1}^{N-1} J(t_k)\delta t = \int_{t_{j+1}}^{T} J(\tau)d\tau,$$

(10.146)

$$I(0) = \sum_{j=0}^{N-1} J(t_j)\delta t$$

$$= \int_0^T J(\tau)d\tau.$$

(10.147)

Then

$$-\frac{1}{2m}\sum_{j=0}^{N-1}\delta t\left(\sum_{k=j+1}^{N-1}J(t_k)\delta t\right)^2$$

$$=-\frac{1}{2m}\sum_{j=0}^{N-1}[I(t_{j+1})]^2\,\delta t=\int_0^T dt I^2(t). \qquad (10.148)$$

Then from Eqs. (10.143) and (10.145–10.148)

$$Z_{FI}(J)=(\frac{im}{2\pi\hbar\delta t})^{\frac{N}{2}}\prod_{i=1}^{N}\int dq_i e^{\frac{i}{\hbar}q_I I(0)}e^{\frac{i}{\hbar}\frac{m}{2\delta t}[\sum_{j=0}^{N-1}[(q_{J+1}-q_j)+\frac{\delta t}{m}I(t_{j+1})]]^2}$$

$$\times\,e^{-\frac{i}{\hbar}\frac{1}{2m}\int_0^t dt I^2(t)}. \qquad (10.149)$$

To deal with the second term in Eq. (10.149), we proceed as in the free particle case and look at

$$\int dq_j e^{\frac{i}{\hbar}\frac{m}{2\delta t}\{(q_{j+1}-q_j+\frac{\delta t}{m}I(t_{j+1}))^2+(q_j-q_{j-1}+\frac{\delta t}{m}I(t_j))^2\}}.$$

We can write the coefficient of $\dfrac{i}{\hbar}\dfrac{m}{2\delta t}$ as

$$2\left[q_j-\frac{1}{2}\left(q_{j+1}+q_j-\frac{\delta t}{m}(I(t_{j+1})+I(t_j))\right)\right]^2$$

$$+\frac{1}{2}\left[(q_{j+1}-q_j)+\frac{\delta t}{m}(I(t_{j+1})+I(t_j))\right]^2.$$

Thus using the Gaussian integration (10.128), the dq_j integration gives

$$\sqrt{\frac{2\pi i\hbar\delta t}{2m}}e^{\frac{im}{4\delta t\hbar}[(q_{j+1}-q_{j-1})+(I(t_{j+1})+I(t_j))]^2}.$$

Therefore after $q_1.....,q_{N-1}$ integrations

$$\left(\frac{m}{2\pi i\hbar\delta t}\right)^{\frac{N}{2}}\prod_{j=1}^{N-1}\left(\frac{2\pi i\delta t\hbar j}{m(j+1)}\right)^{\frac{1}{2}}e^{i\frac{m}{2N\hbar\delta t}[(q_F-q_I)+\frac{\delta t}{m}\sum_{j=0}^{N-1}I(t_j)]}. \qquad (10.150)$$

Taking the continuum limit $\sum_{j=0}^{N-1}I(t_j)\,\delta t=\int_0^t I(t)\,dt$, as in free particle case ($N\delta t=T$), Eq. (10.150) becomes

$$\left(\frac{m}{2\pi\iota\hbar t}\right)^{\frac{1}{2}} e^{\frac{\iota m}{\hbar T}\left[(q_F - q_I) + \frac{1}{m}\int_0^T dt I(t)\right]^2}.$$

(10.151)

Thus finally from Eqs. (10.7.4) and (10.151)

$$Z_{FI}(J) = \left(\frac{m}{2\pi\iota\hbar T}\right)^{\frac{1}{2}} \exp\left[\frac{i}{\hbar}\left[\frac{m(q_F - q_I)^2}{2T} + \frac{q_F - q_I}{T}\int_0^T dt I(t) + \frac{1}{2mT}\right.\right.$$

$$\times \left.\left(\int_0^T dt I(t)\right)^2 + q_I I(0) - \frac{1}{2m}\int_0^T dt I^2(t)\right]\right]$$

(10.152a)

$$= \left(\frac{m}{2\pi\iota\hbar T}\right)^{\frac{1}{2}} \exp[F].$$

where

$$F = \frac{i}{\hbar}\left[\frac{m(q_F - q_I)^2}{2T} + \frac{q_F - q_I}{T}\int_0^T dt I(t)\right.$$

$$+ \frac{1}{2mT}\left.\left(\int_0^T dt I(t)\right)^2 + q_I I(0) - \frac{1}{2m}\int_0^T dt I^2(t)\right].$$

(10.152b)

We note that [see Eq. (10.146)]

$$I(t) = \int_t^T J(\tau)d\tau$$

(10.153)

which can be written as

$$I(t) = \int_0^T \theta(\tau - t) J(\tau)d\tau$$

(10.154)

where

$$\theta(\tau - t) = 1 \quad \text{if} \quad \tau > t,$$
$$= 0 \quad \text{if} \quad \tau < t.$$

Thus

$$I(0) = \int_0^t \theta(\tau) J(\tau) d\tau$$

(10.155)

$$= \int_0^t J(\tau) d\tau.$$

We now show that $Z_{FI}(J)$ serves as the generating functional for time products of operator $\hat{q}_i(t)$ of the free field (zero potential).

For this purpose, we define the functional derivative

$$\frac{\delta f}{\delta g(x)} = \lim_{\epsilon \to 0} \frac{f\left[g(x) + \epsilon(x - y)\right] - f\left[g(x)\right]}{\epsilon}. \tag{10.156}$$

If

$$f\left[g(x)\right] = \int_a^b dx \left[g(x)\right]^n,$$

then

$$\frac{\delta f}{\delta g(x)} = \int_a^b dx \, n\left[g(x)\right]^{n-1} \delta\left(x - y\right) = n[g(x)]^{n-1}.$$

Then from Eq. (10.154)

$$\frac{\delta I(t)}{\delta J(t_1)} = \int_0^T d\tau \theta(\tau - t) \delta\left(\tau - t_1\right) \tag{10.157}$$

$$= \theta\left(t_1 - t\right)$$

and

$$\frac{\delta I(0)}{\delta J(t_1)} = \theta\left(t_1\right) \tag{10.158}$$

$$= 1 \quad \text{for} \quad t_1 > 0.$$

Further, on using Eq. (10.157)

$$\frac{\delta}{\delta J(t_1)} \int_0^T I(t) dt = \int_0^T \theta(t_1 - t) dt = \int_0^{t_1 - t_I} \theta\left(t_1 - t\right) dt$$

$$+ \int_{t_1 - t_I}^T \theta\left(t_1 - t\right) dt = \int_0^{t_1 - t_I} dt = t_1 - t_I. \tag{10.159}$$

Thus

$$-\iota\hbar \frac{\delta Z(J)}{\delta J(t_1)} = -\iota\hbar \left(\frac{m}{2\pi\iota\hbar T}\right)^{\frac{1}{2}} \exp[F].G(t_1) \tag{10.160}$$

where

$$G(t_1) = \left(\frac{\iota}{\hbar}\right) \left[\frac{q_F - q_I}{T} \left(t_1 - t_I\right) + \frac{1}{2mT} 2 \int_0^T dt I(t).(t_1 - t_I) \right.$$

$$\left. + q_I - \frac{1}{2m} \int_0^t dt 2 I(t) \theta(t_1 - t) \right].$$

Hence

$$-\iota \frac{\delta Z}{\delta J(t_1)}\bigg|_{J=0} \equiv \int D_{q(t)} q(t_1) \exp\left[\frac{\iota}{\hbar}\int_0^T dt \frac{1}{2}m\dot{q}^2\right]$$

$$= \left[q_I + \frac{q_F - q_I}{T}(t_1 - t)\right]\left(\frac{m}{2\pi\iota\hbar T}\right)^{\frac{1}{2}} e^{\frac{\iota}{\hbar}\frac{m(q_F-q_I)^2}{2T}} \quad (10.161a)$$

$$= \left[q_I + \frac{q_F - q_I}{T}(t_1 - t)\right] Z(J = 0)$$

$$= \langle q_F, t_F|\hat{q}(t_1)|q_I, t_I\rangle_{J=0}$$

where

$$Z(J = 0) = \left(\frac{m}{2\pi i\hbar T}\right)^{\frac{1}{2}} e^{\frac{i}{\hbar}\frac{m(q_F-q_I)^2}{2T}}. \quad (10.161b)$$

Next we calculate $(-\iota\hbar)^2 \frac{\delta^2 Z(J)}{\delta J(t_1)\delta J(t_2)}$, first for $t_1 > t_2$ and then for $t_1 < t_2$. For $t_1 > t_2$, using Eq. (10.160) $[t_1 \to t_2]$

$$(-\iota\hbar)^2 \frac{\delta^2 Z(J)}{\delta J(t_1)\delta J(t_2)} = (-\iota\hbar)^2 \left(\frac{m}{2\pi\iota\hbar T}\right)^{\frac{1}{2}}$$

$$\times \left[\frac{\delta}{\delta J(t_1)}\exp[F].G(t_2) + \exp[F]\frac{\delta}{\delta J(t_1)}G(t_2)\right]$$

$$= (-\iota\hbar)^2 \left(\frac{m}{2\pi\iota\hbar T}\right)^{\frac{1}{2}} \exp[F].G(t_1)G(t_2) + \exp[F].\left(\frac{\iota}{\hbar}\right)$$

$$\times \left[\frac{2}{2mT}(t_2-t_I)(t_1-t_I) - \frac{2}{2m}\int_0^T dt\theta(t_2-t)\theta(t_1-t)\right]$$

where we have used (10.157), (10.159) and (10.160). Thus

$$(-i\hbar)^2 \frac{\delta^2 Z(J)}{\delta J(t_1)\delta J(t_2)}\bigg|_{J=0}$$

$$= (-\iota\hbar)^2 \left\{\left(\frac{i}{\hbar}\right)^2 \left[q_I + \frac{q_F - q_I}{T}(t_1 - t_I)\right]\left[q_I + \frac{q_F - q_I}{T}(t_2 - t_I)\right]\right.$$

$$\left. + \frac{i}{\hbar}\left[(t_2 - t_I)((t_1 - t_I) - t_F + t_I)\frac{1}{mT}\right]\right\} Z(0)$$

where we have used that [note $T = t_F - t_I$]

$$\int_0^T dt\theta(t_1 - t)\theta(t_2 - t) = \int_{t_I}^{t_2} dt = (t_2 - t_I)\frac{(t_F - t_I)}{T},$$

since $\theta(t_1 - t)$ and $\theta(t_2 - t)$ contribute only when $t < t_1$ as well $< t_2$ and we are taking $t_1 > t_2$.

For $t_1 < t_2$, interchanging $t_2 \leftrightarrow t_1$
Thus using

$$\theta(t_1 - t_2) = 1, \quad t_1 > t_2$$
$$= 0, \quad t_1 < t_2$$

$$(-\iota\hbar)^2 \frac{\delta^2 Z(J)}{\delta J(t_1)\delta J(t_2)}\Big|_{J=0} \equiv \int D_{q(t)} q(t_1) q(t_2) \exp\left[\frac{\iota}{\hbar}\int_0^T dt \frac{1}{2}m\dot{q}^2\right]$$

$$= \left[\iota\hbar\left[\theta(t_1 - t_2)\frac{(t_F - t_1)(t_2 - t_I)}{T}\right.\right.$$

$$\left.+\theta(t_2 - t_1)\frac{(t_F - t_2)(t_1 - t_I)}{T}\right]$$

$$+ \left[q_I + \frac{q_F - q_I}{T}(t_1 - t_I)\right]$$

$$\left.\times \left[q_I + \frac{q_F - q_I}{T}(t_2 - t_I)\right]\right] Z(0)$$

$$= \langle q_F, t_F | T\left(\hat{q}(t_1)\hat{q}(t_2)\right) | q_I, t_I\rangle_{J=0}$$

$Z_{FI}[J]$ serves as the generating functional for time-ordered products of the free (zero potential) theory.

10.8   Problems

10.1 Show that for the simple harmonic oscillator, the Heisenberg equation of motion gives

$$i\hbar\dot{a}(t) = [a(t), H] = \hbar\omega a(t)$$

which has the solution

$$a(t) = a(0)e^{-i\omega t} = ae^{-i\omega t},$$
$$a^\dagger(t) = a^\dagger e^{i\omega t}.$$

Hence show that

$$\hat{q}(t) = \hat{q}(0)\cos\omega t + \frac{\hat{p}(0)}{m\omega}\sin\omega t,$$
$$\hat{p}(t) = [\hat{p}(0)\cos\omega t - m\omega\hat{q}(0)\sin\omega t].$$

Hence show that

$$[\hat{q}(t_1), \hat{q}(t_2)] = \frac{i\hbar}{m\omega} \sin\omega(t_2 - t_1) \neq 0, \quad \text{if } t_2 \neq t_1$$

$$[\hat{q}(t_2), \hat{p}(t_1)] = i\hbar \cos\omega(t_2 - t_1) \neq i\hbar. \quad \text{if } t_2 \neq t_1$$

Thus canonical quantum conditions hold only for equal time $t_2 = t_1 = t$

10.2 Show that for the angular momentum $\boldsymbol{L} = \boldsymbol{x} \times \boldsymbol{p}$, $L_i = \epsilon_{iln}x_l p_n$, the classical PB is

$$(L_i, L_j)_{PB} = \epsilon_{iln}\epsilon_{jrs} \sum_k (\delta_{lk}p_n\delta_{sk}x_r - \delta_{nk}x_l\delta_{rk}p_s)$$

$$= x_i p_j - x_j p_i$$

$$= \epsilon_{ijk}L_k .$$

Thus in quantum mechanics

$$\frac{1}{i\hbar}\left[\hat{L}_i, \hat{L}_j\right] = \epsilon_{ijk}\hat{L}_k,$$

$$\left[\hat{L}_i, \hat{L}_j\right] = i\hbar\epsilon_{ijk}\hat{L}_k .$$

10.3 Show that

$$e^{\hat{A}}\hat{B}e^{-\hat{A}} = \hat{B} + [\hat{A}, \hat{B}] + \frac{1}{2}[\hat{A}, [\hat{A}, \hat{B}]] + \dots.$$

Hint: Consider $\hat{f}(\lambda) = e^{\lambda\hat{A}}\hat{B}e^{-\lambda\hat{A}}$,
and show that

$$\frac{d\hat{f}(\lambda)}{d\lambda} = \left[\hat{A}, \hat{f}(\lambda)\right]$$

$$\frac{d^2 f(\lambda)}{d\lambda^2} = \left[\hat{A}, [\hat{A}, \hat{f}(\lambda)]\right] \qquad \text{etc.}$$

and use the Taylor's series

$$\hat{f}(\lambda) = \hat{f}(0) + \frac{\lambda}{1!}\frac{d\hat{f}(\lambda)}{d\lambda!}\bigg|_{\lambda=0} + \frac{\lambda^2}{2!}\frac{d^2 f}{d\lambda^2}\bigg|_{\lambda=0} + \dots$$

to get the required quantity $f(\lambda)$

10.4 Let x and p_x be the coordinate and linear momentum in one dimension. Evaluate the classical PB

$$(x, F(p_x))_{PB} .$$

Let $\hat{x}$ and $\hat{p}_x$ be corresponding quantum mechanical operators. Then using the rule to go from classical mechanics to quantum mechanics, evaluate the commutator

$$\left[\hat{x}, \exp\left(\frac{ia\hat{p}_x}{\hbar}\right)\right]$$

where a is a real number. Using the above result, prove that $\exp\left(\dfrac{ia\hat{p}_x}{\hbar}\right)|\acute{x}\rangle$ is an eigenstate of the coordinate operator $\hat{x}$ [remember $\hat{x}|\acute{x}\rangle = \acute{x}|\acute{x}\rangle$]. What is the corresponding eigenvalue? Thus $\exp\left(\dfrac{ia\hat{p}_x}{\hbar}\right)$ gives the unitary operator corresponding to the translation.

10.5 Consider the Hamiltonian describing a free particle in an external field ϵ (constant in time)

$$H = \frac{\hat{p}^2}{2m} - e\epsilon\hat{x}.$$

Calculate the operators $\hat{x}(t)$ and $\hat{p}(t)$ in the Heisenberg picture. Suppose at $t = 0$, the particle is in the state $|\psi_0\rangle$ whose wave function in the x-representation is $\psi_0(x) = \langle x|\psi_0\rangle = e^{ikx}\phi(x)$, where $\phi(-x) = \phi(x)$ and $\phi(x)$ is real and $\int |\phi(x)|^2 dx = 1$. Define the uncertainty $(\Delta A)_t$ in the observable $\hat{A}$ at time t as

$$(\Delta A)_t = \left\{ \langle (\hat{A}(t) - \langle \hat{A}(t)\rangle)^2 \rangle \right\}^{\frac{1}{2}},$$

where $\langle \hat{A}(t)\rangle = \langle \psi_0|\hat{A}(t)|\psi_0\rangle$ in the Heisenberg picture. Then prove the following:

(i)

$$\langle \hat{x}(0)\rangle = 0, \quad \langle \hat{p}(0)\rangle = \hbar k, \quad \langle \hat{p}(0)^2\rangle = 2m\langle E\rangle,$$
$$\langle \hat{x}(0)\hat{p}(0) + \hat{p}(0)\hat{x}(0)\rangle = 0.$$

(ii)

$$(\Delta x)_t = \left\{ (\Delta x)_0^2 + \frac{2t^2}{m}\left(\langle E\rangle - \frac{\hbar^2 k^2}{2m}\right) \right\}^{\frac{1}{2}},$$

$$(\Delta p)_t = \left\{ 2m\left(\langle E\rangle - \frac{\hbar^2 k^2}{2m}\right) \right\}^{\frac{1}{2}}$$

$$= (\Delta p)_0.$$

10.6 For the Hamiltonian

$$H = \frac{1}{2}\hat{p}^2 V(\hat{q}),$$

show that the transition amplitude is given by

$$\langle q_F, t_F|q_I, t_I\rangle = \prod_{i=0}^{N-1} \left(\frac{-\iota}{2\pi\delta t}\right)^{\frac{1}{2}} \int dq_i V^{-\frac{1}{2}}(q_i) e^{i\int_{t_I}^{t_F} dt L}$$

where

$$L = \frac{1}{2}V^{-1}\dot{q}^2$$

is the Lagrangian.

10.7 If $T < t_1, t_2 < 0$, $\quad S = \int_0^T L dt$, and

$$\int Dq \quad q(t_1)q(t_2)e^{\iota S} = \langle q_T|T(\hat{q}(t_1)\hat{q}(t_2))|q_0\rangle$$

where $q_T = q(T), q_0 = q(0)$, show that for the Harmonic oscillator

$$\langle q_T|T(\hat{q}(t_1)\hat{q}(t_2))|q_0\rangle$$

$$= \left\{ \frac{\iota\hbar}{m\omega\sin\omega t} \left[\theta(t_1 - t_2)\sin\omega(T - t_1) \right. \right.$$

$$\left. \sin\omega t_2 + \theta(t_2 - t_1)\sin\omega(T - t_2)\sin\omega t_1 \right]$$

$$+ \frac{1}{\sin^2\omega T} \left[q_0\sin\omega(T - t_1) + q_T\sin\omega t_1 \right]$$

$$\left. \left[q_0\sin\omega(T - t_2) + q_T\sin\omega t_2 \right] \right\} \langle q_F|q_I\rangle$$

where

$$\langle q_F|q_I\rangle = \int Dq \; e^{\iota S}$$

with $S = \int_0^T dt \left[\frac{1}{2}m\dot{q}^2 - \frac{1}{2}m\omega^2 q^2 \right]$. Calculate S_{cl} for this case. [Hint: First show that for S.H.O, $\hat{q}(t) = \frac{1}{\sin\omega T} [\hat{q}_T\sin\omega t + \hat{q}_0\sin\omega(T - t)]$.]

10.8 For

$$H = \frac{\hat{p}^2}{2m} - \beta\hat{q}$$

with the boundary conditions

$$q(t) = q_f, \quad q(0) = q_i$$

show that

$$S_{cl} = \frac{\beta q_f}{2}t - \frac{\beta^2}{24m}t^3 + \frac{\beta q_i}{2}t + \frac{m(q_f - q_i)^2}{2t}.$$

Show that the propagator $K(q_f, t; q_i, 0)$ is

$$K(q_f, t; q_i, 0) = \left(\frac{m}{2\pi\hbar\iota t} \right)^{\frac{1}{2}} \exp\left(\frac{\iota S_{cl}}{\hbar} \right).$$

Chapter 11

Angular Momentum and Spin

11.1 The Zeeman Effect

Consider a hydrogen atom in a constant weak magnetic field **B**. Then the Hamiltonian is given by

$$H_B = H + \text{ interaction energy between the atom and}$$
$$\text{the magnetic field,}$$

where H is the Hamiltonian of an unperturbed atom (i.e. in the absence of a magnetic field)

$$H = \frac{p^2}{2m_e} + V(r). \tag{11.1}$$

Let us first calculate the interaction energy between the atom and the magnetic field classically. We may regard the proton as fixed and the electron moving round it to be in a circular orbit. Let the radius of the orbit be r, and the momentum of the electron be p. The velocity of the electron in the orbit is given by

$$v_e = \frac{p}{m_e}. \tag{11.2}$$

The electron orbit can then be regarded as a current loop. Now a current loop which encloses a surface S acts like a shell of strength j/c where j is the current flowing in the loop. Now any element of the shell of area $\mathbf{dS}$ acts like a dipole of moment $j/c \, \mathbf{dS}$. Therefore a circular current loop can be regarded as a dipole of moment

$$\boldsymbol{\mu} = \frac{j}{c} \int \mathbf{dS} = \frac{j}{c} \pi r^2 \mathbf{n} \tag{11.3}$$

where πr^2 is the area enclosed by the circular loop and **n** is a unit vector normal to the plane of the circular loop. Now the circumference of the circular loop $= 2\pi r$, therefore the charge at a fixed point on the orbit $=$ $e/2\pi r$. The current j is then given by

$$
\begin{aligned}
j &= \frac{e}{2\pi r} v_e \\
&= \frac{ep}{2\pi m_e r}.
\end{aligned}
\tag{11.4}
$$

Thus

$$
|\mu| = \frac{erp}{2m_e c}.
\tag{11.5}
$$

Now

$$
|\mathbf{r} \times \mathbf{p}| = rp.
$$

It is conventional to write

$$
\boldsymbol{\mu} = \frac{e}{2m_e c} \mathbf{r} \times \mathbf{p}
\tag{11.6}
$$

where e is the electric charge, of the particle.
But $\mathbf{r} \times \mathbf{p} = \mathbf{L}$, the angular momentum. Therefore

$$
\boldsymbol{\mu} = \frac{e}{2m_e c} \mathbf{L}.
\tag{11.7}
$$

The interaction energy is then given by

$$
\begin{aligned}
V_B &= -\boldsymbol{\mu} \cdot \mathbf{B} \\
&= -\frac{e}{2m_e c} \mathbf{B} \cdot \mathbf{L}.
\end{aligned}
\tag{11.8}
$$

This is a general result, not restricted to a circular orbit. By the correspondence principle, we also regard Eq. (11.8) as the interaction energy of the hydrogen atom with the magnetic field $\mathbf{B}$ in quantum mechanics provided we regard $\mathbf{L}$ as an operator. Thus we write the total Hamiltonian operator of the system

$$
H_B = H - \frac{e}{2m_e c} \mathbf{B} \cdot \mathbf{L}.
\tag{11.9}
$$

Take the z-axis in the direction of $\mathbf{B}$, then

$$
H_B = H - \frac{e}{2m_e c} B L_z.
\tag{11.10}
$$

The extra term $\frac{e}{2m_e c} B L_z$ in H_B only contains the operator L_z. Therefore the simultaneous eigenstates of H, L^2, L_z are also eigenstates of H_B. We have denoted the eigenfunction of H by

$$
u_{nlm}(r, \theta, \phi) \equiv \langle r, \theta, \phi | lmn \rangle
\tag{11.11}
$$

where the eigenvalues of H, L^2, L_z are given respectively by $E_n = \frac{m_e^4}{2\hbar^2 n^2}$, $l(l+1)\hbar^2$ and $m\hbar$. For a given l, m has $(2l+1)$ values

$$-l, -l+1, \ldots, 0, 1, 2, \ldots, l-1, l$$
$$n = 1, 2, \ldots, \text{and} \quad n \geq l+1.$$

Thus

$$H_B u_{nlm}(r, \theta, \phi) = E_n^B u_{nlm}(r, \theta, \phi). \tag{11.12}$$

But

$$H_B u_{nlm}(r, \theta, \phi) = \left(H - \frac{e}{2m_e c} B L_z\right) u_{nlm}(r, \theta, \phi)$$
$$= \left(E_n - \frac{eB}{2m_e c} m\hbar\right) u_{nlm}(r, \theta, \phi). \tag{11.13}$$

Thus

$$E_n^B \equiv E_{n,m} = E_n - \frac{e\hbar}{2m_e c} Bm \tag{11.14}$$

where E_n are the unperturbed hydrogen atom energy levels. Equation (11.14) shows that for given values of n, l, the $(2l+1)$ different states (corresponding to $(2l+1)$ values of m) which previously corresponded to the same level, E_n, are now split into $(2l+1)$ separate levels with spacing according to Eq. (11.14)

$$\Delta E = \frac{e\hbar}{2m_e c} B. \tag{11.15}$$

This effect is observed and shows why m is known as the magnetic quantum number.

According to Eq. (11.13) or Eq. (11.14), there should be no splitting of the ground state ($n = 1, l = 0, m = 0$). However, experimentally the lowest level is observed to split into two. If the splitting has the same physical origin as that derived above, it must be associated with the angular momentum j, which satisfies

$$2j + 1 = 2$$
$$\text{or} \quad j = \frac{1}{2}$$

with $m = -\frac{1}{2}, \frac{1}{2}$.

Since we have shown quite generally that the orbital angular momentum l can only take integral values, this result indicates the necessity for generalization of the formulism. The answer is to introduce a generalized definition of angular momentum and show that half integral eigenvalues are also possible.

11.2   Angular Momentum

As we have discussed previously, orbital angular momentum is defined as

$$\mathbf{L} = \hat{\mathbf{r}} \times \hat{\mathbf{p}}. \tag{11.16}$$

Then using the commutation relation

$$[\hat{x}_i, \hat{p}_j] = i\hbar \delta_{ij}$$

we have shown that $\mathbf{L}$ satisfy the commutation relations

$$\mathbf{L} \times \mathbf{L} = i\hbar \mathbf{L}. \tag{11.17}$$

Also we have shown that for eigenvalues $m\hbar$ of L_z, m is an integer due to the fact that with $\mathbf{L}$ defined as in Eq. (11.16) L_z has the representation

$$-i\hbar \frac{\partial}{\partial \phi}$$

in spherical polar coordinates. We now assume Eq. (11.17) holds whether or not angular momentum can be expressed as in Eq. (11.16). Thus in general for the angular momentum operator $\mathbf{J}$, we take the commutation relations

$$\mathbf{J} \times \mathbf{J} = i\hbar \mathbf{J} \tag{11.18a}$$

as the defining property of the angular momentum $\mathbf{J}$. Equation (11.18a) is equivalent to

$$[J_x, J_y] = i\hbar J_z,$$
$$[J_y, J_z] = i\hbar J_x,$$
$$[J_z, J_x] = i\hbar J_y. \tag{11.18b}$$

From Eq. (11.18b) it follows that

$$[\mathbf{J}^2, J_z] = 0 \quad \text{etc.} \tag{11.18c}$$

Thus we can measure $\mathbf{J}^2$ and any single component of $\mathbf{J}$ which we take to be J_z.

We now show that in general $\mathbf{J}^2$, J_z can have half integral as well as integral eigenvalues. Only when $\mathbf{J} = \mathbf{L}$ given in Eq. (11.16) the eigenvalues of L^2 and L_z are restricted to integers.

Let us denote the simultaneous eigenvectors of $\mathbf{J}^2$ and J_z by $|\lambda m\rangle$:

$$J^2 |\lambda m\rangle = \lambda \hbar^2 |\lambda m\rangle \tag{11.19a}$$
$$J_z |\lambda m\rangle = m\hbar |\lambda m\rangle \tag{11.19b}$$

where $\lambda\hbar^2$ and $m\hbar$ denote the eigenvalues of J^2 and J_z.

We now introduce new operators

$$J_\pm = J_x \pm iJ_y. \tag{11.20}$$

Then it is easy to see using the commutation relations (11.18b) that

$$\begin{aligned}
J_+J_- &= J_x^2 + J_y^2 - i(J_xJ_y - J_yJ_x) \\
&= \mathbf{J}^2 - J_z^2 + \hbar J_z \\
&= J^2 + \frac{1}{4}\hbar^2 - (J_z - \frac{1}{2}\hbar)^2.
\end{aligned} \tag{11.21a}$$

Similarly

$$J_-J_+ = J^2 + \frac{1}{4}\hbar^2 - (J_z + \frac{1}{2}\hbar)^2. \tag{11.21b}$$

Thus

$$[J_+, J_-] = 2\hbar J_z. \tag{11.21c}$$

Also

$$[J_z, J_\mp] = \mp\hbar J_\mp, \tag{11.21d}$$

$$[J^2, J_\pm] = 0. \tag{11.21e}$$

Then using Eqs. (11.21a) and (11.21b)

$$\begin{aligned}
J_zJ_-|\lambda m\rangle &= (-\hbar J_- + J_-J_z)|\lambda m\rangle \\
&= (-\hbar J_- + m\hbar J_-)|\lambda m\rangle \\
&= (m-1)\hbar J_-|\lambda m\rangle,
\end{aligned} \tag{11.22a}$$

$$\begin{aligned}
J^2J_-|\lambda m\rangle &= J_-J^2|\lambda m\rangle \\
&= \lambda\hbar^2 J_-|\lambda m\rangle.
\end{aligned} \tag{11.22b}$$

Thus if $J_-|\lambda m\rangle \neq 0$ then $J_-|\lambda m\rangle$ is an eigenstate of J_z belonging to eigenvalue $(m-1)\hbar$ but to the same eigenvalue $\lambda\hbar^2$ of J^2. Again using Eqs. (11.21) and (11.19b)

$$\begin{aligned}
J_zJ_+|\lambda m\rangle &= (\hbar J_+ + J_+J_z)|\lambda m\rangle \\
&= (m+1)\hbar J_+|\lambda m\rangle,
\end{aligned} \tag{11.23a}$$

$$\begin{aligned}
J^2J_+|\lambda m\rangle &= J_+J^2|\lambda m\rangle \\
&= \lambda\hbar^2 J_+|\lambda m\rangle.
\end{aligned} \tag{11.23b}$$

Thus if $J_+|\lambda m\rangle \neq 0$ then $J_+|\lambda m\rangle$ is an eigenvector of J_z with eigenvalue $(m+1)\hbar$ but same eigenvalue $\lambda\hbar^2$ of J^2. Continuing in this way we see that the eigenvalues of J_z corresponding to eigenvalue $\lambda\hbar^2$ of J^2 are

$$\cdots (m-2)\hbar, (m-1)\hbar, m\hbar, (m+1)\hbar, (m+2)\hbar \cdots, \tag{11.24}$$

i.e. the eigenvalues form a series in which successive terms differ by one unit of $\hbar$. We now show that this series terminates on both sides. Note that if we write

$$|F\rangle = J_-|\lambda m\rangle, \quad \text{then}$$
$$\langle F| = \langle \lambda m|J_+.$$

Thus

$$\langle F|F\rangle = \langle \lambda m|J_+ J_-|\lambda m\rangle \geq 0. \tag{11.25}$$

The equality sign holds when $J_-|\lambda m\rangle = 0$. Using Eq. (11.21a), we have

$$\langle \lambda m|J^2 + \frac{1}{4}\hbar^2 - (J_z - \frac{1}{2}\hbar)^2|\lambda m\rangle \geq 0$$

i.e. on using Eq. (11.19) we get

$$[(\lambda + \frac{1}{4})\hbar^2 - (m - \frac{1}{2})^2\hbar^2]\langle \lambda m|\lambda m\rangle \geq 0.$$

But

$$\langle \lambda m|\lambda m\rangle = 1$$

therefore

$$(\lambda + \frac{1}{4}) \geq (m - \frac{1}{2})^2. \tag{11.26}$$

Thus $(\lambda + \frac{1}{4}) \geq 0$ and we have the inequality

$$-\sqrt{\lambda + 1/4} + 1/2 \leq m \leq \sqrt{\lambda + 1/4} + 1/2. \tag{11.27}$$

Thus m is bounded on both sides. Thus the series (11.24) must terminate on both sides. Therefore m has a minimum and maximum value which we respectively denote by m_2 and m_1. Then we must have

$$J_-|\lambda m_2\rangle = 0 \tag{11.28}$$

since otherwise Eq. (11.22) shows that $J_-|\lambda m_2\rangle$ is an eigenvector of J_z with eigenvalue $(m_2-1)\hbar$ which contradicts that $m_2\hbar$ is the minimum eigenvalue. Thus Eq. (11.28) holds and because of it we must have an equality sign in Eq. (11.26) for $m = m_2$, i.e.

$$(\lambda + 1/4) - (m_2 - 1/2)^2 = 0$$

or

$$m_2 = 1/2 \pm \sqrt{\lambda + 1/4}. \tag{11.29}$$

Similarly we can show that

$$J_+|\lambda m_1\rangle = 0 \tag{11.30}$$

because otherwise, from Eq. (11.23), $J_+|\lambda m_1\rangle$ would be an eigenvector of J_z belonging to eigenvalue $(m_1 + 1)\hbar$ which contradicts that $m_1\hbar$ is the maximum eigenvalue. Thus Eq. (11.30) holds and we have

$$\langle\lambda m_1|J_- J_+|\lambda m_1\rangle = 0$$

or using Eq. (11.21b)

$$\langle\lambda m_1|J^2 + \frac{1}{4}\hbar^2 - (J_z + \frac{1}{2}\hbar)^2|\lambda m_1\rangle = 0$$

i.e. on using Eq. (11.19)

$$\left(\lambda\hbar^2 + \frac{1}{4}\hbar^2 - \hbar^2(m_1 + \frac{1}{2})^2\right)\langle\lambda m_1|\lambda m_1\rangle = 0$$

or

$$(\lambda + \frac{1}{4}) - (m_1 + \frac{1}{2})^2 = 0.$$

Therefore

$$m_1 = -\frac{1}{2} \pm \sqrt{\lambda + 1/4}. \tag{11.31}$$

To avoid min>max we must have a negative sign in Eq. (11.29) and a positive sign in Eq. (11.31). Thus

$$m_2 = (m)_{min} = \frac{1}{2} - \sqrt{\lambda + 1/4} \tag{11.32a}$$

$$m_1 = (m)_{max} = -\frac{1}{2} + \sqrt{\lambda + 1/4} = -m_2 \tag{11.32b}$$

and the series (11.24) for the eigenvalues of J_z becomes

$$m_2\hbar, (m_2 + 1)\hbar, \cdots, (m_1 - 1)\hbar, m_1\hbar. \tag{11.33}$$

Now because the successive terms of the series differ by one unit of $\hbar$, $(m_1 - m_2)$ must be an integer or zero, i.e.

$$m_1 - m_2 = 2\sqrt{\lambda + 1/4} - 1 = 2m_1 \tag{11.34a}$$

or

$$2\sqrt{\lambda + 1/4} - 1 = 2j \tag{11.34b}$$

where $2j$ is an integer or zero. Thus j is restricted to integral or half integral values, including zero. Since from Eq. (11.32b) $m_1 = -m_2$, it follows from Eq. (11.34a) and (11.34b) that

$$m_2 = -j, \; m_1 = j \tag{11.35}$$

and

$$\lambda = j(j+1) \qquad (11.36)$$

where

$$j = 0, 1/2, 1, 3/2, \cdots .$$

Hence the eigenvalues of J^2 and J_z respectively are

$$j(j+1)\hbar^2; j = 0, 1/2, 1, 3/2, \cdots \qquad (11.37)$$
$$m\hbar; m = -j, -j+1, \cdots, j-1, j. \qquad (11.38)$$

The eigenvalue Eq. (11.19) now becomes

$$J^2|jm\rangle = j(j+1)\hbar^2|jm\rangle, \qquad (11.39a)$$
$$J_z|jm\rangle = m\hbar|jm\rangle. \qquad (11.39b)$$

Thus by taking the commutation relations $\mathbf{J} \times \mathbf{J} = i\hbar\mathbf{J}$ as the defining property of angular momentum, we see that both half integral as well as integral eigenvalues of angular momentum are allowed.

11.3   Matrix Representation of Angular Momentum

We take eigenstates $|jm\rangle$ to be normalized. Then the matrix elements of J^2 and J_z in the basis defined by the vectors $|jm\rangle$ can be written as

$$\langle jm|J^2|j'm'\rangle = j'(j'+1)\hbar^2\langle jm|j'm'\rangle$$
$$= j(j+1)\hbar^2\delta_{jj'}\delta_{mm'}, \qquad (11.40a)$$
$$\langle jm|J_z|j'm'\rangle = m\hbar\delta_{jj'}\delta_{mm'}. \qquad (11.40b)$$

Let us now find the matrix representation of $J_{\pm}$. Since

$$[J^2, J_{\pm}] = 0$$

therefore

$$0 = \langle jm|J^2 J_{\pm}|j'm'\rangle - \langle jm|J_{\pm}J^2|j'm'\rangle.$$

Using Eq. (11.39a) we have

$$0 = j(j+1)\hbar^2\langle jm|J_{\pm}|j'm'\rangle - j'(j'+1)\hbar^2\langle jm|J_{\pm}|j'm'\rangle.$$

Hence

$$\langle jm|J_{\pm}|j'm'\rangle = 0 \quad \text{unless} \quad j = j'$$

i.e.

$$\langle jm|J_\pm|j'm'\rangle = \delta_{jj'}\langle jm|J_\pm|j'm'\rangle. \tag{11.41}$$

Now

$$[J_z, J_\pm] = \pm\hbar J_\pm$$

therefore

$$\pm\hbar\langle jm|J_\pm|jm'\rangle = \langle jm|J_z J_\pm|jm'\rangle - \langle jm|J_\pm J_z|jm'\rangle$$
$$= \hbar(m - m')\langle jm|J_\pm|jm'\rangle.$$

Hence

$$\langle jm|J_\pm|jm'\rangle \neq 0, \quad \text{only if}$$
$$(m - m') = \pm 1$$

or

$$m = m' \pm 1 \tag{11.42}$$

i.e. we can write

$$\langle jm|J_\pm|jm'\rangle = C^{m'}_{(\pm)j}\delta_{m,m'\pm 1}. \tag{11.43}$$

But since $(J_\pm)^\dagger = J_\mp$,

$$\langle jm|J_+|jm'\rangle = \langle jm'|J_-|jm\rangle^*$$
$$= C^{*m}_{(-)j}\delta_{m',m-1}. \tag{11.44}$$

Hence

$$C^{m'}_{(+)j}\delta_{m,m'+1} = C^{*m}_{(-)j}\delta_{m',m-1}$$

or

$$C^{m-1}_{(+)j} = C^{*m}_{(-)j}.$$

Now we note from Eq. (11.21a) that

$$\langle jm|J_+J_-|jm'\rangle = \langle jm|(J^2 - J_z^2 + \hbar J_z)|jm'\rangle$$
$$= \hbar^2 j(j+1)\delta_{mm'} - \hbar^2 m^2\delta_{mm'} + \hbar^2 m\delta_{mm'}. \tag{11.45}$$

Now putting a complete set of states:

$$\sum_n \langle jm|J_+|jn\rangle\langle jn|J_-|jm'\rangle = \hbar^2\left(j(j+1) - m(m-1)\right)\delta_{mm'}$$

i.e.

$$\sum_n C^{*m}_{(-)j}\delta_{n,m-1}C^{m'}_{(-)j}\delta_{n,m'-1} = \hbar^2\left(j(j+1) - m(m-1)\right)\delta_{mm'}$$

or

$$|C^m_{(-)j}|^2 = \hbar^2(j+m)(j-m+1). \qquad (11.46)$$

This leaves the phase of the matrix elements of J_- arbitrary. This is not of any physical significance. We therefore choose all phases to be zero and obtain

$$C^m_{(-)j} = \hbar\sqrt{(j+m)(j-m+1)}$$
$$= C^{m-1}_{(+)j}, \qquad (11.47)$$
$$C^m_{(+)j} = \hbar\sqrt{(j+m+1)(j-m)}. \qquad (11.48)$$

From Eqs. (11.43) and (11.47)

$$\langle jm|J_-|jm'\rangle = \hbar\sqrt{(j+m')(j-m'+1)}\delta_{m,m'-1}, \qquad (11.49)$$
$$\langle jm|J_+|jm'\rangle = \hbar\sqrt{(j+m'+1)(j-m')}\delta_{m,m'+1}$$
$$= \hbar\sqrt{(j+m)(j-m+1)}\delta_{m',m-1}. \qquad (11.50)$$

From Eq. (11.43), in view of the fact that $J_\pm|jm\rangle$ is an eigenstate of J_z with eigenvalue $(m \pm 1)\hbar$, it is clear that

$$J_\pm|jm\rangle = C^m_{(\pm)j}|j,m\pm 1\rangle.$$

Hence from Eq. (11.47)

$$J_+|jm\rangle = \hbar\sqrt{(j+m+1)(j-m)}|j,m+1\rangle, \qquad (11.51a)$$
$$J_-|jm\rangle = \hbar\sqrt{(j+m)(j-m+1)}|j,m-1\rangle. \qquad (11.51b)$$

J_+ and J_- are called raising and lowering operators. Rewriting Eqs. (11.40) we have

$$\langle jm|J^2|jm'\rangle = \hbar^2 j(j+1)\delta_{mm'}, \qquad (11.52)$$
$$\langle jm|J_z|jm'\rangle = \hbar m\delta_{mm'}. \qquad (11.53)$$

11.4 Spin

(i) $j = 0$; all above matrices are null.

(ii) spin $(1/2)\hbar$, $j = 1/2$: The vector space is two dimensional with basis vectors $|\frac{1}{2}, \frac{1}{2}\rangle$, $|\frac{1}{2}, \frac{-1}{2}\rangle$:

$$\langle \frac{1}{2}, \pm\frac{1}{2}|J_-|\frac{1}{2}, \pm\frac{1}{2}\rangle = 0,$$
$$\langle \frac{1}{2}, +\frac{1}{2}|J_-|\frac{1}{2}, -\frac{1}{2}\rangle = 0, \qquad (11.54a)$$
$$\langle \frac{1}{2}, -\frac{1}{2}|J_-|\frac{1}{2}, +\frac{1}{2}\rangle = \hbar.$$

Therefore J_- can be represented by the matrix

$$\hbar \begin{pmatrix} 0 & 0 \\ 1 & 0 \end{pmatrix},$$

where the rows and columns correspond to $\langle m|$ and $|m'\rangle$ respectively. Similarly J_+ is represented by

$$\hbar \begin{pmatrix} 0 & 1 \\ 0 & 0 \end{pmatrix}.$$

Now from $J_\pm = J_x \pm iJ_y$, we have

$$J_x = \frac{1}{2}(J_+ + J_-),\ J_y = -\frac{i}{2}(J_+ - J_-).$$

Therefore, J_x and J_y are represented respectively by

$$\frac{\hbar}{2} \begin{pmatrix} 0 & 1 \\ 1 & 0 \end{pmatrix},\quad \frac{\hbar}{2} \begin{pmatrix} 0 & -i \\ i & 0 \end{pmatrix}.$$

Also from Eq. (11.52), J_z is represented by

$$\frac{\hbar}{2} \begin{pmatrix} 1 & 0 \\ 0 & -1 \end{pmatrix},$$

while from Eq. (11.53), J^2 is represented by the matrix

$$\hbar^2 \begin{pmatrix} \frac{1}{2}(\frac{1}{2}+1) & 0 \\ 0 & \frac{1}{2}(\frac{1}{2}+1) \end{pmatrix} = \frac{3}{4}\hbar^2 \begin{pmatrix} 1 & 0 \\ 0 & 1 \end{pmatrix}.$$

For $j = 1/2$, it is convenient to write

$$\mathbf{J} = (1/2)\hbar\boldsymbol{\sigma} \tag{11.54b}$$

so that $\boldsymbol{\sigma}$ satisfies the commutation relations

$$\boldsymbol{\sigma} \times \boldsymbol{\sigma} = 2i\boldsymbol{\sigma}$$

$$\sigma_x\sigma_y - \sigma_y\sigma_x = 2i\sigma_z, \quad \text{etc.} \tag{11.55}$$

It follows from above that σ_x, σ_y, σ_z are represented by the matrices

$$\sigma_x = \begin{pmatrix} 0 & 1 \\ 1 & 0 \end{pmatrix}, \sigma_y = \begin{pmatrix} 0 & -i \\ i & 0 \end{pmatrix}, \sigma_z = \begin{pmatrix} 1 & 0 \\ 0 & -1 \end{pmatrix}. \tag{11.56}$$

These are known as Pauli matrices. We note that

$$\sigma_x^2 = \sigma_y^2 = \sigma_z^2 = 1,$$

$$\sigma_x\sigma_y + \sigma_y\sigma_x = 0, \tag{11.57a}$$

$$\sigma_y\sigma_z + \sigma_z\sigma_y = 0,$$

$$\sigma_z\sigma_x + \sigma_x\sigma_z = 0.$$

Writing

$$\sigma_x = \sigma_1, \sigma_y = \sigma_2, \sigma_z = \sigma_3,$$

we can rewrite Eqs. (11.55) and (11.57) as

$$[\sigma_i, \sigma_j] = 2i\varepsilon_{ijk}\sigma_k,$$

$$[\sigma_i, \sigma_j]_+ = (\sigma_i\sigma_j + \sigma_j\sigma_i) = 2\delta_{ij}. \qquad (11.57\text{b})$$

Since for $j = 1/2$, the eigenvalues of J_z are $\pm(1/2)\hbar$, it follows from Eq. (11.54) that the eigenvalues of σ_z are given by

$$\sigma'_z = \pm 1. \qquad (11.58)$$

Thus there are two eigenvectors of $J_z = (1/2)\hbar\sigma_z$ with eigenvalue equations

$$\frac{1}{2}\sigma_z|\pm 1/2\rangle = \pm 1/2|\pm 1/2\rangle. \qquad (11.59)$$

In the basis defined by the eigenvectors $|+1/2\rangle$, $|-1/2\rangle$, it is very easy to check by direct substitution of Eq. (11.56) into Eq. (11.59) that the eigenvectors of $J_z = \frac{1}{2}\hbar\sigma_z$ are

$$|+1/2\rangle = \begin{pmatrix} 1 \\ 0 \end{pmatrix}, \quad |-1/2\rangle = \begin{pmatrix} 0 \\ 1 \end{pmatrix}. \qquad (11.60)$$

By taking the commutation relations

$$\mathbf{J} \times \mathbf{J} = i\hbar\mathbf{J}$$

for the angular momentum $\mathbf{J}$, we have uncovered the possibility of

$$j = 1/2, m = \pm 1/2.$$

As far as the orbital angular momentum is concerned, the eigenvalues l and m are restricted to integers. Thus $j = 1/2$, $m = \pm 1/2$ cannot be associated with orbital motion, it must be the intrinsic angular momentum or spin of the particle itself. In particular, the electron has spin $(1/2)\hbar$. In fact many elementary particles (e.g. electron, proton, neutron, neutrino) have spin $(1/2)\hbar$. Particles with spin 0, or $1\hbar$ or $(3/2)\hbar$ or even higher values also exist in nature. We shall, however, confine ourselves here to $j = 1/2$ or spin $(1/2)\hbar$. In this case, in addition to a factor $\psi(\mathbf{r})$ specifying its probability distribution in space, the state function of a spin 1/2 particle has a factor $|\chi\rangle$, specifying its spin state. In the basis defined by the eigenvectors $|+1/2\rangle$ and $|-1/2\rangle$ of $(1/2)\hbar\sigma_z$ belonging to eigenvalues $\pm(1/2)\hbar$, we can represent the general spin state $|\chi\rangle$ as

$$\begin{aligned} |\chi\rangle &= |+1/2\rangle\langle+1/2|\chi\rangle + |-1/2\rangle\langle-1/2|\chi\rangle \\ &= a|+1/2\rangle + b|-1/2\rangle, \end{aligned} \qquad (11.61)$$

where

$$a = \langle +1/2|\chi \rangle, \quad b = \langle -1/2|\chi \rangle.$$

Thus the probability of its having $m = \pm 1/2$ is given by

$$P_\chi(+1/2) = |\langle +1/2|\chi \rangle|^2 = |a|^2,$$
$$P_\chi(-1/2) = |\langle -1/2|\chi \rangle|^2 = |b|^2. \tag{11.62a}$$

The normalizing condition is

$$1 = \langle \chi|\chi \rangle = |a|^2 + |b|^2. \tag{11.62b}$$

This ensures that the total probability of one or other of the two possible spin orientations is unity as it should be.

For a particle with spin $1/2$, we have to take into account two additional degrees of freedom due to spin. Thus we may represent the state function in this case by

$$\psi(\mathbf{r}, (1/2)\sigma_z') \quad \text{or} \quad \psi(\mathbf{r})|\chi \rangle \quad , \text{where}$$
$$\psi(\mathbf{r})|\chi \rangle = \psi(\mathbf{r}, +1/2)| +1/2 \rangle + \psi(\mathbf{r}, -1/2)| -1/2 \rangle \tag{11.63}$$

Let us write

$$\psi(\mathbf{r}, +1/2) = \psi_1(\mathbf{r}),$$
$$\psi(\mathbf{r}, -1/2) = \psi_2(\mathbf{r}), \tag{11.64a}$$

where

$$|\psi_{\frac{1}{2}}(\mathbf{r})|^2 = |\psi(\mathbf{r}, \pm 1/2)|^2 \tag{11.64b}$$

is the probability of finding the particle at the point $\mathbf{r}$ with $\sigma_z' = \pm 1$.

Since $|\pm 1/2 \rangle$ are represented by

$$|+1/2 \rangle = \begin{pmatrix} 1 \\ 0 \end{pmatrix}, \quad |-1/2 \rangle = \begin{pmatrix} 0 \\ 1 \end{pmatrix}, \tag{11.65}$$

we can write the state function $\psi(\mathbf{r})|\chi \rangle$ in the form

$$\begin{pmatrix} \psi_1(\mathbf{r}) \\ \psi_2(\mathbf{r}) \end{pmatrix},$$

i.e. it has two components.

(iii) $j = 1$; the vector space is now 3 dimensional with basis vectors $|1, 1 \rangle$, $|1, 0 \rangle$, $|1, -1 \rangle$. The operators are now 3×3 matrices.

$$J^2 = \hbar^2 \begin{pmatrix} 1(1+1) & 0 & 0 \\ 0 & 1(1+1) & 0 \\ 0 & 0 & 1(1+1) \end{pmatrix} = 2\hbar^2 \begin{pmatrix} 1 & 0 & 0 \\ 0 & 1 & 0 \\ 0 & 0 & 1 \end{pmatrix},$$

$$J_x = \frac{1}{\sqrt{2}}\hbar \begin{pmatrix} 0 & 1 & 0 \\ 1 & 0 & 1 \\ 0 & 1 & 0 \end{pmatrix},$$

$$J_y = \frac{1}{\sqrt{2}}\hbar \begin{pmatrix} 0 & -i & 0 \\ i & 0 & -i \\ 0 & i & 0 \end{pmatrix},$$

$$J_z = \hbar \begin{pmatrix} 1 & 0 & 0 \\ 0 & 0 & 0 \\ 0 & 0 & -1 \end{pmatrix}.$$

11.5   Splitting of the Ground State of Hydrogen Atom

The total angular momentum of an electron in an atom is given by

$$\mathbf{J} = \mathbf{L} + \frac{1}{2}\hbar\boldsymbol{\sigma} = \mathbf{L} + \mathbf{S}, \tag{11.66}$$

where $\mathbf{L}$ is the orbital angular momentum and $\mathbf{S} = \frac{1}{2}\hbar\boldsymbol{\sigma}$ is the spin. Then

$$J_z = L_z + \frac{1}{2}\hbar\sigma_z. \tag{11.67}$$

We now show that spin provides a basis for the Zeeman splitting of the ground state of the hydrogen atom. It is natural that the Hamiltonian (11.10) is now replaced by

$$H_\sigma = H - \frac{eB}{2m_e c}L_z - b\frac{\hbar}{2}\sigma_z \tag{11.68}$$

where we determine the constant b from the consideration that the splitting of the ground state agrees with the experiment. Since σ_z commutes with H (as H does not involve any spin variable) and also with L_z, it follows that the eigenstates of H_σ will be those which are simultaneous eigenstates of H, L_z and σ_z.

For the ground state $E_1(l = 0, m_l = 0)$ the wave functions are

$$u_{100}|+1/2\rangle, u_{100}|-1/2\rangle,$$

where

$$Hu_{100}(r) = E_1 u_{100}(r),$$
$$L_z u_{100}(r) = 0,$$
$$\frac{\hbar}{2}\sigma_z|\pm 1/2\rangle = \pm\frac{\hbar}{2}|\pm 1/2\rangle.$$

Thus it follows that

$$H_\sigma u_{100}(r)|\pm 1/2\rangle = (E_1 \mp b\frac{\hbar}{2})u_{100}|\pm 1/2\rangle \qquad (11.69)$$

i.e. $u_{100}(r)|\pm 1/2\rangle$ are eigenstates of H_σ with eigenvalues

$$E_1^\sigma = E_1 \mp b\frac{\hbar}{2}. \qquad (11.70)$$

Thus we see that the ground state splits into two levels, with a separation

$$\Delta E = b\hbar.$$

Experimentally

$$(\Delta E)_{\text{exp}} = \frac{e\hbar B}{m_e c},$$

therefore

$$b = \frac{eB}{m_e c} \quad \text{and} \quad E_1^\sigma = E_1 \mp \frac{e\hbar}{2m_e c}.$$

Thus

$$H_\sigma = H - \frac{eB}{2m_e c}(L_z + \hbar\sigma_z). \qquad (11.71a)$$

Note the curious fact that in order to get agreement with experiment, H_σ is not obtained from H_B by replacing L_z by $J_z = L_z + \frac{1}{2}\hbar\sigma_z = z$-component of the total angular momentum. The factor of $1/2$ in the spin term is missing in Eq. (11.71a). This is known as the "magnetic anomaly of the spin". We may also say that the magnetic moment of the electron is

$$\mu_e = \frac{e\hbar}{2m_e c},$$

the Bohr magneton, since this is the factor by which a magnetic field **B** has to be multiplied to obtain the observed change in the energy. The so called "magnetic anomaly of the spin" naturally arises in Dirac theory (see chapter 20).

11.6 Addition of Spin

We consider the addition of two spin $1/2$ operators $\mathbf{S}_1$ and $\mathbf{S}_2$. Let $\mathbf{S}_1$ and $\mathbf{S}_2$ be spin operators corresponding to two spin $1/2$ particles. We now consider a system of two spin $1/2$ particles and define the total spin operator

$$\mathbf{S} = \mathbf{S}_1 + \mathbf{S}_2. \qquad (11.71b)$$

Now

$$\mathbf{S}^2 = \mathbf{S}_1^2 + \mathbf{S}_2^2 + 2\mathbf{S}_1 \cdot \mathbf{S}_2, \qquad (11.71c)$$

since $\mathbf{S}_1$ and $\mathbf{S}_2$ commute with each other, for they refer to two independent particles. This implies that $\mathbf{S}$ obeys the usual commutation relations of angular momentum.

Now we can express $\mathbf{S}^2$ as

$$\mathbf{S}^2 = S_1^2 + S_2^2 + S_{1+}S_{2-} + S_{1-}S_{2+} + 2S_{1z}S_{2z}. \qquad (11.71d)$$

Further we note that the eigenstates of S_1^2, S_{1z} and S_2^2, S_{2z} are given by

$$S_1^2|\chi_\pm(1)\rangle = \frac{3}{4}\hbar^2|\chi_\pm(1)\rangle \qquad (11.72a)$$

$$S_{1z}|\chi_\pm(1)\rangle = \pm\frac{1}{2}\hbar|\chi_\pm(1)\rangle \qquad (11.72b)$$

and similar equations for S_2^2 and S_{2z}. Further we note (see problem (11.4))

$$S_{1+}|\chi_+(1)\rangle = 0 \qquad (11.73a)$$

$$S_{1-}|\chi_-(1)\rangle = 0 \qquad (11.73b)$$

$$S_{1-}|\chi_+(1)\rangle = \hbar|\chi_-(1)\rangle \qquad (11.73c)$$

$$S_{1+}|\chi_-(1)\rangle = \hbar|\chi_+(1)\rangle \qquad (11.73d)$$

and similar equations for $S_{2\pm}$.

For the system of two spin 1/2 particles, we can form the following states out of $|\chi_\pm(1)\rangle$ and $|\chi_\pm(2)\rangle$

$$|\chi_{++}(1,2)\rangle = |\chi_+(1)\rangle|\chi_+(2)\rangle, \qquad (11.74a)$$

$$|\chi_{+-}(1,2)\rangle = |\chi_+(1)\rangle|\chi_-(2)\rangle, \qquad (11.74b)$$

$$|\chi_{-+}(1,2)\rangle = |\chi_-(1)\rangle|\chi_+(2)\rangle, \qquad (11.74c)$$

$$|\chi_{--}(1,2)\rangle = |\chi_-(1)\rangle|\chi_-(2)\rangle. \qquad (11.74d)$$

We first show that these states are eigenstates of S_z:

$$S_z|\chi_{++}(1,2)\rangle = (S_{1z} + S_{2z})|\chi_{++}(1,2)\rangle$$

$$= (S_{1z} + S_{2z})|\chi_+(1)\rangle|\chi_+(2)\rangle$$

$$= (\frac{1}{2}\hbar + \frac{1}{2}\hbar)|\chi_{++}(1,2)\rangle$$

$$= \hbar|\chi_{++}(1,2)\rangle, \qquad (11.75a)$$

$$S_z|\chi_{+-}(1,2)\rangle = 0, \qquad (11.75b)$$

$$S_z|\chi_{-+}(1,2)\rangle = 0, \qquad (11.75c)$$

$$S_z|\chi_{--}(1,2)\rangle = -\hbar|\chi_{--}(1,2)\rangle. \qquad (11.75d)$$

From Eqs. (11.75), we see that the eigenvalues of S_z are $-\hbar$, 0, and $\hbar$, where the eigenvalue 0 has multiplicity 2.

To find the eigenstates of S^2, we note using Eqs. (11.74), (11.72) and (11.73) that

$$S^2|\chi_{++}(1,2)\rangle = (S_1^2 + S_2^2 + S_{1+}S_{2-} + S_{1-}S_{2+} + 2S_{1z}S_{2z})|\chi_{++}(1,2)\rangle$$

$$= (\frac{3}{4} + \frac{3}{4} + \frac{1}{2})\hbar^2|\chi_{++}(1,2)\rangle$$

$$= 2\hbar^2|\chi_{++}(1,2)\rangle, \tag{11.76a}$$

$$S^2|\chi_{+-}(1,2)\rangle = (\frac{3}{4} + \frac{3}{4} - \frac{1}{2})\hbar^2|\chi_{+-}(1,2)\rangle + \hbar^2|\chi_{-+}(1,2)\rangle,$$

$$= \hbar^2 \left(|\chi_{+-}(1,2)\rangle + |\chi_{-+}(1,2)\rangle\right), \tag{11.76b}$$

$$S^2|\chi_{-+}(1,2)\rangle = \hbar^2 \left(|\chi_{-+}(1,2)\rangle + |\chi_{+-}(1,2)\rangle\right),$$

$$S^2|\chi_{--}(1,2)\rangle = 2\hbar^2|\chi_{--}(1,2)\rangle. \tag{11.76c}$$

We see that above are not the eigenstates of S^2.

From Eqs. (11.75) and (11.76), we see that normalized simultaneous eigenstates of S^2 and S_z are

$$S^2|\chi_+^{+1}(1,2)\rangle = 2\hbar^2|\chi_+^{+1}(1,2)\rangle, \tag{11.77a}$$

$$S^2|\chi_+^0(1,2)\rangle = 2\hbar^2|\chi_+^0(1,2)\rangle, \tag{11.77b}$$

$$S^2|\chi_+^{-1}(1,2)\rangle = 2\hbar^2|\chi_+^{-1}(1,2)\rangle, \tag{11.77c}$$

$$S^2|\chi_-^0(1,2)\rangle = 0, \tag{11.77d}$$

where

$$|\chi_+^{+1}(1,2)\rangle = |\chi_{++}(1,2)\rangle = |\uparrow\uparrow\rangle, \tag{11.78a}$$

$$|\chi_+^0(1,2)\rangle = \frac{1}{\sqrt{2}}\left(|\chi_{+-}(1,2)\rangle + |\chi_{-+}(1,2)\rangle\right) = \frac{|\uparrow\downarrow\rangle + |\downarrow\uparrow\rangle}{\sqrt{2}}, \tag{11.78b}$$

$$|\chi_+^{-1}(1,2)\rangle = |\chi_{--}(1,2)\rangle = |\downarrow\downarrow\rangle, \tag{11.78c}$$

$$|\chi_-^0(1,2)\rangle = \frac{1}{\sqrt{2}}\left(|\chi_{+-}(1,2)\rangle - |\chi_{-+}(1,2)\rangle\right) = \frac{|\uparrow\downarrow\rangle - |\downarrow\uparrow\rangle}{\sqrt{2}}. \tag{11.78d}$$

Hence we have the result that S^2 has eigenvalues $s(s+1)\hbar^2$, where s takes the values

$$s = 1, 0.$$

Sometimes we will write the spin states simply as

$$\chi_+^{+1}(1,2) = \chi_{++}(1,2) = \chi_+(1)\chi_+(2), \tag{11.79a}$$

$$\chi_+^0(1,2) = \frac{1}{\sqrt{2}}(\chi_+(1)\chi_-(2) + \chi_-(1)\chi_+(2)), \tag{11.79b}$$

$$\chi_+^{-1}(1,2) = \chi_{--}(1,2) = \chi_-(1)\chi_-(2), \tag{11.79c}$$

$$\chi_-^0(1,2) = \frac{1}{\sqrt{2}}(\chi_+(1)\chi_-(2) - \chi_-(1)\chi_+(2)). \tag{11.79d}$$

We conclude that the addition of two spin 1/2 angular momenta give the angular momenta with $j = 0$ and 1.

11.7 Addition of Angular Momenta

The normalized eigenstates $|j, m\rangle$ of J^2 and J_z form basis vectors

$$|j, j\rangle, |j, j - 1\rangle \cdots |j, -j\rangle$$

in a space of $(2j + 1)$ dimensions. In the basis defined above, the angular momentum operators J^2 and J_z are represented by $(2j + 1) \times (2j + 1)$ diagonal matrices for all $j = 0, \frac{1}{2}, 1, \frac{3}{2}, ...$

$$J^2 = \hbar^2 \begin{pmatrix} j(j+1) & & & & \\ & j(j+1) & & & \\ & & \cdot & & \\ & & & \cdot & \\ & & & & j(j+1) \end{pmatrix}$$

$$= j(j+1)\hbar^2 I \tag{11.80a}$$

$$J_z = \hbar \begin{pmatrix} j & & & & \\ & j-1 & & & \\ & & \cdot & & \\ & & & \cdot & \\ & & & & -j \end{pmatrix} \tag{11.80b}$$

where I is an unit matrix of $(2j + 1) \times (2j + 1)$ dimensions. j is called the highest weight of the representation. The particles with spin angular momentum $j = 0, 1, 2...$ are called bosons and those with $j = \frac{1}{2}, \frac{3}{2}, ...$ are called fermions. We saw in chapter 10 that components of angular momentum $\mathbf{J}$ are generators of the rotation group

$$U_R(\theta) = e^{-\frac{i}{\hbar}\theta\,\mathbf{n}\cdot\mathbf{J}}.$$

Consider rotating a spin j particle with by an angle θ around an axis, say z-axis: The rotation that does this is

$$g_j(\theta) = e^{-\frac{i}{\hbar}\theta J_3} = \begin{pmatrix} e^{-i\theta j} & 0 & \cdots & 0 \\ 0 & e^{-i\theta(j-1)} & \cdots & 0 \\ \vdots & \vdots & & \vdots \\ 0 & 0 & \cdots & e^{i\theta j} \end{pmatrix}$$

Thus

$$g_j(2\pi) = I, \qquad j = 0, 1, 2,$$
$$= -I, \qquad j = \frac{1}{2}, \frac{3}{2}.$$

Thus bosons come back to themselves after a rotation by 2π but fermions do not. Pauli's exclusion principle holds for fermions but not for bosons (see Chapter 14). Because they behave so differently it seems unlikely that a symmetry can exist which converts bosons into fermions and fermions into bosons. This is so if one only talks about conserved charges coming from symmetries whose generators satisfy commutation relations. The idea of supersymmetry (which unify fermions and bosons) involve a symmetry in which some generators satisfy commutation relations and some anticommutation relations. We already discussed in Chapter 9 a simple example of a supersymmetric system in the form of the supersymmetric oscillator.

Consider two angular momentum operators $\mathbf{J}_1$ and $\mathbf{J}_2$. The basis vectors for $\mathbf{J}_1$ and $\mathbf{J}_2$ are given by

$$|j_1 m_1\rangle, \qquad m_1 = j_1, \cdots, -j_1$$
$$|j_2 m_2\rangle, \qquad m_2 = j_2, \cdots, -j_2$$

in a space R_1 of $(2j_1 + 1)$ dimensions and in a space R_2 of $(2j_2 + 1)$ dimensions respectively. J_1^2, J_{1z} are represented by matrices of $(2j_1 + 1) \times (2j_1 + 1)$ dimensions, whereas J_2^2 and J_{2z} are represented by matrices of $(2j_2 + 1) \times (2j_2 + 1)$ dimensions. The vectors

$$|j_1 j_2 m_1 m_2\rangle = |j_1 m_1\rangle |j_2 m_2\rangle \tag{11.81}$$

form a basis for the new space of dimension $(2j_1 + 1)(2j_2 + 1)$.

Consider the operator

$$\mathbf{J} = \mathbf{J}_1 + \mathbf{J}_2.$$

This is not a matrix addition, because J_1 and J_2 are represented by matrices of different dimensions. The addition is symbolic. The operators $\mathbf{J}_1$ and $\mathbf{J}_2$ commute, for they refer to two independent systems; this implies that the components of $\mathbf{J}$ obey the usual commutation relations of angular momentum.

Now

$$J_z = J_{1z} + J_{2z}, \tag{11.82a}$$
$$J_z |j_1 j_2 m_1 m_2\rangle = (J_{1z} + J_{2z})|j_1 j_2 m_1 m_2\rangle$$
$$= (J_{1z} + J_{2z})|j_1 m_1\rangle |j_2 m_2\rangle$$
$$= (m_1 + m_2)\hbar |j_1 m_1\rangle |j_2 m_2\rangle$$
$$= (m_1 + m_2)\hbar |j_1 j_2 m_1 m_2\rangle. \tag{11.82b}$$

Thus the eigenvalues of J_z are

$$m = m_1 + m_2. \tag{11.83}$$

Therefore the basis $|j_1 j_2 m_1 m_2\rangle$ of the space $R_1 \times R_2$ is a system of orthonormal eigenvectors of J_z with eigenvalues $m = m_1 + m_2$. Since

$$m_1 = j_1, \cdots, -j_1, \tag{11.84a}$$
$$m_2 = j_2, \cdots, -j_2, \tag{11.84b}$$

therefore m have $(2j_1 + 1)(2j_2 + 1)$ eigenvalues which run from

$$j_1 + j_2, \cdots, -(j_1 + j_2). \tag{11.85}$$

What are the possible values of j? Now $(m)_{\max} = j_1 + j_2$. The maximum possible value of j must also be $(j_1 + j_2)$ since otherwise, there must be a corresponding $m = j > j_1 + j_2 = (m)_{\max}$. Thus

$$j \le j_1 + j_2.$$

Now the possible j values differ by an integer and the total number of eigenvalues of J_z are $(2j_1 + 1)(2j_2 + 1)$ for given j_1 and j_2. But for each j, m has $(2j + 1)$ values, therefore

$$\sum_{j_{\min}}^{j_{\max}} (2j + 1) = (2j_1 + 1)(2j_2 + 1).$$

This is an arithmetic progression with common difference 2 and number of terms $(j_{\max} - j_{\min} + 1)$ so that its sum is

$$(j_{\max} - j_{\min} + 1)((2j_{\min} + 1) + (j_{\max} - j_{\min}))$$
$$= (2j_1 + 1)(2j_2 + 1).$$

This gives, on using that $j_{\max} = j_1 + j_2$, $j_{\min}^2 = (j_1 - j_2)^2$, i.e.

$$j_{\min} = |j_1 - j_2|.$$

Thus, the possible j values must obey the inequalities

$$|j_1 - j_2| \le j \le j_1 + j_2. \tag{11.86}$$

As an example consider

$$\mathbf{J} = \mathbf{L} + \mathbf{S}, \quad \mathbf{S} = \frac{1}{2}\hbar\boldsymbol{\sigma},$$

so that $j_1 = l$ and $j_2 = \frac{1}{2}$. Thus, we have

$$\left| l - \frac{1}{2} \right| \le j \le l + \frac{1}{2},$$

i.e. j only has two possible values

$$l \mp \frac{1}{2}, \; l \neq 0$$

and for $l = 0$, $j = \frac{1}{2}$ only. The following table summarizes the situation for some low values of l:

l	Possible value of j	m	$(2j_1 + 1)(2j_2 + 1)$ $= 2(2l + 1)$	
0	$\frac{1}{2}$	$-\frac{1}{2}, \frac{1}{2}$	2	
1	$\frac{3}{2}$	$-\frac{3}{2}, -\frac{1}{2}, \frac{1}{2}, \frac{3}{2}$		
	$\frac{1}{2}$	$-\frac{1}{2}, \frac{1}{2}$	6	$4 \oplus 2$
2	$\frac{5}{2}$	$-\frac{5}{2}, -\frac{3}{2}, -\frac{1}{2}, \frac{1}{2}, \frac{3}{2}, \frac{5}{2}$		
	$\frac{3}{2}$	$-\frac{3}{2}, -\frac{1}{2}, \frac{1}{2}, \frac{3}{2}$	10	$6 \oplus 4$

From Eq. (11.86) we have the result that the angular momentum operator J^2 has the eigenvalues $j(j + 1)\hbar^2$, where j can take the following values

$$j_1 + j_2, \cdots , |j_1 - j_2|.$$

It may be noted that the product space of dimension $((2j_1+1) \times (2j_2 + 1))$ is reducible, that is the space $R_1 \times R_2$ splits into a number of invariant irreducible subspaces, each corresponding to one of the following allowed values of j:

$$j_1 + j_2, j_1 + j_2 - 1, \cdots , |j_1 - j_2|.$$

The total number of independent vectors for the irreducible subspaces, which correspond to possible values of j are given in the following table:

Highest weight of irreducible representation	Number of independent vectors
$j_1 + j_2$	$2j_1 + 2j_2 + 1$
$j_1 + j_2 - 1$	$2j_1 + 2j_2 - 1$
$j_1 + j_2 - 2$	$2j_1 + 2j_2 - 3$
$\cdots$	$\cdots$
$\cdots$	$\cdots$
$|j_1 - j_2|$	$|2j_1 - j_2| + 1$

Clebsch–Gordan Coefficients

Basis vectors of the product space have been labeled as

$$|j_1 m_1\rangle |j_2 m_2\rangle = |j_1 j_2 m_1 m_2\rangle,$$

corresponding to the commuting set J_1^2, J_2^2, J_{1z}, J_{2z}. A second way of labeling a vector of product space is to write it as

$$|j m j_1 j_2\rangle$$

which corresponds to the commuting set J^2, J_z, J_1^2, J_2^2 (note that J^2 does not commute with J_{1z} and J_{2z}). We have

$$J_z |j m j_1 j_2\rangle = m\hbar |j m j_1 j_2\rangle,$$
$$J^2 |j m j_1 j_2\rangle = j(j+1)\hbar^2 |j m j_1 j_2\rangle.$$

Now, the states $|j m j_1 j_2\rangle$ or $|j_1 j_2 m_1 m_2\rangle$ form complete sets so that we can write

$$|j m j_1 j_2\rangle = \sum_{m_1 m_2} |j_1 j_2 m_1 m_2\rangle \langle j_1 j_2 m_1 m_2 | j m j_1 j_2\rangle, \qquad (11.87a)$$

$$|j_1 j_2 m_1 m_2\rangle = \sum_{\substack{j,m \\ m = m_1 + m_2}} |j m j_1 j_2\rangle \langle j m j_1 j_2 | j_1 j_2 m_1 m_2\rangle. \qquad (11.87b)$$

We shall choose the phase factors in such a way that the Clebsch–Gordan coefficients (C.G.), namely, the coefficients in expansion (11.87a) are real

$$\langle j_1 j_2 m_1 m_2 | j m j_1 j_2\rangle = \langle j m j_1 j_2 | j_1 j_2 m_1 m_2\rangle^*$$
$$= \langle j m j_1 j_2 | j_1 j_2 m_1 m_2\rangle \quad \text{(reality)}. \qquad (11.88)$$

Orthogonality relations:

$$\langle j_1 j_2 m_1' m_2' | j_1 j_2 m_1 m_2\rangle = \sum_{jm} \langle j_1 j_2 m_1' m_2' | j m j_1 j_2\rangle \langle j m j_1 j_2 | j_1 j_2 m_1 m_2\rangle$$
$$= \delta_{m_1 m_1'} \delta_{m_2 m_2'} \qquad (11.89a)$$

since the left-hand side vanishes unless $m_1' = m_2$, $m_2' = m_2$. Similarly we have

$$\langle j m j_1 j_2 | j' m' j_1 j_2\rangle$$
$$= \sum_{m_1 m_2} \langle j m j_1 j_2 | j_1 j_2 m_1 m_2\rangle \langle j_1 j_2 m_1 m_2 | j' m' j_1 j_2\rangle$$
$$= \delta_{jj'} \delta_{mm'} \delta(j, j_1, j_2), \qquad (11.89b)$$

since the left-hand side vanishes unless $j = j'$, $m = m'$. Now $\delta(j, j_1, j_2) = 1$ if j has one of the values of the set

$$j_1 + j_2, j_1 + j_2 - 1, \cdots, |j_1 - j_2|$$

i.e.

$$|j_1 - j_2| \le j \le j_1 + j_2$$

and zero otherwise.

The Clebsch–Gordan coefficients can be calculated, but the calculation is quite cumbersome and it is easier to find the C.G. coefficients from books of tables.

Here we give two tables for $j_2 = 1/2$ and $j_2 = 1$. For $j_2 = 1/2$, $m_2 = \pm 1/2$, the C.G. coefficients are given by the following Table I.

Table I

j	$m_2 = 1/2$	$m_2 = -1/2$
$j + 1/2$	$\sqrt{\frac{j_1+m+1/2}{2j_1+1}}$	$\sqrt{\frac{j_1-m+1/2}{2j_1+1}}$
$j - 1/2$	$-\sqrt{\frac{j_1-m+1/2}{2j_1+1}}$	$\sqrt{\frac{j_1+m+1/2}{2j_1+1}}$

For $j_2 = 1$, $m_2 = +1$, 0, -1, the C.G. coefficients $\langle j_1 1 m_1 m_2 | jm j_1 1 \rangle$ are given in Table II.

Table II

j	$m_2 = 1$	$m_2 = 0$	$m_2 = -1$
$j = j_1 + 1$	$\left[\frac{(j_1+m)(j_1+m+1)}{(2j_1+1)(2j_1+2)}\right]^{\frac{1}{2}}$	$\left[\frac{(j_1+1-m)(j_1+m+1)}{(2j_1+1)(j_1+1)}\right]^{\frac{1}{2}}$	$\left[\frac{(j_1-m)(j_1-m+1)}{(2j_1+1)(2j_1+2)}\right]^{\frac{1}{2}}$
$j = j_1$	$-\left[\frac{(j_1+m)(j_1-m+1)}{2j_1(j_1+1)}\right]^{\frac{1}{2}}$	$\frac{m}{[j_1(j_1+1)]^{1/2}}$	$\left[\frac{(j_1-m)(j_1+m+1)}{2j_1(j_1+1)}\right]^{\frac{1}{2}}$
$j = j_1 - 1$	$\left[\frac{(j_1-m)(j_1-m+1)}{2j_1(2j_1+1)}\right]^{\frac{1}{2}}$	$-\left[\frac{(j_1-m)(j_1+m)}{j_1(2j_1+1)}\right]^{\frac{1}{2}}$	$\left[\frac{(j_1+m+1)(j_1+m)}{2j_1(2j_1+1)}\right]^{\frac{1}{2}}$

As an application, consider the case

$$\mathbf{J} = \mathbf{L} + \mathbf{S},$$

where $\mathbf{L}$ is the orbital angular momentum and $\mathbf{S}$ represents the spin $\frac{1}{2}$ operator. As already seen j can only take two values viz. $l + \frac{1}{2}$ and $l - \frac{1}{2}$. Thus

$$J^2|j = l + \tfrac{1}{2}, \, m\rangle = (l + \tfrac{1}{2})(l + \tfrac{3}{2})\hbar^2|j = l + \tfrac{1}{2}, \, m\rangle, \quad (11.90a)$$

$$J^2|j = l - \tfrac{1}{2}, \, m\rangle = (l - \tfrac{1}{2})(l + \tfrac{1}{2})\hbar^2|j = l - \tfrac{1}{2}, \, m\rangle, \quad (11.90b)$$

$$J_z|jm\rangle = m\hbar|jm\rangle, \quad (11.90c)$$

where $j = l + \frac{1}{2}, l - \frac{1}{2}, m = m_l \pm \frac{1}{2}$. Using the C.G. coefficients (Table I) we can write

$$|l + \tfrac{1}{2}, \, m\rangle = \sqrt{\frac{l + m + \frac{1}{2}}{2l + 1}} |l\tfrac{1}{2}m - \tfrac{1}{2}\tfrac{1}{2}\rangle + \sqrt{\frac{l - m + \frac{1}{2}}{2l + 1}} |l\tfrac{1}{2}m + \tfrac{1}{2} - \tfrac{1}{2}\rangle,$$

(11.91)

$$|l - \tfrac{1}{2}, \, m\rangle = -\sqrt{\frac{l - m + \frac{1}{2}}{2l + 1}} |l\tfrac{1}{2}m - \tfrac{1}{2}\tfrac{1}{2}\rangle + \sqrt{\frac{l + m + \frac{1}{2}}{2l + 1}} |l\tfrac{1}{2}m + \tfrac{1}{2} - \tfrac{1}{2}\rangle.$$

(11.92)

Denoting the spin wave functions

$$|\tfrac{1}{2}\tfrac{1}{2}\rangle = \chi_+, |\tfrac{1}{2} - \tfrac{1}{2}\rangle = \chi_-$$

(11.93a)

and noting that

$$\langle \theta\phi | lm_l \rangle = Y_{lm_l}(\theta, \phi),$$

(11.93b)

$$|l\tfrac{1}{2}m - \tfrac{1}{2}\tfrac{1}{2}\rangle = |lm - \tfrac{1}{2}\rangle|\tfrac{1}{2} + \tfrac{1}{2}\rangle$$

$$= |lm - \tfrac{1}{2}\rangle\chi_+,$$

(11.94)

etc. we have

$$\psi_{j=l+\frac{1}{2},m}(\theta, \phi) = \sqrt{\frac{l + m + \frac{1}{2}}{2l + 1}} Y_{lm-\frac{1}{2}}|\chi_+\rangle + \sqrt{\frac{l - m + \frac{1}{2}}{2l + 1}} Y_{lm+\frac{1}{2}}|\chi_-\rangle,$$

(11.95)

$$\psi_{j=l-\frac{1}{2},m}(\theta, \phi) = -\sqrt{\frac{l - m + \frac{1}{2}}{2l + 1}} Y_{lm-\frac{1}{2}}|\chi_+\rangle + \sqrt{\frac{l + m + \frac{1}{2}}{2l + 1}} Y_{lm+\frac{1}{2}}|\chi_-\rangle.$$

(11.96)

Now

$$\mathbf{J}^2 = \mathbf{L}^2 + \mathbf{S}^2 + 2\mathbf{L} \cdot \mathbf{S},$$

(11.97)

since $\mathbf{L}$ and $\mathbf{S}$ commute with each other. Thus we see that $\mathbf{L} \cdot \mathbf{S}$ has the following eigenvalues

$$\frac{1}{2}\left(j(j + 1)\hbar^2 - l(l + 1)\hbar^2 - s(s + 1)\hbar^2\right),$$

(11.98)

where $j = l + \frac{1}{2}, l - \frac{1}{2}, s = \frac{1}{2}$. Hence the eigenvalues of $\mathbf{L} \cdot \mathbf{S}$ are $(1/2) l\hbar^2$ and $(-1/2)(l + 1)\hbar^2$ respectively. It is then clear from Eqs. (11.95)–(11.97) that the corresponding eigenstates are

$$\mathbf{L} \cdot \mathbf{S}\psi_{l+1/2,m} = \frac{1}{2}l\hbar^2\psi_{l+1/2,m},$$

(11.99)

$$\mathbf{L} \cdot \mathbf{S}\psi_{l-1/2,m} = -\frac{1}{2}(l + 1)\hbar^2\psi_{l-1/2,m}.$$

(11.100)

11.8 Rotations: Rotation Matrices

As we have shown in Sec. 10.5, rotations in space are most conveniently described in terms of angular momentum operators $\mathbf{J}$. A rotation of coordinates through angle ω about the direction $\hat{\omega}$ in the positive sense is described by the operator

$$R_\omega = e^{\frac{-i}{\hbar}\omega\cdot\mathbf{J}}. \tag{11.101}$$

Since $\mathbf{J}^2$ commutes with J_x, J_y and J_z, it follows that

$$[R_\omega, \mathbf{J}^2] = 0 \tag{11.102}$$

i.e. a rotation does not change the total angular momentum of the system. Thus

$$\mathbf{J}^2 R_\omega |jm\rangle = R_\omega \mathbf{J}^2 |jm\rangle = j(j+1)\hbar^2 R_\omega |jm\rangle. \tag{11.103}$$

Consider in particular the following rotation. Let $\mathbf{p}_0$ specify the initial direction and $\mathbf{p}$ the desired final direction, $\mathbf{p}$ being obtained from $\mathbf{p}_0$ by a rotation in the positive sense about the direction

$$\mathbf{n} = \mathbf{p}_0 \times \mathbf{p}/|\mathbf{p}_0 \times \mathbf{p}|. \tag{11.104}$$

The rotation angle is determined from the relation

$$\mathbf{p}_0 \cdot \mathbf{p} = \cos\theta, \quad 0 \le \theta \le \pi.$$

In this case the desired rotation is given by

$$R_\theta = e^{-\frac{i}{\hbar}\theta\mathbf{n}\cdot\mathbf{J}}. \tag{11.105}$$

Let us take $\mathbf{p}_0$ along the z-axis ($\mathbf{p}_0$ unit vector along z-axis)

Now θ and ϕ are polar and azimuthal angles of $\mathbf{p}$ in the system in which $\mathbf{p}_0$ is along the z-axis

$$\mathbf{p} = (\sin\theta\cos\phi, \sin\theta\sin\phi, \cos\theta), \tag{11.106a}$$
$$\mathbf{p}_0 \times \mathbf{p} = -\mathbf{i}\sin\theta\sin\phi + \mathbf{j}\sin\theta\cos\phi,$$
$$\mathbf{n} = (-\sin\phi, \cos\phi, 0) = \frac{\mathbf{p}_0 \times \mathbf{p}}{\sin\theta}.$$

Now

$$\mathbf{n} \times \mathbf{p}_0 = \frac{(\mathbf{p}_0 \times \mathbf{p}) \times \mathbf{p}_0}{\sin\theta}$$
$$= \frac{\mathbf{p} - \mathbf{p}_0\cos\theta}{\sin\theta} \tag{11.106b}$$

giving for the infinitesimal rotation θ about direction $\mathbf{n}$

$$\mathbf{p} = \mathbf{p}_0 + \theta\mathbf{n} \times \mathbf{p}_0.$$

Compare it with

$$\mathbf{p}_0' = \mathbf{p}_0 + \mathbf{p}_0 \times \theta \mathbf{n}.$$

In this case we look at $\mathbf{p}_0$ in the rotated frame of reference whereas in the former case we looked at the rotated vector itself. The former can be obtained from the latter by substituting $\theta \to -\theta$. We now show that R_θ can be written in a rather simple form. We have

$$\mathbf{n} \cdot \mathbf{J} = -J_x \sin \phi + J_y \cos \phi. \qquad (11.107)$$

Let us look at

$$e^{-i(\phi/\hbar)J_z} J_y e^{(i\phi/\hbar)J_z} = \left(1 - \frac{i\phi}{\hbar} J_z + \frac{(i\phi)^2}{2!\hbar^2} J_z^2 \cdots\right)$$

$$\times J_y \left(1 + \frac{i\phi}{\hbar} J_z + \frac{(i\phi)^2}{2!\hbar^2} J_z^2 \cdots\right)$$

$$= J_y + (-\frac{i\phi}{\hbar})[J_z, J_y] + \frac{(-i\phi)^2}{2!\hbar^2}[J_z, [J_z, J_y]] + \cdots$$

$$= J_y - \phi J_x - \frac{\phi^2}{2!} J_y + \frac{\phi^3}{3!} J_x \cdots$$

$$= J_y \cos \phi - J_x \sin \phi = \mathbf{n} \cdot \mathbf{J} \qquad (11.108)$$

and

$$e^{-\frac{i\phi}{\hbar} J_z} J_y^2 e^{\frac{i\phi}{\hbar} J_z} = e^{-\frac{i\phi}{\hbar} J_z} J_y e^{\frac{i\phi}{\hbar} J_z} e^{-\frac{i\phi}{\hbar} J_z} J_y e^{\frac{i\phi}{\hbar} J_z}$$

$$= (J_y \cos \phi - J_x \sin \phi)^2 = (\mathbf{n} \cdot \mathbf{J})^2.$$

Generalizing these results, we see that

$$e^{-\frac{i\phi}{\hbar} J_z} e^{-\frac{i\theta}{\hbar} J_y} e^{\frac{i\phi}{\hbar} J_z}$$

$$= e^{-\frac{i\phi}{\hbar} J_z} (1 - \frac{i\theta}{\hbar} J_y + \frac{1}{2!}(-\frac{i\theta}{\hbar})^2 J_y^2 + \cdots) e^{\frac{i\phi}{\hbar} J_z}$$

$$= e^{-\frac{i\theta}{\hbar} \mathbf{n} \cdot \mathbf{J}}. \qquad (11.109)$$

Hence we have the important result

$$R_\theta \equiv R(\phi, \theta, -\phi) = e^{-\frac{i\theta}{\hbar} \mathbf{n} \cdot \mathbf{J}}$$

$$= e^{-\frac{i\phi}{\hbar} J_z} e^{-\frac{i\theta}{\hbar} J_y} e^{\frac{i\phi}{\hbar} J_z}. \qquad (11.110)$$

It follows from Eq. (11.106a) that $R_\theta = R(\phi, \theta, -\phi)$ acting on a state with $|p, \theta = 0, \phi = 0\rangle$ gives the state

$$|p, \theta, \phi\rangle = R(\phi, \theta, -\phi)|p, \theta = 0, \phi = 0\rangle. \qquad (11.111)$$

We have seen that rotation operator $R_\omega = e^{-\frac{i}{\hbar}\boldsymbol{\omega}\cdot\mathbf{J}}$ commutes with $\mathbf{J}^2$ and $R_\omega|jm\rangle$ is also an eigenstate of $\mathbf{J}^2$ with eigenvalue $j(j+1)\hbar^2$. Thus, we can write

$$R_\omega|jm\rangle = \sum_{m'}|jm'\rangle\langle jm'|R_\omega|jm\rangle$$

$$= \sum_{m'=-j}^{j}|jm'\rangle d^{(j)}_{m'm}(\boldsymbol{\omega}) \qquad (11.112)$$

where

$$d^{(j)}_{m'm}(\boldsymbol{\omega}) = \langle jm'|R_\omega|jm\rangle = \langle jm'|e^{-\frac{i}{\hbar}\boldsymbol{\omega}\cdot\mathbf{J}}|jm\rangle. \qquad (11.113)$$

$d^{(j)}(\boldsymbol{\omega})$ is a $(2j+1)$ by $(2j+1)$ matrix whose matrix elements are $d^{(j)}_{m'm}(\boldsymbol{\omega})$; $d^{(j)}(\boldsymbol{\omega})$ is called a rotation matrix.

The rotations form a group, i.e. they satisfy the following four properties
i) Identity rotation operator exists

$$R_{\omega=0} = 1.$$

ii) Inverse of R_ω exists

$$R_\omega = e^{-\frac{i\boldsymbol{\omega}}{\hbar}\cdot\mathbf{J}},$$
$$R_{-\omega} = e^{\frac{i\boldsymbol{\omega}}{\hbar}\cdot\mathbf{J}},$$
$$R_\omega R_{-\omega} = 1.$$

iii) Two successive rotations made in a row say $\boldsymbol{\omega}_1$ followed by $\boldsymbol{\omega}_2$ are equivalent to a single rotation

$$R_\omega = R_{\omega_2}R_{\omega_1} = e^{-\frac{i\boldsymbol{\omega}_2}{\hbar}\cdot\mathbf{J}}e^{-\frac{i}{\hbar}\boldsymbol{\omega}_1\cdot\mathbf{J}}$$
$$\neq R_{\omega_1}R_{\omega_2}, \quad \text{in general,}$$

since $\boldsymbol{\omega}_2\cdot\mathbf{J}$ and $\boldsymbol{\omega}_1\cdot\mathbf{J}$ do not commute in general.
iv) The product operation is associative

$$(R_{\omega_1}R_{\omega_2})R_{\omega_3} = R_{\omega_1}(R_{\omega_2}R_{\omega_3}).$$

We now show that the set of matrices $d^j(\boldsymbol{\omega})$ gives a representation of the rotation group. Now

$$\langle jm'|R_\omega|jm\rangle = \langle jm'|R_{\omega_2}R_{\omega_1}|jm\rangle = \sum_{m''}\langle jm'|R_{\omega_2}|jm''\rangle\langle jm''|R_{\omega_1}|jm\rangle$$

$$= \sum_{m''}d^{(j)}_{m'm''}(\boldsymbol{\omega}_2)d^{(j)}_{m''m'}(\boldsymbol{\omega}_1) \qquad (11.114)$$

or

$$d^{(j)}_{m'm}(\boldsymbol{\omega}) = \sum_{m''} d^{(j)}_{m'm''}(\boldsymbol{\omega}_2)d^{(j)}_{m''m}(\boldsymbol{\omega}_1) \qquad (11.115)$$

or equivalently

$$d^{(j)}(\boldsymbol{\omega}) = d^{(j)}(\boldsymbol{\omega}_2)d^{(j)}(\boldsymbol{\omega}_1). \qquad (11.116)$$

Because of this property the rotation matrix $d^{(j)}(\boldsymbol{\omega})$ gives a representation of the rotation group. For the identity transformation (no rotation), $\boldsymbol{\omega} = 0$ and

$$d^{(j)}_{m'm}(0) = \langle jm'|jm \rangle = \delta_{m'm}. \qquad (11.117)$$

The matrices $d^{(j)}(\boldsymbol{\omega})$ are unitary, since

$$\left[d^{(j)}(\boldsymbol{\omega})^\dagger \right]_{m'm} = \langle jm'|e^{(i/\hbar)\boldsymbol{\omega}\cdot\mathbf{J}}|jm \rangle$$

$$= \langle jm|e^{-(i/\hbar)\boldsymbol{\omega}\cdot\mathbf{J}}|jm' \rangle^*$$

$$= d^{(j)}_{mm'}(\boldsymbol{\omega})^* = d^{(j)}_{m'm}(-\boldsymbol{\omega})$$

$$d^{(j)^\dagger}(\boldsymbol{\omega}) = d^{(j)}(-\boldsymbol{\omega})$$

or

$$d^{(j)^\dagger}(\boldsymbol{\omega})d^{(j)}(\boldsymbol{\omega}) = 1. \qquad (11.118)$$

For the infinitesimal rotation $\boldsymbol{\omega} = \boldsymbol{\epsilon}$, $d^j(\boldsymbol{\omega})$ has a simple representation. Writing $\epsilon_\pm = \epsilon_x \pm i\epsilon_y$, we can write $\boldsymbol{\epsilon} \cdot \mathbf{J}$ in terms of raising and lowering operators $J_\pm$, i.e.

$$\boldsymbol{\epsilon} \cdot \mathbf{J} = \frac{1}{2}(\epsilon_+ J_- + \epsilon_- J_+) + \epsilon_z J_z, \qquad (11.119)$$

so that $d^{(J)}(\boldsymbol{\epsilon})$ has a simple representation

$$d^{(j)}_{m'm}(\boldsymbol{\epsilon}) = \langle jm'|(1 - i\frac{\boldsymbol{\epsilon}}{\hbar} \cdot \mathbf{J})|jm \rangle + O(\epsilon^2)$$

$$= \delta_{mm'} - \frac{i}{2\hbar}\epsilon_+ \langle jm'|J_-|jm \rangle$$

$$- \frac{i}{2\hbar}\epsilon_- \langle jm'|J_+|jm \rangle$$

$$- \frac{i}{\hbar}\epsilon_z \langle jm'|J_z|jm \rangle. \qquad (11.120)$$

The use of Eqs. (11.49) and (11.50) then give

$$d^{(j)}_{m'm}(\boldsymbol{\epsilon}) = \delta_{mm'} - \frac{i\epsilon_+}{2}\sqrt{(j+m)(j-m+1)}\delta_{m',m-1}$$

$$- \frac{i\epsilon_-}{2}\sqrt{(j-m)(j+m+1)}\delta_{m',m+1}$$

$$- i\epsilon_z m\delta_{mm'}, \qquad (11.121)$$

where

$$j = 0, 1/2, 1, \cdots.$$

For the finite rotation $\boldsymbol{\omega} = \theta\mathbf{n}$, $R(\phi, \theta, -\phi) = e^{-\frac{i\theta}{\hbar}\mathbf{n}\cdot\mathbf{J}}$ and we have

$$\begin{aligned}
d^{(j)}_{m'm}(\phi, \theta, -\phi) &= \langle jm'|R(\phi, \theta, -\phi)|jm\rangle \\
&= \langle jm'|e^{-\frac{i\phi}{\hbar}J_z}e^{-\frac{i\theta}{\hbar}J_y}e^{\frac{i\phi}{\hbar}J_z}|jm\rangle \\
&= e^{-im'\phi}e^{im\phi}\langle jm'|e^{-\frac{i\theta}{\hbar}J_y}|jm\rangle \\
&= e^{i(m-m')\phi}d^{(j)}_{m'm}(\theta), \quad\quad\quad (11.122)
\end{aligned}$$

$$d^{(j)}_{m'm}(\theta) = \langle jm'|e^{-i\frac{\theta}{\hbar}J_y}|jm\rangle. \quad\quad\quad (11.123)$$

Thus to construct $d^{(j)}$, we need only the matrix elements for a rotation about the y-axis by an angle θ; ϕ dependence is trivial. We will write $d^{(j)}_{m'm}(\phi, \theta, -\phi) \equiv d^{(j)}_{m'm}(\phi, \theta)$.

The matrix elements $d^{(j)}_{m'm}(\theta)$ turn out to be real and have the simple property

$$d^{(j)}_{-m',-m}(\theta) = (-1)^{2j-m-m'}d^{(j)}_{m'm}(\theta). \quad\quad\quad (11.124)$$

Now

$$R(\phi, \theta, -\phi)|\theta = 0, \phi = 0\rangle = |\theta\phi\rangle,$$

so that

$$\langle jm|R(\phi, \theta, -\phi)|00\rangle = \langle jm|\theta\phi\rangle \quad\quad\quad (11.125)$$

or

$$\sum_{m'}\langle jm|R(\phi, \theta, -\phi)|jm'\rangle\langle jm'|00\rangle = \langle jm|\theta\phi\rangle,$$

i.e.

$$\sum_{m'}d^{(j)}_{mm'}(\phi, \theta)\langle jm'|00\rangle = \langle jm|\theta\phi\rangle. \quad\quad\quad (11.126)$$

For the particular case j being an integer, we have

$$\sum_{m'}d^{(j)}_{mm'}(\phi, \theta)Y^*_{jm'}(0, 0) = Y^*_{jm}(\theta, \phi), \quad\quad\quad (11.127)$$

where

$$Y_{jm}(\theta, \phi) = \langle \theta, \phi|jm\rangle.$$

Now

$$Y_{jm'}(0, 0) = \delta_{m',0}\sqrt{\frac{2j+1}{4\pi}}, \quad\quad\quad (11.128)$$

and we have

$$\sqrt{\frac{2j+1}{4\pi}} \sum_{m'} d_{mm'}^{(j)}(\phi,\theta)\delta_{m',0} = Y_{jm}^*(\theta,\phi) \qquad (11.129)$$

or

$$d_{m0}^{(j)}(\phi,\theta) = \sqrt{\frac{4\pi}{2j+1}}Y_{jm}^*(\theta,\phi) \qquad (11.130)$$

or

$$e^{-im\phi}d_{m0}^{(j)}(\theta) = \sqrt{\frac{4\pi}{2j+1}}Y_{jm}^*(\theta,\phi). \qquad (11.131)$$

An application of this result is given in problem (11.18).

Rotation Matrices for the Combination of Two Angular Momenta

Consider $\mathbf{J} = \mathbf{J}_1 + \mathbf{J}_2$, then we show that using the C.G. coefficients, we can express the rotation matrix corresponding to $\mathbf{J}$ in terms of products of those for $\mathbf{J}_1$ and $\mathbf{J}_2$.
Now

$$e^{-\frac{i}{\hbar}\boldsymbol{\omega}\cdot\mathbf{J}} = e^{-\frac{i}{\hbar}\boldsymbol{\omega}\cdot(\mathbf{J}_1+\mathbf{J}_2)}$$

$$= e^{-\frac{i}{\hbar}\boldsymbol{\omega}\cdot\mathbf{J}_1}e^{-\frac{i}{\hbar}\boldsymbol{\omega}\cdot\mathbf{J}_2}, \qquad (11.132)$$

since $\mathbf{J}_1$ and $\mathbf{J}_2$ commute with each other. Therefore, we have

$$e^{-(i/\hbar)\boldsymbol{\omega}\cdot\mathbf{J}}|j_1j_2m_1m_2\rangle$$
$$= e^{-(i/\hbar)\boldsymbol{\omega}\cdot\mathbf{J}_1}|j_1m_1\rangle e^{-(i/\hbar)\boldsymbol{\omega}\cdot\mathbf{J}_2}|j_2m_2\rangle$$
$$= \sum_{m_1'}|j_1m_1'\rangle d_{m_1'm_1}^{(j_1)}(\boldsymbol{\omega})\sum_{m_2'}|j_2m_2'\rangle d_{m_2'm_2}^{(j_2)}(\boldsymbol{\omega})$$
$$= \sum_{m_1'm_2'}|j_1j_2m_1'm_2'\rangle d_{m_1'm_1}^{(j_1)}(\boldsymbol{\omega})d_{m_2'm_2}^{(j_2)}(\boldsymbol{\omega}) \qquad (11.133)$$

or

$$\langle j_1j_2m_1''m_2''|e^{-(i/\hbar)\boldsymbol{\omega}\cdot\mathbf{J}}|j_1j_2m_1m_2\rangle$$
$$= \sum_{m_1'm_2'}\langle j_1j_2m_1''m_2''|j_1j_2m_1'm_2'\rangle d_{m_1'm_1}^{(j_1)}(\boldsymbol{\omega})d_{m_2'm_2}^{(j_2)}(\boldsymbol{\omega})$$
$$= \sum_{m_1'm_2'}\delta_{m_1''m_1'}\delta_{m_2''m_2'}d_{m_1'm_1}^{(j_1)}(\boldsymbol{\omega})d_{m_2'm_2}^{(j_2)}(\boldsymbol{\omega})$$
$$= d_{m_1''m_1}^{(j_1)}(\boldsymbol{\omega})d_{m_2''m_2}^{(j_2)}(\boldsymbol{\omega}).$$

Hence

$$\langle j_1 j_2 m_1' m_2' | e^{-(i/\hbar)\boldsymbol{\omega}\cdot\mathbf{J}} | j_1 j_2 m_1 m_2 \rangle = d^{(j_1)}_{m_1' m_1}(\boldsymbol{\omega}) d^{(j_2)}_{m_2' m_2}(\boldsymbol{\omega}). \quad (11.134)$$

But

$$\langle j_1 j_2 m_1' m_2' | e^{-(i/\hbar)\mathbf{J}\cdot\boldsymbol{\omega}} | j_1 j_2 m_1 m_2 \rangle$$
$$= \sum_{jj'm'm} \langle j_1 j_2 m_1' m_2' | j'm' j_1 j_2 \rangle \langle j'm' j_1 j_2 | e^{-(i/\hbar)\mathbf{J}\cdot\boldsymbol{\omega}} | jm j_1 j_2 \rangle$$
$$\times \langle jm j_1 j_2 | j_1 j_2 m_1 m_2 \rangle. \quad (11.135)$$

Now since $e^{-(i/\hbar)\mathbf{J}\cdot\boldsymbol{\omega}}|jm j_1 j_2\rangle$ is an eigenstate of J^2, we can write [c.f. Eq. (11.113)]

$$\langle j'm' j_1 j_2 | e^{-(i/\hbar)\mathbf{J}\cdot\boldsymbol{\omega}} | jm j_1 j_2 \rangle = \delta_{jj'} d^{(j)}_{m'm}(\boldsymbol{\omega}). \quad (11.136)$$

Hence

$$\langle j_1 j_2 m_1' m_2' | e^{-(i/\hbar)\mathbf{J}\cdot\boldsymbol{\omega}} | j_1 j_2 m_1 m_2 \rangle$$
$$= \sum_{jmm'} \langle j_1 j_2 m_1' m_2' | jm' j_1 j_2 \rangle \langle jm j_1 j_2 | j_1 j_2 m_1 m_2 \rangle d^{(j)}_{m'm}(\boldsymbol{\omega})$$

$$(11.137)$$

and on using Eq. (11.134) and the reality of C.G. coefficient, we have

$$d^{(j_1)}_{m_1' m_1}(\boldsymbol{\omega}) d^{(j_2)}_{m_2' m_2}(\boldsymbol{\omega})$$
$$= \sum_{jmm'} \langle j_1 j_2 m_1' m_2' | jm' j_1 j_2 \rangle \langle j_1 j_2 m_1 m_2 | jm j_1 j_2 \rangle d^{(j)}_{m'm}(\boldsymbol{\omega}).$$

$$(11.138)$$

Note that the factors multiplying $d^{(j)}_{m'm}(\boldsymbol{\omega})$ are C.G. coefficients. On the other hand, the inverse relation is given by

$$\sum_{m_1 m_1'} \sum_{m_2 m_2'} d^{(j_1)}_{m_1' m_1}(\boldsymbol{\omega}) d^{(j_2)}_{m_2' m_2}(\boldsymbol{\omega}) \langle j_1 j_2 m_1 m_2 | jm j_1 j_2 \rangle \langle j'm' j_1 j_2 | j_1 j_2 m_1 m_2 \rangle$$
$$= \delta_{jj'} d^{(j)}_{m'm}(\boldsymbol{\omega}). \quad (11.139)$$

Now we use Eq. (11.138) to prove some of the important results.

For j_1 and j_2 both integers and m_1 and m_2 both zero, m must also be zero. In this case, from Eq. (11.138), we have

$$d^{(j_1)}_{m_1' 0}(\boldsymbol{\omega}) d^{(j_2)}_{m_2' 0}(\boldsymbol{\omega})$$
$$= \sum_{jm'} \langle j_1 j_2 m_1' m_2' | jm' j_1 j_2 \rangle \langle j0 j_1 j_2 | j_1 j_2 00 \rangle d^{(j)}_{m'0}(\boldsymbol{\omega}). \quad (11.140)$$

Note that for $\boldsymbol{\omega} = \theta\mathbf{n}$ and j an integer [c.f. Eq. (11.130)]

$$d_{m0}^{(j)}(\boldsymbol{\omega}) = \sqrt{\frac{4\pi}{2j+1}} Y_{jm}^*(\theta, \phi),$$

we have from Eq. (11.140) removing the prime and taking the complex conjugate

$$Y_{j_1 m_1}(\theta, \phi) Y_{j_2 m_2}(\theta, \phi)$$
$$= \Sigma_{jm} \sqrt{\frac{(2j_1 + 1)(2j_2 + 1)}{4\pi(2j+1)}} \langle j_1 j_2 m_1 m_2 | jm j_1 j_2 \rangle \langle j0 j_1 j_2 | j_1 j_2 00 \rangle Y_{jm}(\theta, \phi).$$

$$(11.141)$$

Multiplying both sides by $Y_{j'm'}^*(\theta, \phi)$ and integrating over the angles and using the orthogonality of spherical harmonics, and replacing $j'm'$ by jm at the end, we have often used the integral

$$\int Y_{jm}^*(\theta, \phi) Y_{j_1 m_1}(\theta, \phi) Y_{j_2 m_2}(\theta, \phi) d\Omega$$
$$= \left[\frac{(2j_1 + 1)(2j_2 + 1)}{4\pi(2j+1)} \right]^{1/2} \langle j_1 j_2 m_1 m_2 | jm j_1 j_2 \rangle \langle j0 j_1 j_2 | j_1 j_2 00 \rangle.$$

$$(11.142)$$

11.9 Vector and Tensor Operators

The set of $(2j+1)$ states $|jm\rangle$, for fixed j, transform into themselves under rotation according to the transformation law

$$R_\omega |jm\rangle = \sum_{m'=-j}^{j} |jm'\rangle d_{m'm}^j(\boldsymbol{\omega}).$$

Consider an operator $\hat{A}$. Then under rotation, $\hat{A}$ will transform to a new operator

$$\hat{A}_\omega = R_\omega \hat{A} R_\omega^{-1}.$$

To see this we note that

$$\langle \psi | \hat{A} | \psi \rangle$$
$$= \langle \psi | R_\omega^{-1} R_\omega \hat{A} R_\omega^{-1} R_\omega | \psi \rangle$$
$$= \langle \psi_\omega | R_\omega \hat{A} R_\omega^{-1} | \psi_\omega \rangle$$

therefore $$\hat{A}_\omega = R_\omega \hat{A} R_\omega^{-1}.$$ $$(11.143)$$

Now we study the transformation properties of operators under rotation.
1. Scalar operator: A scalar operator is invariant under rotation, i.e.

$$R_\omega S R_\omega^{-1} = S.$$

For an infinitesimal rotation $\boldsymbol{\omega}$

$$(1 - \frac{i}{\hbar}\boldsymbol{\omega} \cdot \mathbf{J})S(1 + \frac{i}{\hbar}\boldsymbol{\omega} \cdot \mathbf{J}) = S$$

or

$$[J_i, S] = 0$$

i.e. a scalar operator commutes with the angular momentum operator.
2. Vector operator: A vector operator $\mathbf{V}$ should transform in the same way as $\mathbf{J}$, i.e. a vector operator obeys the following commutation relations.

$$[J_i, V_j] = i\hbar\varepsilon_{ijk}V_k. \tag{11.144}$$

For infinitesimal rotation, using Eq. (10.103)

$$R_\omega V_i R_\omega^{-1} = (1 - \frac{i}{\hbar}\boldsymbol{\omega} \cdot \mathbf{J})V_i(1 + \frac{i}{\hbar}\boldsymbol{\omega} \cdot \mathbf{J})$$

$$= V_i - \frac{i}{\hbar}\omega_k[J_k, V_i]$$

$$= V_i + \varepsilon_{kil}\omega_k V_l$$

$$= V_i + (\mathbf{V} \times \boldsymbol{\omega})_i. \tag{11.145}$$

On the other hand under infinitesimal rotation, we know that a vector operator transforms as

$$V_i' = a_{ij}V_j = (\delta_{ij} + \omega_{ij})V_j. \tag{11.146}$$

Noting that

$$\omega_{ij} = -\omega_{ji},$$

$$\omega_{12} = \omega_3, \omega_{23} = \omega_1, \omega_{31} = \omega_2,$$

i.e. $\omega_{ij} = \epsilon_{ijk}\omega_k$, we have

$$V_i' = V_i + \epsilon_{ijk}V_j\omega_k = [\mathbf{V} + \mathbf{V} \times \boldsymbol{\omega}]_i.$$

We see that using Eq. (11.144) we get the correct transformation law for the vector operator.

Let us go to the "spherical" or "canonical" base in which instead of J_x and J_y, we consider $J_+ = J_x + iJ_y$, $J_- = J_x - iJ_y$. Then we have

$$[J_z, J_\pm] = \pm\hbar J_\pm,$$

$$[J_+, J_-] = 2\hbar J_z.$$

These commutation relations can be written compactly as

$$[J_z, J_q] = \hbar q J_q, \quad q = 0, \pm 1$$
$$[J_\pm, J_q] = \hbar \sqrt{1(1+1) - q(q \pm 1)} J_{q\pm 1}, \quad q = 0, \pm 1$$
$$J_0 = J_z,$$
$$J_1 = -\frac{1}{\sqrt{2}}(J_x + iJ_y),$$
$$J_{-1} = \frac{1}{\sqrt{2}}(J_x - iJ_y).$$

Thus in a "canonical" or a "spherical" base we will write the component of a vector as

$$V_1 = -\frac{V_x + iV_y}{\sqrt{2}},$$
$$V_0 = V_z, \qquad (11.147)$$
$$V_{-1} = \frac{V_x - iV_y}{\sqrt{2}}.$$

Then we can write the commutation relations of V_q as

$$[J_z, V_q] = \hbar q V_q, \qquad (11.148)$$
$$[J_\pm, V_q] = \hbar \sqrt{1(1+1) - q(q \pm 1)} V_{q\pm 1}.$$

Now we generalize; an irreducible tensor operator $T_q^{(k)}$ has $q = -k, \cdots, k$, so it has $2k + 1$ components. A vector is a tensor for which $k = 1$ and thus it has three components $V_q = T_q^{(1)}$. We will always be working in the representation in which J_z is diagonal. An irreducible tensor operator of order k will satisfy the following commutation relations with angular momentum

$$[J_z, T_q^{(k)}] = \hbar q T_q^{(k)}, \qquad (11.149)$$
$$[J_\pm, T_q^{(k)}] = \hbar \sqrt{k(k+1) - q(q \pm 1)} T_{q\pm 1}^{(k)}.$$

Noting that

$$\langle jm' | J_\mp | jm \rangle = \hbar \sqrt{j(j+1) - m(m \mp 1)} \delta_{m', m\mp 1},$$

we can write

$$\langle kq' | J_\pm | kq \rangle = \hbar \sqrt{k(k+1) - q(q \pm 1)} \delta_{q', q\pm 1},$$
$$\langle kq' | J_z | kq \rangle = \hbar q \langle kq' | kq \rangle = \hbar q \delta_{qq'}.$$

Hence

$$\sum_{q'} T_{q'}^{(k)} \langle kq'|J_\pm|kq\rangle$$

$$= \hbar\sqrt{k(k+1) - q(q \pm 1)}\Sigma_{q'} T_{q'}^{(k)} \delta_{q',q\pm1}$$

$$= \hbar\sqrt{k(k+1) - q(q \pm 1)}T_{q\pm1}^{(k)}, \tag{11.150}$$

so that

$$[J_\pm, T_q^{(k)}] = \hbar\sqrt{k(k+1) - q(q \pm 1)}T_{q\pm1}^{(k)}$$

$$= \Sigma q' T_{q'}^{(k)} \langle kq'|J_\pm|kq\rangle \tag{11.151a}$$

$$[J_z, T_q^{(k)}] = \hbar q T_q^{(k)} = \Sigma_{q'} T_{q'}^{(k)} \langle kq'|J_z|kq\rangle \tag{11.151b}$$

or

$$[\mathbf{J}, T_q^{(k)}] = \Sigma_{q'} T_{q'}^{(k)} \langle kq'|\mathbf{J}|kq\rangle. \tag{11.151c}$$

Under infinitesimal rotation

$$R_\omega T_q^{(k)} R_\omega^{-1} = T_q^{(k)} - \frac{i}{\hbar}\boldsymbol{\omega} \cdot [\mathbf{J}, T_q^{(k)}]$$

$$= T_q^{(k)} - \frac{i}{\hbar}\Sigma_{q'} T_{q'}^{(k)} \langle kq'|\boldsymbol{\omega} \cdot \mathbf{J}|kq\rangle$$

$$= \Sigma_{q'} T_{q'}^{(k)} \langle kq'|R_\omega|kq\rangle. \tag{11.152}$$

The above result also holds for finite rotation since a finite rotation can be built from infinitesimal rotations. Hence, we have the result that under rotation an irreducible tensor operator $T_q^{(k)}$ transforms as

$$R_\omega T_q^{(k)} R_\omega^{-1} = \sum_{q'} T_{q'}^{(k)} \langle kq'|R_\omega|kq\rangle$$

$$= \sum_{q'=-k}^{k} T_{q'}^{(k)} d_{q'q}^{(k)}(\boldsymbol{\omega}). \tag{11.153}$$

Conversely we can start from this definition of the tensor operator and can obtain results about the commutation relation of $T_q^{(k)}$ with $\mathbf{J}$.

Taking the Hermitian conjugate of Eq. (11.153), we have

$$\left(R_\omega T_q^{(k)} R_\omega^{-1}\right)^\dagger = \sum_{q'=-k}^{k} \left(T_{q'}^{(k)}\right)^\dagger d_{q'q}^{(k)*}(\boldsymbol{\omega}),$$

$$R_\omega T_q^{(k)\dagger} R_\omega^{-1} = \sum_{q'=-k}^{k} T_{q'}^{(k)\dagger} d_{q'q}^{(k)*}(\boldsymbol{\omega}).$$

But

$$d_{q'q}^{(k)*}(\boldsymbol{\omega}) = (-1)^{2k-q-q'} d_{-q'-q}^{(k)}(\boldsymbol{\omega})$$

therefore

$$R_\omega T_q^{(k)\dagger} R_\omega^{-1} = \sum_{q'} T_q^{(k)\dagger} (-1)^{2k-q-q'} d_{-q'-q}^{(k)}(\boldsymbol{\omega}).$$

Now q' being a dummy index, we can write $q' \to -q'$. Changing $q \to -q$, multiplying by $(-1)^q$ on both sides of the above equation and noting that $(k+q)$ is always an integer so that $(-1)^{2k+2q} = 1$, we have

$$R_\omega \left((-1)^q T_{-q}^{(k)\dagger} \right) R_\omega^{-1} = \sum_{q'=-k}^{k} T_{-q'}^{(k)\dagger} (-1)^{q'} d_{q'q}^{(k)}(\boldsymbol{\omega}).$$

Hence $(-1)^q T_{-q}^{(k)\dagger}$ transforms under rotation in the same way as $T_q^{(k)}$. If $T_q^{(k)}$ is to represent a physical quantity, it should return to itself after a rotation through 2π so that k is an integer.

11.10a   Wigner–Eckart Theorem

First we note that $T_q^{(k)}|\alpha j_1 m_1\rangle$ is an eigenstate of J_z with eigenvalues $(m_1 + q)\hbar$. Note that α is an eigenvalue of another observable which commutes with J^2 and J_z. To see this, we consider

$$J_z T_q^{(k)}|\alpha j_1 m_1\rangle = (\hbar q T_q^{(k)} + T_q^{(k)} J_z)|\alpha j_1 m_1\rangle$$
$$= \hbar(q + m_1) T_q^{(k)}|\alpha j_1 m_1\rangle. \qquad (11.154)$$

Let us consider how the state $T_q^{(k)}|\alpha j_1 m_1\rangle$ transforms under rotation [using Eq.(11.112)]

$$R_\omega T_q^{(k)}|\alpha j_1 m_1\rangle = R_\omega T_q^{(k)} R_\omega^{-1} R_\omega |\alpha j_1 m_1\rangle$$
$$= \sum_{q'} T_{q'}^{(k)} d_{q'q}^{(k)}(\boldsymbol{\omega}) \sum_{m_1'} |\alpha j_1 m_1'\rangle d_{m_1' m_1}^{(j_1)}(\boldsymbol{\omega})$$
$$= \sum_{q'} \sum_{m_1'} T_{q'}^{(k)}|\alpha j_1 m_1'\rangle d_{q'q}^{(k)}(\boldsymbol{\omega}) d_{m_1' m_1}^{(j_1)}(\boldsymbol{\omega}). \quad (11.155)$$

$T_q^{(k)}|\alpha j_1 m_1\rangle$ transforms under rotation in the same way as the state $|k j_1 q m_1\rangle$ [cf. Eq.(11.133)]. Hence we can write

$$T_q^{(k)}|\alpha j_1 m_1\rangle$$
$$\equiv |\alpha k j_1 q m_1\rangle = \sum_{jm} |\alpha j m\rangle \langle j m k j_1 | k j_1 q m_1 \rangle, \qquad (11.156)$$

where

$$J^2|\alpha jm\rangle = j(j+1)\hbar^2|\alpha jm\rangle$$
$$J_z|\alpha jm\rangle = m\hbar|\alpha jm\rangle.$$

Conversely we can write

$$|\alpha jm\rangle = \sum_{qm_1} T_q^{(k)}|\alpha j_1 m_1\rangle\langle kj_1 qm_1|jmkj_1\rangle. \tag{11.157}$$

From Eq. (11.156), changing $j_1 m_1 \to jm$ and replacing $\sum_{jm}$ by $\sum_{j''m''}$, we have

$$T_q^{(k)}|\alpha jm\rangle = \sum_{j''m''} |\alpha j''m''\rangle\langle j''m''kj|kjqm\rangle, \tag{11.158}$$

$$\langle \alpha'j'm'|T_q^{(k)}|\alpha jm\rangle = \sum_{j''m''} \langle \alpha'j'm'|\alpha j''m''\rangle\langle j''m''kj|kjqm\rangle. \tag{11.159}$$

Now from the orthogonality relation

$$\langle \alpha'j'm'|\alpha j''m''\rangle = \delta_{j'j''}\delta_{m'm''}\langle \alpha'j'm'|\alpha j'm'\rangle. \tag{11.160}$$

Hence

$$\langle \alpha'j'm'|T_q^{(k)}|\alpha jm\rangle = (\langle \alpha'j'm'|\alpha j'm'\rangle)\langle j'm'kj|kjqm\rangle. \tag{11.161}$$

Now we prove that $\langle \alpha'j'm'|\alpha j'm'\rangle$ is independent of m'. To see this, consider the transformation

$$|\alpha jm\rangle = \sum_\beta |\beta jm\rangle\langle \beta jm|\alpha jm\rangle,$$

so that

$$|\alpha jm+1\rangle = \sum_\beta |\beta jm+1\rangle\langle \beta jm+1|\alpha jm+1\rangle$$

and

$$J_+|\alpha jm\rangle = \hbar\sqrt{(j+m+1)(j-m)}|\alpha jm+1\rangle$$
$$= \hbar\sqrt{(j+m+1)(j-m)}\sum_\beta |\beta jm+1\rangle\langle \beta jm+1|\alpha jm+1\rangle. \tag{11.162a}$$

But

$$J_+|\alpha jm\rangle = \left(J_+\sum_\beta |\beta jm\rangle\right)\langle \beta jm|\alpha jm\rangle$$
$$= \hbar\sqrt{(j+m+1)(j-m)}\sum_\beta |\beta jm+1\rangle\langle \beta jm|\alpha jm\rangle. \tag{11.162b}$$

Comparing Eqs. (11.162a) and (11.162b), we have

$$\langle \beta j m | \alpha j m \rangle = \langle \beta j m + 1 | \alpha j m + 1 \rangle,$$

i.e. it must be independent of m. Hence we have the result that $\langle \alpha' j' m' | \alpha j' m' \rangle$ is independent of m'.

This is a remarkable result, it shows that the matrix elements of a tensor operator Eq. (11.161) between two angular momentum states factorize into a C.G. coefficient containing all the angular momentum details but independent of the dynamics and a coefficient independent of m, m' and q but containing all the dynamics.

Thus it is possible to write Eq. (11.161) as

$$\langle \alpha' j' m' | T_q^{(k)} | \alpha j m \rangle = \frac{\langle \alpha' j' || T^{(k)} || \alpha j \rangle}{\sqrt{2j' + 1}} \langle j' m' k j | k j q m \rangle, \qquad (11.163)$$

where it is customary to write

$$\langle \alpha' j' m' | \alpha j' m' \rangle = \frac{\langle \alpha' j' || T^{(k)} || \alpha j \rangle}{\sqrt{2j' + 1}}.$$

The quantity $\langle \alpha' j' || T^{(k)} || \alpha j \rangle$ is known as the reduced matrix element and the result (11.163) is the Wigner–Eckart Theorem. Because of the reality of C.G. coefficient

$$\langle j' m' k j | k j q m \rangle^* = \langle k j q m | j' m' k j \rangle = \langle j' m' k j | k j q m \rangle$$

$$\langle \alpha' j' m' | T_q^{(k)} | \alpha j m \rangle = \frac{\langle \alpha' j' || T^{(k)} || \alpha j \rangle}{\sqrt{2j' + 1}} \langle k j q m | j' m' k j \rangle. \qquad (11.164)$$

11.10b   Application of Wigner–Eckart Theorem

As a simple application of Wigner–Eckart theorem, we derive the angular momentum selection rules for the electric dipole ($E1$) transition.

First we note that (for j integer)

$$\langle j' m' | Y_{kq}(\theta, \phi) | j m \rangle$$

$$= \int \langle j' m' | \theta \phi \rangle Y_{kq}(\theta, \phi) \langle \theta \phi | j m \rangle d\Omega$$

$$= \int Y_{j'm'}^*(\theta \phi) Y_{kq}(\theta \phi) Y_{jm}(\theta \phi) d\Omega$$

$$= \sqrt{\frac{(2k + 1)(2j + 1)}{4\pi(2j' + 1)}} \langle k j q m | j' m' k j \rangle \langle j' 0 k j | k j 0 0 \rangle, \qquad (11.165)$$

on using Eq. (11.142). By the Wigner–Eckart Theorem,

$$\langle j'm'|Y_{kq}|jm\rangle = \langle j'||Y^{(k)}||j\rangle \frac{\langle kjqm|j'm'kj\rangle}{\sqrt{2j'+1}}. \tag{11.166}$$

Hence we have the result

$$\langle j'||Y^{(k)}||j\rangle = \sqrt{\frac{(2k+1)(2j+1)}{4\pi}}\langle j'0kj|kj00\rangle. \tag{11.167}$$

For the $E1$ transition one comes across (see Sec. 17.6) the matrix elements of the form $\langle l'm'|\mathbf{r}|lm\rangle$. Now $\mathbf{r}$ in the canonical base can be written as

$$\mathbf{r} \equiv (r_1, r_0, r_{-1}),$$

where

$$r_1 = -\frac{x+iy}{\sqrt{2}} = -\frac{r}{\sqrt{2}}\sin\theta e^{i\phi}$$
$$r_0 = z = r\cos\theta \tag{11.168}$$
$$r_{-1} = \frac{x-iy}{\sqrt{2}} = \frac{r}{\sqrt{2}}\sin\theta e^{-i\phi}$$

or

$$r_q = \sqrt{\frac{4\pi}{3}}rY_{1q}(\theta,\phi), q = 1, 0, -1.$$

Now using Eq. (11.165)

$$\langle l'm'|Y_{1q}(\theta\phi)|lm\rangle$$
$$= \sqrt{\frac{3}{4\pi}}\sqrt{\frac{2l+1}{2l'+1}}\langle 1lqm|l'm'1l\rangle\langle l'01l|1l00\rangle. \tag{11.169}$$

The C.G. coefficient

$$\langle 1lqm|l'm'1l\rangle \neq 0,$$

for $l' = l+1, l, l-1$ and for $m' = m+q$. Hence we have the selection rules

$$\Delta l \equiv l' - l = \pm 1, 0, \tag{11.170}$$
$$\Delta m \equiv m' - m = q = \pm 1, 0.$$

The parity selection rule forbids $\Delta l = 0$ transitions.

Furthermore, using the C.G. Table II, and Eq. (11.169) we have

$$\langle l+1m|Y_{10}|lm\rangle = \sqrt{\frac{3}{4\pi}}\sqrt{\frac{(l+1)^2-m^2}{(2l+3)(2l+1)}},$$

$$\langle l-1m|Y_{10}|lm\rangle = \sqrt{\frac{3}{4\pi}}\sqrt{\frac{l^2-m^2}{(2l+1)(2l-1)}},$$

$$\langle l+1m+1|Y_{11}|lm\rangle = \sqrt{\frac{3}{4\pi}}\sqrt{\frac{(l+m+1)(l+m+2)}{2(2l+1)(2l+3)}}, \qquad (11.171)$$

$$\langle l-1m+1|Y_{11}|lm\rangle = \sqrt{\frac{3}{4\pi}}\sqrt{\frac{(l-m-1)(l-m)}{2(2l+1)(2l+3)}},$$

$$\langle l+1m-1|Y_{1-1}|lm\rangle = \sqrt{\frac{3}{4\pi}}\sqrt{\frac{(l-m+1)(l-m+2)}{2(2l+1)(2l+3)}},$$

$$\langle l-1m-1|Y_{1-1}|lm\rangle = \sqrt{\frac{3}{4\pi}}\sqrt{\frac{(l+m)(l+m-1)}{2(2l+1)(2l-1)}}.$$

Thus we know completely the angular momentum part of the matrix elements for the $E1$ transition.

11.11   Problems

11.1 Consider an operator $\hat{u}$, which obeys the commutation relations

$$[\hat{u}, J_z] = \frac{1}{2}\hbar\hat{u},$$

$$[\hat{u}, J_+] = 0.$$

Show that

$$\hat{u}|j,j\rangle = \text{constant} \quad |j-1/2, j-1/2\rangle.$$

11.2 Show that
(i) $\sigma_x\sigma_y\sigma_z = i$.
(ii) $(\boldsymbol{\sigma}\cdot\mathbf{a})(\boldsymbol{\sigma}\cdot\mathbf{b}) = (\mathbf{a}\cdot\mathbf{b})I + i\boldsymbol{\sigma}\cdot(\mathbf{a}\times\mathbf{b})$, where $\mathbf{a}$ and $\mathbf{b}$ are ordinary vectors and I is a 2×2 unit matrix.
(iii) $(\boldsymbol{\sigma}\cdot\mathbf{n})$ has eigenvalues ± 1, where $\mathbf{n}$ is a unit vector.

11.3 The eigenvectors of σ_z corresponding to eigenvalues ± 1 are denoted by $|\pm 1/2\rangle$. Show that
(i) $\frac{1}{\sqrt{2}}(|+1/2\rangle \pm |-1/2\rangle)$ are normalized eigenvectors of σ_x with eigenvalues ± 1.

(ii) $\frac{1}{\sqrt{2}}(|+1/2\rangle \pm i|-1/2\rangle)$ are normalized eigenvectors of σ_y with eigenvalues ± 1.

Write the above eigenvectors as column vectors.

11.4 S_x, S_y and S_z are components of spin $1/2$ operator $\mathbf{S}$ and obey the commutation relations

$$[S_x, S_y] = i\hbar S_z, \text{etc.}$$

Show that for $S_\pm = S_x \pm i S_y$

$$S_+|+1/2\rangle = 0,$$
$$S_-|-1/2\rangle = 0,$$
$$S_-|+1/2\rangle = \hbar|-1/2\rangle,$$
$$S_+|-1/2\rangle = \hbar|+1/2\rangle,$$

where $|\pm 1/2\rangle$ are eigenstates of S_z belonging to eigenvalues $\pm 1/2\,\hbar$.

11.5 If a particle is in an eigenstate of σ_x, find the probability of finding it in the eigenstate of σ_z belonging to eigenvalue $+1$.

11.6 Consider an electron in a uniform magnetic field $\mathbf{B}$ in the positive z-direction. The result of a measurement has shown that the electron spin is along the positive x-direction at $t = 0$. An arbitrary spin state at time t can be written as

$$|\chi(t)\rangle = a(t)|+1/2\rangle + b(t)|-1/2\rangle,$$

where $|\chi(t)\rangle$ satisfies the Schrödinger equation

$$i\hbar \frac{d}{dt}|\chi(t)\rangle = H|\chi(t)\rangle.$$

Find the probability at $t > 0$, for finding the electron in the spin state (a) $S_x = (1/2)\hbar$, (b) $S_x = -(1/2)\hbar$, and (c) $S_z = (1/2)\hbar$. Hint: As far as spin is concerned, the Hamiltonian $H = \mu_0 B\sigma_z$, where μ_0 is the magnetic moment of the electron. Use matrix representation for S_z etc.

11.7 Consider an operator $\hat{u}$ which obeys the commutation relation

$$[\hat{u}, J_z] = (1/2)\hbar\hat{u},$$
$$[[\hat{u}, J^2], J^2] = \frac{1}{2}(\hat{u}J^2 + J^2\hat{u}) + \frac{3\hbar^4}{16}\hat{u}.$$

Consider the representation in which the basis vectors are $|jm\rangle$, the simultaneous eigenvectors of J^2 and J_z belonging to the eigenvalues $j(j+1)\hbar^2$ and $m\hbar$ respectively. Show that if

$$\langle jm|\hat{u}|j'm'\rangle \neq 0$$

(i) $m' = m + 1/2$,

(ii) $(j'(j'+1) - j(j+1))^2 = \frac{1}{2}(j'(j'+1) + j(j+1)) + \frac{3}{16}$.

11.8 Consider a system of two distinguishable particles, each with spin $1/2$ and magnetic moments $\boldsymbol{\mu}_1 = \mu_1 \boldsymbol{\sigma}_1$, $\boldsymbol{\mu}_2 = \mu_2 \boldsymbol{\sigma}_2$ in an external magnetic field B in the z-direction. The spin–spin interaction of the particles is $b \boldsymbol{\sigma}_1 \cdot \boldsymbol{\sigma}_2$, where b is a constant. Find the exact energy eigenvalues of this system.

11.9 In the normal Zeeman effect, an energy level characterized by quantum number n is split into $(2l+1)$ different levels:

$$E_{n,m} = E_n - \frac{e\hbar}{2m_e c} Bm,$$

where $m = -l, \cdots, l$. The equal spacing between the levels is given by

$$\Delta E = \frac{e\hbar}{2m_e c} B.$$

Show that

$$\Delta E = \left(\frac{1}{2}\alpha^2 m_e c^2\right)\alpha \frac{B}{e/a^2}$$

$$= \frac{B}{(2.4 \times 10^9 \text{Gauss})} \times 13.6\text{eV},$$

where $\alpha = \frac{e^2}{\hbar c} = \frac{1}{137}$ is the fine structure constant and $E_1 = \frac{1}{2}m_e c^2 \alpha^2 = 13.6$ eV is the binding energy of electron in the first Bohr orbit of hydrogen atom. a is the radius of first Bohr orbit. Draw the energy level diagram for the Zeeman effect for p and d level. In atomic transitions $\Delta m = m_f - m_i = -1, 0, 1$. Draw 9 possible transitions on the energy level diagram.

11.10 An electron can be regarded as a magnetic dipole of magnetic moment $\boldsymbol{\mu} = \mu_0 \boldsymbol{\sigma}$, where $\mu_0 = \frac{e\hbar}{2m_e c}$ is the electron magnetic moment. The interaction energy between magnetic dipoles is given by

$$V = \frac{1}{r^3}\left\{(\boldsymbol{\mu}_1 \cdot \boldsymbol{\mu}_2) - 3\frac{(\boldsymbol{\mu}_1 \cdot \mathbf{r})(\boldsymbol{\mu}_2 \cdot \mathbf{r})}{r^2}\right\}.$$

Find the dipole–dipole magnetic interaction energy of an electron and a positron at a fixed distance a, in eigenstates of total spin.

11.11 Show explicitly that the state $|j = l - 1/2, m\rangle$ as given by Eq. (11.92) is an eigenstate of the operator $(\mathbf{L} + \mathbf{S})^2$ with eigenvalue $j(j+1)\hbar^2$.

11.12 Consider an electron in an external magnetic field $\mathbf{B}$. The energy of the spinning electron is given by

$$H_{\text{spin}} = -\boldsymbol{\mu} \cdot \mathbf{B} = \frac{eg}{2mc}\mathbf{S} \cdot \mathbf{B},$$

show that

$$\frac{d\mathbf{S}}{dt} = -\frac{eg}{2mc}\mathbf{S} \times \mathbf{B}.$$

Further show that the expectation value of the spin operator $\mathbf{S}$ at times t, when initially the particle is in the spin state with $S_x = \frac{1}{2}\hbar$, is given by

$$\langle \mathbf{S} \rangle = \frac{\hbar}{2}\cos\omega t \mathbf{e}_x + \frac{\hbar}{2}\sin\omega t \mathbf{e}_y,$$

where $\mathbf{e}_x$ and $\mathbf{e}_y$ are the unit vectors along the x- and y-axis respectively and $\omega = \frac{egB}{2mc}$.

11.13 The Hamiltonian for a particle of mass m is given by

$$H = \frac{\hat{p}^2}{2m} + V(r) + U\hbar\mathbf{L} \cdot \boldsymbol{\sigma},$$

where U is some constant. Show that $\mathbf{L}$ is not a constant of motion, but $\mathbf{J} = \mathbf{L} + \frac{1}{2}\hbar\boldsymbol{\sigma}$ is a constant of motion.

11.14 Show that for the spin $\frac{1}{2}\hbar$ case $(\mathbf{J} = \mathbf{S} = \frac{1}{2}\hbar\boldsymbol{\sigma})$, the rotation operator can be written as

$$U_R(\theta) = e^{-(i/2)\theta\mathbf{n}\cdot\boldsymbol{\sigma}}$$

$$= \cos\frac{\theta}{2} - i\sin\frac{\theta}{2}\mathbf{n} \cdot \boldsymbol{\sigma},$$

where θ is the angle of rotation about $\mathbf{n}(n^2 = 1)$. In general a spin state can be written as

$$|\chi\rangle = C_1|+\frac{1}{2}\rangle + C_2|-\frac{1}{2}\rangle$$

$$= \begin{pmatrix} C_1 \\ C_2 \end{pmatrix},$$

called a two component spinor. Its transformation law under the rotation is

$$|\chi'\rangle = e^{-(i/2)\theta\mathbf{n}\cdot\boldsymbol{\sigma}}|\chi\rangle.$$

Show that
(i) for $\theta = 2\pi$, $|\chi\rangle = -|\chi\rangle$,
(ii) for rotation about x-axis, for which $\hat{\mathbf{n}} = \mathbf{e}_x$, $\boldsymbol{\sigma} \cdot \hat{\mathbf{n}} = \sigma_x$,

$$\begin{pmatrix} C_1' \\ C_2' \end{pmatrix} = \begin{pmatrix} \cos\frac{\theta}{2} & -i\sin\frac{\theta}{2} \\ i\sin\frac{\theta}{2} & \cos\frac{\theta}{2} \end{pmatrix} \begin{pmatrix} C_1 \\ C_2 \end{pmatrix}.$$

Compare it with the transformation law for a vector under the above rotation. For the same rotation, show that

$$\boldsymbol{\sigma}' = e^{-i(\theta/2)\mathbf{n}\cdot\boldsymbol{\sigma}}\boldsymbol{\sigma}\,e^{i(\theta/2)\mathbf{n}\cdot\boldsymbol{\sigma}}$$

gives

$$\sigma_x' = \sigma_x,$$
$$\sigma_y' = \cos\theta\,\sigma_y + \sin\theta\,\sigma_z,$$
$$\sigma_z' = -\sin\theta\,\sigma_y + \cos\theta\,\sigma_z,$$

i.e. $\boldsymbol{\sigma}$ transforms like a vector under the rotation.

11.15 Show that

$$\boldsymbol{\sigma}\cdot\mathbf{p}\,e^{-(i/2)\theta\mathbf{n}\cdot\boldsymbol{\sigma}} = e^{-(i/2)\theta\mathbf{n}\cdot\boldsymbol{\sigma}}\boldsymbol{\sigma}\cdot\mathbf{e}_z = e^{(i/2)\theta\mathbf{n}\cdot\boldsymbol{\sigma}}\sigma_z.$$

Hence, show that for the states

$$|\mathbf{p}\uparrow\rangle = e^{-(i/2)\theta\mathbf{n}\cdot\boldsymbol{\sigma}}|+\tfrac{1}{2}\rangle$$
$$= \cos\tfrac{\theta}{2}|+\tfrac{1}{2}\rangle + e^{i\phi}\sin\tfrac{\theta}{2}|-\tfrac{1}{2}\rangle,$$
$$|\mathbf{p}\downarrow\rangle = e^{-(i/2)\theta\mathbf{n}\cdot\boldsymbol{\sigma}}|-\tfrac{1}{2}\rangle$$
$$= \cos\tfrac{\theta}{2}|-\tfrac{1}{2}\rangle - e^{-i\phi}\sin\tfrac{\theta}{2}|+\tfrac{1}{2}\rangle,$$

we have

$$\boldsymbol{\sigma}\cdot\mathbf{p}|\mathbf{p}\uparrow\rangle = |\mathbf{p}\uparrow\rangle,$$
$$\boldsymbol{\sigma}\cdot\mathbf{p}|\mathbf{p}\downarrow\rangle = -|\mathbf{p}\downarrow\rangle.$$

$\boldsymbol{\sigma}\cdot\mathbf{p}$ is called the Helicity Operator.

11.16 Show that the rotation matrix for the operator $R_\omega = e^{-(i/\hbar)\boldsymbol{\omega}\cdot\mathbf{J}}$ for $\boldsymbol{\omega} = \theta\mathbf{n}$ and $\mathbf{J} = \tfrac{1}{2}\hbar\boldsymbol{\sigma}$ (spin $\tfrac{1}{2}$ operator),

$$d^{(\frac{1}{2})}(\phi,\theta,-\phi) = \begin{pmatrix} \cos\tfrac{\theta}{2} & -e^{-i\phi}\sin\tfrac{\theta}{2} \\ e^{i\phi}\sin\tfrac{\theta}{2} & \cos\tfrac{\theta}{2} \end{pmatrix}.$$

11.17 (a) Starting from Eq. (11.138), show that

$$\sum_{\substack{m_1 m_2 \\ m_1+m_2=m}} d^{(j_1)}_{m_1' m_1}(\boldsymbol{\omega})d^{(j_2)}_{m_2' m_2}(\boldsymbol{\omega})\langle j_1 j_2 m_1 m_2|jmj_1 j_2\rangle$$

$$= \sum_{m'=m_1'+m_2'} \langle j_1 j_2 m_1' m_2'|jm'j_1 j_2\rangle d^{(j)}_{m'm}(\boldsymbol{\omega}).$$

[Hint: operate Eq. (11.138) on $|j''m''j_1 j_2\rangle$ and then take matrix elements with $\langle j_1 j_2 m_1 m_2|$, use the orthogonality property of C.G.

coefficients and replace at the end j'', m'' by j, m.]

(b) Use the above relation and a table of C.G. coefficients to show that

$$\sqrt{\frac{j+m}{2j}}\,d^{j-\frac{1}{2}}_{m'-\frac{1}{2},m-\frac{1}{2}}(\boldsymbol{\omega})d^{\frac{1}{2}}_{\frac{1}{2}\frac{1}{2}}(\boldsymbol{\omega}) + \sqrt{\frac{j-m}{2j}}\,d^{j-\frac{1}{2}}_{m'-\frac{1}{2},m+\frac{1}{2}}(\boldsymbol{\omega})d^{\frac{1}{2}}_{\frac{1}{2}-\frac{1}{2}}(\boldsymbol{\omega}),$$

$$= \sqrt{\frac{j+m'}{2j}}\,d^{j}_{m'm}(\boldsymbol{\omega}).$$

11.18 Show that

$$P_j(\cos\theta') = \frac{4\pi}{2j+1}\sum_m Y^*_{jm}(\beta,\alpha)Y_{jm}(\theta,\phi),$$

where j is an integer; θ, ϕ and θ', ϕ' are the spherical polar coordinates of the same physical point in the old and new coordinate systems while β, α are the spherical polar coordinates of z' axis in the old coordinate system.

[Hint: consider the eigenstates $|jm\rangle$ and $|jm\rangle'$ respectively of J_z and J'_z where

$$|jm\rangle' = R_\omega|jm\rangle.$$

Take the **r**-representation and make use of $d^{(j)}_{m0}(\alpha,\beta) = \sqrt{\frac{4\pi}{2j+1}}Y^*_{jm}(\beta,\alpha)$ (c.f. Eq. (11.130)).]

11.19 Consider the 4 state system consisting of two spin-$\frac{1}{2}$ particles. The vector space of the system is spanned by the 4 orthonormal states:

$$|\uparrow\uparrow\rangle \equiv |\uparrow\rangle_1|\uparrow\rangle_2 \quad |\uparrow\downarrow\rangle \equiv |\uparrow\rangle_1|\downarrow\rangle_2 \quad |\downarrow\uparrow\rangle \equiv |\downarrow\rangle_1|\uparrow\rangle_2 \quad |\downarrow\downarrow\rangle \equiv |\downarrow\rangle_1|\downarrow\rangle_2$$

where the arrows refer to the direction of the spin along the z-axis and the subscript 1 and 2 refer to the particle. Suppose that the Hamiltonian of this system is given by

$$H = \gamma\,(S_{1,z} + S_{2,z}) + \frac{\gamma}{\hbar}\mathbf{S}_1 \cdot \mathbf{S}_2$$

a. Write the above Hamiltonian in terms of $S_{1,\pm}, S_{1,z}, S_{2,\pm}, S_{2,z}$.

b. Using the form of the Hamiltonian found in part (a) find the matrix of H in the basis given above.

c. Write the Hamiltonian in terms of the $\mathbf{S}_{total}$ where $\mathbf{S}_{total} = \mathbf{S}_1 + \mathbf{S}_2$.

d. Find the energies and stationary states of the Hamiltonian.

e. If the system is in the state $|\uparrow\downarrow\rangle$ at time $t = 0$ what is the probability of finding the system in the single state at time t?

11.20 Consider a system of three particles. Particle 1 has spin $\frac{1}{2}$, particle 2 has spin $\frac{1}{2}$ and particle 3 has spin 1. This system has 12 states:

$$| \uparrow\uparrow 1 \rangle \; | \uparrow\uparrow 0 \rangle \; | \uparrow\uparrow -1 \rangle$$
$$| \uparrow\downarrow 1 \rangle \; | \uparrow\downarrow 0 \rangle \; | \uparrow\downarrow -1 \rangle$$
$$| \downarrow\uparrow 1 \rangle \; | \downarrow\uparrow 0 \rangle \; | \downarrow\uparrow -1 \rangle$$
$$| \downarrow\downarrow 1 \rangle \; | \downarrow\downarrow 0 \rangle \; | \downarrow\downarrow -1 \rangle$$

a. What are the possible eigenvalues of J^2_{total}?

b. For each of the eigenvalues found in part (a) what are possible eigenvalues of $J_{total,z}$?

c. Determine $(J_{1,x} + J_{2,x})^2 | \uparrow\uparrow 1 \rangle$ and $J_{1,x} J_{3,y} | \uparrow\uparrow 1 \rangle$.

d. Write down the normalized state with total angular momentum eigenvalue 0 in terms of the individual spin states given above.

11.21 A particle of spin $\frac{1}{2}$ is in a D-state of the orbital angular momentum:

a. What are its possible states of total angular momentum?

b. For the Hamiltonian

$$H = a + b\mathbf{L} \cdot \mathbf{S} + c\mathbf{L}^2$$

where a, b and c are numbers, find the values of the energy for each of the different states of total angular momentum (express your answer in terms of a, b, c).

11.22 Show that:

a)

$$e^{\frac{i}{\hbar}\frac{\pi}{2}J_y} J_x e^{\frac{-i}{\hbar}\frac{\pi}{2}J_y} = J_z,$$

b)

$$e^{\frac{i}{\hbar}\frac{\pi}{2}J_y} e^{\frac{i}{\hbar}\alpha J_x} e^{\frac{-i}{\hbar}\frac{\pi}{2}J_y} = e^{\frac{i}{\hbar}\alpha J_z},$$

c) For any vector operators $\mathbf{A}$:

$$e^{\frac{i}{\hbar}\alpha J_y} A_x e^{\frac{-i}{\hbar}\alpha J_y} = A_x \cos\alpha + A_z \sin\alpha.$$

11.23 Show that:

a)

$$[J_x, [J_x, T_q^{(k)}]] = \sum_{q'} T_{q'}^{(k)} \langle kq' | J_x^2 | kq \rangle,$$

b)

$$[J_x, [J_x, T_q^{(k)}]] + [J_y, [J_y, T_q^{(k)}]] + [J_z, [J_z, T_q^{(k)}]] = k(k+1)\hbar^2 T_q^{(k)}.$$

Chapter 12

Time Independent Perturbation Theory

12.1 Introduction

There are very few problems which can be solved exactly. One has to resort to approximate methods, which are of two types:
(i) Perturbative
(ii) Non-perturbative
We first discuss perturbative methods. Suppose we have some problem which cannot be solved exactly. We look at a similar problem that can be solved exactly. Then see what is the deviation between the problem we want to solve and the problem we have solved. In perturbation theory this deviation is small. So we have a main effect which we know and the small effect which we get as an approximation.

12.2 Stationary Perturbation Theory

Suppose we want to find eigenvalues and eigenfunctions of the Hamiltonian H for stationary states :

$$H\psi_s = E_s\psi_s. \tag{12.1}$$

Assume that H can be decomposed as

$$H = H_0 + \lambda V, \tag{12.2}$$

such that we can solve exactly the eigenvalue equation

$$H_0 u_n(\mathbf{x}) = \epsilon_n u_n(\mathbf{x}). \tag{12.3}$$

λV is called the perturbation. λ is some parameter which is small. We assume that V does not depend explicitly on time (the same is always assumed for H_0). Since the eigenfunctions $u_n(\mathbf{x})$ form a complete set, we can write

$$\psi_s(\mathbf{x}) = \sum_n C_{ns} u_n(\mathbf{x}). \qquad (12.4)$$

We insert Eq. (12.4) in Eq. (12.1) with H as given in Eq. (12.2). Thus

$$(H_0 + \lambda V) \sum_n C_{ns} u_n(\mathbf{x}) = E_s \sum_n C_{ns} u_n(\mathbf{x}).$$

Making use of Eq. (12.3), we get

$$\sum_n C_{ns}(E_s - \varepsilon_n) u_n(\mathbf{x}) = \sum_n \lambda V C_{ns} u_n(\mathbf{x}).$$

We multiply both sides by $u_m^*(\mathbf{x})$ and integrate over $\mathbf{x}$:

$$\sum_n C_{ns}(E_s - \varepsilon_n) \int u_m^*(\mathbf{x}) u_n(\mathbf{x}) d^3x$$

$$= \sum_n \lambda C_{ns} \int u_m^*(\mathbf{x}) V u_n(\mathbf{x}) d^3x.$$

This gives

$$\sum_n C_{ns}(E_s - \varepsilon_n) \delta_{mn} = \lambda \sum_n C_{ns} V_{mn}$$

or

$$C_{ms}(E_s - \varepsilon_m) = \lambda \sum_n C_{ns} V_{mn}, \qquad (12.5a)$$

where

$$V_{mn} = \int u_m^*(\mathbf{x}) V u_n(\mathbf{x}) d^3x$$

$$= \langle m|V|n \rangle. \qquad (12.5b)$$

Note that V_{mn} are matrix elements of the perturbation with respect to the eigenfunctions of the unperturbed Hamiltonian. Equations (12.5a) are an infinite set of linear equations for C_{ms}. When $\lambda = 0$, perturbation is zero, then

$$E_s = \varepsilon_s,$$

$$C_{ms} = \delta_{ms}. \qquad (12.6)$$

For $\lambda \neq 0$, we make an expansion of C's

$$C_{ms} = \delta_{ms} + \lambda C_{ms}^{(1)} + \lambda^2 C_{ms}^{(2)} + \cdots \qquad (12.7a)$$

$$E_s = \varepsilon_s + \lambda E_s^{(1)} + \lambda^2 E_s^{(2)} + \cdots \qquad (12.7b)$$

$$\psi_s = u_s + \lambda \psi_s^{(1)} + \lambda^2 \psi_s^{(2)} \cdots . \qquad (12.7c)$$

We substitute the above equations back in Eq. (12.5a) and solve these equations for a certain power of λ. In the first order (keeping terms up to λ only)

$$[\delta_{ms} + \lambda C_{ms}^{(1)}][\varepsilon_s + \lambda E_s^{(1)} - \varepsilon_m] = \lambda \sum_n \delta_{ns} V_{mn}.$$

Now $\delta_{ms}(\varepsilon_s - \varepsilon_m) = 0$, thus to order λ, we have

$$\lambda E_s^{(1)} \delta_{ms} + \lambda C_{ms}^{(1)}(\varepsilon_s - \varepsilon_m) = \lambda V_{ms}.$$

When $m = s$, $\delta_{ms} = 1$, we get

$$E_s^{(1)} = V_{ss}. \tag{12.8}$$

When $m \neq s$, $\delta_{ms} = 0$, we get

$$C_{ms}^{(1)} = \frac{V_{ms}}{\varepsilon_s - \varepsilon_m}, \quad m \neq s. \tag{12.9}$$

Note that $C_{ss}^{(1)}$ remains arbitrary. Up to first order

$$E_s = \varepsilon_s + \lambda V_{ss}, \tag{12.10}$$

$$\psi_s = \sum_m (\delta_{ms} + \lambda C_{ms}^{(1)}) u_m$$

$$= u_s + \lambda \sum_m C_{ms}^{(1)} u_m. \tag{12.11}$$

We choose $C_{ss}^{(1)}$ such that

$$\int \psi_s^*(\mathbf{x})\psi_s(\mathbf{x}) d^3 x = 1, \quad \text{to first order in } \lambda.$$

Since

$$\int u_s^* u_m d^3 x = \delta_{sm}$$

it follows

$$C_{ss}^{(1)*} + C_{ss}^{(1)} = 0$$

that is the real part of $C_{ss}^{(1)} = 0$. There is no loss of generality (because we can select a phase such that $C_{ss}^{(1)}$ is real) in making the simple choice Im $C_{ss}^{(1)} = 0$, so that

$$C_{ss}^{(1)} = 0.$$

Thus using Eq. (12.9), we have from Eq. (12.11) up to first order in λ

$$\psi_s = u_s + \lambda \sum_{m \neq s} \frac{V_{ms} u_m}{\varepsilon_s - \varepsilon_m}. \tag{12.12}$$

For an observable $\hat{F}$ we have matrix elements of the form

$$f_{sr} = \int \psi_s^* \hat{F} \psi_r d^3 x. \tag{12.13a}$$

Up to first order, we have

$$f_{sr} = \int u_s^* \hat{F} u_r d^3 x + \lambda \sum_{m \neq r} \frac{V_{mr}}{\varepsilon_r - \varepsilon_m} \int u_s^* \hat{F} u_m d^3 x$$

$$+ \lambda \sum_{m \neq s} \frac{V_{ms}^*}{\varepsilon_s - \varepsilon_m} \int u_m^* \hat{F} u_r d^3 x. \tag{12.13b}$$

We write

$$\int u_s^* \hat{F} u_r d^3 x = f_{sr}^{(0)}. \tag{12.13c}$$

Also note

$$V_{ms}^* = \langle m|V|s \rangle^* = \langle s|V|m \rangle = V_{sm}.$$

Thus up to first order

$$f_{sr} = f_{sr}^{(0)} + \lambda \sum_{m \neq s} \frac{V_{sm} f_{mr}^{(0)}}{\varepsilon_s - \varepsilon_m} + \lambda \sum_{m \neq r} \frac{V_{mr} f_{sm}^{(0)}}{\varepsilon_r - \varepsilon_m}. \tag{12.13d}$$

We also note from Eq. (12.12) that the condition for the applicability of the perturbation theory is that

$$\lambda |V_{ms}| \ll |\varepsilon_s - \varepsilon_m|. \tag{12.14}$$

Let us calculate E_s to second order in λ. Substituting Eq. (12.7) in Eq. (12.5a) and retaining terms up to order λ^2, we have

$$[\delta_{ms} + \lambda C_{ms}^{(1)} + \lambda^2 C_{ms}^{(2)}][\varepsilon_s + \lambda E_s^{(1)} + \lambda^2 E_s^{(2)} - \varepsilon_m]$$

$$= \lambda \sum_n (\delta_{ns} + \lambda C_{ns}^{(1)}) V_{mn}.$$

Equating terms of order λ^2

$$C_{ms}^{(2)}(\varepsilon_s - \varepsilon_m) + E_s^{(2)} \delta_{ms} + C_{ms}^{(1)} E_s^{(1)} = \sum_n C_{ns}^{(1)} V_{mn}. \tag{12.15}$$

Thus for $m = s$ [since $C_{ss}^{(1)} = 0$ and $\delta_{ss} = 1$]

$$E_s^{(2)} = \sum_n C_{ns}^{(1)} V_{sn}.$$

Using Eq. (12.9), we get

$$E_s^{(2)} = \sum_{n \neq s} \frac{V_{ns} V_{sn}}{\varepsilon_s - \varepsilon_n}. \tag{12.16}$$

Now

$$V_{ns} = V_{sn}^*.$$

Thus to second order in λ

$$E_s = \varepsilon_s + \lambda V_{ss} + \lambda^2 \sum_{n \neq s} \frac{|V_{sn}|^2}{\varepsilon_s - \varepsilon_n}. \qquad (12.17)$$

Similarly we can calculate $C_{ms}^{(2)}$ from Eq. (12.15) and hence ψ_s to second order. From Eq. (12.15), for $m \neq s$, we have

$$C_{ms}^{(2)}(\epsilon_s - \epsilon_m) + \frac{V_{ms}}{\epsilon_s - \epsilon_m} V_{ss} = \sum_{n \neq s} \frac{V_{ns}}{\epsilon_s - \epsilon_n} V_{mn}, \qquad s \neq m,$$

$$C_{ms}^{(2)} = \sum_{n \neq s} \frac{V_{ns} V_{mn}}{(\epsilon_s - \epsilon_n)(\epsilon_s - \epsilon_m)} - \frac{V_{ms} V_{ss}}{(\epsilon_s - \epsilon_m)^2}, \qquad s \neq m.$$

From $\int \psi_s^* \psi_s dx = 1$, to order λ^2, we have $[\psi_s = u_s + \sum_m \lambda C_{ms}^{(1)} u_m + \sum_m \lambda^2 C_{ms}^{(2)} u_m]$

$$\lambda C_{ss}^{(1)*} + \lambda C_{ss}^{(1)} + \lambda^2 \sum_n C_{ns}^{(1)*} C_{ns}^{(1)} + \lambda^2 C_{ss}^{(2)} + \lambda^2 C_{ss}^{(2)*} = 0.$$

Therefore $C_{ss}^{(2)} + C_{ss}^{(2)*} = -\sum_{n \neq s} \frac{V_{ns}^* V_{ns}}{(\epsilon_s - \epsilon_n)^2} = 2\mathrm{Re}\, C_{ss}^{(2)}$.

We take $\mathrm{Im}\, C_{ss}^{(2)} = 0$, hence

$$C_{ss}^{(2)} = -\frac{1}{2} \sum_{n \neq s} \frac{V_{ns}^* V_{ns}}{(\epsilon_s - \epsilon_n)^2},$$

$$\psi_s = u_s + \lambda \sum_{m \neq s} C_{ms}^{(1)} u_m + \lambda^2 \sum_{m \neq s} C_{ms}^{(2)} u_m + \lambda^2 C_{ss}^{(2)} u_s$$

$$= u_s + \lambda \sum_{m \neq s} \frac{V_{ms} u_m}{\epsilon_s - \epsilon_m} + \lambda^2 \sum_{m \neq s} (\{ \sum_{n \neq s} \frac{V_{ns} V_{mn}}{(\epsilon_s - \epsilon_n)(\epsilon_s - \epsilon_m)}$$

$$- \frac{V_{ms} V_{ss}}{(\epsilon_s - \epsilon_m)^2} \}) u_m + \lambda^2 (\frac{1}{2} \sum_{n \neq s} \frac{|V_{ns}|^2}{(\epsilon_s - \epsilon_n)^2}) u_s. \qquad (12.18)$$

12.3 Degeneracy

We now consider the case when H_0 has degenerate eigenvalues i.e. corresponding to eigenvalue ϵ_s, there are more than one eigenfunctions. Let us consider the case of two fold degeneracy i.e.

$$H_0 u_{s,1} = \epsilon_s u_{s,1}$$

$$H_0 u_{s,2} = \epsilon_s u_{s,2}.$$

They form a complete set, so that we can write

$$\psi_s(\mathbf{x}) = \sum_{j=1}^{2} \sum_n C_{ns,j} u_{nj}(\mathbf{x}). \tag{12.19}$$

Substituting Eq. (12.19) in the eigenvalue equation

$$(H_0 + \lambda V)\psi_s = E_s \psi_s \tag{12.20}$$

we get

$$\sum_j \sum_n C_{ns,j}(E_s - \varepsilon_n) u_{nj} = \lambda \sum_j \sum_n C_{ns,j} V u_{nj}. \tag{12.21}$$

Multiply both sides by u_{mi}^* and make use of orthonormality relations

$$\int u_{mi}^*(\mathbf{x}) u_{nj}(\mathbf{x}) d^3 x = \delta_{mn} \delta_{ij}, \tag{12.22}$$

then we get

$$C_{ms,i}(E_s - \varepsilon_m) = \lambda \sum_j \sum_n C_{ns,j} \int u_{mi}^* V u_{nj} d^3 x. \tag{12.23}$$

We now make the expansions

$$C_{ms,i} = \delta_{ms} d_i + \lambda C_{ms,i}^{(1)} + \lambda^2 C_{ms,i}^{(2)} + \cdots \tag{12.24a}$$

$$E_s = \varepsilon_s + \lambda E_s^{(1)} + \lambda^2 E_s^{(2)} + \cdots \tag{12.24b}$$

and correspondingly we write

$$\psi_s = \psi_s^0 + \lambda \psi_s^{(1)} + \lambda^2 \psi_s^{(2)} \cdots . \tag{12.24c}$$

Note that on the right-hand side in the first term of Eq. (12.24a), we cannot simply have δ_{ms} as in the non-degenerate case since the index i is not then balanced. Therefore we have to introduce d_i which we determine later.

Substitute Eq. (12.24) in Eq. (12.23) and keep terms up to order λ:

$$(\delta_{ms} d_i + \lambda C_{ms,i}^{(1)})(\varepsilon_s + \lambda E_s^{(1)} - \varepsilon_m) = \lambda \sum_j \sum_n \delta_{ns} d_j \int u_{mi}^* V u_{nj} d^3 x.$$

Now $d_i \delta_{ms}(\varepsilon_s - \varepsilon_m)$ is zero because when $m \neq s$, $\delta_{ms} = 0$ and when $m = s$, $\varepsilon_s = \varepsilon_m$. Therefore up to terms of order λ, we have

$$\lambda d_i E_s^{(1)} \delta_{ms} + \lambda C_{ms,i}^{(1)}(\varepsilon_s - \varepsilon_m) = \lambda \sum_j d_j \int u_{mi}^* V u_{sj} d^3 x. \tag{12.25}$$

Hence for $m = s$

$$d_i E_s^{(1)} = \sum_j d_j V_{ij}, \qquad (12.26a)$$

where

$$V_{ij} = \int u_{si}^* V u_{sj} d^3 x. \qquad (12.26b)$$

Also from Eqs. (12.24a,c) and Eq. (12.19) , up to zero'th order

$$\psi_s^0 = \sum_j \sum_n \delta_{ns} d_j u_{nj}$$

$$= \sum_j d_j u_{sj} . \qquad (12.27)$$

Now from Eq. (12.26)

$$d_1 E_s^{(1)} = d_1 V_{11} + d_2 V_{12},$$

$$d_2 E_s^{(1)} = d_1 V_{21} + d_2 V_{22}. \qquad (12.28)$$

A solution of these homogeneous simultaneous equations exists only if

$$\begin{bmatrix} V_{11} - E_s^{(1)} & V_{12} \\ V_{21} & V_{22} - E_s^{(1)} \end{bmatrix} = 0. \qquad (12.29)$$

Now

$$V_{12} = V_{21}^* \quad \text{or} \quad V_{21} = V_{12}^*.$$

Then Eq. (12.29) gives

$$E_{s\pm}^{(1)} = \frac{1}{2} \left((V_{11} + V_{22}) \pm \sqrt{(V_{11} - V_{22})^2 + 4|V_{12}|^2} \right). \qquad (12.30)$$

Thus from Eq. (12.24b) to order λ

$$E_{s\pm} = \varepsilon_s + \lambda E_{s\pm}^{(1)}$$

or

$$E_{s\pm} = \varepsilon_s + \frac{\lambda}{2}(V_{11} + V_{22}) \pm \frac{\lambda}{2}\sqrt{(V_{11} - V_{22})^2 + 4|V_{12}|^2}. \qquad (12.31)$$

Thus due to perturbation the level ε_s (which is degenerate with respect to the unperturbed Hamiltonian H_0) is split up into two levels given in Eq. (12.31) when we turn on the perturbation.

Thus we can remove the degeneracy by turning on a small perturbation. Also we note from Eq. (12.28) that $\frac{d_2}{d_1}$ has two values corresponding to two values $E_{s\pm}^{(1)}$ given in Eq. (12.30). Thus it follows that the zero'th order wave functions are $\psi_{s\pm}^{(0)}$ which are linear combinations of the unperturbed degenerate wave functions u_{s_1} and u_{s_2}

$$\psi_{s+}^{(0)} = d_1 \left(u_{s_1} + \frac{E_{s+}^{(1)} - V_{11}}{V_{12}} u_{s_2} \right) : E_{s+},$$

$$\psi_{s-}^{(0)} = d_1 \left(u_{s_1} + \frac{E_{s-}^{(1)} - V_{11}}{V_{12}} u_{s_2} \right) : E_{s-}. \qquad (12.32)$$

12.4   The Stark Effect

Consider an electron bound in an atom in electric field

$$\mathbf{E} = -\nabla \phi = -\frac{1}{r}\frac{\partial \phi}{\partial r}\mathbf{r} \qquad (12.33)$$

where ϕ is the electrostatic potential thus

$$\phi(\mathbf{r}) = -\mathbf{E} \cdot \mathbf{r}.$$

Origin is taken at the position of the nucleus. Perturbation potential is then

$$eV = -e\phi = e\mathbf{E} \cdot \mathbf{r}.$$

Energy of the system to second order in e is given by

$$E_s = \varepsilon_s + eV_{ss} + e^2 \sum_{n \neq s} \frac{|V_{sn}|^2}{\varepsilon_s - \varepsilon_n}$$

where ε_s denotes the energy in the absence of the electric field. Now

$$
\begin{aligned}
V_{sn} &= \langle s|V|n\rangle \\
&= \mathbf{E} \cdot \langle s|\mathbf{r}|n\rangle \qquad (12.34)\\
&= \mathbf{E} \cdot \mathbf{r}_{sn},
\end{aligned}
$$

$$
\begin{aligned}
\mathbf{r}_{sn} &= \langle s|\mathbf{r}|n\rangle \\
&= \int u_s^* \mathbf{r} u_n d^3 r \\
&= \langle n|\mathbf{r}|s\rangle^* \\
&= \mathbf{r}_{ns}^*. \qquad (12.35)
\end{aligned}
$$

Thus to second order

$$E_s = \varepsilon_s + e\mathbf{E} \cdot \mathbf{r}_{ss} + e^2 \sum_{n \neq s} \frac{(\mathbf{E} \cdot \mathbf{r}_{sn})(\mathbf{E} \cdot \mathbf{r}_{ns})}{\varepsilon_s - \varepsilon_n}. \qquad (12.36)$$

The shift of energy levels in the electric field is known as the Stark effect. The above expansion to order e^2 holds to a good accuracy when $|E|$ is small compared to the internal electric field of the atom.

The energy eigenfunction of the perturbed Hamiltonian $H = H_0 + eV$ to first order in e is given by

$$\psi_s = u_s + e\mathbf{E} \cdot \sum_{n \neq s} \frac{\mathbf{r}_{ns} u_n}{\varepsilon_s - \varepsilon_n}. \qquad (12.37)$$

We now calculate the static electric dipole moment of the system in one of its stationary states. In quantum mechanics, the electric charge density is given by

$$\rho(\mathbf{r}) = -e\psi_s^* \psi_s.$$

Thus to order e^2

$$\rho(\mathbf{r}) = -eu_s^* u_s - e^2 \mathbf{E} \cdot \sum_{n \neq s} \frac{(u_s^* \mathbf{r}_{ns} u_n + u_n^* \mathbf{r}_{sn} u_s)}{\varepsilon_s - \varepsilon_n} \qquad (12.38)$$

and the dipole moment is given by

$$\mathbf{D} = \int \rho(\mathbf{r}) \mathbf{r} d^3 r$$

$$= -e \int u_s^* \mathbf{r} u_s d^3 r$$

$$\quad -e^2 \mathbf{E} \cdot \sum_{n \neq s} \frac{1}{\varepsilon_s - \varepsilon_n} \left(\mathbf{r}_{ns} \int u_s^* \mathbf{r} u_n d^3 r + \mathbf{r}_{sn} \int u_n^* \mathbf{r} u_s d^3 r \right)$$

$$= -e\mathbf{r}_{ss} - e^2 \sum_{n \neq s} \frac{\mathbf{r}_{ns} \mathbf{r}_{sn} + \mathbf{r}_{sn} \mathbf{r}_{ns}}{\varepsilon_s - \varepsilon_n} \cdot \mathbf{E}$$

$$= \mathbf{D}_0 + \mathbf{D}_1, \qquad (12.39)$$

where

$$\mathbf{D}_0 = -e\mathbf{r}_{ss} = -e \int u_s^* \mathbf{r} u_s d^3 r \qquad (12.40)$$

$$\mathbf{D}_1 = -e^2 \sum_{n \neq s} \frac{\mathbf{r}_{ns} \mathbf{r}_{sn} + \mathbf{r}_{sn} \mathbf{r}_{ns}}{\varepsilon_s - \varepsilon_n} \mathbf{E} \qquad (12.41)$$

$$= \vec{\vec{\alpha}} \cdot \mathbf{E},$$

where

$$\vec{\vec{\alpha}} = -e^2 \sum_{n \neq s} \frac{\mathbf{r}_{ns} \mathbf{r}_{sn} + \mathbf{r}_{sn} \mathbf{r}_{ns}}{\varepsilon_s - \varepsilon_n}. \qquad (12.42)$$

$\mathbf{D}_0$ is called the permanent electric dipole moment of the system as it is determined by the unperturbed state of the system and is entirely independent of the applied field. $\mathbf{D}_1$ is called the induced dipole moment for the states. The tensor $\vec{\vec{\alpha}}$ is called the polarisability for the state s.

Theorem: If the unperturbed Hamiltonian H_0 is invariant with respect to the reflection of coordinates, i.e. $H_0(\mathbf{r}) = H_0(-\mathbf{r})$, then $r_{ss} = 0$ if the non-degenerate eigenstate $u_s(\mathbf{r})$ of H_0 has definite parity.

This can be seen as follows:

$$\hat{P}u_s(\mathbf{r}) = \pm u_s(\mathbf{r})$$

where $\hat{P}$ denotes the parity operator. But

$$\hat{P}u_s(\mathbf{r}) = u_s(-\mathbf{r}).$$

Therefore

$$u_s(-\mathbf{r}) = \pm u_s(\mathbf{r}).\qquad(12.43)$$

Now

$$\mathbf{r}_{ss} = \int u_s^*(\mathbf{r})\mathbf{r}u_s(\mathbf{r})d^3r.$$

In the integration change

$$\mathbf{r} \to -\mathbf{r},$$

so that

$$\mathbf{r}_{ss} = \int u_s^*(-\mathbf{r})(-\mathbf{r})u_s(-\mathbf{r})d^3r$$

$$= \int u_s^*(\mathbf{r})(-\mathbf{r})u_s(\mathbf{r})d^3r,$$

$$= -\mathbf{r}_{ss}.\qquad(12.44)$$

Hence $r_{ss} = 0$. For H atom,

$$H_0 u_{nlm}(\mathbf{r}) = \epsilon_n u_{nlm}(\mathbf{r})$$

where

$$\epsilon_n = -\frac{me^4}{2\hbar^2 n^2}, \quad n = 1, 2, \cdots, \quad l+1 \le n, \quad -l \le m \le l$$

and the parity of $u_{nlm}(\mathbf{r})$ is $(-1)^l$.

Due to Coulomb interaction, all energy levels are degenerate except the ground state since only the ground state has definite parity $+1$. Thus to first order, the Stark effect vanishes for the ground state of H atom and we have to go to the second order. However the radially excited states are degenerate, hence the first order Stark effect does not vanish for such states.

First order Stark Effect
Consider the first excited state which correspond to $n = 2$:

$$l = 0 \qquad m = 0 \qquad 2s \qquad u_{200}$$
$$l = 1 \qquad m = -1, 0, 1 \qquad 2p \qquad u_{21-1}, u_{210}, u_{211}$$

Take the electric field along the z-axis so that

$$\mathbf{E} = |\mathbf{E}|e_z, V = e|\mathbf{E}|z, \tag{12.45a}$$

$$V_{ij} = \int u_{s,i}^* V u_{s,j} d^3 r \quad i, j = 1, \ldots, 4. \tag{12.45b}$$

Instead of i, j we can use l, m; which specify the above matrix elements. Thus we have to calculate the matrix elements

$$\langle 2, l, m \ |V| 2, l', m' \rangle = e|\mathbf{E}| \int u_{2lm}^* z u_{2l'm'} d^3 r.$$

Now

$$L_z = -i\hbar(x \frac{\partial}{\partial y} - y \frac{\partial}{\partial x}). \tag{12.46a}$$

Therefore

$$[L_z, V] = 0. \tag{12.46b}$$

Hence [$n = 2$ is understood]

$$\langle l, m \ |[L_z, V]| l', m' \rangle = 0 \tag{12.46c}$$

or

$$(m\hbar - m'\hbar)\langle lm|V|l'm' \rangle = 0. \tag{12.46d}$$

Thus

$$\langle lm|V|lm' \rangle = 0 \quad \text{unless } m = m'. \tag{12.46e}$$

Parity of states $|lm\rangle$ and $|l'm'\rangle$ are $(-1)^l$ and $(-1)^{l'}$. Thus

$$\hat{P}|lm\rangle = (-1)^l|lm\rangle, \tag{12.47a}$$
$$\hat{P}|l'm'\rangle = (-1)^{l'}|l'm'\rangle. \tag{12.47b}$$

Now

$$\hat{P}V\langle r|\psi \rangle = e|\mathbf{E}|\hat{P}\{z\langle r|\psi \rangle\}$$
$$= -e|\mathbf{E}|z\hat{P}\langle r|\psi \rangle = -V\hat{P}\langle r|\psi \rangle. \tag{12.48}$$

Therefore

$$\hat{P}V + V\hat{P} = 0. \tag{12.49}$$

Hence

$$\left((-1)^l + (-1)^{l'}\right) \langle lm|V|l'm'\rangle = 0. \tag{12.50}$$

Thus

$$\langle lm|V|l'm'\rangle = 0 \quad \text{for} \quad l = l'. \tag{12.51}$$

Hence using Eqs. (12.46e) and (12.51), the 4×4 determinant equation which determines $E_2^{(1)}$, the first order correction, is

$$\begin{bmatrix} -E_2^{(1)} & 0 & \langle 00|V|10\rangle & 0 \\ 0 & -E_2^{(1)} & 0 & 0 \\ \langle 10|V|00\rangle & 0 & -E_2^{(1)} & 0 \\ 0 & 0 & 0 & -E_2^{(1)} \end{bmatrix} = 0. \tag{12.52}$$

Now

$$\langle 10|V|00\rangle = \langle 00|V|10\rangle^* = e|\mathbf{E}| \int u_{210}^*(\mathbf{r})zu_{200}(\mathbf{r})d^3r. \tag{12.53}$$

But [here a is the Bohr radius]

$$z = r\cos\theta, \tag{12.54a}$$

$$u_{210} = \frac{1}{4(2\pi)^{1/2}a^{5/2}}re^{-r/2a} \cdot \cos\theta, \tag{12.54b}$$

$$u_{200} = \frac{1}{4(2\pi)^{1/2}a^{3/2}}(2 - \frac{r}{a})e^{-r/2a}. \tag{12.54c}$$

Therefore

$$\langle 10|V|00\rangle = e|\mathbf{E}|\frac{1}{16(2\pi)a^4}2\pi \int_0^\pi \int_0^\infty r^2\cos^2\theta\left(2 - \frac{r}{a}\right)e^{-r/a}\sin\theta d\theta r^2 dr$$

$$= e|\mathbf{E}|\frac{1}{16a^4}\frac{2}{3}\left(-72a^5\right)$$

$$= -3e|\mathbf{E}|a. \tag{12.55}$$

Equations (12.52) and (12.55) give

$$-E_2^{(1)}\left(-E_2^{(1)}(E_2^{(1)2}) - 3e|\mathbf{E}|a(-3e|\mathbf{E}|aE_2^{(1)})\right) = 0. \tag{12.56}$$

The four roots are

$$E_2^{(1)} = 0, \quad E_2^{(1)} = 0, \quad E_2^{(1)} = \pm3e|\mathbf{E}|a. \tag{12.57}$$

Thus half of the four fold degeneracy is removed in the first order. The third and fourth values of $E_2^{(1)}$ correspond to

$$\frac{1}{\sqrt{2}}\left(u_{200} - u_{210}\right) \quad \text{and} \quad \frac{1}{\sqrt{2}}\left(u_{200} + u_{210}\right)$$

and the first two correspond to any linearly independent combination of u_{211} and u_{21-1}. Thus a hydrogen atom in its excited state $(n = 2)$ behaves as though it has a permanent electric dipole moment of magnitude $3ea$ that can be oriented in three different ways: one state parallel to the external field, one state anti-parallel to the field and two states with zero component along the field.

12.5 Non-Perturbative Method (Variational Principle)

In the perturbative approach we have a series expansion in terms of a small parameter. In the non-perturbative approach we have no series expansion. A well known example of this is the variational principle. This is based on the observation that[1]

$$\frac{\langle \psi | H | \psi \rangle}{\langle \psi | \psi \rangle} \geq E_0,$$

where E_0 is the ground state energy and $|\psi\rangle$ is an arbitrary state. This can be seen as follows: In the basis $|E_n\rangle$, E_n being energy eigenvalues of H,

$$H = \sum_m \sum_n |E_m\rangle\langle E_m | H | E_n\rangle\langle E_n|$$

$$= \sum_m \sum_n |E_m\rangle E_n \delta_{E_m E_n} \langle E_n|$$

$$= \sum_n |E_n\rangle E_n \langle E_n|.$$

Thus

$$\langle \psi | H | \psi \rangle = \sum_n \langle \psi | E_n\rangle\langle E_n | \psi\rangle E_n$$

$$= \sum_n |\langle \psi | E_n\rangle|^2 E_n.$$

Now $|\langle \psi | E_n\rangle|^2 > 0$ and $E_n \geq E_0$, E_0 being the lowest possible energy, i.e. ground state energy corresponding to $|\psi_0\rangle$. Thus

$$\langle \psi | H | \psi \rangle \geq \sum_n \langle \psi | E_n\rangle\langle E_n | \psi\rangle E_0 = \langle \psi | \psi\rangle E_0. \qquad (12.58)$$

Hence $\frac{\langle \psi | H | \psi \rangle}{\langle \psi | \psi \rangle} \geq E_0$, the equality sign holds when $|\psi\rangle$ is the ground state $|\psi_0\rangle$. In other words the absolute minimum of $\frac{\langle \psi | H | \psi \rangle}{\langle \psi | \psi \rangle}$ is the ground state energy and corresponds to $|\psi\rangle = |\psi_o\rangle$.

In applying the variational principle, one has to pick up a state $|\psi(\alpha)\rangle$, which depends on a parameter α. Then choose the value of α which minimizes $\frac{\langle \psi(\alpha) | H | \psi(\alpha)\rangle}{\langle \psi(\alpha) | \psi(\alpha)\rangle}$, the $|\psi\rangle$ for that value of α is the closest of all the $|\psi(\alpha)\rangle$ to the ground state and as such the least value of $\frac{\langle \psi | H | \psi \rangle}{\langle \psi | \psi \rangle}$ has the most accurate approximation to the ground state energy and wave function.

---

[1]See G. Baym, Lectures on Quantum Mechanics, Benjamen, 1969, p. 242

We illustrate this approach for He atom. The Hamiltonian of two electron atom is

$$H = \frac{\hat{\mathbf{p}}_1^2}{2m} + \frac{\hat{\mathbf{p}}_2^2}{2m} - \frac{Ze^2}{r_1} - \frac{Ze^2}{r_2} + \frac{e^2}{|\mathbf{r}_1 - \mathbf{r}_2|} \qquad (12.59)$$

where the last term is the electron Coulomb energy; $\mathbf{r}_1$ and $\mathbf{r}_2$ are the position vectors of two electrons measured from the nucleus. It is not possible to solve this problem exactly for energy eigenvalues and eigenfunctions.

The simplest approximation is to consider the electron–electron interaction as perturbation

$$H = H_0 + H_1. \qquad (12.60)$$

The ground state is

$$\psi_0(1,2) = \psi_{1s}(\mathbf{r}_1)\psi_{1s}(\mathbf{r}_2)\chi_{00} \qquad (12.61)$$

where χ_{00} is spin singlet. This is required by the Pauli principle as the spatial part of $\psi_0(1,2)$ is symmetric under $\mathbf{r}_1 \rightarrow \mathbf{r}_2$ and therefore the spin wave function must be antisymmetric in spin coordinates to make $\psi_0(1,2)$ antisymmetric. Here

$$\begin{aligned}\psi_{1s}(\mathbf{r}) &= \psi_{100}(\mathbf{r}) \\ &= Y_{00}(\theta,\phi)R_{10}(r) \\ &= \frac{1}{\pi}(Z/a_0)^{3/2}e^{-Zr/a_0},\end{aligned} \qquad (12.62)$$

for $\mathbf{r} = \mathbf{r}_1$ or $\mathbf{r} = \mathbf{r}_2$. Now the energy of each electron in the Coulomb potential $(-Ze^2/r)$ is [see Hydrogen atom sec. 7.2, and 7.3]:

$$\begin{aligned}\left(-\frac{Z^2e^2}{2a_0}\right) &= -Z^2(-13.605\text{eV}) \\ &= -Z^2 Ry\end{aligned} \qquad (12.63)$$

so that the ground state energy to the zeroth order

$$\begin{aligned}E_0^0 &= -\frac{Z^2e^2}{2a_0} + \frac{-Z^2e^2}{2a_0} \\ &= -\frac{Z^2e^2}{a_0}.\end{aligned} \qquad (12.64)$$

The first order change in the energy is the expectation value of the H_1 in the ground state, on using the wave functions in Eq. (12.62) is

$$\begin{aligned}\Delta E &= \langle H_1 \rangle \\ &= \langle 1(100)2(100)|H_1|1(100)2(100)\rangle \\ &= \frac{1}{\pi^2}(Z/a_0)^6 e^2 \int \int d^3\mathbf{r}_1 d^3\mathbf{r}_2 \frac{1}{|\mathbf{r}_1 - \mathbf{r}_2|}e^{-\frac{2Z}{a_0}(r_1+r_2)}.\end{aligned} \qquad (12.65)$$

Now using

$$\frac{1}{|\mathbf{r}_1 - \mathbf{r}_2|} = \int \frac{d^3k}{(2\pi)^3} e^{i\mathbf{k}\cdot(\mathbf{r}_1-\mathbf{r}_2)} \frac{4\pi}{k^2}, \tag{12.66}$$

we get

$$\Delta E = \frac{1}{2\pi^4}(Z/a_0)^6 e^2 \int \frac{d^3k}{k^2}\left(\int d^3r_1 e^{i\mathbf{k}\cdot\mathbf{r}_1 - \frac{2Z}{a_0}r_1}\int d^3r_2 e^{-i\mathbf{k}\cdot\mathbf{r}_2 - \frac{2Z}{a_0}r_2}\right)$$

$$= \frac{1}{2\pi^4}(Z/a_0)^6 e^2 \int \frac{d^3k}{k^2}\left|\int d^3r\, e^{-i\mathbf{k}\cdot\mathbf{r} - \frac{2Z}{a_0}r_0}\right|^2. \tag{12.67}$$

Then

$$\int d^3r\, e^{-i\mathbf{k}\cdot\mathbf{r} - \frac{2Z}{a_0}r_0} = \frac{16\pi Z a_0}{[k^2 + (2Z/a_0)^2]^2}, \tag{12.68}$$

$d^3k = 4\pi k^2 dk$, and putting $a_0 k/2Z = x$, one obtains

$$\Delta E = \frac{4}{\pi}\frac{Ze^2}{a_0}\int_0^\infty \frac{dx}{(x^2+1)^4}$$

$$= \frac{4}{\pi}\frac{Ze^2}{a_0}\left(\frac{5\pi}{32}\right)$$

$$= \frac{5Ze^2}{8a_0}. \tag{12.69}$$

One can determine experimentally the ground state energy for a two electron atom by measuring the ionization energy E_i i.e. the minimum energy required to remove one of the electrons to ∞. Then

$$E_i = -(E_0 - E'_{j0}) \tag{12.70}$$

where E'_{j0} is the ground state energy of the remaining electron and is given in Eq. (12.63). For He ($Z = 2$), $E_i = 1.807R_y$ experimentally so that

$$E_0^{\text{exp}} = E'_{j0} - E_i$$

$$= [-4 - (1.807)]R_y$$

$$= -5.807R_y$$

$$\approx -79eV \tag{12.71}$$

while the perturbative result, using Eq. (12.64) and Eq. (12.68) gives ($Z = 2$)

$$E_0^{\text{per}} = E_0^0 + \Delta E$$

$$= -(Z - \frac{5}{8})\frac{Ze^2}{a_0}$$

$$= -(Z - \frac{5}{8})Z.2R_y$$

$$= -5.5R_y$$

$$\approx -74.8eV. \tag{12.72}$$

Thus there is a discrepancy of a about $0.31 Ry(5.45\%)$ between the perturbative result and the experimental value of the ground state energy of He. One reason is that the wave function (12.62) ignores the fact that one electron tends to screen the charge on the nucleus as seen by the other electron. That is to say, when one electron looks at the nucleus it sees not only positive charge of the nucleus but also some negative charge density around the nucleus from the other electron. This suggests that one can improve the accuracy by choosing a trial wave function of the form (12.62) with an effective Z' value in ψ_{1s} :

$$\psi_{1s} = \frac{1}{\sqrt{\pi}} (Z'/a_0)^{3/2} e^{-Z'r/a_0} \tag{12.73}$$

and treat Z' as a variational parameter. Then

$$\begin{aligned} E_0(Z') &= \langle H \rangle \\ &= \langle H_0 \rangle + \langle H_1 \rangle \\ &= \langle H_0 \rangle + \frac{5}{8} \frac{Z'e^2}{a_0}. \end{aligned} \tag{12.74}$$

The expectation value $\langle H_0 \rangle$ is with respect to the ground state of a Coulombic potential $Z'e^2/r$, so that

$$\langle \text{K.E of each electron} \rangle = \frac{Z'e^2}{2a_0} \tag{12.75}$$

and

$$\begin{aligned} \langle \text{nuclear Coulomb P.E for each electron} \rangle &= \langle -\frac{Ze^2}{r} \rangle \\ &= \frac{Z}{Z'} \langle -\frac{Z'e^2}{r} \rangle \\ &= \frac{Z}{Z'} (-\frac{Z'^2 e^2}{a_0}). \end{aligned} \tag{12.76}$$

Thus

$$E_0(Z') = (Z'^2 - 2ZZ' + \frac{5}{8} Z') \frac{e^2}{a_0}. \tag{12.77}$$

To find the minimum of $E_0(Z')$,

$$\frac{\partial E_0(Z')}{\partial Z'} = 0 \quad \text{gives} \quad Z' = Z - \frac{5}{16} \tag{12.78}$$

while $\frac{\partial^2 E_0(Z')}{\partial Z'^2} = 2 > 0$. Thus $E_0(Z')$ has a minimum at $Z' = Z - \frac{5}{16}$. Thus as expected the effective charge is somewhat less than Z. The minimum value of $E_0(Z')$ is the ground state energy $(Z = 2)$

$$E_0^{var} = -(Z - \frac{5}{16})^2 \frac{e^2}{a^2}$$
$$= -5.695 R_y$$
$$\approx -77.46 eV \qquad (12.79)$$

which is great improvement on the perturbative value as the discrepancy is now reduced to $0.115 R_y(\sim 2\%)$.

12.6 Problems

12.1 Consider a harmonic oscillator of mass m on which acts a uniform time-independent external force F. Call its ground state

$$\tilde{u}_0(x) = \langle x|\tilde{0}\rangle.$$

If

$$u_1(x) = \langle x|1\rangle$$

is the first excited state of the same oscillator when $F = 0$, show that to the first order in perturbation the probability of finding it in the first excited state is

$$|\langle 1|\tilde{0}\rangle|^2 = |\int u_1^*(x)\tilde{u}_0(x)dx|^2$$
$$= \frac{F^2}{2m\hbar} \frac{1}{\omega^3}.$$

Also calculate the shift in the ground state energy to second order in F.

12.2 A hydrogen atom is placed in an external constant field F in the z-direction. For the ground state s of the atom, write

$$e^2 F^2 \sum_{n\neq s} \frac{|V_{sn}|^2}{\varepsilon_s - \varepsilon_n} = -\frac{1}{2}F^2\alpha.$$

Show that

$$4a^3 < \alpha < \frac{16}{3}a^3$$

where a is the Bohr radius $\frac{\hbar^2}{me^2}$.

12.3 If the perturbation potential is given by

$$eV = e\mathbf{F} \cdot \mathbf{r} = eFr\cos\theta$$

where F is a constant, show that for the hydrogen atom

$$\langle 100|V|nlm\rangle = 0 \text{ for } m \neq 0$$

and $\langle 100|V|nl0\rangle$ is non-zero only for $l = 1$ and in this case it is given by

$$eF\frac{1}{\sqrt{3}} \int_0^\infty R_{10}^*(r) R_{nl}(r) r^3 \, dr.$$

Hence show that for the ground state

$$E_1 = \epsilon_1 + \frac{e^2 F^2}{3} \sum_{n \neq 1} \frac{|\int_0^\infty R_{10} R_{n1} r^3 dr|^2}{\epsilon_1 - \epsilon_n}.$$

12.4 Consider an atom which has a nucleus of charge Z and one electron. Using first order perturbation theory, calculate the energy shift of the $1s$, $2s$ and $2p$ states of ^{1}H and ^{235}U atoms, assuming that the nucleus is a uniformly charged sphere of radius $R = r_0 A^{1/3}$, where $r_0 \approx 1.2 \times 10^{-13}$ cm and A is the atomic number. Note that $R \ll a$, where a is the Bohr radius.

Hint: The Coulomb potential is given by

$$V(r) = -\frac{3Ze^2}{2R^3}\left(r^2 - \frac{1}{3}r^2\right) \quad r < R,$$

$$= -\frac{Ze^2}{r} \qquad\qquad r > R.$$

12.5 Zeeman Effect:

The Hamiltonian for a hydrogen like atom can be written as

$$H_0 = \frac{\hat{\mathbf{p}}^2}{2m_e} - \frac{Ze^2}{r} + \frac{Ze^2}{2m_e^2 c^2 r^3}\mathbf{S}\cdot\mathbf{L}.$$

The last term represents the interaction of the magnetic moment of the electron $\boldsymbol{\mu} = -\frac{e}{m_e c}\mathbf{S}$ with the magnetic field $\mathbf{B} = \frac{1}{2c}\mathbf{E}\times\mathbf{v} = \frac{1}{2}\frac{Ze}{e}\frac{r^3}{r^3}\mathbf{r}\times\mathbf{v} = \frac{1}{2m_e c}\frac{Ze}{r^3}\mathbf{r}\times\mathbf{p}$ which arises due to the motion of an electron in the electric field E of nucleus of charge Ze.

Consider the atom in an external weak homogeneous magnetic field B along the z-axis so that the magnetic interaction energy is given by

$$V = -\boldsymbol{\mu}_L\cdot\mathbf{B} - \boldsymbol{\mu}\cdot\mathbf{B} = \frac{e}{2m_e c}BL_z + \frac{e}{m_e c}BS_z$$

$$= \frac{eB}{2m_e c}(L_z + 2S_z).$$

It is clear that the eigenstates of H_0 are $|njm\rangle$,
$$H_0|njm\rangle = E_{nj}|njm\rangle$$
Treating V as a perturbation, and using Eqs. (11.95,96), show that the energy shift
$$\Delta E_{nl}^B = \langle njm|V|njm\rangle$$
$$= \frac{eB}{2m_ec}\langle njm|(L_z + 2S_z)|njm\rangle$$
$$= \frac{eB\hbar}{2m_ec}m\left(1 \pm \frac{1}{2l+1}\right), \quad j = l \pm 1/2.$$

12.6 If $H(\lambda)$ depends on a parameter λ, and $|\psi_k(\lambda)\rangle$ is an eigenvector of $H(\lambda)$ which is normalized to unity, prove that
$$\frac{dE_k(\lambda)}{d\lambda} = \langle \psi_k(\lambda)|\frac{\partial H(\lambda)}{\partial \lambda}|\psi_k(\lambda)\rangle$$
[Helmann–Feynman Theeorem]

12.7 Estimate the ground state energy of the hydrogen atom by using the three dimensional harmonic oscillator ground state function as a trial function. How this estimate compares with the exact value, which is $-\frac{me^4}{2\hbar^2}$?
Three dimensional harmonic oscillator ground state function is:
$$\left(\frac{4\alpha^3}{\sqrt{\pi}}\right)^{1/2}e^{-\frac{1}{2}\alpha^2r^2}, \quad \alpha = \sqrt{\mu\omega}.$$
Take α as variational parameter. Useful integral
$$\int_0^\infty x^{2n}e^{\alpha^2x^2}dx = \frac{1}{2\alpha^{2n+1}}\frac{\sqrt{\pi}}{n!}\frac{(2n)!}{2^{2n}}.$$
Repeat the same problem with the following trial function:
$$\phi_\alpha(r) = 1 - \frac{r}{\alpha}, \quad r \le \alpha$$
$$= 0, \quad r > \alpha$$
which one is the better trial function?

12.8 Use the variational principle to estimate the ground state energy of the anharmonic oscillator
$$H = \frac{\hat{p}^2}{2m} + \lambda x^4.$$
Compare your result with the exact result
$$E_0 = 1.060\lambda^{1/3}\left(\frac{\hbar^2}{2m}\right)^{2/3}.$$
You may use the trial function
$$Ae^{-\alpha^2x^2/2}.$$

12.9 Use the variational principle to estimate the ground state energy
of a particle in the potential

$$V = \infty \quad \text{for} \quad x < 0,$$
$$V = cx \quad \text{for} \quad x > 0.$$

Take Axe^{iax} as the trial function. Why can't one select Ae^{-ax} as
the trial function? Useful integral:

$$\int_0^\infty x^n e^{-bx} dx = \frac{n!}{b^{n+1}}.$$

Chapter 13

Time Dependent Perturbation Theory

13.1 Transition Probability

Let us discuss how the perturbation method works for the time dependent Schrödinger equation

$$ i\hbar \frac{\partial \psi}{\partial t} = H\psi \qquad (13.1a) $$

where we assume that

$$ H = H_0 + \lambda V(\mathbf{r}; \mathbf{p}, t) \qquad (13.1b) $$

i.e. the perturbation V may depend on time while the unperturbed Hamiltonian H_0 does not explicitly depend on t.

The Schrödinger equation involving H_0 is

$$ i\hbar \frac{\partial \phi}{\partial t} = H_0\phi \qquad (13.2) $$

which can be solved exactly. λV is assumed to be small compared to H_0.

In general, solution of Eq. (13.2) is

$$ \phi(\mathbf{r}, t) = \sum C_n u_n(\mathbf{r}) e^{-i\varepsilon_n t/\hbar} \qquad (13.3) $$

where $u_n(\mathbf{r})$ are eigenfunctions of H_0 belonging to eigenvalues ε_n.

We now assume that solution ψ of Eq. (13.1) can be written as

$$ \psi(\mathbf{r}, t) = \sum_n C_n(t) u_n(\mathbf{r}) e^{-i\varepsilon_n t/\hbar} \qquad (13.4) $$

where the C_n are in general dependent on time. Everything in Eq. (13.4) is known except $C_n(t)$. We substitute Eq. (13.4) in Eq. (13.1a) with H as given in Eq. (13.1b) and make use of

$$ H_0 u_n(\mathbf{r}) = \varepsilon_n u_n(\mathbf{r}). $$

301

Then we obtain

$$\sum_n (i\hbar\dot{C}_n(t) + \varepsilon_n C_n(t))u_n(\mathbf{r})e^{-i\varepsilon_n t/\hbar}$$

$$= \sum_n \varepsilon_n e^{-i\varepsilon_n t/\hbar}u_n(\mathbf{r})C_n(t) + \lambda\sum_n C_n(t)e^{-i\varepsilon_n t/\hbar}V u_n(\mathbf{r})$$

or

$$i\hbar\sum_n \dot{C}_n(t)u_n(\mathbf{r})e^{-i\varepsilon_n t/\hbar} = \lambda\sum_n C_n(t)e^{-i\varepsilon_n t/\hbar}V u_n(\mathbf{r}). \tag{13.5}$$

We multiply both sides by $u_m^*(\mathbf{r})e^{i\varepsilon_m t/\hbar}$, integrate over $\mathbf{r}$ and make use of

$$\int u_m^*(\mathbf{r})u_n(\mathbf{r})d^3r = \delta_{mn}.$$

Then we get

$$i\hbar\dot{C}_m(t) = \lambda\sum_n C_n(t)e^{-i(\varepsilon_n-\varepsilon_m)t/\hbar}V_{mn}(t), \tag{13.6a}$$

where

$$V_{mn}(t) = \int u_m^*(\mathbf{r})V(t)u_n(\mathbf{r})d^3r$$

$$= \langle m|V(t)|n\rangle. \tag{13.6b}$$

These are coupled equations.

Suppose that for $t < t_0$, $V = 0$, so that our system is unperturbed and is in an eigenstate of H_0. Then

$$\psi(t < t_0) = \phi(t < t_0)$$

$$= u_s(\mathbf{r})e^{-i\varepsilon_s t/\hbar}. \tag{13.7a}$$

Comparing Eq. (13.7a) with (13.3), we get

$$C_s = 1, C_n = 0, n \neq s \text{ for } t < t_0. \tag{13.7b}$$

Suppose that the perturbation is effective from t_0 to t. Under the action of the perturbation, the system may pass from its initial stationary state u_s to another state. According to the general rule, the probability of a transition from the original eigenstate s of H_0 to another state n is given by

$$P(s,n) = |C_n(t)|^2. \tag{13.8}$$

We now rewrite Eq. (13.6a) as

$$i\hbar\dot{C}_m(t) = \lambda C_s(t)e^{i(\varepsilon_m-\varepsilon_s)t/\hbar}V_{ms}(t) + \lambda\sum_{n\neq s} C_n(t)$$

$$\times e^{i(\varepsilon_m-\varepsilon_n)t/\hbar}V_{mn}(t). \tag{13.9}$$

In view of Eq. (13.7b), in the first approximation we may replace on the right-hand side of Eq. (13.9), $C_n(t)$ by 0 for $n \neq s$ and equal to 1 for $n = s$. Thus we get in the first approximation

$$i\hbar \dot{C}_m(t) = \lambda e^{i(\varepsilon_m - \varepsilon_s)t/\hbar} V_{ms}(t), \quad m \neq s \tag{13.10}$$

and for $m = s$

$$i\hbar \dot{C}_s(t) = \lambda C_s(t) V_{ss}(t). \tag{13.11}$$

The solutions of Eqs. (13.10) and (13.11) are

$$C_m(t) = -\frac{i}{\hbar}\lambda \int_{t_0}^t e^{i(\varepsilon_m - \varepsilon_s)t'/\hbar} V_{ms}(t')dt', \quad m \neq s, \tag{13.12}$$

$$C_s(t) = e^{-i\lambda/\hbar \int_{t_0}^t V_{ss}(t')dt'}. \tag{13.13}$$

Note that $|C_s| = 1$, only its phase has changed. Now

$$\psi(\mathbf{r}, t) = \sum_n C_n(t) u_n(\mathbf{r}) e^{-i\varepsilon_n t/\hbar}. \tag{13.14}$$

But from Eq. (13.13)

$$C_s(t) = e^{-i(\lambda/\hbar)\bar{V}_{ss}(t - t_0)}, \tag{13.15a}$$

where

$$\bar{V}_{ss} = \frac{1}{t - t_0} \int_{t_0}^t V_{ss}(t')dt'. \tag{13.15b}$$

Thus separating out the $n = s$ term in Eq. (13.14) and using Eqs. (13.15) and (13.12) (for $m = n$), we have

$$\psi(\mathbf{r}, t) \approx u_s(\mathbf{r}) e^{-i(\varepsilon_s + \lambda \bar{V}_{ss})\frac{t - t_0}{\hbar}}$$

$$+ \lambda \sum_{n \neq s} \left(-\frac{i}{\hbar}\right) \int_{t_0}^t e^{i(\varepsilon_n - \varepsilon_s)t'/\hbar} V_{ms}(t')dt'$$

$$\times u_n(\mathbf{r}) e^{-i\varepsilon_n t/\hbar}. \tag{13.16}$$

Originally ψ is given by

$$\phi = u_s(\mathbf{r}) e^{-i\varepsilon_s t/\hbar}.$$

Time dependence of the s-th state after the perturbation to the lowest order is

$$e^{-iE_s \frac{t - t_0}{\hbar}}, \tag{13.17a}$$

where

$$E_s = \varepsilon_s + \lambda \bar{V}_{ss}, \tag{13.17b}$$

i.e. energy is changed slightly and is given above in the first approximation.

We are interested in $C_n(t)$ for $n \neq s$, since then $|C_n(t)|^2$ gives the probability of transition from state s to another state n as given in Eq. (13.8). Hence using Eq. (13.12) to first order

$$
\begin{aligned}
P(s,n) &= |C_n(t)|^2 \\
&= \frac{\lambda^2}{\hbar^2} |\int_{t_0}^{t} e^{i(\varepsilon_n - \varepsilon_s)t'/\hbar} V_{ns}(t')dt'|^2.
\end{aligned}
\tag{13.18}
$$

For second order, we put Eq. (13.12), replacing m by n, back into Eq. (13.6a); so that we get

$$\dot{C}_m(t') = (\frac{-i}{\hbar}\lambda)^2 \sum_n \int_{t_0}^{t'} dt'' V_{ns}(t'')e^{i\omega_{ns}t''} V_{mm}(t')e^{i\omega_{mm}t'}$$

$$C_m(t) = (\frac{-i}{\hbar}\lambda)^2 \sum_n \int_{t_0}^{t} dt' V_{mn}(t')e^{i\omega_{mn}t'} \int_{t_0}^{t'} V_{ns}(t'')e^{i\omega_{ns}t''}dt'',$$

where

$$\omega_{mn} = \frac{\epsilon_m - \epsilon_n}{\hbar}. \tag{13.19}$$

Special cases:
(A) Sudden Approximation

$$t < t_0 \quad \text{no perturbation}$$
$$t \geq t_0 \quad \text{there is perturbation}$$

Thus

$$V_{ms}(t) = 0 \qquad t < t_0$$

and assume that

$$V_{ms}(t) = V_{ms} \qquad t \geq t_0 \,,$$

i.e. we consider transitions caused by a perturbation independent of t. Then from Eq. (13.12)

$$
\begin{aligned}
C_m(t) &= -\frac{i}{\hbar}\lambda V_{ms} \int_{t_0}^{t} e^{i(\varepsilon_m - \varepsilon_s)t/\hbar} dt \\
&= -\lambda V_{ms} \frac{e^{i(\varepsilon_m - \varepsilon_s)t/\hbar}}{\varepsilon_m - \varepsilon_s} \left(1 - e^{-i(\varepsilon_m - \varepsilon_s)\frac{t-t_0}{\hbar}}\right).
\end{aligned}
\tag{13.20}
$$

Hence

$$P(s,n) = |C_n(t)|^2$$
$$= \frac{4\lambda^2 |V_{ns}|^2}{(\varepsilon_n - \varepsilon_s)^2} \sin^2 \left(\frac{(\varepsilon_n - \varepsilon_s)(t - t_0)}{2\hbar} \right). \tag{13.21}$$

This shows that if the perturbation is turned on, the transition probability oscillates between 0 and some number.

From Eq. (13.21)

$$P(s,n) = \frac{\lambda^2}{\hbar^2} |V_{ns}|^2 \frac{\sin^2 \frac{1}{2} \omega_{ns} t}{(\omega_{ns}/2)^2}, \tag{13.22}$$

where $\omega_{ns} = \frac{\varepsilon_n - \varepsilon_s}{\hbar}$ and we have taken $t_0 = 0$. We now show that for large t,

$$P(n,s) \rightarrow t.$$

Now

$$\lim_{t \to \infty} \frac{\sin^2 \frac{1}{2} \omega_{ns} t}{(\omega_{ns}/2)^2 t} = \lim_{t \to \infty} \frac{\sin^2 \frac{1}{2} \omega_{ns} t}{(\omega_{ns} t/2)^2} t$$
$$= \lim_{t \to \infty} t, \quad \text{for } \omega_{ns} = 0. \tag{13.23}$$

Further

$$\int_{-\infty}^{\infty} dt \frac{\sin^2 \frac{1}{2} \omega_{ns} t}{(\omega_{ns}/2)^2 t} d(\frac{\omega_{ns} t}{2}) = \int_{-\infty}^{\infty} dx \frac{\sin^2 x}{x^2} = \pi.$$

Thus

$$\lim_{t \to \infty} \frac{\sin^2 \frac{1}{2} \omega_{ns} t}{(\omega_{ns}/2)^2 t} = \pi \delta(\frac{1}{2} \omega_{ns} t). \tag{13.24}$$

Hence for large t, we may replace

$$\frac{\sin^2 \frac{1}{2} \omega_{ns} t}{(\frac{\omega_{ns}}{2})^2} = \delta \left(\frac{1}{2} \omega_{ns} t \right) \pi t$$
$$= 2\pi \hbar \delta(\varepsilon_n - \varepsilon_s). \tag{13.25}$$

Thus Eq. (13.21) becomes for large t

$$P(s,n) = \frac{2\pi \lambda^2}{\hbar} |V_{ns}|^2 \delta(\varepsilon_n - \varepsilon_s) t. \tag{13.26}$$

Hence the transition probability per unit time (in the limit of $t \to \infty$) is given by

$$W = \frac{2\pi}{\hbar} \lambda^2 |V_{ns}|^2 \delta(\varepsilon_n - \varepsilon_s). \tag{13.27}$$

This explicitly exhibits the fact that energy is conserved. If the transition is to a continuum of states about state n, then $\rho_f(\varepsilon_n)d\varepsilon_n$ is the number of final states with energies between ε_n and $\varepsilon_n + d\varepsilon_n$, where $\rho_f(\varepsilon_n)$ is the density of final states, i.e. the number of states per unit energy interval. In this case the transition probability per unit time is given by

$$
\begin{aligned}
\mathcal{W}_{ns} &= \frac{2\pi}{\hbar}\lambda^2 \int |V_{ns}|^2 \delta(\varepsilon_n - \varepsilon_s)\rho_f(\varepsilon_n)d\varepsilon_n \\
&= \frac{2\pi}{\hbar}\lambda^2 |V_{ns}|^2 \rho_f(\varepsilon_n).
\end{aligned}
\tag{13.28}
$$

The above formula represents a constant rate of transition. The formula (13.28) is a fundamental result and has been called the Golden Rule by Fermi. For an application of the Golden Rule to potential scattering, see Sec. 18.7.

We end this section by the following remarks: we note from the above analysis if we plot $4\frac{\sin^2 \frac{1}{2}\omega_{ns}t}{\omega_{ns}^2}$ as a function of ω_{ns} for fixed t, we get the curve shown in Fig 13.1.

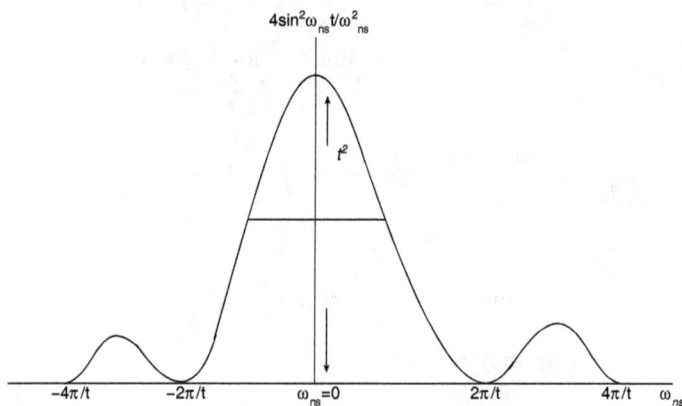

Fig. 13.1   Plot of $4\frac{\sin^2 \omega_{ns}t}{\omega_{ns}^2}$ verses ω_{ns}

We note that width is proportional to $\frac{1}{t}$. Height of the middle peak: t^2. $P(s,n)$ is large only for those final states that satisfy

$$
\omega_{ns} \sim \frac{2\pi}{t}
$$

$$
t \sim \frac{2\pi}{\omega_{ns}} = \frac{2\pi\hbar}{|\epsilon_n - \epsilon_s|}
\tag{13.29}
$$

i.e. $\Delta E \Delta t \sim \hbar$.

13.2 Problems

13.1 A harmonic oscillator is perturbed by a weak, time dependent but spatially uniform force $F(t)$, where $F(t) = 0$ for $t < 0$. Initially, (i.e. $t < 0$) the system is in its ground state $|0\rangle$. Up to first order in F, show that the probability of finding the system in state $|1\rangle$ at time t is given by

$$P(0,1) = \frac{1}{2m\hbar\omega} \left| \int_0^t e^{i\omega t} F(t)dt \right|^2 .$$

For $t \geq 0$, take $F(t) = F_0(1 - e^{-t/\tau})$. Show that in this case, the probability of finding the system in $|1\rangle$ for large times is

$$P(0,1) \approx \frac{F_0^2}{2m\hbar\omega^3},$$

if $\omega\tau \gg 1$. If $\tau = 0$, show that this probability oscillates indefinitely.

13.2 A hydrogen atom in its ground state is placed in an electric field $\mathbf{E}(t)$, which is spatially uniform and has the time dependence

$$\mathbf{E}(t) = 0 \qquad t < 0,$$
$$= \mathbf{E}_0 e^{-t/\tau} \qquad t \geq 0.$$

What is the first-order probability of finding the atom in the $2p$ state after a long time?

Chapter 14

Statistics and the Exclusion Principle

14.1 Introduction

Identical particles cannot be distinguished by means of any inherent property; however in classical mechanics, they can be labelled in some way so that they do not lose their individuality despite the identity of their physical properties. This is because in classical mechanics the existence of sharply definable trajectories for individual particles makes it possible, in principle, to distinguish them by their paths and each particle can be followed during the course of an experiment. In quantum mechanics, it is assumed that for identical quantum particles such as electrons, no such distinction is possible. Consider a quantum mechanical system of two identical particles described by the state function

$$\psi(1, 2) \tag{14.1}$$

where each of the numbers represents all the coordinates (position and spin) of one of the particles. The Hamiltonian of the system $H(1, 2)$ must be symmetric for exchange of particles

$$H(1, 2) = H(2, 1). \tag{14.2}$$

This is because an energy measurement cannot distinguish between the particles. Also $|\psi(1, 2)|^2$ determines the probability for finding one particle here and the other there, but can make no distinction as to which particle is which. Thus the probability distribution is unchanged by exchange of the particles' coordinates implying

$$|\psi(1, 2)|^2 = |\psi(2, 1)|^2. \tag{14.3}$$

Hence

$$\psi(1, 2) = \pm\psi(2, 1) , \tag{14.4}$$

i.e. the wave function is either symmetric or anti-symmetric under the interchange of the particles' coordinates. It is found that the choice of positive or negative sign is a property of the type of particle. For an electron (in general for half integral spin particles, called fermions), the sign must always be negative. As will be seen below this leads to the Pauli Exclusion Principle. The state function for two electrons (in general for two identical fermions) is always anti-symmetric for exchange of their coordinates. Electrons are then said to satisfy Fermi–Dirac statistics. On the other hand, a state function for two photons (or in general for integral spin particles, called bosons) must always have positive sign in Eq. (14.2) or be symmetric if 1 and 2 are interchanged. Photons are said to satisfy Bose–Einstein statistics. Thus there is an intimate connection between spin and statistics.

14.2   Permutation Operator and Exclusion Principle

We now introduce an operator $\hat{P}_{12}$, which interchanges all the coordinates of particles 1 and 2.

$$\hat{P}_{12}\psi(1,2) = \psi(2,1),$$
$$\hat{P}_{12}^2\psi(1,2) = \psi(1,2). \tag{14.5}$$

Thus $\hat{P}_{12}$ has two eigenvalues ± 1. Let the corresponding eigenstates be $\psi_\pm$

$$\psi_+ = \tfrac{1}{2}(1 + \hat{P}_{12})\psi = \tfrac{1}{2}(\psi(1,2) + \psi(2,1)), \tag{14.6}$$
$$\psi_- = \tfrac{1}{2}(1 - \hat{P}_{12})\psi = \tfrac{1}{2}(\psi(1,2) - \psi(2,1)). \tag{14.7}$$

Now

$$\hat{P}_{12}H(1,2)\psi(1,2) = H(2,1)\psi(2,1)$$
$$= H(1,2)\hat{P}_{12}\psi(1,2). \tag{14.8}$$

Hence

$$\left[\hat{P}_{12}, H\right] = 0. \tag{14.9}$$

In other words, $\hat{P}_{12}$ is a constant of motion. It means the symmetry which a state initially has, does not change with time.

The situation with more than 2 particles is more complicated. Suppose there are 3 particles described by an arbitrary wave function $\psi(1,2,3)$.

P_{ij} is an operator that interchanges the variables of the ith and jth particles:

$$P_{12}\psi(1,2,3) = \psi(2,1,3),$$
$$P_{13}\psi(1,2,3) = \psi(3,2,1), \text{ etc .} \tag{14.10}$$

Then, for example

$$P_{12}P_{13}\psi(1,2,3) = P_{12}\psi(3,2,1) = \psi(2,3,1),$$
$$P_{13}P_{12}\psi(1,2,3) = P_{13}\psi(2,1,3) = \psi(3,1,2), \tag{14.11}$$

so that

$$P_{12}P_{13} \neq P_{13}P_{12} , \text{ i.e. } [P_{12}, P_{13}] \neq 0. \tag{14.12}$$

Thus we cannot find a complete set of eigenfunctions belonging to all the P's simultaneously. But this does not mean that there are no simultaneous eigenfunctions at all. For example, even though angular momentum operators $\hat{L}_x$, $\hat{L}_y$ and $\hat{L}_z$ do not commute, the function $Y_{00}(\theta,\psi)R_{n0}(r) \equiv f(r)$, which is spherically symmetric is an eigenfunction of all three of them with eigenvalue zero. In the present case, there exist exactly 2 kinds of ψ's that are eigenfunctions of all the P's, one totally symmetric and one totally anti-symmetric:

$$P_{ij}\psi_{s,a}(1,2,3) = \pm\psi_{s,a}(1,2,3) \tag{14.13}$$

where

$$\psi_{s,a}(1,2,3) = \{\psi(1,2,3) \pm \psi(2,1,3) \pm \psi(1,3,2)$$
$$\pm\psi(3,2,1) + \psi(3,1,2) + \psi(2,3,1)\}. \tag{14.14}$$

Out of $3! = 6$ independent states, there are 4 further states with more complicated permutation properties than the above two states.

We need not concern ourselves with such states because of the following experimental fact. Experimentally for each system of n identical particles, the states are either completely anti-symmetric or symmetric under the interchange of coordinates of any two particles:

$$P_{ij}\psi(1, ..., i, ...j, ...n) = \psi(1, ..., j, ...i, ...n)$$
$$= -\psi(1, ..., i, ...j, ...n), \quad n \text{ identical fermions,}$$
$$= \psi(1, ..., i, ...j, ...n), \quad n \text{ identical bosons.} \tag{14.15}$$

This property depends only on the species of particles and not on the number n.

The anti-symmetry of the wave function for identical fermions implies, for example

$$\psi(2,1,3,\ldots) = -\psi(1,2,3,\ldots) \tag{14.16}$$

so that if $1 = 2$, i.e. if 2 identical fermions have the same values of their coordinates then ψ vanishes. This shows that identical fermions obey Pauli's exclusion principle, which states: Two identical fermions cannot occupy the same point if they have the same spin orientation; nor can they have the same value of momentum if they have the same spin orientation; in general two (or more) identical fermions cannot occupy the same state.

14.3   Non-interacting Particles and Exclusion Principle

We first consider a system of 2 *non-interacting* identical particles so that we write the Hamiltonian as

$$H_0 = H_0(1) + H_0(2). \tag{14.17}$$

We now write a solution in the form

$$\psi_{\alpha,\beta} = \psi_\alpha(1)\psi_\beta(2), \tag{14.18}$$

$$E = E_\alpha + E_\beta, \tag{14.19}$$

where $\psi_\alpha(1)$ and $\psi_\beta(2)$ are solutions of

$$\begin{aligned} H_0(1)\psi_\alpha(1) &= E_\alpha\psi_\alpha(1), \\ H_0(2)\psi_\beta(2) &= E_\beta\psi_\beta(2). \end{aligned} \tag{14.20}$$

We can also write the solution as

$$\psi_{\beta,\alpha} = \psi_\beta(1)\psi_\alpha(2). \tag{14.21}$$

Here particle 1 is in state β and particle 2 is in state α, but energy $E = E_\alpha + E_\beta$ while $\psi_{\alpha,\beta}$ and $\psi_{\beta,\alpha}$ are two different wave functions belonging to the same energy eigenvalue. Therefore, we have degeneracy, called exchange degeneracy.

We now consider a system of n particles. We write the Hamiltonian as

$$H_0 = H_0(1) + H_0(2) + \cdots H_0(n). \tag{14.22}$$

Then a possible eigenfunction of H_0 is

$$\psi_\alpha(1)\psi_\beta(2)\cdots\psi_\nu(n) \tag{14.23}$$

with eigenvalue

$$E = E_\alpha + E_\beta + \cdots + E_\nu. \tag{14.24}$$

The other possible eigenfunctions can be obtained by permutation of the n numbers $1, \ldots, n$. There are $n!$ possible wave functions corresponding to the same eigenvalue E. We thus have $n!$ fold degeneracy.

In general Eq. (14.23) is not an admissible solution for n identical particles since it lacks symmetry under the interchange of any two particles.

For particles which obey Fermi–Dirac statistics (e.g. electrons) the antisymmetric state functions must be constructed from the products of the above type. This is most easily expressed in the form of the determinant:

$$\psi_A(1, 2, \ldots) = \left(\frac{1}{n!}\right)^{1/2} \begin{vmatrix} \psi_\alpha(1) & \psi_\beta(1) & \cdots & \psi_\nu(1) \\ \psi_\alpha(2) & \psi_\beta(2) & \cdots & \psi_\nu(2) \\ \vdots & \vdots & & \vdots \\ \psi_\alpha(n) & \psi_\beta(n) & \cdots & \psi_\nu(n) \end{vmatrix}. \tag{14.25}$$

This has the required property of anti-symmetry, since exchanging any pair of particle coordinates is equivalent to interchanging two rows of the determinant which changes its sign. Equation (14.25) has the important property that it vanishes if two or more of the ψ's are the same since then two or more columns of the determinant are the same e.g.

$$\psi_A(1, 2, \ldots, n) = 0 \text{ if } \psi_\alpha = \psi_\beta. \tag{14.26}$$

Thus no two (or more) electrons can be in the same state. In other words, there cannot be more than one electron in any of the states α, $\beta, \cdots, \nu$. This is known as the Pauli exclusion principle which was postulated as an explanation of the periodic system of the chemical elements.

For a system of n photons (or any particle which satisfies Bose–Einstein statistics), one must construct symmetric state functions from products of type Eq. (14.23). In this case any number of the particles may be in the same state.

Illustration:

Consider a system of two particles 1 and 2 with two possible states ψ_α and ψ_β.

Fermi–Dirac statistics:

$$\psi_A(1, 2) = \frac{1}{\sqrt{2}}\left(\psi_\alpha(1)\psi_\beta(2) - \psi_\beta(1)\psi_\alpha(2)\right) \text{ (1 state)} \tag{14.27}$$

Bose–Einstein statistics:

Possible symmetric wave functions are

$\psi_\alpha(1)\psi_\alpha(2)$, $\frac{1}{\sqrt{2}}(\psi_\alpha(1)\psi_\beta(2) + \psi_\beta(1)\psi_\alpha(2))$ and $\psi_\beta(1)\psi_\beta(2)$ (3 states)

In the classical situation, one can think of the different ψ_α, as specifying possible orbits of the particles. In this case, each of the following represents a classically distinguishable situation

$\psi_\alpha(1)\psi_\alpha(2)$, $\psi_\alpha(1)\psi_\beta(2)$, $\psi_\beta(1)\psi_\alpha(2)$ and $\psi_\beta(1)\psi_\beta(2)$ (4 states)

This way of counting the number of different states of the system is known as classical or Boltzmann statistics. The above considerations for non-interacting identical particles will be useful where we can treat the interaction between the identical particles as a perturbation. This is illustrated in the next section.

14.4   Two Electrons System (Helium Atom)

The Hamiltonian of the system can be written as

$$H = \frac{\hat{\mathbf{p}}_1^2}{2m} + \frac{\hat{\mathbf{p}}_2^2}{2m} - \frac{2e^2}{r_1} - \frac{2e^2}{r_2} + \frac{e^2}{r_{12}}, \qquad (14.28)$$

where r_{12} is the distance between the two electrons. r_1 and r_2 are the distances of electrons 1 and 2 from the nucleus which is taken to be very heavy. We write

$$H = H_0 + V \qquad (14.29\text{a})$$

where

$$\begin{aligned} H_0 &= H_0(1) + H_0(2) \\ &= \frac{\hat{p}_1^2}{2m} - \frac{2e^2}{r_1} + \frac{\hat{p}_2^2}{2m} - \frac{2e^2}{r_2}, \end{aligned} \qquad (14.29\text{b})$$

$$V = \frac{e^2}{r_{12}}.$$

We treat V as a perturbation. For the unperturbed part the solution can be written in the form

$$u_1 \equiv u_{\alpha,\beta} = u_\alpha(\mathbf{r}_1)u_\beta(\mathbf{r}_2), \qquad (14.30\text{a})$$

$$E = E_\alpha + E_\beta, \qquad (14.30\text{b})$$

where $u_\alpha(\mathbf{r}_1)$ is a solution of the eigenvalue equation

$$\left(\frac{\hat{p}_1^2}{2m} - \frac{2e^2}{r_1} \right) u_\alpha(\mathbf{r}_1) = E_\alpha u_\alpha(\mathbf{r}_1). \tag{14.31}$$

We can also write the solution

$$u_2 \equiv u_{\beta,\alpha} = u_\beta(\mathbf{r}_1) u_\alpha(\mathbf{r}_2). \tag{14.32}$$

Here electron 1 is in state β and electron 2 is in state α, but the energy $E = E_\alpha + E_\beta$. Thus we have two fold exchange degeneracy corresponding to two eigenfunctions $u_{\alpha,\beta}$ and $u_{\beta,\alpha}$ with eigenvalue E. This is for the unperturbed Hamiltonian. When the perturbation is turned on, the wave functions become different; there is a shift in the energy and degeneracy is removed. The matrix elements V_{ij} are given by

$$\lambda V_{11} = e^2 \int u_1^* \frac{1}{r_{12}} u_1 d^3 r_1 d^3 r_2, \tag{14.33a}$$

$$\lambda V_{22} = e^2 \int u_2^* \frac{1}{r_{12}} u_2 d^3 r_1 d^3 r_2, \tag{14.33b}$$

$$\lambda V_{12} = e^2 \int u_1^* \frac{1}{r_{12}} u_2 d^3 r_1 d^3 r_2, \tag{14.33c}$$

$$\lambda V_{21} = e^2 \int u_2^* \frac{1}{r_{12}} u_1 d^3 r_1 d^3 r_2. \tag{14.33d}$$

Now

$$V_{11} = V_{22}; V_{12} = V_{21}; V_{12} = V_{21}^*, \tag{14.34}$$

and therefore V_{12} and V_{21} are real. To order λ

$$E_\pm = E + \frac{\lambda}{2}(V_{11} + V_{22}) \pm \frac{\lambda}{2}\sqrt{(V_{11} - V_{22})^2 + 4|V_{12}|^2}$$

$$= E + \lambda V_{11} \pm \lambda V_{12} = E + \lambda E_\pm^{(1)}, \tag{14.35}$$

where

$$E_\pm^{(1)} = V_{11} \pm V_{12}. \tag{14.36}$$

The zeroth order wave function is then given by

$$\psi_\pm = \frac{1}{\sqrt{2}} \left(u_1 + \frac{E_\pm^{(1)} - V_{11}}{V_{12}} u_2 \right)$$

$$= \frac{1}{\sqrt{2}} (u_1 \pm u_2) \tag{14.37}$$

$$= \frac{1}{\sqrt{2}} \left(u_\alpha(\mathbf{r}_1) u_\beta(\mathbf{r}_2) \pm u_\beta(\mathbf{r}_1) u_\alpha(\mathbf{r}_2) \right). \tag{14.38}$$

The energy eigenvalues corresponding to eigenfunctions $\psi_\pm$ are as given in Eq. (14.36). Note that ψ_+ is symmetric with respect to interchange of electrons 1 and 2 while the other ψ_- is anti-symmetric. If the electrons are in the same state, only ψ_+ is possible. Also if the electrons are at the same place, the anti-symmetric wave function vanishes. The electrons represented by the symmetric wave function ψ_+ are said to be in the para state and electrons represented by anti-symmetric wave function ψ_- are said to be in the ortho state. So far we have neglected the spin of electrons. We now take into consideration the spin of electrons. Electrons are fermions, thus the total wave function must be anti-symmetric under interchange of electrons 1 and 2. For para state, the spin wave function must be anti-symmetric [spin singlet]:

$$\chi_-(\sigma_1, \sigma_2) = \frac{1}{\sqrt{2}} \left(\chi_+(1)\chi_-(2) - \chi_+(2)\chi_-(1) \right)$$

$$= -\chi_-(\sigma_2, \sigma_1). \tag{14.39}$$

Thus for para state the wave function is given

$$\psi(1,2) = \psi_+(\mathbf{r}_1, \mathbf{r}_2)\chi_-(\sigma_1, \sigma_2). \tag{14.40}$$

For ortho state, the spin wave function must be symmetric [spin triplet]. In this case there are three linearly independent states viz.

$$\chi_+^{+1}(\sigma_1, \sigma_2) = \chi_+(1)\chi_+(2), \tag{14.41a}$$

$$\chi_+^{0}(\sigma_1, \sigma_2) = \frac{1}{\sqrt{2}} \left(\chi_+(1)\chi_-(2) + \chi_+(2)\chi_-(1) \right), \tag{14.41b}$$

$$\chi_+^{-1}(\sigma_1, \sigma_2) = \chi_-(1)\chi_-(2). \tag{14.41c}$$

For ortho state the wave function is given by

$$\psi(1,2) = \psi_-(\mathbf{r}_1, \mathbf{r}_2)\chi_+(\sigma_1, \sigma_2). \tag{14.42}$$

When the two electrons are in the same state $\alpha = \beta$, then ψ_- vanishes, only para state is possible and the spins of the two electrons are anti-parallel. Thus the ground state for the helium atom is a para state. Also we note that for the para state electrons can be very close to each other, whereas for the ortho state two electrons tend to stay away from each other, since in this case ψ_- vanishes as $\mathbf{r}_1 \to \mathbf{r}_2$. We have repulsion when the spins of two electrons are parallel to each other and we have attraction when their spins are anti-parallel. This is how a hydrogen molecule is formed, viz. we get a chemical bond between two hydrogen atoms when two electrons have opposite spins, because they can then come close to each other.

14.5   Scattering of Identical Particles

Consider the scattering of two particles a and b in the centre of mass system.

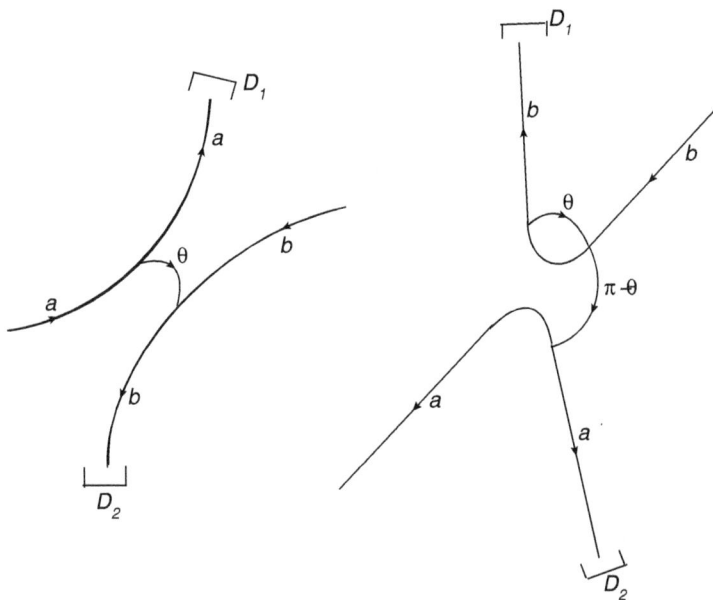

Fig. 14.1   Scattering of particles at an angle θ and the angle $\pi - \theta$.

Let $f(\theta)$ be the scattering amplitude for scattering of particle a at an angle θ. Let $f(\pi - \theta)$ be the scattering amplitude for scattering of particle a at an angle $\pi - \theta$. Now $f(\pi - \theta) =$ the amplitude for particle b scattering through the angle θ. Thus the probability of some particle detected at D_1 is given by

$$\sigma(\theta) = |f(\theta)|^2 + |f(\pi - \theta)|^2. \qquad (14.43)$$

The two amplitudes $f(\theta)$ and $f(\pi - \theta)$ are distinguishable in principle when particles a and b are not identical.

Consider the scattering of two identical particles. First consider the case of scattering of two identical Bose particles of spin zero, for example, α particles. In this case there are two ways to get an α-particle into the counter D_1: by scattering the bombarding α-particle by an angle θ and or by scattering it by an angle $\pi - \theta$. Since the particles are indistinguishable, we cannot tell whether the bombarding particle or the target particle entered the counter. Thus the two amplitudes $f(\theta)$ and $f(\pi - \theta)$ can interfere. The wave function must be symmetric. Thus

$$\sigma(\theta) = |f(\theta) + f(\pi - \theta)|^2, \tag{14.44}$$

or

$$\sigma(\theta) = |f(\theta)|^2 + |f(\pi - \theta)|^2 + 2\mathrm{Re}\ f(\theta)f^*(\pi - \theta). \tag{14.45}$$

This differs from the classical result by the interference term $2\mathrm{Re}\ f(\theta)f^*(\pi - \theta)$. For example, for $\theta = \pi/2$, the classical result gives

$$\sigma(\pi/2) = 2|f(\pi/2)|^2 \tag{14.46}$$

compared to the result obtained using Bose statistics

$$\sigma(\pi/2) = 4|f(\pi/2)|^2. \tag{14.47}$$

Now consider the case of scattering of two identical Fermi particles of spin $1/2$, for example an electron or proton. Here the total wave function must be anti-symmetric. If the two particles are in a spin singlet state (spin wave function anti-symmetric), then the spatial wave function is symmetric and

$$\sigma_s(\theta) = |f(\theta) + f(\pi - \theta)|^2. \tag{14.48}$$

When the two particles are in a spin triplet state then the spatial wave function must be anti-symmetric

$$\sigma_t(\theta) = |f(\theta) - f(\pi - \theta)|^2. \tag{14.49}$$

For unpolarised particles, noting that there is one singlet state and three triplet states

$$
\begin{aligned}
\sigma(\theta) &= \frac{1}{4}\sigma_s(\theta) + \frac{3}{4}\sigma_t(\theta) \\
&= \frac{1}{4}|f(\theta) + f(\pi - \theta)|^2 + \frac{3}{4}|f(\theta) - f(\pi - \theta)|^2 \\
&= \frac{1}{2}|f(\theta) - f(\pi - \theta)|^2 + \frac{1}{2}|f(\theta)|^2 + \frac{1}{2}|f(\pi - \theta)|^2 \tag{14.50} \\
&= |f(\theta)|^2 + |f(\pi - \theta)|^2 - \mathrm{Re}\ f(\theta)f^*(\pi - \theta). \tag{14.51}
\end{aligned}
$$

Again we see that this result for scattering of two identical Fermi particles is different from the classical result.

14.6   Problems

14.1 A deuteron consists of a neutron and a proton. Consider a system of 2 deuterons as a composite of 2 neutrons and 2 protons with the wave function

$$\psi(n_2, p_2; n_1, p_1)$$

where n and p denote *all* the coordinates of a neutron and a proton respectively. How does the above wave function behave under the interchange of 2 deuterons, i.e. under

$$(n_1, p_1) \leftrightarrow (n_2, p_2)?$$

14.2 For a system of 2 identical particles, is

$$\psi(1, 2) = \psi_{100}(r_1)\psi_{200}(r_2)$$

where

$$\psi_{100}(r) = \frac{1}{\sqrt{2\pi}}\beta^{3/2}e^{-\beta r},$$

$$\psi_{200}(r) = \frac{1}{4\sqrt{2\pi}}\beta^{3/2}(2 - \beta r)e^{-\beta r},$$

an acceptable wave function? If not, find an acceptable wave function. If the particles have spin $\frac{1}{2}$, write down acceptable wave functions which include the spin wave functions also.

14.3 For a helium atom $(Z = 2)$, show that the ground state energy to first order in perturbation is

$$E_0 = E_0^{(0)} + \Delta E$$

where

$$E_0^{(0)} = -\frac{Z^2 e^2}{a_0}, \quad \Delta E = (5/8)\frac{Z e^2}{a_0}.$$

Hint: Ground state wave function is

$$\psi(1, 2) = u_{100}(\mathbf{r}_1)u_{100}(\mathbf{r}_2)\chi_-(\sigma_1, \sigma_2)$$

where

$$u_{100}(\mathbf{r}) = \frac{1}{\sqrt{\pi}}(\frac{Z}{a_0})^{3/2}e^{-Zr/a_0}.$$

Apply Eq. (14.33a) with the above wave function.

14.4 Consider an electron confined in a cubical box of length L. Show that the energy eigenvalues and normalised eigenfunctions are given by

$$E = \frac{\hbar^2 \pi^2}{2mL^2}(n_1^2 + n_2^2 + n_3^2),$$

$$u_n(\mathbf{r}) = \left(\frac{2}{L}\right)^{2/3} \sin \frac{n_1 \pi}{L} x \sin \frac{n_2 \pi}{L} y \sin \frac{n_3 \pi}{L} z.$$

From this it follows that the number of quantum states between momentum $\mathbf{p}$ and $\mathbf{p} + d\mathbf{p}$ is given by [$\mathbf{p} = \hbar \mathbf{k}$ and $\frac{2\pi n_1}{L} = k_x$ etc.]

$$2\left(\frac{L}{2\pi}\right)^3 dk_x dk_y dk_z = 2\left(\frac{L}{2\pi}\right)^3 d^3k$$

$$= 2\left(\frac{L}{2\pi}\right)^3 4\pi k^2 dk,$$

where the factor 2 is due to two spin orientation. Consider now a non-interacting electron gas in the box. A quantum state is specified by three quantum numbers (n_1, n_2, n_3). Since an electron has spin $1/2$, only two electrons can be in the same state specified by (n_1, n_2, n_3). Now the Pauli principle forces the electrons to occupy all the states between 0 and ϵ_F, where ϵ_F is the Fermi energy and for non-relativistic case is given by $\epsilon_F = \frac{P_F^2}{2m}$. Show that

$$\epsilon_F = \frac{\pi^2 \hbar^2}{2m}\left(\frac{3\rho}{m\pi}\right)^{2/3}$$

where $\rho = \frac{N}{L^3}m$, is the density of electrons in the box. Find the pressure as a function of ρ viz. the equation of state for the degenerate gas. Obtain also the equation of state for a relativistic degenerate electron gas ($\epsilon_F \approx cp_F$).

14.5 Consider two electrons confined in a box whose sides are of length L. There exists an attractive potential of strength V_0 between pairs of electrons whenever they are very close to each other. Using perturbation theory, calculate the ground state energy and the wave function. Hint: Take the potential

$$V(\mathbf{r}_1 - \mathbf{r}_2) = -V_0 \qquad r < a,$$
$$= 0 \qquad r > a, a \ll L.$$

Then approximate this potential as

$$\int V(\mathbf{r}_1 - \mathbf{r}_2)d^3r_2 = -V_0 \frac{4\pi}{3}a^3$$

$$V(\mathbf{r}_1 - \mathbf{r}_2) = -V_0 \frac{4\pi}{3} a^3 \Delta(\mathbf{r}_1 - \mathbf{r}_2).$$

Treat this potential as a perturbation. Then using equations similar to Eqs. (14.33), (14.34), (14.36), (14.37), (14.39) and (14.41) in the text and the result of the first part of problem (14.4), calculate energy levels and energy eigenfunctions. For a numerical estimate take $L = 10^{-8}$cm, $a < 10^{-10}$cm and $V_0 = 10^{-3}$eV.

Chapter 15

Two State Systems

15.1 Introduction

A two state system, i.e. when two states form a complete set, is simple but rich in content. It provides a simple framework for quantum mixing; a consequence of superposition principle of quantum mechanics. This leads to quantum mechanical phenomena of interferometry which provides a sensitive method to probe extremely small effects. Consider a system with orthonormal basis $\{|1\rangle, |2\rangle\}$:

$$\langle i|j \rangle = \delta_{ij}, \quad i, j = 1, 2. \tag{15.1}$$

Let $\hat{H}$ be the Hamiltonian of the system which in the above basis is a 2×2 hermitian matrix

$$H = \begin{pmatrix} H_{11} & H_{12} \\ H_{21} & H_{22} \end{pmatrix}, \quad H_{21} = H_{12}^*; \quad H_{11}, H_{22} \quad \text{real.} \tag{15.2}$$

In general this is not a diagonal matrix and therefore the basis states $|1\rangle$ and $|2\rangle$ are not eigenstates of H. Its diagonalization

$$\det |H - \lambda I| = 0$$

gives two eigenvalues

$$E_{\pm} = \frac{1}{2}\{H_{11} + H_{22} \pm \sqrt{(H_{11} - H_{22})^2 + 4H_{12}H_{21}}\} \tag{15.3}$$

which are possible energies of the system. We denote the corresponding eigenstates by $|+\rangle$ and $|-\rangle$, which are the stationary states of the system if H does not depend explicitly on time.

Now the set of four matrices 1, σ^i $(i = 1, 2, 3)$, the Pauli matrices, form the basis in which one can express H as

$$H = n_0 1 + \mathbf{n} \cdot \boldsymbol{\sigma} \tag{15.4}$$

where

$$n_0 = \frac{1}{2} Tr(H) = \frac{H_{11} + H_{22}}{2},$$

$$\mathbf{n} = \frac{1}{2} Tr(H\boldsymbol{\sigma}) = \frac{1}{2}((H_{11} + H_{22}), i(H_{12} - H_{21}), (H_{11} - H_{22})). \tag{15.5}$$

Since $(\mathbf{n} \cdot \boldsymbol{\sigma})^2 = n^2 1$, the eigenvalues of $\mathbf{n} \cdot \boldsymbol{\sigma}$ are $\pm |\mathbf{n}|$. This implies that the eigenvalues of H are

$$E_\pm = n_0 \pm |\mathbf{n}| \tag{15.6}$$

with the eigenvalue equation

$$\begin{pmatrix} n_0 + n_3 & n_1 - in_2 \\ n_1 + in_2 & n_0 - n_3 \end{pmatrix} \begin{pmatrix} u_\pm \\ v_\pm \end{pmatrix} = (n_0 \pm |\mathbf{n}|) \begin{pmatrix} u_\pm \\ v_\pm \end{pmatrix}. \tag{15.7}$$

This gives $[n = |\mathbf{n}|]$

$$u_\pm(n_3 \mp n) = \mp(n_1 - in_2)v_\pm \tag{15.8}$$

which fixes only $u_\pm/v_\pm$ i.e. the relative phase between $u_\pm$ and $v_\pm$. Using the normalizing condition

$$|u_\pm|^2 + |v_\pm|^2 = 1, \tag{15.9}$$

then apart from an overall phase, the normalized eigenstates are

$$u_+ = \frac{(n + n_3)^{1/2}}{(2n)^{1/2}}, \quad v_+ = \frac{(n - n_3)^{1/2}}{(2n)^{1/2}} \frac{n_1 + in_2}{\sqrt{n_1^2 + n_2^2}},$$

$$u_- = -\frac{(n - n_3)^{1/2}}{(2n)^{1/2}} \frac{n_1 + in_2}{\sqrt{n_1^2 - n_2^2}}, \quad v_- = \frac{(n + n_3)^{1/2}}{(2n)^{1/2}}. \tag{15.10}$$

We can express the vector $\mathbf{n}$ in terms of spherical polar coordinates:

$$\mathbf{n} = n(\sin\theta\cos\phi, \sin\theta\sin\phi, \cos\theta) :$$

$$n = |\mathbf{n}|$$

$$= \frac{1}{2}\sqrt{(H_{11} - H_{22})^2 + 4H_{12}H_{21}},$$

$$\cos\theta = \frac{H_{11} - H_{22}}{2n},$$

$$\cos\phi = \frac{H_{12} + H_{21}}{2|H_{12}|}. \tag{15.11}$$

Then the normalized eigenstates of H are given by $\begin{pmatrix} \cos\frac{\theta}{2} \\ e^{i\phi}\sin\frac{\theta}{2} \end{pmatrix}$ and $\begin{pmatrix} -e^{-i\phi}\sin\frac{\theta}{2} \\ \cos\frac{\theta}{2} \end{pmatrix}$ with the eigenvalues $E_\pm = n_0 \pm n$ respectively.

It is easy to see that above states are eigenstates of helicity operator $\frac{\vec{\sigma}\cdot\vec{n}}{|\vec{n}|}$ with eigenvalues ± 1.

Since $|1\rangle$ is $\begin{pmatrix}1\\0\end{pmatrix}$ and $|2\rangle = \begin{pmatrix}0\\1\end{pmatrix}$,

$$|+\rangle = \cos\frac{\theta}{2}|1\rangle + e^{i\phi}\sin\frac{\theta}{2}|2\rangle,$$
$$|-\rangle = -e^{-i\phi}\sin\frac{\theta}{2}|1\rangle + \cos\frac{\theta}{2}|2\rangle, \tag{15.12}$$

or we can invert them

$$|1\rangle = \cos\frac{\theta}{2}|+\rangle - e^{i\phi}\sin\frac{\theta}{2}|-\rangle,$$
$$|2\rangle = e^{-i\phi}\sin\frac{\theta}{2}|+\rangle + \cos\frac{\theta}{2}|-\rangle. \tag{15.13}$$

Now the Schrödinger equation of motion is

$$i\hbar\frac{\partial}{\partial t}|\psi(t)\rangle = \hat{H}|\psi(t)\rangle. \tag{15.14}$$

In the basis provided by $\{|1\rangle, |2\rangle\}$, $|1\rangle\langle 1| + |2\rangle\langle 2| = 1$, we can express $|\psi(t)\rangle$ as

$$|\psi(t)\rangle = |1\rangle\langle 1|\psi(t)\rangle + |2\rangle\langle 2|\psi(t)\rangle$$
$$= \psi_1(t)|1\rangle + \psi_2(t)|2\rangle. \tag{15.15}$$

Thus

$$i\hbar\frac{\partial}{\partial t}\begin{pmatrix}\psi_1(t)\\\psi_2(t)\end{pmatrix} = H\begin{pmatrix}\psi_1(t)\\\psi_2(t)\end{pmatrix} \tag{15.16}$$

where H is given in Eq. (15.2). Using Eq. (15.3), we can also express $|\psi\rangle$ in the orthonormal basis $\{|+\rangle, |-\rangle\}$:

$$|\psi(t)\rangle = \psi_+(t)|+\rangle + \psi_-(t)|-\rangle \tag{15.17}$$

where

$$\psi_+(t) = \langle +|\psi(t)\rangle$$
$$= \cos\frac{\theta}{2}|\psi_1(t)\rangle + e^{-i\phi}\sin\frac{\theta}{2}|\psi_2(t)\rangle,$$
$$\psi_-(t) = \langle -|\psi(t)\rangle$$
$$= -e^{i\phi}\sin\frac{\theta}{2}|\psi_1(t)\rangle + \cos\frac{\theta}{2}|\psi_2(t)\rangle. \tag{15.18}$$

In other words

$$\begin{pmatrix} \psi_+(t) \\ \psi_-(t) \end{pmatrix} = U \begin{pmatrix} \psi_1(t) \\ \psi_2(t) \end{pmatrix} \tag{15.19}$$

where U is the Unitary matrix

$$U = \begin{pmatrix} \cos\frac{\theta}{2} & e^{-i\phi}\sin\frac{\theta}{2} \\ -e^{i\phi}\sin\frac{\theta}{2} & \cos\frac{\theta}{2} \end{pmatrix}, \quad UU^\dagger = U^\dagger U = 1. \tag{15.20}$$

Thus the two representations are related by a unitary transformation. Then from Eqs. (15.16) and (15.19)

$$i\hbar\frac{\partial}{\partial t}\begin{pmatrix} \psi_+(t) \\ \psi_-(t) \end{pmatrix} = UHU^{-1}\begin{pmatrix} \psi_+(t) \\ \psi_-(t) \end{pmatrix}$$

$$-i\hbar U\frac{\partial}{\partial t}U^{-1}\begin{pmatrix} \psi_+(t) \\ \psi_-(t) \end{pmatrix} \tag{15.21}$$

where

$$UHU^{-1} = \begin{pmatrix} E_+ & 0 \\ 0 & E_- \end{pmatrix} \tag{15.22}$$

i.e. in the basis $\{|+\rangle, |-\rangle\}$ the Hamiltonian is diagonal.

If U is independent of t, then

$$\psi_\pm(t) = \psi_\pm(0)e^{-iE_\pm t/\hbar}. \tag{15.23}$$

Then from Eq. (15.19)

$$\begin{pmatrix} \psi_1(t) \\ \psi_2(t) \end{pmatrix} = U^{-1}\begin{pmatrix} \psi_+(0)e^{-iE_+ t/\hbar} \\ \psi_-(0)e^{-iE_- t/\hbar} \end{pmatrix}. \tag{15.24}$$

Then from Eq. (15.17)

$$|\psi(t)\rangle = [e^{-iE_+ t/\hbar}\psi_+(0)|+\rangle + e^{-iE_- t/\hbar}\psi_-(0)|-\rangle]$$

$$= e^{-iEt/\hbar}[e^{-i\Delta Et/\hbar}\psi_+(0)|+\rangle + e^{+i\Delta Et/\hbar}\psi_-(0)|-\rangle] \tag{15.25}$$

where $E = \frac{E_+ + E_-}{2}$, $\Delta E = \frac{E_+ - E_-}{2}$. Suppose at $t = 0$ $\psi_1(0) = 1$, $\psi_2(0) = 0$, then from Eq. (15.18)

$$\psi_+(0) = \cos\frac{\theta}{2}, \quad \psi_-(0) = -e^{i\phi}\sin\frac{\theta}{2}. \tag{15.26}$$

Then using the Eq. (15.12), in the basis $\{|1\rangle, |2\rangle\}$,

$$|\psi(t)\rangle = e^{-iEt/\hbar}\{(\cos^2\frac{\theta}{2}e^{-i\Delta Et/\hbar} + \sin^2\frac{\theta}{2}e^{i\Delta Et/\hbar})|1\rangle$$

$$+e^{i\phi}\sin\frac{\theta}{2}\cos\frac{\theta}{2}(e^{-i\Delta Et/\hbar} - e^{i\Delta Et/\hbar})|2\rangle\}$$

$$= e^{-iEt/\hbar}\{(\cos\frac{\Delta Et}{\hbar} - i\cos\theta\sin\frac{\Delta Et}{\hbar})|1\rangle$$

$$-ie^{i\phi}\sin\theta\sin\frac{\Delta Et}{\hbar}|2\rangle\}. \tag{15.27}$$

The important conclusion one draws from this equation is that although at $t = 0$ the system was in state $|1\rangle$, after time t it develops the component $|2\rangle$. The probability of finding the system in state $|2\rangle$ after time t is

$$P_{1-2}(t) = |\langle 2|\phi(t)\rangle|^2 = \sin^2 \theta \sin^2 \frac{\Delta Et}{\hbar}. \tag{15.28}$$

The survival probability is

$$P_{1-1}(t) = 1 - \sin^2 \theta \sin^2 \frac{\Delta Et}{\hbar}. \tag{15.29}$$

The above equations illustrate the quantum mechanical phenomenon of interferometry which provides a sensitive method to probe extremely small effects. Eqs. (15.28) and (15.29) are plotted in Fig. 15.1.

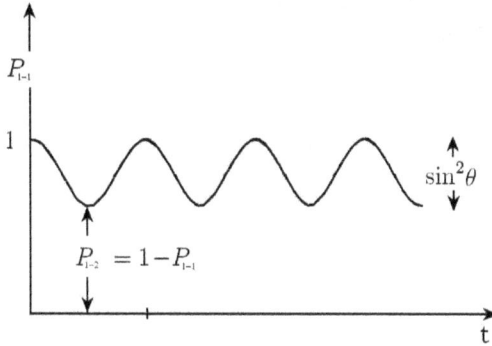

Fig. 15.1   The probability for spin flip.

This shows that a state produced as pure $|1\rangle$ state at $t = 0$ continuously oscillates between $|1\rangle$ and state $|2\rangle$ with a period of oscillation $\tau = \frac{2\pi}{\Delta E/\hbar}$. Thus by measuring τ, one can measure $\Delta E/\hbar$.

15.2   Magnetic Spin Resonance

The spin system is a two state system viz. the states

$$|1\rangle = |\frac{1}{2}\rangle, \quad \sigma_z|\frac{1}{2}\rangle = |\frac{1}{2}\rangle$$

$$|2\rangle = |-\frac{1}{2}\rangle, \quad \sigma_z|-\frac{1}{2}\rangle = |-\frac{1}{2}\rangle \tag{15.30}$$

form a complete set. In electron spin resonance experiments, the spin moves under the joint action of a large uniform magnetic field $\mathbf{B}$ in the z-direction which is constant in time and a small oscillatory magnetic field $\mathbf{b}(t)$ in the x–y plane:

$$\mathbf{B} = B\mathbf{e}_z,$$
$$\mathbf{b}(t) = b(\mathbf{e}_x \cos \omega t) + \mathbf{e}_y \sin \omega t). \tag{15.31}$$

Spin Hamiltonian

$$H = -\boldsymbol{\mu} \cdot [\mathbf{B} + \mathbf{b}(t)]$$
$$\boldsymbol{\mu} = g_s \frac{e\hbar}{2mc} \frac{1}{2} \boldsymbol{\sigma} \tag{15.32}$$

where e is the charge of the particle and g_s is the spin gyromagnetic ratio of the electron, Dirac equation gives $g_s = 2$ (see Ch. (20)). Thus for the electron $(-e, e > 0)$

$$H = g_s \frac{e\hbar}{2mc} \frac{1}{2} \boldsymbol{\sigma} \cdot (\mathbf{B} + \mathbf{b}(t))$$
$$= g_s \frac{e\hbar}{2mc} \frac{1}{2} (B\sigma_z + \boldsymbol{\sigma} \cdot \mathbf{b}(t))$$
$$= \frac{\hbar}{2} \begin{pmatrix} \omega_s & \omega_b e^{-i\omega t} \\ \omega_b e^{i\omega t} & -\omega_s \end{pmatrix} \tag{15.33}$$

where

$$\omega_s = g_s \frac{e\hbar}{2mc} B, \quad \omega_b = g_s \frac{e\hbar}{2mc} b. \tag{15.34}$$

Note that here the Hamiltonian explicitly depends on time. The Schrödinger equation of motion for the spin state $|\chi(t)\rangle$ is

$$i\hbar \frac{d}{dt} |\chi(t)\rangle = H |\chi(t)\rangle. \tag{15.35}$$

The spin state $|\chi(t)\rangle$ can be written in terms of the basis provided by $|\frac{1}{2}\rangle$ and $|-\frac{1}{2}\rangle$, the eigenstate of $S_z = \frac{\hbar}{2}\sigma_z$ given in Eq. (15.30). Since H explicitly depends on t, to deal with it we write it as

$$H = \hbar V^\dagger(t) \begin{pmatrix} \frac{\omega_s}{2} & \frac{\omega_b}{2} \\ \frac{\omega_b}{2} & -\frac{\omega_s}{2} \end{pmatrix} V(t) = V^\dagger(t) \tilde{H} V(t) \tag{15.36}$$

where

$$V(t) = \begin{pmatrix} e^{\frac{i\omega t}{2}} & 0 \\ 0 & e^{\frac{-i\omega t}{2}} \end{pmatrix} \tag{15.37}$$

and

$$V(t)V^\dagger(t) = 1, \quad V(0) = 1 = V^\dagger(0)$$

and

$$\tilde{H} = \hbar \begin{pmatrix} \frac{\omega_s}{2} & \frac{\omega_b}{2} \\ \frac{\omega_b}{2} & \frac{-\omega_s}{2} \end{pmatrix}$$

$$= \hbar(\frac{\omega_s}{2}\sigma_z + \frac{\omega_b}{2}\sigma_x). \tag{15.38}$$

Then the Eq. (15.35) takes the form

$$i\hbar\frac{d}{dt}|\chi(t)\rangle = V^\dagger(t)\tilde{H}V(t)|\chi(t)\rangle \tag{15.39}$$

or

$$i\hbar\frac{d}{dt}|\chi'(t)\rangle = H'|\chi'(t)\rangle \tag{15.40}$$

where

$$|\chi'(t)\rangle = V(t)|\chi(t)\rangle, \quad |\chi'(0)\rangle = |\chi(0)\rangle, \tag{15.41}$$

$$H' = i\hbar\frac{dV(t)}{dt}V^\dagger(t) + \tilde{H}$$

$$= i^2\hbar\frac{\omega}{2}\sigma_z + \tilde{H} = \hbar(\frac{\omega_s - \omega}{2}\sigma_z + \frac{\omega_b}{2}\sigma_x). \tag{15.42}$$

and is independent of t. H' is not diagonal and the unitary matrix which diagonalizes it is given in Eq. (15.20), where now [cf Eq. (15.11)]

$$\cos\theta = \frac{H'_{11} - H'_{22}}{2n'} = \frac{\omega_s - \omega}{\Omega}, \quad \cos\phi = \frac{H'_{12} + H'_{21}}{2|H'_{12}|} = 1, \quad \phi = 0,$$

$$n' = \frac{\hbar}{2}[(\omega_s - \omega)^2 + \omega_b^2]^{\frac{1}{2}} = \frac{\Omega}{2}\hbar, \quad \sin\theta = \frac{\omega_b}{\Omega}, \tag{15.43}$$

$$E_\pm = \pm n'\hbar = \pm\hbar\frac{\Omega}{2},$$

$$\Delta E = \Omega\hbar, \quad E = 0. \tag{15.44}$$

Then from Eq. (15.27), using boundary conditions $|\chi(0)\rangle = |\chi'(0)\rangle = |1\rangle = |+\frac{1}{2}\rangle$

$$|\chi'(t)\rangle = (\cos\frac{\Omega}{2}t - i\cos\theta\sin\frac{\Omega}{2}t)|1\rangle - i\sin\theta\sin\frac{\Omega}{2}t|2\rangle. \tag{15.45}$$

Then

$$|\chi(t)\rangle = V^\dagger(t)|\chi'(t)\rangle \tag{15.46}$$

$$= e^{-i\omega t}(\cos\frac{\Omega}{2}t - i\cos\theta\sin\frac{\Omega}{2}t)|1\rangle - ie^{i\omega t}\sin\theta\sin\frac{\Omega}{2}t|2\rangle.$$

$$\tag{15.47}$$

Therefore the probability that spin has flipped by time t i.e. the particle has spin down is given by

$$P_{-\frac{1}{2}} = |\langle -\frac{1}{2}|\chi(t)\rangle|^2 = \sin^2\theta \sin^2\frac{\Omega}{2}t$$

$$= \frac{\omega_b^2}{(\omega - \omega_s)^2 + \omega_b^2}\sin^2\left[\left(\frac{(\omega - \omega_s)^2 + \omega_b^2}{2}\right)^{\frac{1}{2}}t\right]. \qquad (15.48)$$

At resonance $\omega = \omega_s$

$$P_{-\frac{1}{2}} = \sin^2\omega_b t, \quad (P_{-\frac{1}{2}})_{max} = 1 \qquad (15.49)$$

ω_b is known as the Rabi frequency since Rabi was the first person who solved this problem. Above is a model for spin resonance experiments. If the field B is known, a measurement of resonance frequency $\omega = \omega_s$, measures the parameter g_s. The interest in g_s is to detect any deviation from the value $g_s = 2$ predicted by Dirac equation. The experimental value for the parameter:

$$a = \frac{1}{2}(g_s - 2) \quad \text{is} \quad a_{exp} = (1159652.4 \pm 2) \times 10^{-9}.$$

This small deviation was expected and can be explained by radioactive corrections to Dirac theory.

15.3 Particle Mixing

15.3.1 *Introduction*

In this section, we discuss two state problems involving particle mixing. Particles are divided into two classes; leptons and hadrons. Leptons are spin $\frac{1}{2}$ particles. There are three generations of leptons; the lightest one is the electron (e^-) which is stable. The experimental limit on the lifetime of electron is $\tau_e^- > 4.6 \times 10^{26} yr$. The second and third generation of leptons are muon (μ^-) and τ-lepton (τ^-); each type of charged lepton has a corresponding neutral partner called neutrino, ν_e, ν_μ, ν_τ, which are stable. The heavy charged leptons decay into lighter leptons by β-decay:

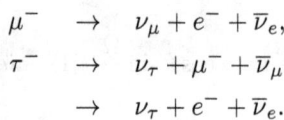

$$\mu^- \rightarrow \nu_\mu + e^- + \bar{\nu}_e,$$
$$\tau^- \rightarrow \nu_\tau + \mu^- + \bar{\nu}_\mu$$
$$\rightarrow \nu_\tau + e^- + \bar{\nu}_e.$$

Corresponding to each lepton, there is an antilepton with the same mass, same spin but opposite charge. Each lepton is assigned a quantum number called lepton number $L = 1$; an antilepton has $L = -1$. Thus corresponding to leptons (e^+, ν_e), (μ^+, ν_μ), (τ^+, ν_τ), antileptons are $(e^-, \bar{\nu}_e)$, $(\mu^-, \bar{\nu}_\mu)$ and $(\tau^-, \bar{\nu}_\tau)$. In any process lepton number and charge is conserved.

Hadrons are specified by intrinsic spin and intrinsic parity J^P, J denoting the spin and P intrinsic parity. Hadrons are divided into two classes: baryons and mesons. Low lying baryons have $J^P = \frac{1}{2}^+$ and are assigned a quantum number $B = 1$; antibaryons have $B = -1$, but same mass, spin and negative intrinsic parity. The lightest baryon is the proton (p), which is stable and positively charged. The experimental limit on proton life time: $\tau_p > 10^{31} yr$. It has a neutral partner, called neutron (n) which is unstable and decays into a proton by β-decay with a life time $\tau_{\frac{1}{2}} \simeq 4624 sec$.

$$n \quad \rightarrow \quad p + e^- + \bar{\nu}_e$$

conserving both charge, baryon number and lepton number. Each hadron member is assigned an internal quantum number, such as isospin I and hypercharge Y, which are related to electric charge of each hadron:

$$Q = I_3 + \frac{Y}{2}.$$

The neutron and proton are an isospin doublet $I = \frac{1}{2}$, $I_3 = \frac{1}{2}$ (proton), $I_3 = -\frac{1}{2}$ (neutron) and have hypercharge $Y = 1$. Heavy baryons decay to lighter baryons by weak interaction. Low lying mesons with spin and parity $J^P = 0^-$ are π^-, π^0, π^+ called pions:

$$(\pi^-, \pi^0, \pi^+), \quad I = 1, \quad I_3 = -1, 0, 1, \quad Y = 0$$

and kaons:

$$(K^+, K^0) : \quad I = \frac{1}{2}, \quad I_3 = \frac{1}{2}, \frac{-1}{2}, \quad Y = 1,$$

$$(\bar{K}^0, K^-) : \quad I = \frac{1}{2}, \quad I_3 = \frac{1}{2}, \frac{-1}{2}, \quad Y = -1.$$

They are unstable; for example

$$\pi^\pm \quad \rightarrow \quad \mu^\pm + \nu_\mu(\bar{\nu}_\mu),$$
$$\pi^0 \quad \rightarrow \quad 2\gamma,$$
$$K^0 \quad \rightarrow \quad \pi^+ \pi^-.$$

π^- is antiparticle of π^+; $(\overline{K}^0, K^-)$ are antiparticles of (K^0, K^+). Mesons and antimesons have same mass, spin and parity, but opposite I_3 and hypercharge. In weak decays, isospin and hypercharge are not conserved.

We conclude this section with the remark that baryon number and electric charge are absolutely conserved; we can consider particle mixing only for K^0 and $\overline{K}^0$ and heavy neutral mesons $B_d^0 - \overline{B}_d^0$ and $B_s^0 - \overline{B}_s^0$. It is also possible to consider mixing between neutrinos which are electrically neutral.

In a production process involving strong hadronic (or electromagnetic) interaction hypercharge Y is conserved and K^0 and $\overline{K}^0$ appear as two distinctly different particles. In the presence of weak interaction Y is no longer conserved and transitions between K^0 and $\overline{K}^0$ can occur. Both K^0 and $\overline{K}^0$ decay into $\pi^+\pi^-$:

$$K^0 \quad \rightarrow \quad \pi^+\pi^- \quad \leftarrow \quad \overline{K}^0, \qquad \Delta Y = \mp 1,$$

$$K^0 \quad \rightarrow \quad \pi^+\pi^- \quad \rightarrow \quad \overline{K}^0, \qquad |\Delta Y| = 2.$$

Hence weak interaction can mix K^0 and $\overline{K}^0$:

$$\langle K^0 | H | \overline{K}^0 \rangle \neq 0$$

i.e. off diagonal matrix elements are not zero. Thus K^0 and $\overline{K}^0$ cannot be mass eigenstates, when weak interaction Hamiltonian is included in H.

15.3.2 *General Formalism for Particle Mixing*

We now discuss particle mixing for the $(K^0, \overline{K}^0)$ complex or in general $(\chi^0, \overline{\chi}^0)$. For an unstable particle, the complex mass $M = m - \frac{i}{2}\Gamma$, where m is the mass of the particle and Γ is decay width of the decaying particle. In the $|\chi^0\rangle, |\overline{\chi}^0\rangle$ basis ($\hbar = c = 1$), we can express a state $|\psi(t)\rangle$ as

$$|\psi(t)\rangle = a(t)|\chi^0\rangle + \overline{a}(t)|\overline{\chi}^0\rangle \qquad (15.50)$$

where t is measured in the rest system of the particle χ^0. Then the time evolution of the state

$$\psi(t) = \begin{pmatrix} a(t) \\ \overline{a}(t) \end{pmatrix}$$

is given by

$$i\frac{\partial}{\partial t}\psi(t) = M\psi(t). \tag{15.51}$$

The mass matrix M given in Eq. (15.51) is not diagonal in this basis and is given by

$$M = m - \frac{i}{2}\Gamma = \begin{pmatrix} m_{11} - \frac{i}{2}\Gamma_{11} & m_{12} - \frac{i}{2}\Gamma_{12} \\ m_{21} - \frac{i}{2}\Gamma_{21} & m_{22} - \frac{i}{2}\Gamma_{22} \end{pmatrix}. \tag{15.52}$$

Hermiticity of the matrices $m_{\alpha\alpha'}$ and $\Gamma_{\alpha\alpha'}$ gives $(\alpha, \alpha' = 1, 2)$

$$(m)_{\alpha\alpha'} = (m^{\dagger})_{\alpha\alpha'} = (m^{*})_{\alpha\alpha'}; \quad \Gamma_{\alpha\alpha'} = \Gamma_{\alpha\alpha'}^{*},$$
$$m_{21} = m_{12}^{*}, \quad \Gamma_{21} = \Gamma_{12}^{*}. \tag{15.53}$$

Since $\overline{\chi}^0$ is the antiparticle of χ^0, χ^0 and $\overline{\chi}^0$ have same mass and decay width:

$$\langle \overline{\chi}^0 | M | \overline{\chi}^0 \rangle = \langle \chi^0 | M | \chi^0 \rangle$$
$$m_{11} = m_{22}, \quad \Gamma_{11} = \Gamma_{22}. \tag{15.54}$$

Diagonalization of mass matrix M, gives the eigenvalues [c.f Eq. (15.3)]

$$m_{\mp} - \frac{i}{2}\Gamma_{\mp} = \left[(m_{11} - \frac{i}{2}\Gamma_{11}) \mp \sqrt{(m_{12} - \frac{i}{2}\Gamma_{12})(m_{12}^{*} - \frac{i}{2}\Gamma_{12}^{*})} \right]$$
$$= (m_{11} - \frac{i}{2}\Gamma_{11}) \mp \sqrt{M_{12}M_{21}}. \tag{15.55}$$

We can write

$$M = n_0 + \boldsymbol{\sigma} \cdot \mathbf{n} \tag{15.56}$$

where

$$n_0 = m_{11} - \frac{i}{2}\Gamma_{11}$$
$$\mathbf{n} = \left(M_{12} + M_{21}, i(M_{12} - M_{21}), 0 \right) \tag{15.57}$$

so that the magnitude of $\mathbf{n}$ is

$$n = (M_{12}M_{21})^{1/2}. \tag{15.58}$$

and from Eq. (15.10), eigenvectors corresponding to eigenvalues (15.55) are

$$\frac{1}{\sqrt{2}} \begin{pmatrix} 1 \\ \sqrt{\frac{M_{21}}{M_{12}}} \end{pmatrix}, \quad \frac{1}{\sqrt{2}} \begin{pmatrix} -\sqrt{\frac{M_{12}}{M_{21}}} \\ 1 \end{pmatrix}. \tag{15.59}$$

Thus the mass eigenstates are [c.f Eq.(15.13)]

$$|\chi_\mp\rangle = \frac{1}{\sqrt{2}}[|\chi^0\rangle \mp \frac{q}{p}|\overline{\chi}^0\rangle] \tag{15.60}$$

where

$$\frac{q}{p} = \sqrt{\frac{M_{21}}{M_{12}}} = \sqrt{\frac{m_{12}^* - \frac{i}{2}\Gamma_{12}^*}{m_{12} - \frac{i}{2}\Gamma_{12}}} = (\frac{p}{q})^*. \tag{15.61}$$

The mass eigenstates $|\chi_\pm\rangle$ form a complete set:

$$|\psi(t)\rangle = a_-(t)|\chi_-\rangle + a_+(t)|\chi_+\rangle. \tag{15.62}$$

In this basis, the mass matrix M is diagonal so that:

$$i\frac{d}{dt}\begin{pmatrix} a_-(t) \\ a_+(t) \end{pmatrix} = \begin{pmatrix} m_- - \frac{i}{2}\Gamma_- & 0 \\ 0 & m_+ - \frac{i}{2}\Gamma_+ \end{pmatrix} \begin{pmatrix} a_-(t) \\ a_+(t) \end{pmatrix}. \tag{15.63}$$

The solution is

$$a_\pm(t) = a_\pm(0)e^{-im_\pm t - \frac{1}{2}\Gamma_\pm t}. \tag{15.64}$$

If we start with state $|\chi^0\rangle$, i.e. $|\psi(0)\rangle = |\chi^0\rangle$ so that from Eq (15.50) $a(0) = 1$, $\overline{a}(0) = 0$, then from Eqs (15.60) and (15.62):

$$|\chi^0\rangle = \frac{1}{\sqrt{2}}\left[a_+(0)(|\chi^0\rangle + \frac{q}{p}|\overline{\chi}^0\rangle) + a_-(0)(|\chi^0\rangle - \frac{q}{p}|\overline{\chi}^0\rangle)\right].$$

This gives

$$a_\pm(0) = \frac{1}{\sqrt{2}}. \tag{15.65}$$

Hence from Eq. (15.62) after time t,

$$\begin{aligned} |\psi(t)\rangle &= \frac{1}{\sqrt{2}}[\exp(-im_- t - \frac{1}{2}\Gamma_- t)|\chi_-\rangle \\ &\quad + \exp(-im_+ t - \frac{1}{2}\Gamma_+ t)|\chi_+\rangle] \\ &= \frac{1}{2}[\exp(-im_- t - \frac{1}{2}\Gamma_- t) + \exp(-im_+ t - \frac{1}{2}\Gamma_+ t)]|\chi^0\rangle \\ &\quad - \frac{q}{p}[\exp(-im_- t - \frac{1}{2}\Gamma_- t) - \exp(-im_+ t - \frac{1}{2}\Gamma_+ t)]|\overline{\chi}^0\rangle. \end{aligned} \tag{15.66}$$

where we have used Eq. (15.60). Thus although at time $t = 0$, the system was in the state $|\chi^0\rangle$, after time t it develops the component $|\overline{\chi}^0\rangle$. The probability of finding the particles $|\chi^0\rangle$ or $|\overline{\chi}^0\rangle$ at time t is represented by

$$P(\chi^0 \to \chi^0, t) = |\langle\chi^0|\psi(t)\rangle|^2 = \frac{1}{4}[e^{-\Gamma_- t} + e^{-\Gamma_+ t} + 2e^{-\Gamma t}\cos\Delta mt] \tag{15.67}$$

$$P(\chi^0 \to \overline{\chi}^0, t) = |\langle \overline{\chi}^0 | \psi(t) \rangle|^2 = \frac{1}{4} \left| \frac{q}{p} \right|^2 [e^{-\Gamma_- t} + e^{-\Gamma_+ t}$$

$$- 2e^{-\Gamma_- t} \cos \Delta mt], \quad (15.68)$$

where

$$\Delta m = m_+ - m_-, \qquad \Gamma = \frac{\Gamma_+ + \Gamma_-}{2}. \qquad (15.69)$$

One may make the following remark here: under the operation (called charge conjugation) which takes a particle state into its antiparticle state,

$$|\chi^0\rangle \xrightarrow{C} |\overline{\chi}^0\rangle$$

while under parity

$$|\chi^0\rangle \xrightarrow{P} -|\chi^0\rangle$$

so that under the combined operator CP

$$|\chi^0\rangle \xrightarrow{CP} \eta_{CP} |\overline{\chi}^0\rangle \qquad (15.70)$$

where the CP phase η_{CP} is arbitrary ($|\eta_{CP}|^2 = 1$) but we select it to be -1. Then we see from Eq. (15.60) that if $\frac{q}{p} = 1$ the mass eigenstate $|\chi_{\mp}\rangle$ will be the eigenstates of CP with eigenvales ± 1 respectively, which we designate as $|\chi_{1,2}\rangle$. From Eq. (15.61) it is clear that $\frac{q}{p}$ is unity if m_{12} and Γ_{12} are real. Writing $\frac{q}{p} = 1 + \epsilon$, then ϵ is a measure of CP violation and arises due to mismatch between mass and flavor eigenstates:

$$|\chi_-\rangle = |\chi_1^0\rangle + \epsilon|\chi_2^0\rangle$$
$$|\chi_+\rangle = |\chi_2^0\rangle + \epsilon|\chi_1^0\rangle \qquad (15.71)$$

where $CP|\chi_{1,2}\rangle = \pm|\chi_{1,2}\rangle$.

15.3.3 $K^0 - \overline{K}^0$ Complex

We now apply the general formalism developed in Sec. 15.3.2 to $K^0 - \overline{K}^0$ system. Here we denote K_- and K_+ as K_S and K_L. The subscripts S and L stand for short and long lived since the particles K_S and K_L which have definite mass and life time have $\tau_S = (0.8935 \pm 0.0008)10^{-10}s$ and $\tau_L = (5.17 \pm 0.14)10^{-8}s$ respectively so that $\tau_S \ll \tau_L$. Suppose that at $t = 0$, K^0 ($Y = 1$) is produced by the reaction $\pi^- p \to K^0 \Lambda^0$. The initial state is then pure $Y = 1$. It is clear from Eq. (15.66) [with $\chi^0 = K^0$] that a kaon beam which has been produced in a pure $Y = 1$ state has changed into one containing both the parts with $Y = 1$ and $Y = -1$. Experimentally

$\overline{K}^0$ can be verified through the observation of hadronic signature such as $\overline{K}^0 p \to \pi^+ \Lambda^0$ since $\pi^+ \Lambda^0$ can only be produced by $\overline{K}^0$ and not by K^0. The probability of finding $Y = -1$ component at t in the kaon produced at $t = 0$ in a pure $Y = 1$ state is given by Eq. (15.68) ($|\frac{q}{p}| \ll 1$, see below)

$$P(K^0 \to \overline{K}^0, t) = |\langle \overline{K}^0 | \psi(t) \rangle|^2$$
$$= \frac{1}{4} \{ \exp(-\frac{t}{\tau_S}) + \exp(-\frac{t}{\tau_L}) - 2\exp[-\frac{1}{2}(\frac{t}{\tau_S} + \frac{t}{\tau_L})] \cos(\Delta m)t \}$$

where $\Delta m = m_L - m_S$ and since τ_L is much larger than τ_S

$$P(K^0 \to \overline{K}^0, t) = \frac{1}{4}(1 + e^{-\frac{t}{\tau_S}} - 2e^{-\frac{1}{2}\frac{t}{\tau_S}} \cos(\Delta m)t). \qquad (15.72)$$

If kaons were stable ($\tau_L, \tau_S \to \infty$), then

$$P(K^0 \to \overline{K}^0, t) = \frac{1}{2}(1 - \cos(\Delta m)t) \qquad (15.73)$$

which shows that the state produced as pure $Y = 1$ state at $t = 0$ continuously oscillates between $Y = 1$ and $Y = -1$ states with frequency $\omega = \frac{\Delta m}{\hbar}$ and period of oscillation,

$$\tau = \frac{2\pi}{\Delta m / \hbar}. \qquad (15.74)$$

Kaons, however, decay and their oscillations are damped. By measuring the period of oscillation, Δm can be determined:

$$\Delta m = m_L - m_S \approx (3.5)10^{-12} MeV. \qquad (15.75)$$

Such a small number is measured as a consequence of quantum phenomena of interferometry.

Finally from Eq. (15.71)

$$|K_S\rangle = |K_1^0\rangle + \epsilon|K_2^0\rangle$$
$$|K_L\rangle = |K_2^0\rangle + \epsilon|K_1^0\rangle \qquad (15.76)$$

where $CP|K_{1,2}^0\rangle = \pm|K_{1,2}^0\rangle$. In the decay $K \to 2\pi$, (2π) state has $CP = +1$. Thus the CP non conservation manifests itself by the ratio

$$\eta_\pm = \frac{A(K_L \to \pi^+\pi^-)}{A(K_S \to \pi^+\pi^-)} = \epsilon$$

which experimentally is of order 2×10^{-3}, showing that $|\epsilon| \ll 1$.

15.3.4 $B_q^0 - \overline{B}_q^0$ Complex $(q = d, s)$

For $B^0 - \overline{B}^0$ complex (suppressing the subscript q) it is known that Γ_{12} and m_{12} have the same phase and $|\Gamma_{12} \ll |m_{12}|$. Thus from Eq. (15.61)

$$q/p = e^{-2i\beta}$$

where $m_{12} = |m_{12}|e^{2i\beta}$. The mass eigenstates $B_\mp$, which we now denote as $B_{L,H}$ are:

$$|B_L^0\rangle = \frac{1}{\sqrt{2}}[|B^0\rangle - e^{-2i\beta}|\overline{B}^0\rangle],$$

$$|B_H^0\rangle = \frac{1}{\sqrt{2}}[|B^0\rangle + e^{-2i\beta}|\overline{B}^0\rangle]. \tag{15.77}$$

In the limit $\beta \to 0$

$$|B_L^0\rangle \to |B_1^0\rangle = \frac{1}{\sqrt{2}}[|B^0\rangle - |\overline{B}^0\rangle], \quad CP = +1,$$

$$|B_H^0\rangle \to |B_2^0\rangle = \frac{1}{\sqrt{2}}[|B^0\rangle + |\overline{B}^0\rangle], \quad CP = -1. \tag{15.78}$$

The mismatch between mass and CP eigenstates is a possible source of CP-violation in B^0 decays. For $B^0 - \overline{B}^0$, complex:

$$\Delta\Gamma = \Gamma_1 - \Gamma_2$$

$$\ll \Delta m = m_2 - m_1 \tag{15.79}$$

If we start with $|B^0\rangle$, then from Eq (15.66)

$$|\psi(t)\rangle = e^{-imt}e^{-\frac{1}{2}\Gamma t}[\cos(\frac{\Delta m}{2})t|B^0\rangle - ie^{-2i\beta}\sin(\frac{\Delta m}{2}t)|\overline{B}^0\rangle] \tag{15.80}$$

where $m = \frac{m_1+m_2}{2}$.

The probabilities of finding $|\overline{B}^0\rangle$ and $|B^0\rangle$ at time t are given by

$$P(B^0 \to \overline{B}^0, t) = |\langle\overline{B}^0|\psi(t)\rangle|^2$$

$$= e^{-\Gamma t}\sin^2(\frac{\Delta m}{2}t)$$

$$= \frac{1}{2}e^{-\Gamma t}(1 - \cos(\Delta m)t), \tag{15.81}$$

$$P(B^0 \to B^0, t) = |\langle B^0|\psi(t)\rangle|^2$$

$$= e^{-\Gamma t}\cos^2(\frac{\Delta m}{2}t)$$

$$= \frac{1}{2}e^{-\Gamma t}(1 + \cos(\Delta m)t). \tag{15.82}$$

These equations are like those of a damped harmonic oscillator, the angular frequency of which is given by

$$\omega = \frac{\Delta m}{\hbar}, \quad \tau = \frac{2\pi\hbar}{\Delta m}.$$

The mixing parameter

$$r = \frac{\int_0^T |\langle \overline{B}^0 | B(t) \rangle|^2 dt}{\int_0^T |\langle B^0 | B(t) \rangle|^2 dt} \xrightarrow{T \to \infty} \frac{(\frac{\Delta m}{\Gamma})^2}{2 + (\frac{\Delta m}{\Gamma})^2}$$

$$= \frac{x^2}{2 + x^2}. \tag{15.83}$$

Experimentally, for B_d^0, B_s^0

$$\Delta m_{B_d^0} \approx 3.34 \times 10^{-10} MeV, \quad \tau_{B_d^0} = 1.52 \times 10^{-12} s,$$

$$\Delta m_{B_s^0} \approx 117 \times 10^{-10} MeV, \quad \tau_{B_s^0} = 1.47 \times 10^{-12} s,$$

$$x_d = \left(\frac{\Delta m_{B_d^0}}{\Gamma_{B_d^0}} \right) = \frac{(\Delta m_{B_d^0})}{\hbar} \tau_{B_d^0} \approx 0.77,$$

$$x_s = \frac{(\Delta m_{B_s^0})}{\hbar} \tau_{B_s^0} \approx 26.2.$$

Non-zero values of x_d and x_s clearly show mixing between B^0, $\overline{B}^0$; a proof of quantum interference.

15.3.5 *Neutrino Oscillations*

The neutrinos occur in three flavors, ν_e, ν_μ, ν_τ which are produced in β-decays of charged leptons. If neutrinos are massless, then these are also the mass eigenstates. If any of them have a mass, then mass eigenstates which we denote by ν_i ($i = 1, 2, 3$) are in general different from the flavor eigenstates ν_α ($\alpha = e, \mu, \tau$). We note that two set of states are connected with each other by a unitary transformation

$$|\nu_\alpha\rangle = \sum_i U_{\alpha i} |\nu_i\rangle, \quad \alpha = e, \mu, \tau \tag{15.84}$$

where

$$\sum_i U_{\alpha i} U_{j\alpha}^* = \delta_{ij}. \tag{15.85}$$

The mismatch between flavor and mass eigenstates gives rise to neutrino oscillations.

For example, for the conversion of ν_e to ν_x ($x = \mu, \tau$), we have from Eq. (15.27), where states $|\pm\rangle$ are mass eigenstates $|\nu_i\rangle, i = 1, 2$ and states $|1\rangle$ and $|2\rangle$ are flavor eigenstates $|\nu_e\rangle$ and $|\nu_x\rangle$ and we change $\frac{\theta}{2} \to \theta$ to conform to the usual convention used in neutrino physics $[\phi = 0]$:

$$|\psi(t)\rangle = e^{\frac{-iEt}{\hbar}} \left[\left(\cos \frac{\Delta E t}{\hbar} - i \cos 2\theta \sin \frac{\Delta E t}{\hbar} \right) |\nu_e\rangle - i \sin 2\theta \sin \frac{\Delta E t}{\hbar} |\nu_x\rangle \right]$$

(15.86)

where $E = \frac{E_+ + E_-}{2} = \frac{E_1 + E_2}{2}$ and $\Delta E = \frac{E_1 - E_2}{2}$. Then from Eq. (15.28) and (15.29), the probability of conversion from ν_e to ν_x and survival probability are respectively

$$P_{\nu_e \to \nu_x} = \sin^2 2\theta \sin^2 \frac{\Delta E t}{\hbar},$$

$$P_{\nu_e \to \nu_e} = 1 - P_{\nu_e \to \nu_x}.$$

(15.87)

Here θ is the mixing angle. Now

$$E_i = (k^2 + m_i^2)^{\frac{1}{2}} \simeq |k| + \frac{m_i^2}{2k},$$

(15.88)

since $k \gg m_i$ and we take the extreme relativistic limit. Then $\Delta E = \frac{m_1^2 - m_2^2}{2|k|} = -\frac{\Delta m_{12}^2}{2k}$ where $\Delta m_{12}^2 = m_2^2 - m_1^2$, it is customary to take $m_2 > m_1$ and write $|k| \simeq E_\nu$. Define

$$\lambda = \frac{2\pi}{E_2 - E_1} = \frac{4\pi E_\nu}{\Delta m^2} = 4\pi \frac{E_\nu}{MeV} \frac{eV^2}{\Delta m^2} 10^{12} \frac{1}{MeV}.$$

(15.89)

Putting c and $\hbar$ back

$$\frac{1}{MeV} = \frac{c\hbar}{MeV} = 1.97 \times 10^{-13} m$$

$$\lambda = \frac{4\pi \frac{E_\nu}{MeV}}{\frac{\Delta m^2}{eV^2}} \times 10^{12} \times 1.97 \times 10^{-13} m$$

$$\approx 2.47 m \frac{E_\nu}{MeV} \frac{eV^2}{\Delta m^2}.$$

(15.90)

Now $L = ct, (c = 1, t = L)$ is the distance travelled after which ν_e is converted into ν_x. Hence

$$P(\nu_e \to \nu_e, t) = \sin^2 2\theta \sin^2 \left(\frac{\Delta m^2}{1.24 E_\nu} L \right)$$

$$= \sin^2 2\theta \sin^2 \left(\frac{\pi L}{\lambda} \right),$$

(15.91)

$$P(\nu_e \to \nu_e, t) = 1 - \sin^2 2\theta \sin^2 \left(\frac{\pi L}{\lambda} \right).$$

(15.92)

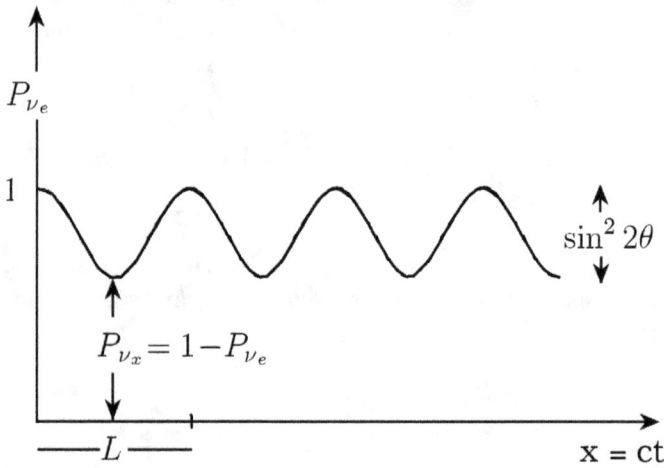

Fig. 15.2   Neutrino oscillations.

The probability P oscillates and frequency $\frac{c}{\lambda}$ and amplitude $\sin^2 2\theta$. The amplitude of the oscillation $\sin^2 2\theta$ is given by the mixing angle. These oscillations are shown in Fig. 15.2.

It is clear from above analysis, that oscillations are between neutrinos of same momentum but different masses. To look for oscillations, one needs factors, which enhance tiny effects: a coherent source (there are many, the sun, cosmic rays, reactor etc), low energy neutrinos, large base line (size of sun, and that of the earth), large mixing angle and large flux.

For 3 flavors, it is customary to parameterize mixing matrix elements $U_{\alpha i}$ (not all independent) by three mixing angles $\theta_{12}, \theta_{13}, \theta_{23}$ and one complex phase δ

$$\frac{|U_{e2}|^2}{|U_{e1}|^2} = \tan^2 \theta_{12}, \quad \frac{|U_{\mu3}|^2}{|U_{\tau3}|^2} = \tan^2 \theta_{23}, \quad \frac{|U_{e3}|^2}{|U_{e1}|^2} = \tan^2 \theta_{13}.$$

The mass eigenstate $|\nu_i\rangle$ $(i = 1, 2, 3)$ has a well defined mass m_i and it is customary to order the mass eigenvalues such that

$$\Delta m_{12}^2 = m_2^2 - m_1^2 > 0$$

and

$$\Delta m^2 = m_3^2 - \frac{m_2^2 + m_1^2}{2}$$

where $\Delta m^2 > 0$ (< 0) correspond to normal (inverted) mass spectrum hierarchy. We also summarize the various sources which provide neutrinos

(1) Reactor: $n \to p + e^- + \bar{\nu}_e$.
(2) Solar: in pp cycle $4p \to^4 He + 2e + 2\nu_e + 26.7 MeV$.
   The depletion of ν_e flux from the sun would indicate the conversion of ν_e to another species of neutrinos, e.g. ν_μ. This is observed experimentally.
(3) Atmosphere: Interaction of Cosmic rays with atmosphere

$$p + A \to \pi^\pm + A'$$
$$\to \mu^\pm \nu_\mu (\bar{\nu}_\mu)$$
$$\to e^\pm \nu_e (\bar{\nu}_e) \nu_\mu (\bar{\nu}_\mu).$$

   Thus one would expect $\frac{N(\nu_\mu)}{N(\nu_e)} \simeq 2$, if there is no osillation i.e. no flavor change. But experimentally it is substantially reduced from ~ 2, providing an evidence for oscillation.
(4) Accelerator.

Typically, a single experiment is mainly sensitive to only one of the above mass gaps and to one mixing parameter, although subleading effects driven by remaining parameters may become relevant in precision oscillation searches.

So far, solar and long-baseline reactor neutrino experiments have measured the mass-mixing parameters $(\Delta m_{12}^2, \theta_{12})$ in the $\nu_e \to \nu_e$ channel, while atmospheric and long-baseline accelerator (LBL) experiments have measured $(\Delta m^2, \theta_{23})$ in the $\nu_\mu \to \nu_\mu$ channel. Conversely short-baseline reactor experiments are mainly sensitive to $(\Delta m^2, \theta_{13})$.

To summarize detailed combined analysis of all neutrino data give best fit values:

$$\Delta m_{12}^2 = 7.65 \times 10^{-5} eV^2,$$
$$\Delta m^2 = 2.40 \times 10^{-3} eV^2,$$
$$\theta_{12} \simeq 34^\circ,$$
$$\theta_{23} \simeq 45^\circ,$$
$$\theta_{13} \simeq 9^\circ.$$

Currently, there is no constraint on the sign of Δm^2. To conclude there is compelling evidence that neutrinos change flavor, have non-zero masses and that neutrino mass eigenstates are different from weak eigenstates. Note again that the quantum mechanical phenomena has enabled us to measure differences of order 10^{-3} to $10^{-4}eV^2$.

To conclude this chapter. We have provided a unified treatment of quantum interference and have shown that this have been observed in variety of cases with no connection with each other, except that all are two state systems. In the next chapter we discuss another application of two state systems in quantum computation.

15.4   Problems

15.1 Show that $\begin{pmatrix} \cos\frac{\theta}{2} \\ e^{i\phi}\sin\frac{\theta}{2} \end{pmatrix}$ and $\begin{pmatrix} -e^{i\phi}\sin\frac{\theta}{2} \\ \cos\frac{\theta}{2} \end{pmatrix}$ are eigenstates of the helicity operator $\frac{\sigma\cdot n}{|n|}$ with eigenvalues ± 1 respectively. Here $\frac{n}{|n|} = (\sin\theta\cos\phi, \sin\theta\sin\phi, \cos\theta)$.

15.2 Show that the probability of finding the particle in eigenstates of σ_x with eigenvalues ± 1 is

$$P_{x\uparrow\downarrow} = \frac{1}{2}\{\frac{1}{2}1 \pm 2(\sin^2\frac{\Omega}{2}t\sin\theta\cos\theta\cos\omega t + \sin\frac{\Omega}{2}t\cos\frac{\Omega}{2}t\sin\theta\sin\omega t)\}.$$

What is the corresponding result for $P_{y\uparrow\downarrow}$?

Hint: Eigenstates of σ_x and σ_y are respectively

$$\frac{1}{\sqrt{2}}[|+\frac{1}{2}\rangle \pm |-\frac{1}{2}\rangle], \quad \frac{1}{\sqrt{2}}[|+\frac{1}{2}\rangle \pm i|-\frac{1}{2}\rangle].$$

Chapter 16

Quantum Computation

16.1 Introduction

Theoretical and conceptual understanding of the fundamental principles of quantum mechanics, as explained in earlier chapters, lead to the development of a tool box to engineer computers and their algorithms based on quantum laws. The discussion begins with the basic elements of computers, such as, bits and logic gates, ultimately leads to new concepts like super dense coding, quantum teleportation and quantum networks.

At present the consistent development in the subject has led to the development of sophisticated quantum algorithms, fundamentally newer quantum computers, quantum clocks and quantum satellites for communication between distant parties. Secure communication, that is, cryptography is conventionally based on factorization of large numbers, such as in RSA. It is challenged by quantum ciphers for factorization such as Shor's Algorithm, and rescued by the quantum cryptographic techniques, such as **BB84** and **Ekert92**. In recent years secure data communication between distant parties based on quantum laws has crossed the barrier of 100 kilometers.

The infinite dimension Hilbert space simplifies drastically as we require only two dimensions to develop a prototype of a classical bit. A quantum mechanical bit or a qubit, therefore, becomes a simplest example of Hilbert space and provides understanding of the superposition principle in quantum mechanics. The understanding helps us to think of more and more physically controllable systems to engineer quantum computers.

16.2   Qubit

In conventional computation the information storage and data manipulation is done by means of the basic building block of hardware, that is, a bit. Hence, a data written in binary form is mapped on a series of bits as a sequence of zero and one, that conventionally corresponds to low and high voltages, respectively. A quantum mechanical bit, that is a qubit or qbit, requires a two state quantum system which may be spin of electrons, internal states of an atom or two modes of an electromagnetic field. Hence, the system can be found in either of the two states. In Dirac notation we label the two state as $|0\rangle$ and $|1\rangle$. As an example, we consider them indicating ground and excited state of an atom, that is, $|0\rangle = |g\rangle$ and $|1\rangle = |e\rangle$, respectively. However in contrast to the conventional classical bit the qubit appears in a superposition state. This feature marks a major difference in classical and quantum computers and introduces a lot of philosophical and conceptual debate. Hence, in general we express the qubit as

$$|\psi\rangle = c_1|0\rangle + c_2|1\rangle \tag{16.1}$$

where c_1 and c_2 are complex numbers and define probability amplitudes, such that

$$|c_1|^2 + |c_2|^2 = 1.$$

Physically the superposition state may arise as the atom is performing a transition from its ground state, that is, state vector $|0\rangle$ to $|1\rangle$.

Since the probability amplitudes, c_1 and c_2, are complex numbers we may write then as

$$c_1 = |c_1|e^{i\gamma_1},$$
$$c_2 = |c_2|e^{i\gamma_2}.$$

Hence, equation(16.1) becomes

$$|\psi\rangle = |c_1|e^{i\gamma_1}|0\rangle + |c_2|e^{i\gamma_2}|1\rangle$$

where γ_1 and γ_2 are phases corresponding to state vector $|0\rangle$ and $|1\rangle$, respectively. We may further simplify the above expression

$$|\psi\rangle = e^{i\gamma_1}\left(|c_1||0\rangle + |c_2|e^{i\gamma}|1\rangle\right).$$

Hence, we identify $\gamma = \gamma_1 - \gamma_2$ as the relative phase and γ_1 expressing a global phase on the qubit. Without loss of generality we may express the qubit as

$$|\psi\rangle = |c_1||0\rangle + |c_2|e^{i\gamma}|1\rangle. \tag{16.2}$$

Another useful representation of a qubit comes as we write the real valued $|c_1|$ and $|c_2|$ in the parametric relation, that is,

$$|c_1| = \cos\left(\tfrac{\theta}{2}\right)$$
$$|c_2| = \sin\left(\tfrac{\theta}{2}\right).$$

This helps us to express the qubit in the most general form; we may write them as

$$|\psi\rangle = \cos\left(\tfrac{\theta}{2}\right)|0\rangle + \sin\left(\tfrac{\theta}{2}\right)e^{i\gamma}|1\rangle. \tag{16.3}$$

Hence, the definition of the qubit is based on γ and θ, which makes the qubit a point on a sphere of unit radius r in spherical coordinates, (r, γ, θ). It is known as the Bloch sphere and shown in Fig. 16.1. From the above equation, it is evident that for $\theta = 0$ and for $\theta = \pi$ the qubit, respectively, is in the $|0\rangle$ and $|1\rangle$ state, whereas for any other value θ it appears as another point on the Bloch sphere, shown in Fig. 16.1. It is important

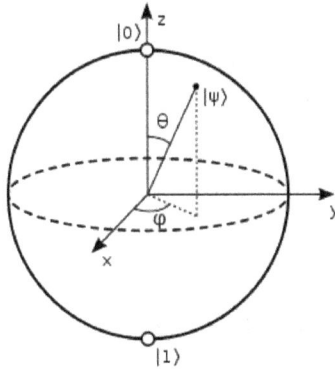

Fig. 16.1

to note that the qubit expressed as a general vector on the basis of two dimensional Hilbert space is defined on a three dimensional sphere. From where one extra dimension is appearing? The riddle is answered as we note that the qubit is effectively defined on the two dimensional surface of the Bloch sphere as its radius is always fixed to unity.

16.3 Multiple Qubit Systems

In order to express a physical system which consists of a large setup of many qubits, we express the state of the system as a tensor product of each individual system. For this reason if $|\psi_1\rangle, |\psi_2\rangle, ... |\psi_n\rangle$ express n numbers of qubits, the combined state of the system becomes

$$|\psi\rangle = |\psi_1\rangle \otimes |\psi_2\rangle \otimes \cdots \otimes |\psi_n\rangle. \qquad (16.4)$$

Hence, Eq. (16.1) becomes

$$
\begin{aligned}
|\psi_{1,2}\rangle &= |\psi\rangle \otimes |\psi_2\rangle \\
|\psi_{1,2}\rangle &= (|c_1||0\rangle + |c_2||1\rangle) \otimes (|d_1||0\rangle + |d_2||1\rangle) \\
|\psi_{1,2}\rangle &= a|0,0\rangle + b|0,1\rangle + c|1,0\rangle + d|1,1\rangle
\end{aligned} \qquad (16.5)
$$

where $|0,0\rangle = |0\rangle \otimes |0\rangle$ and same for the other terms. Interestingly, in conventional computers for a two bit case, we have the same four options, which are: $00, 01, 10, 11$. However, in the quantum domain in addition to these four options we have their superposition state as well. Furthermore, the probability amplitudes are complex numbers which relate a phase to each of the four options. This particular feature makes quantum mechanics different from statistical mechanics and gives rise to the interference phenomenon. It is important to note that as a special case if any one of the amplitudes becomes zero, the two systems are not independent. Hence, we cannot write the final state as a tensor product of the two independent states as we did in Eq.(16.5). In this situation two states are correlated, or entangled to each other.

16.4   Matrix Representation of Qubits

A very useful expression of the qubit comes from matrix representation. We define the basis vectors $|0\rangle$ and $|1\rangle$ in matrix form, such that:

$$|0\rangle = \begin{pmatrix} 1 \\ 0 \end{pmatrix}, |1\rangle = \begin{pmatrix} 0 \\ 1 \end{pmatrix}$$

which leads us to write the qubit as

$$|\psi\rangle = \begin{pmatrix} c_1 \\ c_2 \end{pmatrix}.$$

This implies that the definition of a particular qubit requires the definition of complex probability amplitudes, c_1 and c_2. Alternatively we say that for a particular choice of c_1 and c_2 we have one qubit. We consider a two qubits system, say, $|\psi_1\rangle$ and $|\psi_2\rangle$. The corresponding combined state of the system shall be a tensor product of these two states. As we define tensor products as a matrix of matrices, the combined state of the two qubits, $|\chi\rangle$, becomes

$$
\begin{aligned}
|\chi\rangle &= |\psi_1\rangle \otimes |\psi_2\rangle \\
&= \begin{pmatrix} c_1 \\ c_2 \end{pmatrix} \otimes |\psi_2\rangle \\
&= \begin{pmatrix} c_1|\psi_2\rangle \\ c_2|\psi_2\rangle \end{pmatrix}.
\end{aligned}
\tag{16.6}
$$

Here, state vectors $|\psi_1\rangle$ and $|\psi_2\rangle$ are themselves matrices which appear as elements of another matrix. The above expression can further be simplified as:

$$
|\chi\rangle = \begin{pmatrix} c_1 d_1 \\ c_1 d_2 \\ c_2 d_1 \\ c_2 d_2 \end{pmatrix}.
\tag{16.7}
$$

We conclude that a single qubit or a multiple qubit system is described as a column matrix.

16.5   Single Qubit Logic Gates

In conventional computation, manipulation of data is made by means of classic logic gates. For this reason we define logic gates as operators and transforms which change the state of a system of qubits according to requirement. The most simplest among them is the NOT gate which transforms a particular binary code into its complement.

In the quantum domain, we may define a Quantum NOT gate as an operator, X, such that

$$
X|0\rangle \equiv |1\rangle \qquad X|1\rangle \equiv |0\rangle.
$$

We define it figuratively as

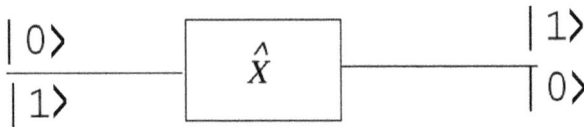

Fig. 16.2

The corresponding definition as the outer product in Dirac notation becomes

$$\hat{X} = |0\rangle\langle 1| + |1\rangle\langle 0|. \tag{16.8}$$

We take the basis vector as discussed above and we obtain

$$\hat{X} = \begin{pmatrix} 0 & 1 \\ 1 & 0 \end{pmatrix}. \tag{16.9}$$

It is worth mentioning here that the square matrix defines NOT gate operation. Generalizing the statement we say that any operator can be defined as a square matrix.

In contrast to one single bit logic gate in conventional computational, in quantum domain we have more. In some applications it is needed to introduce a particular phase on a qubit. For that matter we require a phase gate $\hat{Y}$. The phase gate has the basis vectors $|0\rangle$ and $|1\rangle$ as its eigenstates with eigenvalues $+1$ and -1, respectively, as shown in Fig. 16.3.

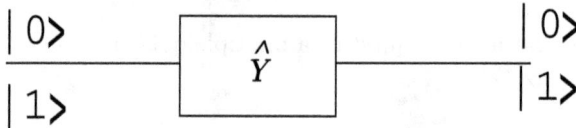

Fig. 16.3

We define the phase gate in the Dirac notations as:

$$\hat{Y} \equiv |0\rangle\langle 0| - |1\rangle\langle 1|.$$

As we consider the matrix representation for the basis vectors, the phase gate is

$$\hat{Y} = \begin{pmatrix} 1 & 0 \\ 0 & -1 \end{pmatrix}. \tag{16.10}$$

Engineering a superposition on a qubit is performed by means of the superposition gate $\hat{Z}$, which is referred some times in literature as the Hadamard gate, $\hat{H}$. The logic gate operates on either of the basis vectors $|0\rangle$ or $|1\rangle$, and generates superposition between the two with equal probability, such that [see Fig 16.4]

$$\hat{H}|0\rangle = \frac{1}{\sqrt{2}}|0\rangle + \frac{1}{\sqrt{2}}|1\rangle,$$

$$\hat{H}|1\rangle = \frac{1}{\sqrt{2}}|0\rangle - \frac{1}{\sqrt{2}}|1\rangle.$$

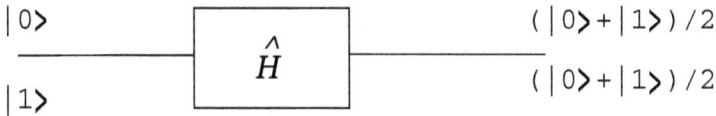

$\lvert 0\rangle$		$(\lvert 0\rangle + \lvert 1\rangle)/2$
	$\hat{H}$	
$\lvert 1\rangle$		$(\lvert 0\rangle + \lvert 1\rangle)/2$

Fig. 16.4

In the outer product we define the Hadamard gate as

$$H = \frac{\lvert 0\rangle + \lvert 1\rangle}{\sqrt{2}} \lvert 0\rangle + \frac{\lvert 0\rangle - \lvert 1\rangle}{\sqrt{2}} \lvert 1\rangle,$$

whereas the corresponding matrix notation for the Hadamard gate is

$$\hat{H} = \frac{1}{\sqrt{2}} \begin{pmatrix} 1 & 1 \\ 1 & -1 \end{pmatrix}. \tag{16.11}$$

16.6 Quantum Logic Gates at Work

In order to develop a physical insight in the operation of quantum single bit logic gates we consider the simple example of atom–field interaction in cavity QED (quantum electrodynamics). The atom transitions between ground and excited state in the presence of field. The atom, therefore, acts like a qubit so that $\lvert \psi \rangle = c_1 \lvert 0 \rangle + c_2 \lvert 1 \rangle$, where $\lvert 0 \rangle$ corresponds to ground state and $\lvert 1 \rangle$ defines excited state. Under the condition of resonance between the atomic transition frequency and field frequency, the square of the probability amplitudes $\lvert c_1(t) \rvert^2$ and $\lvert c_2(t) \rvert^2$ change in time periodically and alternatively, as shown in Figure 16.5. Hence, choosing a time of interaction provides the desired operation of a NOT gate and H gate. Non-dispersive interaction of the atom with the field leads to the phase gate, which does not change the state but the phase in a controlled fashion.

16.7 Two Qubit Logic Gates

Control NOT logic gate (CNOT), $\hat{u}_{cont}$, is a transformation which acts on two qubits, one is the controlled qubit and the other is the target qubit. In its operation the CNOT gate acts on a set of two qubits, such that it does not change the controller qubit but modifies the target qubit using modulus two operation. We define the CNOT gate mathematically as:

$$\hat{u}_{cont}[\lvert q_1 \rangle \otimes \lvert q_2 \rangle] = \lvert q_1 \rangle \otimes \lvert q_1 \oplus q_2 \rangle,$$

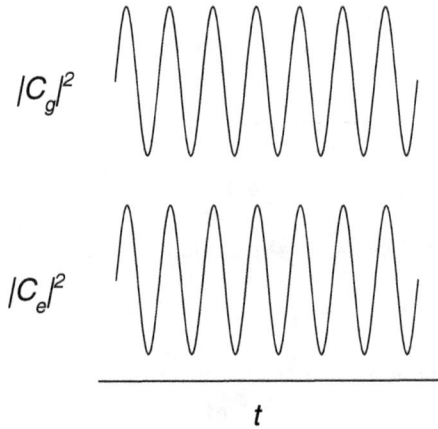

Fig. 16.5

here notation $\oplus$ represent modulus two operation, moreover, $|q_1\rangle$ is the controller qubit and $|q_2\rangle$ is the target qubit [see Fig. 16.6]. Thus, for

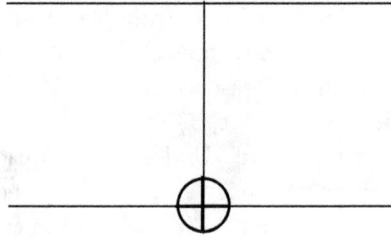

Fig. 16.6

different input we have following results:

$$\hat{u}_{cont}[|0\rangle \otimes |1\rangle] = |0\rangle \otimes |0 \oplus 1\rangle = |0\rangle \oplus |1\rangle,$$
$$\hat{u}_{cont}[|0\rangle \otimes |0\rangle] = |0\rangle \otimes |0 \oplus 0\rangle = |0\rangle \oplus |0\rangle,$$
$$\hat{u}_{cont}[|1\rangle \otimes |0\rangle] = |1\rangle \otimes |1 \oplus 0\rangle = |1\rangle \oplus |1\rangle,$$
$$\hat{u}_{cont}[|1\rangle \otimes |1\rangle] = |1\rangle \otimes |1 \oplus 1\rangle = |1\rangle \oplus |0\rangle. \tag{16.12}$$

Further analysis shows that the CNOT gate flips the target qubit if the controller qubit is 1. The CNOT gate together with a single qubit logic gate acts as a universal logic gate. In quantum computation we perform various different tasks by using the quantum logic gates and developing corresponding circuits.

16.8   Quantum Circuits

Logic circuits are created from conventional logic devices to perform specific tasks. Here our aim is to bring into light these tasks using quantum logic gates, as discussed above. For this purpose, we consider two simple applications, one is the swapping operation, and the other is quantum cloning in a special situation.

We consider a specific circuit shown below [Fig. 16.7]. The above cir-

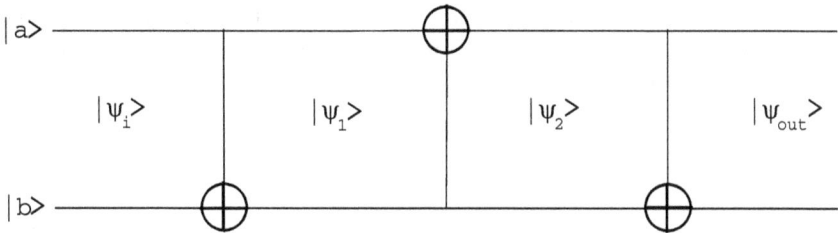

Fig. 16.7

cuitry is a schematic diagram of series of three Control NOT gates in a certain combination. Let's see that what is the output generated at each stage. We consider the initial state $|\psi_1\rangle$ which is in fact a combined state of $|a\rangle$, controller bit, and $|b\rangle$, target bit. As we operate Control NOT operation on the initial state $|\psi_i\rangle$, to generate state $|\psi_1\rangle$, as follows

$$
\begin{aligned}
|\psi_1\rangle &= \hat{u}_{cont}[|\psi_i\rangle], \\
|\psi_1\rangle &= \hat{u}_{cont}[|a\rangle \otimes |b\rangle], \\
|\psi_1\rangle &= |a\rangle \otimes |a \oplus b\rangle.
\end{aligned}
\tag{16.13}
$$

Now the control bit for the next stage $|\psi_2\rangle$ is $|a \oplus b\rangle$ and target bit is $|a\rangle$, thus,

$$
\begin{aligned}
|\psi_2\rangle &= \hat{u}_{cont}[|\psi_1\rangle], \\
|\psi_2\rangle &= \hat{u}_{cont}[|a\rangle \otimes |a \oplus b\rangle], \\
|\psi_2\rangle &= |a \oplus a \oplus b\rangle \otimes |a \oplus b\rangle, \\
|\psi_2\rangle &= |b\rangle \otimes |a \oplus b\rangle.
\end{aligned}
\tag{16.14}
$$

First term in tensor product results in state $|b\rangle$. Next considering $|b\rangle$ as control bit and $|a \oplus b\rangle$ as a target bit for the output state, $|\psi_0\rangle$, we proceed as follows:

$$|\psi_{out}\rangle = \hat{u}_{cont}[|\psi_2\rangle],$$
$$|\psi_{out}\rangle = \hat{u}_{cont}[|b\rangle \otimes |a \oplus b\rangle],$$
$$|\psi_{out}\rangle = |b\rangle \otimes |a \oplus b \oplus b\rangle,$$
$$|\psi_{out}\rangle = |b\rangle \otimes |a\rangle.$$

Hence this circuit performs the *swapping* operation as it swaps the state of two qubits.

Here we discuss *possible quantum cloning* circuitry using quantum logic. The generalized schematic diagram for cloning operation is as follows. Simply we take the Control NOT gate and specify the target bit as zero.

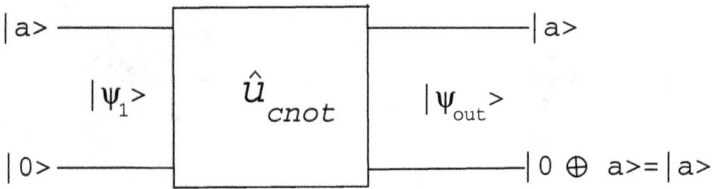

Fig. 16.8

Hence, the output state of the system becomes $|\psi_{out}\rangle$, such that [see Fig. 16.8]

$$|\psi_{out}\rangle = \hat{u}_{cont}[|\psi_1\rangle],$$
$$|\psi_{out}\rangle = \hat{u}_{cont}[|a\rangle \otimes |0\rangle],$$
$$|\psi_{out}\rangle = |a\rangle \otimes |0 \oplus a\rangle,$$
$$|\psi_{out}\rangle = |a\rangle \otimes |a\rangle.$$

This generates a clone of the controller qubit state, $|a\rangle$, at the target qubit. A physical example of quantum cloning phenomena is observed in the stimulated emission process as an excited atom, in the presence of a resonant photon, emits an identical photon. The interaction, therefore, transforms a single photon in the system to two identical photons.

16.9   EPR Entanglement

A simple modification in the circuit for quantum cloning justifies that quantum cloning in general is not possible. In addition, the outcome reveals another purely quantum phenomenon, that is, entanglement. We consider that the controller qubit of the above circuit now passes through a single bit Hadamard gate, before the application of two qubit controlled NOT gate operation. The corresponding schematic diagram is shown below [Fig. 16.9]: Considering $|a\rangle = |0\rangle$, then

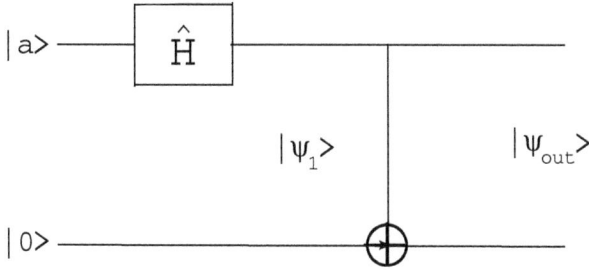

Fig. 16.9

$$|\psi_1\rangle = \left[\frac{|0\rangle + |1\rangle}{\sqrt{2}}\right]$$

and output state is

$$|\psi_{out}\rangle = \hat{u}_{cont}\left[|\psi_1\rangle \otimes |0\rangle\right],$$

$$|\psi_{out}\rangle = \hat{u}_{cont}\left[\left[\frac{|0\rangle + |1\rangle}{\sqrt{2}}\right] \otimes |0\rangle\right],$$

$$|\psi_{out} = \frac{1}{\sqrt{2}}(|0\rangle \otimes |0\rangle + |1\rangle \otimes |1\rangle).$$

In the above equation we have controller qubit before tensor notation and target qubit after it in each of the two terms. At this stage we have an important and surprising effect. The control and target qubit states cannot be factored out and discussed independently, they are *entangled*. If the first qubit is in state zero the second is also in state zero, whereas if the first qubit is in state one, second shall also be in state one, and the probability of occurrence of the two events is equal.

From the above discussion we infer that in the process of quantum cloning if in case the controller qubit is in a superposition state, in the output state $|\psi_{out}\rangle$, we shall not have its clone at the target qubit. In contrast we have engineered an entanglement between the controller qubit and the target qubit. This implies that in general cloning is not possible, which is established as *no-cloning theorem*.

It is important to note that for two different values of controller qubit $|a\rangle = |0\rangle$ and $|a\rangle = |1\rangle$, we have two output states $|\alpha_+\rangle$ and $|\alpha_-\rangle$, viz.

$$|\alpha_\pm\rangle = \frac{|0,0\rangle \pm |1,1\rangle}{\sqrt{2}}.$$

Furthermore, as we apply a NOT gate to the target qubit before CNOT operation, we obtain two other entangled states $|\beta_+\rangle$ and $|\beta_-\rangle$, viz.

$$|\beta_\pm\rangle = \frac{|0,1\rangle \pm |1,0\rangle}{\sqrt{2}}.$$

The four entangled states $|\alpha_\pm\rangle$ and $|\beta_\pm\rangle$ are orthogonal and normalized to unity. These are known as EPR states or EPR pairs. Therefore they define a complete set of the basis vectors for a two qubit system. This implies that we can write any arbitrary two qubit state in the vector space spanned by the four EPR entangled states.

Historically, EPR entangled states initiated a hefty philosophical discussion as they were used primarily to question the foundations of quantum mechanics by Einstein, Podolsky and Rosen. In later years, the discussion between Einstein, Schrodinger and Max Born, led to comprehend the newly developed mechanics at a very fine scale. Interestingly, in 1964, John Bell with the help of the same entangled states settled the long-standing discussion on the validity of the classical laws or the quantum laws in understanding the physical world, by calculating upper limit of strength of correlations, known as Bell's inequalities. He showed that the upper limit of correlation between two parties would differ when calculated using classical mechanics and quantum mechanics. The experimentally verifiable Bell's inequalities led a name for these states, Bell's states.

16.10 Problems

16.1 Show that the CNOT quantum logic gate is a reversible two-bit gate.

16.2 Show that Bell states, given as

$$|\alpha_\pm\rangle = \frac{|0,0\rangle \pm |1,1\rangle}{\sqrt{2}},$$

$$|\beta_\pm\rangle = \frac{|0,1\rangle \pm |1,0\rangle}{\sqrt{2}},$$

define a complete set of the basic vectors for a two qubit system, that is, the four entangled states are orthogonal and normalized to unity.

16.3 Show that the tensor product of two unit operators is a unit operator.

16.4 Show that the tensor product of two unitary operators is a unitary operator.

Chapter 17

Perturbation Induced by Electromagnetic Field

17.1   Interaction of Electromagnetic Field with Electrons

Classically the electromagnetic field is described by Maxwell's equations, which in vacuum are given by

$$\nabla \cdot \mathbf{B}(\mathbf{r}, t) = 0, \tag{17.1a}$$

$$\nabla \cdot \mathbf{E}(\mathbf{r}, t) = 4\pi\rho(\mathbf{r}, t), \tag{17.1b}$$

$$\nabla \times \mathbf{E}(\mathbf{r}, t) = -\frac{1}{c}\frac{\partial \mathbf{B}(\mathbf{r}, t)}{\partial t}, \tag{17.1c}$$

$$\nabla \times \mathbf{B}(\mathbf{r}, t) = \frac{1}{c}\frac{\partial \mathbf{E}(\mathbf{r}, t)}{\partial t} + \frac{4\pi}{c}\mathbf{j}(\mathbf{r}, t), \tag{17.1d}$$

where $\rho(\mathbf{r}, t)$ and $\mathbf{j}(\mathbf{r}, t)$ are the charge and current densities, that is, sources of electromagnetic fields $\mathbf{E}(\mathbf{r}, t)$ and $\mathbf{B}(\mathbf{r}, t)$. Charge conservation is expressed by the continuity equation

$$\frac{\partial \rho(\mathbf{r}, t)}{\partial t} + \nabla \cdot \mathbf{j}(\mathbf{r}, t) = 0. \tag{17.2}$$

It is convenient to introduce a scalar potential $\phi(\mathbf{r}, t)$ and a vector potential $\mathbf{A}(\mathbf{r}, t)$, such that

$$\mathbf{E} = -\nabla\phi - \frac{1}{c}\frac{\partial \mathbf{A}}{\partial t}, \tag{17.3a}$$

$$\mathbf{B} = \nabla \times \mathbf{A}. \tag{17.3b}$$

We note that physically observable quantities $\mathbf{E}$ and $\mathbf{B}$ do not determine ϕ and $\mathbf{A}$ uniquely. For, if we change

$$\phi \to \phi' = \phi - \frac{1}{c}\frac{\partial}{\partial t}\Lambda(\mathbf{r}, t), \tag{17.4a}$$

$$\mathbf{A} \to \mathbf{A}' = \mathbf{A} + \nabla\Lambda(\mathbf{r}, t), \tag{17.4b}$$

Eqs. (17.3a) and (17.3b) remain unchanged. This is known as gauge transformation.

In terms of potentials $\mathbf{A}$ and ϕ, Eqs. (17.1) can be written as

$$\nabla^2 \mathbf{A} - \frac{1}{c^2}\frac{\partial^2 \mathbf{A}}{\partial t^2} - \nabla\left(\frac{1}{c}\frac{\partial \phi}{\partial t} + \nabla\cdot\mathbf{A}\right) = -\frac{4\pi}{c}\mathbf{j}(\mathbf{r},t), \qquad (17.5)$$

$$\nabla^2\phi + \frac{1}{c}\frac{\partial}{\partial t}(\nabla\cdot\mathbf{A}) = -4\pi\rho(\mathbf{r},t). \qquad (17.6)$$

For static Coulomb field for which $\rho(\mathbf{r},t) = \rho(\mathbf{r})$ and $\frac{\partial\phi}{\partial t} = 0$, we can select the gauge such that

$$\nabla\cdot\mathbf{A} = 0.$$

This is called the Coulomb gauge. For this case, we have

$$\nabla^2\phi = -4\pi\rho(\mathbf{r}), \qquad (17.7a)$$

$$\nabla^2\mathbf{A} - \frac{1}{c^2}\frac{\partial^2\mathbf{A}}{\partial t^2} = -\frac{4\pi}{c}\mathbf{j}(\mathbf{r},t). \qquad (17.7b)$$

For the non-static case, it is more convenient to select the gauge such that

$$\frac{1}{c}\frac{\partial\phi}{\partial t} + \nabla\cdot\mathbf{A} = 0. \qquad (17.8)$$

This is called the Lorentz gauge. For this gauge Eq. (17.7b) remains unaltered, but Eq. (17.7a) becomes

$$\nabla^2\phi - \frac{1}{c^2}\frac{\partial^2\phi}{\partial t^2} = \rho(\mathbf{r},t). \qquad (17.9)$$

For a source-free region, Eqs. (17.7b) and (17.9) become

$$\nabla^2\mathbf{A} - \frac{1}{c^2}\frac{\partial^2\mathbf{A}}{\partial t^2} = 0, \qquad (17.10a)$$

$$\nabla^2\phi - \frac{1}{c^2}\frac{\partial^2\phi}{\partial t^2} = 0. \qquad (17.10a)$$

For a free particle, the Hamiltonian is given by

$$H = \frac{\mathbf{p}^2}{2m}. \qquad (17.11)$$

For a particle of mass m and charge e moving in an electromagnetic field, the classical Hamiltonian is given by

$$H = \frac{1}{2m}\left(\mathbf{p} - \frac{e}{c}\mathbf{A}(\mathbf{r},t)\right)^2 + e\phi(\mathbf{r}). \qquad (17.12)$$

Here we confine ourselves to static scalar potential.

By the correspondence principle, the Hamiltonian in quantum mechanics is given by

$$H = \frac{1}{2m}\left(\hat{\mathbf{p}} - \frac{e}{c}\mathbf{A}(\hat{\mathbf{r}}, t)\right)^2 + e\phi(\hat{\mathbf{r}}), \tag{17.13}$$

where in the Schrödinger representation

$$\hat{\mathbf{r}} \to \mathbf{r}, \quad \hat{\mathbf{p}} \to -i\hbar\nabla.$$

Hence

$$\begin{aligned}
H &= \frac{\hat{\mathbf{p}}^2}{2m} e\phi(\mathbf{r}) - \frac{e}{2mc}(\hat{\mathbf{p}} \cdot \mathbf{A} + \mathbf{A} \cdot \hat{\mathbf{p}}) + \frac{e^2}{2mc^2}\mathbf{A}^2 \\
&= -\frac{\hbar^2}{2m}\nabla^2 - \frac{e}{2mc}(2\mathbf{A} \cdot (-i\hbar\nabla) - i\hbar\nabla \cdot \mathbf{A})e\phi + \frac{e^2}{2mc^2}\mathbf{A}^2,
\end{aligned} \tag{17.14}$$

where we have used the fact that

$$\hat{\mathbf{p}} \cdot \mathbf{A} = \mathbf{A} \cdot \hat{\mathbf{p}} - i\hbar\nabla \cdot \mathbf{A}. \tag{17.15}$$

This follows from the commutator

$$\begin{aligned}
[\hat{p}_i, A_i(\mathbf{r}, t)] &= -i\hbar\frac{\partial A_i(\mathbf{r}, t)}{\partial x_i} \\
&= -i\hbar\nabla \cdot \mathbf{A}.
\end{aligned} \tag{17.16}$$

Since ϕ is independent of time, it follows from the Lorentz gauge that

$$\nabla \cdot \mathbf{A} = 0.$$

Hence

$$H = -\frac{\hbar^2}{2m}\nabla^2 - \frac{e}{2mc}(2\mathbf{A} \cdot (-i\hbar\nabla)) + e\phi + \frac{e^2}{2mc^2}\mathbf{A}^2. \tag{17.17}$$

We note from Eq. (17.13) that the Schrödinger equation for a free particle

$$-\frac{\hbar^2}{2m}\nabla^2\psi = i\hbar\frac{\partial\psi}{\partial t} \tag{17.18}$$

goes over to

$$-\frac{\hbar^2}{2m}\left(\nabla - \frac{i}{\hbar}\frac{e}{c}\mathbf{A}\right)^2\psi = i\hbar\left(\frac{\partial}{\partial t} + \frac{ie}{\hbar}\phi\right)\psi \tag{17.19}$$

for a particle of charge e in an electromagnetic field.

We also note that $\rho = \psi^*\psi = |\psi|^2$ gives the probability per unit volume of finding the particle at $\mathbf{r}$ and when we find the particle at some place the entire charge is there. The probability is conserved and we have the equation

$$\frac{\partial \rho}{\partial t} + \nabla \cdot \mathbf{j} = 0, \qquad (17.20)$$

where

$$\mathbf{j} = \frac{i\hbar}{2mc} \left(\psi^* \nabla \psi - (\nabla \psi^*) \psi \right) + \frac{e}{2mc} \mathbf{A} \psi^* \psi \qquad (17.21)$$

is the probability current. If we have a situation in which ψ is the wave function for each of a large number of particles with the same charge which are all in the same state, $\psi^* \psi$ can be interpreted as the density of particles, then we can regard $\psi^* \psi$ as electric charge density and $\mathbf{j}$ as electric current.

We note from Eqs. (17.18) and (17.19) that if in the free particle Schrödinger equation (17.18), we make the substitutions

$$\nabla \to \nabla - \frac{i}{\hbar} \frac{e}{c} \mathbf{A} \qquad (17.22a)$$

$$\frac{\partial}{\partial t} \to \frac{\partial}{\partial t} + \frac{ie}{\hbar} \phi, \qquad (17.22b)$$

we get the Schrödinger equation (17.19) for a particle of charge e moving in an electromagnetic field. Also we note that Eq. (17.19) is invariant under gauge transformation (17.4) (with $\nabla^2 \Lambda = 0$) provided that simultaneously we make the transformation

$$\psi \to \psi'(\mathbf{r}, t) = e^{+(ie/\hbar c)\Lambda(\mathbf{r},t)} \psi(\mathbf{r}, t). \qquad (17.23)$$

The last transformation is necessary in order to maintain the gauge invariance. Lastly, we note that the electromagnetic field energy

$$\frac{1}{8\pi} (\mathbf{E}^2 + \mathbf{B}^2)$$

is also invariant under gauge transformation (17.4) and (17.23).

17.2 Gauge Principle: Aharanov–Bohm Effect

In the previous section, we started from classical electromagnetic theory and wrote down the classical Hamiltonian for a charged particle in an electromagnetic field. We then went over to quantum mechanics by the correspondence principle. In order that physical laws as expressed by the Schrödinger equation be invariant under gauge transformations, the wave function ψ must undergo the space-time dependent phase transformation (17.23).

Suppose we did not know how to write the interaction of a charged particle described by wave function ψ with the electromagnetic field $(\mathbf{A}, \phi)$. Can we then formulate a physical principle which would lead to the correct law for this case? The answer is obviously yes from the previous discussion.

Therefore let us demand now that physical laws as expressed by the Schrödinger equation be invariant under the space-time dependent phase transformation (gauge transformation of the second kind) as given in Eq. (17.23) for an electrically charged system.

We see that the free-particle Schrödinger equation (17.18) is not invariant under the gauge transformation (17.23). In order to restore gauge invariance, we have to postulate the fields $(\mathbf{A}, \phi)$ and replace

$$\nabla \to \nabla - \frac{i}{\hbar}\frac{e}{c}\mathbf{A} \equiv \overrightarrow{D} \tag{17.24}$$

$$\frac{\partial}{\partial t} \to \frac{\partial}{\partial t} + \frac{ie}{\hbar}\phi \equiv D_0 \tag{17.25}$$

(Here $\mathbf{D}$ and D_0 are called co-variant derivatives) in Eq. (17.18). Then the resulting Schrödinger equation

$$-\frac{\hbar^2}{2m}\left(\nabla - \frac{i}{\hbar}\frac{e}{c}\mathbf{A}\right)^2 \Psi = i\hbar\left(\frac{\partial}{\partial t} + \frac{ie}{\hbar}\phi\right)\Psi \tag{17.26}$$

is left invariant under gauge transformation (17.23) provided that we assume $\mathbf{A}$ and ϕ undergo transformations simultaneously given in Eq. (17.4). This is because

$$D_0\Psi \to D_0 e^{i\frac{e}{c\hbar}\Lambda}\Psi = e^{i\frac{e}{c\hbar}\Lambda}D_0\Psi,$$

$$\mathbf{D}\Psi \to e^{i\frac{e}{c\hbar}\Lambda}\mathbf{D}\Psi,$$

$$\mathbf{D}^2\Psi \to e^{i\frac{e}{c\hbar}\Lambda}\mathbf{D}\cdot\mathbf{D}\Psi = e^{i\frac{e}{c\hbar}\Lambda}\mathbf{D}^2\Psi.$$

This is the gauge principle. It gives the form of the interaction of a non-relativistic particle with an electromagnetic field correctly. This has now been generalised that all particle interactions at the fundamental level are of gauge character.

Taking the vector potential $\mathbf{A}$ to be independent of time and putting $V = e\phi$, note that if $\Psi^0(\mathbf{r}, t)$ is the solution of the free particle Schrödinger equation, then

$$\Psi(\mathbf{r}, t) = \Psi^0(\mathbf{r}, t)e^{i\gamma(\mathbf{r})} \tag{17.27}$$

where

$$\gamma(\mathbf{r}) = \frac{e}{\hbar c}\int^r A(\mathbf{r}') \cdot dl', \tag{17.28}$$

is a solution of Eq. (17.26) This can be seen by noting that

$$\mathbf{D}\Psi \equiv (\nabla - \frac{ie}{\hbar c}\mathbf{A}))$$

$$= e^{i\gamma(\mathbf{r})}\nabla\Psi^0,$$

$$\mathbf{D}^2\Psi = e^{i\gamma(\mathbf{r})}\nabla^2\Psi^0. \tag{17.29}$$

Thus (17.27) is a solution of Eq. (17.26) where $\mathbf{A}(\mathbf{r}) \neq 0$ if $\Psi^0(\mathbf{r},t)$ satisfies

$$-\frac{(\hbar^2)}{2m}\nabla^2\Psi^0 + V\Psi^0 - i\hbar\frac{\partial\Psi^0}{\partial t}. \tag{17.30}$$

Now $\Psi(\mathbf{r},t)$ can be regarded as a wave function of a particle that goes from one place to another along a certain route where there is a field $\mathbf{A}$ present. $\Psi^0(\mathbf{r},t)$ is the wave function for the same particle going along the same route but with $\mathbf{A} = 0$.The solution (17.27) has some striking physical consequences as shown in the two slit electron interferometry experiment proposed by Aaharonov and Bohm [Fig 17.1]. In this experiment

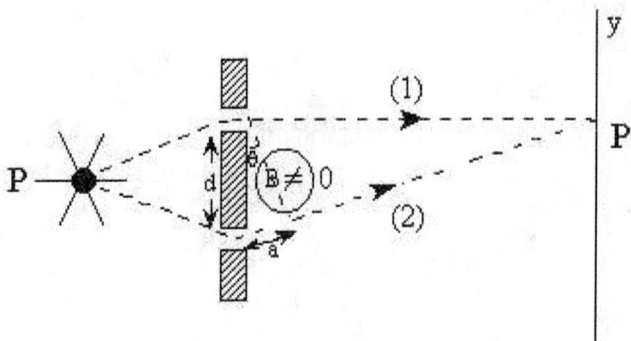

Fig. 17.1   Two paths enclosing a non-zero $\mathbf{B}$ field.

the magnetic field $\mathbf{B}$ (pointing in a horizontal direction out of the paper) is produced by a long solenoid of small cross-section and is confined to the interior of the solenoid so that the two electron beams (1) and (2) can go above and below the $\mathbf{B} \neq 0$ region but stay within the $\mathbf{B} = 0$ region and finally meet in the interference region P'. In the interference region, the wave function of the electron is

$$\Psi = \Psi_1 + \Psi_2 \tag{17.31}$$

so that

$$|\Psi|^2 = |\Psi_1^0|^2 + |\Psi_2^0|^2 + 2|\Psi_1^0||\Psi_2^0|\cos[\gamma_1(\mathbf{r}) - \gamma_2(\mathbf{r})) \tag{17.32a}$$

where

$$\gamma_1 = \gamma_1^0 + \frac{e}{\hbar c} \int_{(1)P}^{P'} \mathbf{A}(\mathbf{r}') \cdot d\mathbf{l}', \tag{17.32b}$$

$$\gamma_2 = \gamma_2^0 + \frac{e}{\hbar c} \int_{(1)P}^{P'} \mathbf{A}(\mathbf{r}') \cdot d\mathbf{l}'. \tag{17.32c}$$

Here γ_1^0 and γ_2^0 are the phases of the wave functions Ψ_1^0 and Ψ_2^0 in the absence of $\mathbf{A}$. The interference pattern is determined by the phase difference

$$\delta(\mathbf{B} \neq 0) = \gamma_1 - \gamma_2$$

$$= \gamma_1^0 - \gamma_2^0 + e \oint_C \mathbf{A}(\mathbf{r}') \cdot d\mathbf{l}'$$

$$= \delta(\mathbf{B} = 0) + \mathbf{\Delta}, \tag{17.33a}$$

where C is the closed path $PP'P$ and

$$\mathbf{\Delta} = e \oint_C \mathbf{A}(\mathbf{r}') \cdot d\mathbf{l}' = e \int_S \mathbf{B} \cdot d\boldsymbol{\sigma} = \frac{e}{\hbar c} \Phi. \tag{17.33b}$$

In Eq. (17.33b) we have used Stokes theorem and put $\mathbf{B} = \nabla \times \mathbf{A}$ and Φ is the magnetic flux through the surface S bounded by the closed path C. Note the important fact that the phase difference is gauge invariant. This is because

$$\oint \nabla \Lambda \cdot d\mathbf{l}' = \oint d\Lambda = 0$$

while the individual phases γ_1 and γ_2 are not. Note also the remarkable fact that the amount of interference can be controlled by varying magnetic flux even though in the idealized experimental arrangement, electrons never enter the region $\mathbf{B} \neq 0$.

Now referring to Fig. 17.1

$$\frac{\text{Phase difference}}{2\pi} = \frac{\text{Path difference}}{\lambda}$$

$$\frac{\delta}{2\pi} = \frac{a}{\lambda} = \frac{1}{\lambda} d \sin\theta \simeq \frac{d}{\lambda} \frac{y}{L} \tag{17.34}$$

where L is the distance of the screen from the slits. Thus from Eq. (17.33a) we see that the diffraction maximum of the interference pattern for $\mathbf{B} \neq 0$ is shifted from that for $\mathbf{B} = 0$ by the amount Δy given by

$$\Delta y = \frac{L\lambda}{d2\pi}[\delta(\mathbf{B} \neq 0) - \delta(\mathbf{B} = 0)]$$

$$= \frac{e\Phi}{\hbar c}\left(\frac{L}{d}\frac{\lambda}{2\pi}\right) \tag{17.35}$$

where we have used Eq. (17.34). This shift in the diffraction maximum, being gauge invariant, should be measurable. In fact the existence and magnitude of Aharanov–Bhom effect has been confirmed to within 5% of the theoretical prediction (17.35) by two qualitatively different experimental arrangements—one involving an electron biprism interferometer while the second used a Josephson-junction interferometer. The following comments are in order.

(i) Measurement of Aharanov–Bhom effect not only verifies the gauge principle in electromagnetism but also quantum mechanics itself since classically the dynamical behavior of electrons is controlled by Lorentz force which is zero when the electrons go through magnetic field free region; yet in quantum mechanics observable effects are seen and depend on the magnetic field in a region inaccessible to the electrons.

(ii) The vector potential $\mathbf{A}$ rather than the fields plays a crucial role as the basic dynamical variable in quantum mechanics.

(iii) By varying $\mathbf{B}$ (and hence Φ) we change the relation phase between the contributions from the 2 paths and move the interference pattern up and down. When $\frac{e}{\hbar c}\Phi = 2n\pi$ or $\Phi = n\phi_0$ [$\phi_0 = 2\pi\hbar c/e = 4.135 \times 10^{-7}$ gauss cm^2], the pattern will return to its initial form, as if there were no field. In other words, an integral multiple of the flux quanta $\phi_0 = \frac{2\pi}{e}\hbar c$ will not make any observable difference to the quantum mechanics of the particle.

The quantization of the magnetic flux [cf. remark (iii) above]

$$\Phi = \frac{2\pi\hbar c}{e}n, \quad n = 0, \pm 1, \pm 2, \cdots. \tag{17.36}$$

was first predicted by F. London. This was confirmed experimentally in superconductors. In a superconductor, due to interactions of the electrons with the vibrations of the atoms in the lattice, there is a small net effective attraction between the electrons. As a result of this attraction, the electrons form bound pairs; such a correlated pair of electrons behave like a quasi Bose particle of charge $2e$. When a superconducting ring is placed in a magnetic field $\mathbf{B}$, inside the material of the ring there is no magnetic field (the absence of the magnetic field in a superconductor is called the Meissner effect). The flux enclosed by the superconducting ring is quantised and instead of Eq. (17.36), we have

$$\Phi = \frac{2\pi\hbar c}{2e}n, \quad n = 0, \pm 1, \pm 2, \cdots. \tag{17.37}$$

Thus we see that the magnetic flux trapped by a superconducting ring should be quantised.

17.3 Landau Levels

These are of great theoretical interest in the study of quantum Hall effect (QHE). Consider a particle of mass m and charge e in a uniform magnetic field along the z-axis for which the Hamiltonian is given in Eq. (17.13) with $\phi = 0$.

First we note that the vector potential $\mathbf{A}$ is given by [see problem 17.1]

$$\mathbf{A} = \frac{1}{2}\mathbf{B} \times \mathbf{r}. \tag{17.38}$$

This satisfies the relation $\mathbf{B} = \nabla \times \mathbf{A}$ and the gauge condition $\nabla \cdot \mathbf{A} = 0$. For $\mathbf{B} \equiv (0, 0, B)$, $\mathbf{A} = \frac{1}{2}(-By, Bx, 0)$.

The selection of $\mathbf{A}$ is not unique. A new vector potential

$$\mathbf{A}' = \mathbf{A} + \nabla\Lambda$$

with

$$\nabla \cdot \mathbf{A}' = \nabla \cdot \mathbf{A} + \nabla^2\Lambda = 0$$

or

$$\nabla^2\Lambda = 0,$$

also gives the same B. This gauge freedom can be exploited to eliminate y-dependence. Select $\Lambda = \frac{B}{2}xy$ so that

$$\nabla\Lambda = \left(\mathbf{i}\frac{\partial}{\partial x} + \mathbf{j}\frac{\partial}{\partial y} + \mathbf{k}\frac{\partial}{\partial Z} \right) \frac{1}{2}Bxy$$

$$= [y\mathbf{i} + x\mathbf{j}]\frac{B}{2}$$

and then,

$$\mathbf{A}' = \mathbf{A} + \nabla\Lambda = Bx\mathbf{j}.$$

Removing the prime, we have

$$\mathbf{A} \equiv (0, Bx, 0). \tag{17.39}$$

Then, the Hamiltonian (17.13) is given by

$$H = \frac{\hat{\mathbf{p}}^2}{2m} - \omega x\hat{p}_y + \frac{1}{2}m\omega^2 x^2 \tag{17.40}$$

where $\omega = \frac{eB}{mc}$. Now

$$[H, \hat{p}_z] = 0, \quad [H, \hat{p}_y] = 0.$$

Thus it is possible to construct wave functions which are simultaneous eigenfunctions of $\hat{p}_y$, $\hat{p}_z$ and H so that we can write

$$\psi(x, y, z) = e^{ip_z z/\hbar} e^{ip_y y/\hbar} u(x) \tag{17.41}$$

where

$$H\psi = \left(\frac{\hat{p}_x^2}{2m} + \frac{p_y^2}{2m} + \frac{p_z^2}{2m} - \omega x p_y + \frac{1}{2} m\omega^2 x^2 \right) \psi$$
$$= E\psi \tag{17.42}$$

or

$$\left(\frac{\hat{p}_x^2}{2m} - \omega x p_y + \frac{1}{2} m\omega^2 x^2 \right) u(x) = \left(E - \frac{p_y^2}{2m} - \frac{p_z^2}{2m} \right) u(x)$$

or

$$\left(\frac{\hat{p}_x^2}{2m} + \frac{1}{2} m\omega^2 (x - \frac{p_y}{m\omega})^2 \right) u(x) = \left(E - \frac{p_z^2}{2m} \right) u(x). \tag{17.43}$$

This is an eigenvalue equation for a simple harmonic oscillator. The energy eigenvalues are given by

$$E - \frac{p_z^2}{2m} = (n + \frac{1}{2})\hbar\omega. \tag{17.44}$$

For the motion in $x - y$ plane, we can put $p_z = 0$. The energy levels labeled by n are called Landau levels. Now a charged particle of charge e in a magnetic field moves in a Larmer circle of radius r

$$\frac{mv^2}{r} = \frac{evB}{c}. \tag{17.45}$$

Classically radius is not fixed. But if we quantize the angular momentum, then

$$mvr = n^* h = n^* (2\pi\hbar) \tag{17.46}$$

so that

$$\frac{eB}{c} r^2 = 2\pi n^* \hbar. \tag{17.47}$$

Thus a quantum electron takes up an area A of order

$$A = \pi r^2 = \frac{2\pi n^* \hbar}{eBc}. \tag{17.48}$$

Hence

$$n^* = \frac{A}{2\pi l_B^2} \tag{17.49}$$

where

$$l_B = \sqrt{\frac{\hbar c}{eB}} \tag{17.50}$$

is the magnetic length and A is the area of the system, which can accommodate one electron per energy level so that there is a total of $n^* = \frac{A}{2\pi l_B^2}$ electrons that fill the lowest Landau level. This has relevance to the Integral Quantum Hall Effect.

17.4 Quantization of Radiation Field

We confine ourselves to a pure radiation field so that we put $\phi = 0$ and the Lorentz condition then becomes $\nabla \cdot \mathbf{A} = 0$. The vector field $\mathbf{A}(\mathbf{r}, t)$ then satisfies the equation

$$\nabla^2 \mathbf{A} - \frac{1}{c^2}\frac{\partial^2 \mathbf{A}}{\partial t^2} = 0. \tag{17.51}$$

The plane wave solution of Eq. (17.51) can be written as

$$\mathbf{C}^\lambda(\mathbf{k}, t)e^{\pm i\mathbf{k}\cdot\mathbf{r}}, \tag{17.52}$$

where λ specifies the polarisation. $\lambda = 1, 2$ for linear polarisation and $\lambda = 1 \pm i2$ for circular polarisation.
Substituting Eq. (17.37) in Eq. (17.36), we get

$$\mathbf{C}^\lambda(\mathbf{k}, t) = -c^2 k^2 \mathbf{C}^\lambda(\mathbf{k}, t). \tag{17.53}$$

With $\omega = c|\mathbf{k}|$, we get the solution

$$\mathbf{C}^\lambda(\mathbf{k}, t) = \mathbf{C}_1^\lambda(\mathbf{k})e^{-i\omega t} + \mathbf{C}_2^\lambda(\mathbf{k})e^{+i\omega t}. \tag{17.54}$$

Hence

$$\mathbf{A}(\mathbf{r}, t) = e^{\pm i\mathbf{k}\cdot\mathbf{r}}\left(\mathbf{C}_1^\lambda(\mathbf{k})e^{-i\omega t} + \mathbf{C}_2^\lambda(\mathbf{k})e^{+i\omega t}\right). \tag{17.55}$$

Since $A(\mathbf{r}, t)$ is a real field, we can write the plane wave solution of Eq. (17.51) as

$$\mathbf{A}(\mathbf{r}, t) = \mathbf{C}^\lambda(\mathbf{k})e^{i\mathbf{k}\cdot\mathbf{r}}e^{-i\omega t} + \mathbf{C}^{*\lambda}(\mathbf{k})e^{-i\mathbf{k}\cdot\mathbf{r}}e^{+i\omega t}. \tag{17.56}$$

Let us write

$$\mathbf{C}^\lambda(\mathbf{k}) = \boldsymbol{\epsilon}^\lambda(\mathbf{k})C_\lambda(\mathbf{k}), \tag{17.57}$$

where $\epsilon^\lambda(\mathbf{k})$ is called the polarisation vector. In what follows we will write simply ϵ^λ instead of $\epsilon^\lambda(\mathbf{k})$.

Suppose that the field $\mathbf{A}(\mathbf{r}, t)$ is enclosed in a cube of volume $V = L^3$, then the functions $\frac{1}{\sqrt{V}} e^{i\mathbf{k}\cdot\mathbf{r}}$, with $k_i = n_i(2\pi/L)$ $(i = 1,2,3)$ form a complete set $(n_i =$ an integer). Thus we can expand $\mathbf{A}(\mathbf{r}, t)$ as

$$\mathbf{A}(\mathbf{r}, t) = \frac{1}{\sqrt{V}} \sum_{\mathbf{k}} \sum_{\lambda=1,2} (\epsilon^\lambda C_\lambda(\mathbf{k}) e^{i\mathbf{k}\cdot\mathbf{r}} e^{-i\omega t} + \epsilon^{\lambda^*} C_\lambda^*(\mathbf{k}) e^{-i\mathbf{k}\cdot\mathbf{r}} e^{i\omega t}),$$

$$(17.58)$$

where we have also summed over polarisations.

We note that

$$\epsilon^\lambda \cdot \epsilon^{\lambda'^*} = \delta_{\lambda\lambda'}, \tag{17.59}$$

$$\frac{1}{V} \int e^{i\mathbf{k}\cdot\mathbf{r}} e^{-i\mathbf{k'}\cdot\mathbf{r}} d^3 r = \delta_{\mathbf{k},\mathbf{k'}}, \tag{17.60}$$

$$\frac{1}{V} \int e^{i\mathbf{k}\cdot\mathbf{r}} e^{i\mathbf{k'}\cdot\mathbf{r}} d^3\mathbf{r} = \delta_{\mathbf{k},-\mathbf{k'}}, \tag{17.61}$$

$$\mathbf{k} \cdot \epsilon^\lambda = \mathbf{k} \cdot \epsilon^{\lambda^*} = 0. \tag{17.62}$$

We now calculate the electric and magnetic fields in terms of $C_\lambda(\mathbf{k})$ and $C_\lambda^*(\mathbf{k})$.

$$\mathbf{E} = -\frac{1}{c} \frac{\partial \mathbf{A}}{\partial t}$$

$$= -\frac{1}{c} \cdot \frac{1}{\sqrt{V}} \sum_{\mathbf{k}} \sum_{\lambda} (\epsilon^\lambda C_\lambda(\mathbf{k}) e^{i\mathbf{k}\cdot\mathbf{r}} (-i\omega) e^{-i\omega t} + \epsilon^{\lambda^*} C_\lambda^*(\mathbf{k})$$

$$\times e^{-i\mathbf{k}\cdot\mathbf{r}} (+i\omega) e^{+i\omega t}). \tag{17.63}$$

$$\mathbf{B} = \mathrm{curl} \mathbf{A}$$

$$= \frac{i}{\sqrt{V}} \sum_{\mathbf{k}} \sum_{\lambda} (\mathbf{k} \times \epsilon^\lambda) C_\lambda(\mathbf{k}) e^{i\mathbf{k}\cdot\mathbf{r}} e^{-i\omega t} - (\mathbf{k} \times \epsilon^{\lambda^*}) C_\lambda^*(\mathbf{k})$$

$$\times e^{-i\mathbf{k}\cdot\mathbf{r}} e^{+i\omega t}). \tag{17.64}$$

Hence the Hamiltonian of the radiation field H is given by

$$H = \frac{1}{8\pi} \int (\mathbf{E}^2 + \mathbf{B}^2) d^3 x$$

$$= \frac{1}{8\pi} \sum_{\mathbf{k}} \sum_{\lambda} (\frac{\omega^2}{c^2} (C_\lambda(\mathbf{k}) C_\lambda^*(\mathbf{k}) + C_\lambda^*(\mathbf{k}) C_\lambda(\mathbf{k}))$$

$$+ \mathbf{k}^2 (C_\lambda(\mathbf{k}) C_\lambda^*(\mathbf{k}) + C_\lambda^*(\mathbf{k}) C_\lambda(\mathbf{k}))$$

$$+ \text{oscillating terms}), \tag{17.65}$$

where we have used Eqs. (17.57)–(17.60) and the fact that

$$(\mathbf{k} \times \boldsymbol{\epsilon}^\lambda) \cdot (\mathbf{k}' \times \boldsymbol{\epsilon}^{\lambda'*}) = \mathbf{k} \cdot \mathbf{k}' \boldsymbol{\epsilon}^\lambda \cdot \boldsymbol{\epsilon}^{\lambda'*} - \mathbf{k} \cdot \boldsymbol{\epsilon}^{\lambda'*} \cdot \mathbf{k}' \cdot \boldsymbol{\epsilon}^\lambda. \qquad (17.66)$$

If we average over time so that oscillating terms drop out, we have

$$H = \frac{1}{4\pi} \sum_{\mathbf{k}} \sum_{\lambda} \frac{\omega^2}{c^2} \left(C_\lambda(\mathbf{k}) C_\lambda^*(\mathbf{k}) + C_\lambda^*(\mathbf{k}) C_\lambda(\mathbf{k}) \right). \qquad (17.67)$$

If we write

$$C_\lambda(\mathbf{k}) = \frac{\sqrt{4\pi c}}{2\omega} \left(P_\lambda(\mathbf{k}) - i\omega Q_\lambda(\mathbf{k}) \right) \qquad (17.68)$$

$$C_\lambda^*(\mathbf{k}) = \frac{\sqrt{4\pi c}}{2\omega} \left(P_\lambda(\mathbf{k}) + i\omega Q_\lambda(\mathbf{k}) \right) \qquad (17.69)$$

then

$$H = \frac{1}{2} \sum_{\mathbf{k}} \sum_{\lambda} \left(P_\lambda^2(\mathbf{k}) + \omega^2 Q_\lambda^2(\mathbf{k}) \right). \qquad (17.70)$$

Comparing Eq. (17.70) with the Hamiltonian of an oscillator $\frac{1}{2m}(p^2 + \omega^2 q^2)$, we see that Hamiltonian for radiation field behaves like a collection of oscillators [cf. Sec. 9.5].

The comparison with Sec. 9.5 suggests that, in order to quantise the radiation field, we replace

$$C_\lambda(\mathbf{k}) \rightarrow \frac{\sqrt{4\pi c}}{2} \sqrt{2\hbar\omega} a_\lambda(\mathbf{k}), \qquad (17.71)$$

$$C_\lambda^*(\mathbf{k}) \rightarrow \frac{\sqrt{4\pi c}}{2\omega} \sqrt{2\hbar\omega} a_\lambda^\dagger(\mathbf{k}), \qquad (17.72)$$

so that

$$H = \frac{1}{2} \sum_{\mathbf{k}} \sum_{\lambda} \hbar\omega \left(a_\lambda(\mathbf{k}) a_\lambda^\dagger(\mathbf{k}) + a_\lambda^\dagger(\mathbf{k}) a_\lambda(\mathbf{k}) \right). \qquad (17.73)$$

Note that $a_\lambda(\mathbf{k})$ and $a_\lambda^\dagger(\mathbf{k})$ are now operators which satisfy the commutation relation

$$\left[a_\lambda(\mathbf{k}), a_{\lambda'}^\dagger(\mathbf{k}) \right] = \delta_{\mathbf{k}\mathbf{k}'}\delta_{\lambda\lambda'}. \qquad (17.74)$$

In analogy with Eqs. (9.120a,b) of Sec. 9.5, we have

$$a_\lambda(\mathbf{k})|0\rangle = 0, \qquad (17.75)$$

$$a_\lambda^\dagger(\mathbf{k})|n_\lambda(\mathbf{k})\rangle = \sqrt{n_\lambda(\mathbf{k}) + 1}|n_\lambda(\mathbf{k}) + 1\rangle, \qquad (17.76)$$

$$a_\lambda(\mathbf{k})|n_\lambda(\mathbf{k})\rangle = \sqrt{n_\lambda(\mathbf{k})}|n_\lambda(\mathbf{k}) - 1\rangle, \qquad (17.77)$$

$$a_\lambda^\dagger(\mathbf{k})a_\lambda(\mathbf{k})|n_\lambda(\mathbf{k})\rangle = n_\lambda(\mathbf{k})|n_\lambda(\mathbf{k})\rangle, \qquad (17.78)$$

where

$$|n_\lambda(\mathbf{k})\rangle = \frac{1}{\sqrt{n_\lambda(\mathbf{k})!}} \left(a_\lambda^\dagger(\mathbf{k})\right)^{n_\lambda(\mathbf{k})} |0\rangle. \tag{17.79}$$

Now using the commutation relation (17.74), we can write

$$H = \sum_{\mathbf{k}} \sum_{\lambda} \hbar\omega \left(N_\lambda(\mathbf{k}) + \frac{1}{2}\right), \tag{17.80}$$

where

$$N_\lambda(\mathbf{k}) = a_\lambda^\dagger(\mathbf{k})a_\lambda(\mathbf{k}), \tag{17.81}$$

is called the number operator, because it has eigenvalue $n_\lambda(\mathbf{k})$, where $n_\lambda(\mathbf{k})$ is an integer. The n number of photons, of wave number $\mathbf{k}$ and polarisation λ is represented by a state $|n_\lambda(\mathbf{k})\rangle$:

$$H|n_\lambda(\mathbf{k})\rangle = \hbar\omega \left(n_\lambda(\mathbf{k}) + \frac{1}{2}\right) |n_\lambda(\mathbf{k})\rangle. \tag{17.82}$$

It is thus clear that we have the following interpretation. The state

$$a_\lambda^\dagger(\mathbf{k})|0\rangle \tag{17.83}$$

represents one photon with polarisation λ and wave number $\mathbf{k}$. The state (17.79) represents n photons each with polarisation λ and wave number $\mathbf{k}$. The operators $a_\lambda^\dagger(\mathbf{k})$ and $a_\lambda(\mathbf{k})$ act as creation and annihilation operators respectively. We note further that

$$a_\lambda^\dagger(\mathbf{k})a_{\lambda'}^{\ \dagger}(\mathbf{k})'|0\rangle \tag{17.84}$$

represents a state with two photons, one with wave number $\mathbf{k}$ and polarisation λ and the other with wave number $\mathbf{k}'$ and polarisation λ'. Also from Eqs. (17.76) and (17.77), we have

$$\langle n_\lambda(\mathbf{k}) + 1|a_\lambda^\dagger(\mathbf{k})|n_\lambda(\mathbf{k})\rangle = \sqrt{n_\lambda(\mathbf{k}) + 1}, \tag{17.85a}$$

$$\langle n_\lambda(\mathbf{k}) - 1|a_\lambda(\mathbf{k})|n_\lambda(\mathbf{k})\rangle = \sqrt{n_\lambda(\mathbf{k})}. \tag{17.85b}$$

Finally Eq. (17.58) in terms of creation and annihilation operators is given by

$$\mathbf{A}(\mathbf{r}, t) = \frac{\sqrt{4\pi\hbar c}}{\sqrt{V}} \sum_{\mathbf{k}} \sum_{\lambda} \frac{1}{\sqrt{2\omega}} \Big(\epsilon^\lambda a_\lambda(\mathbf{k})e^{-i\mathbf{k}\cdot\mathbf{r}}e^{-i\omega t}$$
$$+\epsilon^{\lambda*} a_\lambda^\dagger(\mathbf{k})e^{-i\mathbf{k}\cdot\mathbf{r}}e^{i\omega t}\Big). \tag{17.86}$$

17.5 Perturbation Induced by Electromagnetic Field

Let the Hamiltonian of the atomic electron (charge $e < 0$) in a static potential $V(r)$ be

$$H_0 = \frac{\hat{\mathbf{p}}^2}{2m} + V(r). \tag{17.87}$$

Then the Hamiltonian of the atomic electron in an electromagnetic field $\mathbf{A}(\mathbf{r}, t)$ (pure radiation field) is given by (see Eq. (17.17))

$$H = \frac{\hat{\mathbf{p}}^2}{2m} + V(r) - \frac{e}{mc}\mathbf{A} \cdot \hat{\mathbf{p}} + \frac{e^2}{2mc^2}\mathbf{A}^2$$

$$= H_0 - \frac{e}{mc}\mathbf{A} \cdot \hat{\mathbf{p}} + \frac{e^2}{2mc^2}\mathbf{A}^2. \tag{17.88}$$

We call H_0 the unperturbed Hamiltonian. The other two terms in Eq. (17.86), we regard as a perturbation. Since $\frac{e^2}{2mc^2}\mathbf{A}^2$ is of higher order as compared with the term $\frac{e}{mc}\mathbf{A} \cdot \hat{\mathbf{p}}$ (because e^2 is small) we neglect this term. Thus the perturbation potential is given by

$$\lambda V(t) = H_{int} = -\frac{e}{mc}\mathbf{A} \cdot \hat{\mathbf{p}}. \tag{17.89}$$

Substituting Eq. (17.86) in Eq. (17.89), we have

$$H_{int} = -\frac{ec\sqrt{4\pi\hbar}}{mc\sqrt{V}} \sum_{\mathbf{k}'} \sum_{\lambda'} \frac{1}{\sqrt{2\omega'}} (a_{\lambda'}(\mathbf{k}')e^{i\mathbf{k}'\cdot\mathbf{r}}e^{-i\omega't}$$

$$\times \boldsymbol{\epsilon}^{\lambda'} \cdot \hat{\mathbf{p}} + a_{\lambda'}^{\dagger}(\mathbf{k}')e^{-i\mathbf{k}'\cdot\mathbf{r}}e^{i\omega't}\boldsymbol{\epsilon}^{\lambda'*} \cdot \hat{\mathbf{p}}). \tag{17.90}$$

We take the matrix elements of (17.89), between initial state $|a, n_\lambda(\mathbf{k})\rangle$ and final state $|b, n_\lambda(\mathbf{k})-1\rangle$. The initial state contains n photons of wave number $\mathbf{k}$ and polarisation λ and the final state contains $(n-1)$ photons with the same wave number and polarisation. We thus have, using Eq. (17.90)

$$T_{ba}(t) = \langle b, n_\lambda(\mathbf{k}) - 1|H_{int}|b\rangle)$$

$$= -\frac{e}{mc}\langle b, n_\lambda(\mathbf{k}) - 1|\mathbf{A} \cdot \hat{\mathbf{p}}|a, n_\lambda(\mathbf{k})\rangle$$

$$= -\frac{e}{mc}\frac{c\sqrt{4\pi\hbar}}{\sqrt{V}} \sum_{\mathbf{k}'} \sum_{\lambda'} \frac{1}{\sqrt{2\omega'}}\langle b, n_\lambda(\mathbf{k}) - 1|$$

$$\times (a_{\lambda'}(\mathbf{k}')e^{i\mathbf{k}'\cdot\mathbf{r}}e^{-i\omega't}\boldsymbol{\epsilon}^{\lambda'} \cdot \hat{\mathbf{p}} + a_{\lambda'*}^{\dagger}(\mathbf{k}')$$

$$\times e^{-i\mathbf{k}'\cdot\mathbf{r}}e^{+i\omega't}\boldsymbol{\epsilon}^{\lambda'*} \cdot \hat{\mathbf{p}})|a, n_\lambda(\mathbf{k})\rangle. \tag{17.91}$$

Using Eq. (17.77), we have

$$\langle n_\lambda(\mathbf{k}) - 1|a_{\lambda'}^\dagger(\mathbf{k}) = \delta_{\lambda\lambda'}\delta_{\mathbf{kk}'}\sqrt{n_\lambda(\mathbf{k}) - 1}\langle n_\lambda(\mathbf{k}) - 2|$$
$$a_{\lambda'}(\mathbf{k}')|n_\lambda(\mathbf{k})\rangle = \delta_{\lambda\lambda'}\delta_{\mathbf{kk}'}\sqrt{n_\lambda(\mathbf{k})}|n_\lambda(\mathbf{k}) - 1\rangle,$$

so that Eq. (17.91) gives

$$T_{ba}(t) = -\frac{e}{m}\sqrt{\frac{4\pi\hbar\, n_\lambda(\mathbf{k})}{V}\frac{}{2\omega}}\langle b|e^{i\mathbf{k}\cdot\mathbf{r}}\boldsymbol{\epsilon}^\lambda\cdot\hat{\mathbf{p}}|a\rangle e^{-i\omega t}. \qquad (17.92)$$

These are the matrix elements for the absorption of a single photon of frequency ω by a charged particle from an initial state that already has n photons of frequency ω and polarisation λ.

Similarly, using $\langle n_\lambda(\mathbf{k}) + 1|a_{\lambda'}^\dagger(\mathbf{k}') = \delta_{\lambda\lambda'}\delta_{\mathbf{kk}'}\sqrt{n_\lambda(\mathbf{k}) + 1}\langle n_\lambda(\mathbf{k})|$, etc,

$$T_{ba}(t) = \frac{e}{m}\langle b, n_\lambda(\mathbf{k}) + 1|\mathbf{A}\cdot\hat{\mathbf{p}}|a, n_\lambda(\mathbf{k})\rangle$$
$$= \frac{e}{m}\sqrt{\frac{4\pi\hbar(n_\lambda(\mathbf{k}) + 1)}{V2\omega}}\langle b|e^{-i\mathbf{k}\cdot\mathbf{r}}\boldsymbol{\epsilon}^{\lambda*}\cdot\hat{\mathbf{p}}|a\rangle e^{i\omega t}, \qquad (17.93)$$

gives the matrix elements for emission of a single photon of frequency ω by a charged particle into a final state that has $n + 1$ photons, that is, from an initial state with n photons of frequency ω.

Thus we see that the vector potential $\mathbf{A}(\mathbf{r}, t)$ acts as an effective potential

$$\sqrt{\frac{2\pi c^2\hbar n_\lambda(\mathbf{k})}{V\omega}}\boldsymbol{\epsilon}^\lambda e^{i(\mathbf{k}\cdot\mathbf{r} - \omega t)} \qquad (17.94\text{a})$$

for absorption of a photon of frequency ω and for emission of a photon it acts as

$$\sqrt{\frac{2\pi c^2\hbar(n_\lambda(\mathbf{k}) + 1)}{V\omega}}\boldsymbol{\epsilon}^{\lambda*} e^{-i(\mathbf{k}\cdot\mathbf{r} - \omega t)}. \qquad (17.94\text{b})$$

17.6 Stimulated Emission and Absorption of Radiation

The transition probability from the state a to state b is given by

$$|C_b(\infty)|^2,$$

where [see Eq. (13.12)]

$$C_b(\infty) = -i\frac{\lambda}{\hbar}\int_{-\infty}^{\infty} e^{i(E_b - E_a)t/\hbar}V_{ba}(t)dt. \qquad (17.95\text{a})$$

Using the formula

$$\int_{-\infty}^{\infty} e^{i(E_b - E_a \mp \hbar\omega)t/\hbar}dt = 2\pi\hbar\delta(E_b - E \mp \hbar\omega), \qquad (17.95\text{b})$$

we have from Eqs. (17.95)(with $\lambda = -e$), (17.94a) and (17.94b), for absorption or emission of a photon of frequency ω and polarisation λ:

$$C_b^{\text{absorption}}(\infty) = \frac{ie}{m\hbar}\sqrt{\frac{n_\lambda(\mathbf{k})2\pi\hbar}{\omega V}}2\pi\hbar\delta(E_b - E_a - \hbar\omega)$$
$$\times \langle b|e^{i\mathbf{k}\cdot\mathbf{r}}\boldsymbol{\epsilon}^\lambda \cdot \hat{\mathbf{p}}|a\rangle, \tag{17.96a}$$

$$C_b^{\text{emission}}(\infty) = \frac{ie}{m\hbar}\sqrt{\frac{(n_\lambda(\mathbf{k}) + 1)2\pi\hbar}{\omega V}}2\pi\hbar\delta(E_b - E_a + \hbar\omega)$$
$$\times \langle b|e^{-i\mathbf{k}\cdot\mathbf{r}}\boldsymbol{\epsilon}^{\lambda^*} \cdot \hat{\mathbf{p}}|a\rangle. \tag{17.96b}$$

The transition probability per unit time from the state $|a\rangle$ to the state $|b\rangle$ with the absorption of a photon is given by

$$W_{ba}^{abs} = \frac{1}{t}|C_b^{abs}(\infty)|^2$$

$$= \frac{e^2}{m^2\hbar^2}\frac{n_\lambda(\mathbf{k})}{\omega V}\frac{2\pi\hbar}{t}[2\pi\hbar\delta(E_b - E_a - \hbar\omega)]^2$$
$$\times |\langle b|e^{i\mathbf{k}\cdot\mathbf{r}}\boldsymbol{\epsilon}^\lambda \cdot \hat{\mathbf{p}}|a\rangle|^2. \tag{17.97a}$$

Now

$$[\delta(E_b - E_a \mp \hbar\omega)]^2 = \frac{1}{\hbar}\delta(0)\delta(E_b - E_a \mp \hbar\omega)$$

$$= \lim_{t\to\infty}\frac{t}{2\pi\hbar}\delta(E_b - E_a \mp \hbar\omega).$$

Thus

$$W_{ba}^{abs} = [\frac{n_\lambda(\mathbf{k})c}{V}]\frac{4\pi^2e^2}{m^2\omega c}|\langle b|e^{i\mathbf{k}\cdot\mathbf{r}}\boldsymbol{\epsilon}^\lambda \cdot \hat{\mathbf{p}}|a\rangle|^2\delta(\epsilon_b - \epsilon_a - \hbar\omega). \tag{17.97b}$$

Now $\frac{n_\lambda(\mathbf{k})c}{V}$ is the incident proton flux so that absorption cross section is given by

$$\sigma_{ba}^{abs}(\omega) = W_{ba}^{abs}/(\text{Flux})_{\text{in}}$$

$$= \frac{4\pi^2e^2}{m^2\omega c}|\langle b|e^{i\mathbf{k}\cdot\mathbf{r}}\boldsymbol{\epsilon}^\lambda \cdot \hat{\mathbf{p}}|a\rangle|^2\delta(E_b - E_a - \hbar\omega). \tag{17.97c}$$

Similarly

$$W_{bu}^{\text{emission}} = \frac{[n_\lambda(\mathbf{k}) + 1]}{V}\frac{4\pi^2e^2}{m^2\omega}\langle b|e^{-i\mathbf{k}\cdot\mathbf{r}}\boldsymbol{\epsilon}^{\lambda^*} \cdot \hat{\mathbf{p}}|a\rangle|^2\delta(E_b - E_a + \hbar\omega). \tag{17.97d}$$

In actual physical situations, the photon momentum $\hbar\mathbf{k}$ does not uniquely specify the photon state. The photon is in general emitted or absorbed in some momentum interval $\hbar\mathbf{k}$ and $\hbar(\mathbf{k} + d\mathbf{k})$. The number of states in this interval is given by

$$\frac{V}{(2\pi)^3}d^3k = \frac{V}{(2\pi\hbar)^3}d^3p = \frac{V}{(2\pi\hbar)^3}\left(p^2\frac{dp}{dE}\right)dEd\Omega$$

$$= \rho(E)dE = \rho(\omega)\hbar d\omega, \tag{17.98a}$$

where

$$\rho(E) \equiv \rho(\omega) = \frac{V}{(2\pi\hbar)^3}\left(p^2\frac{dp}{dE}\right)d\Omega = \frac{V}{(2\pi c)^3\hbar}\omega^2 d\Omega \qquad (17.98b)$$

is called the number of states per unit energy interval.

Hence using Eqs. (17.97) and (17.98), the total transition probability per unit time from an atomic state a to a state b with the emission of a photon is given by

$$W_{ba}^{\text{emission}} = \frac{4\pi^2 e^2}{m^2}\int \frac{(n_\lambda(\mathbf{k})+1)}{\omega V}\frac{V}{(2\pi c)^3}\omega^2 d\omega d\Omega$$

$$\times\delta(E_b - E_a + \hbar\omega)|\langle b|e^{-i\mathbf{k}\cdot\mathbf{r}}\boldsymbol{\epsilon}^{\lambda^*}\cdot\hat{\mathbf{p}}|a\rangle|^2$$

$$= \frac{e^2\omega}{2\pi m^2\hbar c^3}(n_\lambda(\mathbf{k})+1)|\langle b|e^{-i\mathbf{k}\cdot\mathbf{r}}E^{\lambda^*}\cdot\hat{\mathbf{p}}|a\rangle|^2 d\Omega \qquad (17.99a)$$

with

$$\omega = (E_a - E_b)/\hbar. \qquad (17.99b)$$

Similarly the total transition probability from an atomic state a to a state b with the absorption of a photon is given by

$$W_{ba}^{\text{absorption}} = \frac{e^2\omega}{2\pi m^2\hbar c^3}n_\lambda(\mathbf{k})|\langle b|e^{i\mathbf{k}\cdot\mathbf{r}}\boldsymbol{\epsilon}^{\lambda}\cdot\hat{\mathbf{p}}|a\rangle|^2 d\Omega \qquad (17.100a)$$

with

$$\omega = (E_b - E_a)/\hbar, \qquad (17.100b)$$

which is the Bohr frequency condition for the absorption of radiation of frequency $\hbar\omega$. Using the above formulation, we can derive Planck's radiation law from the point of view of quantum field theory. Suppose we have atoms and a radiation field which can freely exchange energy by the reversible process

$$a \rightleftarrows \gamma + b \qquad (17.101)$$

in such a way that thermal equilibrium is established. In the above transition we have taken the atom to have two states of energy, E_a and E_b with $E_a > E_b$. If the population of the upper and lower atomic levels are denoted by $N(a)$ and $N(b)$ respectively, we have the equilibrium condition

$$N(a)W_{ba}^{\text{emission}} = N(b)W_{ab}^{\text{absorption}}. \qquad (17.102)$$

Now from Eqs. (17.99) and (17.100), we have

$$\frac{W_{ba}^{\text{emission}}}{W_{ab}^{\text{absorption}}} = \frac{(n_\lambda(\mathbf{k})+1)|\langle b|e^{-i\mathbf{k}\cdot\mathbf{r}}\boldsymbol{\epsilon}^{\lambda^*}\cdot\hat{\mathbf{p}}|a\rangle|^2}{n_\lambda(\mathbf{k})|\langle a|e^{i\mathbf{k}\cdot\mathbf{r}}\boldsymbol{\epsilon}^{\lambda}\cdot\hat{\mathbf{p}}|b\rangle^*|^2}. \qquad (17.103)$$

But

$$\langle b|e^{-i\mathbf{k}\cdot\mathbf{r}}\boldsymbol{\epsilon}^{\lambda*}\cdot\hat{\mathbf{p}}|a\rangle = \langle a|\hat{\mathbf{p}}^\dagger\cdot\boldsymbol{\epsilon}^\lambda e^{i\mathbf{k}\cdot\mathbf{r}}|b\rangle^*$$
$$= \langle a|e^{i\mathbf{k}\cdot\mathbf{r}}\boldsymbol{\epsilon}^\lambda\cdot\hat{\mathbf{p}}|b\rangle^*. \tag{17.104}$$

From statistical mechanics, we know

$$\frac{N(b)}{N(a)} = \frac{e^{-E_b/kT}}{e^{-E_a/kT}} = e^{\hbar\omega/kT}. \tag{17.105}$$

But using Eq. (17.104), we have from Eqs. (17.102) and (17.103)

$$\frac{N(b)}{N(a)} = \frac{W_{ba}^{\text{emission}}}{W_{ab}^{\text{absorption}}} = \frac{n_\lambda(\mathbf{k})+1}{n_\lambda(\mathbf{k})}. \tag{17.106}$$

Using Eqs. (17.105) and (17.106), we obtain

$$n_\lambda(\mathbf{k}) = \frac{1}{e^{\hbar\omega/kT}-1}. \tag{17.107}$$

The above result is for photon states for which $\hbar\omega = E_a - E_b$. Suppose the radiation field is enclosed in a cavity. The walls of the cavity are made up of various kinds of atoms and are capable of absorbing and re-emitting photons of any energy. The energy of the radiation field, per unit volume, in the angular frequency interval ω and $\omega + d\omega$ is given by

$$U(\omega)d\omega = \frac{1}{V}\frac{V}{(2\pi)^3}d^3k\,2\frac{1}{e^{\hbar\omega/kT}-1}\hbar\omega$$
$$= \frac{1}{(2\pi c)^3}\omega^2 d\omega\,4\pi\frac{2\hbar\omega}{e^{\hbar\omega/kT}-1}$$
$$= \frac{8\pi\hbar}{c^3}\left(\frac{\omega}{2\pi}\right)^3\left(\frac{1}{e^{\hbar\omega/kT}-1}\right)d\omega. \tag{17.108}$$

The factor of 2 in the above expression comes from the two polarisation states and in deriving the above formula we have used Eq. (17.98a). The energy distribution per unit frequency per unit volume ($\omega = 2\pi\nu$)

$$U(\nu) = U(\omega)\frac{d\omega}{d\nu}$$
$$= \frac{8\pi h\nu^3}{c^3}\frac{1}{e^{h\nu/kT}-1}. \tag{17.109}$$

This is Planck's law.

We note from Eqs. (17.99) and (17.100), that in the presence of a large number of photons of a given wave length ($n_\lambda(\mathbf{k})$ large) transition rates (stimulated emission of light) corresponding to that wave length will be enormously enhanced. Thus if many atoms can be raised to a given excited state in the environment of the right kind of photons, then they will decay in a very short time, thus giving rise to an intense, coherent, and monochromatic pulse of radiation. In a laser, we have just this. But at equilibrium, we have

$$N(a) = N(b)e^{-\hbar\omega/kT} \qquad (17.110)$$

and it is not possible to have $N(a) > N(b)$. Some special techniques are to be used for population inversion in which the number $N(a)$ in the excited state is very much greater than the number $N(b)$ in the lower state so that the number in the lower state is practically zero. Then light which has the frequency $\hbar\omega = E_a - E_b$ will not be strongly absorbed, because there are not many atoms in state b to absorb it. On the other hand, when that light is present, it will induce the emission from this upper state. So, if we have many atoms in the upper state, the moment the atoms begin to emit, more would be caused to emit and we will have a strong stimulated emission.

17.7 Spontaneous Emission of Radiation: Dipole Approximation

From Eq. (17.99a), we see that even in the absence of radiation field $(n_\lambda(\mathbf{k}) = 0)$, the probability for radiative transition from an atomic state a to a state b is not zero. This is called spontaneous emission of radiation. From Eq. (17.99), the total transition probability per unit time for spontaneous emission from an atomic state a to a state b is given by

$$W_{ba} = \frac{e^2\omega}{2\pi m^2\hbar c^3}|\langle b|e^{-i\mathbf{k}\cdot\mathbf{r}}\boldsymbol{\epsilon}^{(\lambda^*)}\cdot\hat{\mathbf{p}}|a\rangle|^2 d\Omega. \qquad (17.111)$$

We now expand the exponential $e^{-i\mathbf{k}\cdot\mathbf{r}}$

$$e^{-i\mathbf{k}\cdot\mathbf{r}} = 1 - i\mathbf{k}\cdot\mathbf{r}\cdots. \qquad (17.112)$$

If $|\mathbf{k}\cdot\mathbf{r}| \sim kr$ is small, i.e. if

$$|kr| \ll 1, \qquad (17.113)$$

then it is a good approximation to use only the first term. This is called the dipole approximation and it holds if Eq. (17.113) is satisfied. Now r is of the order of the atomic radius (Z is the atomic number)

$$r \sim a = \frac{\hbar^2}{mZe^2} = \frac{\hbar}{mcZ\alpha} \qquad (17.114)$$

and $\hbar\omega$ is given by

$$\hbar\omega = E_a - E_b = -\frac{mc^2 Z^2 \alpha^2}{2}\left(\frac{1}{n_a^2} - \frac{1}{n_b^2}\right)$$

$$\approx \frac{1}{2}mc^2 Z^2 \alpha^2, \tag{17.115a}$$

so that

$$|k| = \frac{\omega}{c} \approx \frac{\frac{1}{2}mcZ^2\alpha^2}{\hbar}. \tag{17.115b}$$

Hence

$$kr \sim \frac{1}{2}Z\alpha$$

$$\sim \frac{1}{2}Z\frac{1}{137}, \tag{17.115c}$$

where $\alpha = \frac{e^2}{\hbar c} = \frac{1}{137}$ is the fine structure constant. Thus the dipole approximation is valid, unless Z is very high.

The transition probability W_{ba} in the electric dipole $(E1)$ approximation is given by

$$W_{ba} = \frac{\alpha\omega}{2\pi m^2 c^2}|\langle b|\boldsymbol{\epsilon}^{\lambda^*}\cdot\hat{\mathbf{p}}|a\rangle|^2 d\Omega. \tag{17.116}$$

Now

$$i\hbar\frac{d\hat{\mathbf{r}}}{dt} = [\hat{\mathbf{r}}, H]$$

$$= [\hat{\mathbf{r}}, H_0 + \lambda V]$$

$$= [\hat{\mathbf{r}}, H_0] + \text{terms of order } \lambda \text{ or } e. \tag{17.117}$$

We can neglect these terms since we are calculating W_{ba} to order e. Thus we have

$$\langle b|\hat{\mathbf{p}}|a\rangle = \frac{m}{i\hbar}\langle b|[\mathbf{r}, H_0]|a\rangle + O(e)$$

$$\approx -\frac{im}{\hbar}(\epsilon_a - \epsilon_b)\langle b|\mathbf{r}|u\rangle$$

$$= -im\omega\langle b|\mathbf{r}|a\rangle. \tag{17.118}$$

Hence

$$W_{ba} = \frac{\alpha\omega^3}{2\pi c^2}|\langle b|\mathbf{r}|a\rangle\cdot\boldsymbol{\epsilon}^{\lambda^*}|^2 d\Omega$$

$$= \frac{\alpha E_\gamma^3}{2\pi\hbar^3 c^2}|\langle b|\mathbf{r}|a\rangle\cdot\boldsymbol{\epsilon}^{\lambda^*}|^2 d\Omega, \tag{17.119}$$

where

$$E_\gamma = \hbar\omega = (E_a - E_b). \tag{17.120}$$

Since $\boldsymbol{\epsilon}^{(1)}$, $\boldsymbol{\epsilon}^{(2)}$, $e_k (= \frac{\mathbf{k}}{|\mathbf{k}|})$ form an orthonormal set, satisfying the relations:

$$\sum_{\lambda=1}^{2} \boldsymbol{\epsilon}^{(\lambda)} \cdot \boldsymbol{\epsilon}^{(\lambda)} = 2,$$

$$\sum_{\lambda=1}^{2} \epsilon_i^{(\lambda)} \cdot \epsilon_j^{(\lambda)} = \delta_{ij} - \frac{k_i k_j}{k^2}, \qquad (17.121)$$

we can express $\mathbf{r}$ as

$$\mathbf{r} = r[\sin\theta \cos\phi \, \boldsymbol{\epsilon}^{(1)} + \sin\theta \sin\phi \, \boldsymbol{\epsilon}^{(2)} + \cos\theta \, \mathbf{e}_k].$$

Thus

$$\mathbf{r} \cdot \boldsymbol{\epsilon}^{*(+)} = -\mathbf{r} \cdot \boldsymbol{\epsilon}^{(-)} = -\frac{r}{\sqrt{2}} \sin\theta e^{-i\phi} = -\sqrt{\frac{4\pi}{3}} r Y_{1,-1}$$

$$\mathbf{r} \cdot \boldsymbol{\epsilon}^{*(-)} = -\mathbf{r} \cdot \boldsymbol{\epsilon}^{(+)} = \frac{r}{\sqrt{2}} \sin\theta e^{i\phi} = -\sqrt{\frac{4\pi}{3}} r Y_{1,1} \qquad (17.122a)$$

for circularly polarized light, $\boldsymbol{\epsilon}^{(\pm)} = \mp \frac{\boldsymbol{\epsilon}^{(1)} \pm i\boldsymbol{\epsilon}^{(2)}}{\sqrt{2}}$; for plane polarized light, we take $\boldsymbol{\epsilon}$ along z-axis and $\mathbf{k}$ in $x-y$ plane and

$$\mathbf{r} \cdot \boldsymbol{\epsilon}^{\lambda^*} = r\cos\theta = \sqrt{\frac{4\pi}{3}} r Y_{10}. \qquad (17.122b)$$

Since Y_{lm} has parity $(-1)^l$ and $l=1$ in Eqs. (17.122), we can say that dipole photon carries an angular momentum of one unit, and negative parity. Thus parity conservation give:

$$\Pi_a = -\Pi_b \qquad (17.123)$$

(The initial and final states have opposite parities).

On the other hand from the rules of addition of angular momentum in quantum mechanics

$$|J_a - J_b| \leq 1 \leq J_a + J_b. \qquad (17.124)$$

The above relation implies that the change

$$\Delta J = J_b - J_a \qquad (17.125)$$

must be restricted to $\Delta J = 0, \pm 1$; $J_a = J_b = 0$ is strictly forbidden since the emitted or absorbed photon (spin 1) cannot have zero angular momentum. The above are the parity and angular momentum selection rules for an electric dipole transition or $E1$ transition.

The above selection rules also follow directly from the electron dipole matrix elements given in Eq. (17.119). We note that initial and final states

$|a\rangle$ and $|b\rangle$ for an atomic electron for a hydrogen-like atom can be specified by

$$|a\rangle = |n_i l_i m_i\rangle, \tag{17.126a}$$

$$|b\rangle = |n_f l_f m_f\rangle, \tag{17.126b}$$

$$\langle \mathbf{r}|a\rangle = u_{n_i l_i m_i}(\mathbf{r}) = R_{n_i l_i}(r) Y_{l_i m_i}(\theta, \phi), \tag{17.126c}$$

$$\langle \mathbf{r}|b\rangle = u_{n_f l_f m_f}(\mathbf{r}) = R_{n_f l_f}(r) Y_{l_f m_f}(\theta, \phi). \tag{17.126d}$$

Hence, using Eqs. (15.122),

$$\langle b|\mathbf{r} \cdot \boldsymbol{\epsilon}_\lambda^*|a\rangle = \langle n_f l_f m_f|\mathbf{r} \cdot \boldsymbol{\epsilon}_\lambda^*|n_i l_i m_i\rangle$$

$$= \int dr\, r^2 R_{n_f l_f}(\mathbf{r}) R_{n_i l_i}(\mathbf{r}) \int d\Omega Y_{l_f m_f}^*(\theta, \phi) Y_{1m}(\theta, \phi) Y_{l_i m_i}(\theta, \phi). \tag{17.127}$$

Now from Eq. (11.165), the integral involving spherical harmonics is proportional to the Clebsch–Gordon coefficient $\langle 1l_i m m_i | l_f m_f 1 l_i \rangle$ which vanishes unless

$$l_f = l_i \pm 1$$

$$m_f = m + m_i \tag{17.128}$$

i.e. the selection rules for EI transition are

$$\Delta l = l_f - l_i = \pm 1$$

$$\Delta m = m_f - m_i = m = \pm 1, 0, \tag{17.129}$$

and parity must change.

Sometimes, the selection rules do not allow $E1$ transition between certain states. This occurs when $\langle b|\mathbf{r}|a\rangle = 0$, between states a and b. In this case, we have to take the term $-i\mathbf{k} \cdot \mathbf{r}$ in Eq. (17.112) seriously, so that, transition matrix elements for the radiative transition are given by

$$\langle b|(\mathbf{k} \cdot \mathbf{r})(\boldsymbol{\epsilon}^{\lambda^*} \cdot \hat{\mathbf{p}})|a\rangle.$$

We can write

$$(\mathbf{k} \cdot \mathbf{r})(\boldsymbol{\epsilon}^{\lambda^*} \cdot \hat{\mathbf{p}}) = \langle b|\frac{1}{2}\left((\mathbf{k} \cdot \mathbf{r})(\boldsymbol{\epsilon}^{\lambda^*} \cdot \hat{\mathbf{p}}) + (\boldsymbol{\epsilon}^{\lambda^*} \cdot \mathbf{r})(\mathbf{k} \cdot \hat{\mathbf{p}})\right)$$

$$+ \frac{1}{2}\left((\mathbf{k} \cdot \mathbf{r})(\boldsymbol{\epsilon}^{\lambda^*} \cdot \hat{\mathbf{p}}) - (\boldsymbol{\epsilon}^{\lambda^*} \cdot \mathbf{r})(\mathbf{k} \cdot \hat{\mathbf{p}})\right)|a\rangle. \tag{17.130}$$

The second term can be written as (since $\mathbf{k} \cdot \boldsymbol{\epsilon}^{\lambda^*} = 0$) follows

$$\frac{1}{2}\langle b|(\mathbf{k} \cdot \mathbf{r})(\boldsymbol{\epsilon}^{\lambda^*} \cdot \hat{\mathbf{p}}) - (\boldsymbol{\epsilon}^{\lambda^*} \cdot \mathbf{r})(\mathbf{k} \cdot \hat{\mathbf{p}})|a\rangle = \langle b|(\mathbf{k} \times \boldsymbol{\epsilon}^{\lambda}) \cdot (\mathbf{r} \times \hat{\mathbf{p}})|a\rangle$$
$$= \frac{1}{2}\langle b|(\mathbf{k} \times \boldsymbol{\epsilon}^{\lambda}) \cdot \mathbf{L}|a\rangle,$$

$$(17.131)$$

where $\mathbf{L}$ is the orbital angular momentum operator. Now $\boldsymbol{\mu}_L = \frac{e}{2mc}\mathbf{L}$, gives the orbital magnetic moment, hence the radiative transition due to this term is called a magnetic dipole ($M1$) transition. From Eq. (17.115), noting that kr is of order $Z\alpha$ we see that the matrix elements for $M1$ transitions are suppressed by a factor of $Z\alpha$ as compared with $E1$ transitions. The selection rules for a $M1$ transition are $\Delta J = \pm 1, 0$, no change in parity of states. For circularly polarised light this can be seen as follows for a hydrogen-like atom.

For $M1$ transition (see Eq. (17.131)), for circularly polarised light, we have the matrix elements of the form

$$\langle n_f l_f m_f|(\mathbf{k} \times \boldsymbol{\epsilon}^{(\pm)}) \cdot \mathbf{L}|n_i l_i m_i\rangle.$$

But

$$\mathbf{k} \times \boldsymbol{\epsilon}^{(\pm)} = \mp i|\mathbf{k}|\boldsymbol{\epsilon}^{\pm},$$

since $\boldsymbol{\epsilon}^{(1)}, \boldsymbol{\epsilon}^{(2)}$ and $\frac{\mathbf{k}}{|\mathbf{k}|}$ form an orthogonal coordinate system. Also we have

$$(\mathbf{k} \times \boldsymbol{\epsilon}^{\pm}) \cdot \mathbf{L} = \frac{i}{\sqrt{2}}|\mathbf{k}|(L_x \pm iL_y)$$
$$= i\frac{|\mathbf{k}|}{\sqrt{2}}L\pm.$$

Thus

$$\langle n_f l_f m_f|(\mathbf{k} \times \boldsymbol{\epsilon}^{(\pm)}) \cdot \mathbf{L}|n_i l_i m_i\rangle$$
$$= i\frac{|\mathbf{k}|}{\sqrt{2}}\langle n_f l_f m_f|L\pm|n_i l_i m_i\rangle \sim \langle n_f l_f m_f|n_i l_i m_i \pm 1\rangle.$$

These matrix elements are non-zero only for

$$\Delta l \equiv l_f - l_i = 0$$
$$\Delta m \equiv m_f - m_i = \pm 1$$

giving us the required selection rules. Further since $(\mathbf{k} \times \boldsymbol{\epsilon}^{-*})$ is invariant under parity, then parity conservation gives $(-1)^{l_i} = (-1)^{l_f}$ or $(-1)^{\Delta l} = 1$ i.e. no parity change.

The first term in Eq. (17.130) can be written as

$$\mathbf{k} \cdot (\mathbf{r}\hat{\mathbf{p}} + \hat{\mathbf{p}}\mathbf{r}) \cdot \boldsymbol{\epsilon}^{\lambda}.$$

Thus we can write

$$
\begin{aligned}
\langle b|\mathbf{k} \cdot (\mathbf{r}\hat{\mathbf{p}} + \hat{\mathbf{p}}\mathbf{r}) \cdot \boldsymbol{\epsilon}^{\lambda}|a\rangle &= \frac{m}{i\hbar}\langle b|\mathbf{k} \cdot \{\mathbf{r}[\mathbf{r}, H_0] + [\mathbf{r}, H_0]\mathbf{r}\} \cdot \boldsymbol{\epsilon}^{\lambda}|a\rangle + O(e) \\
&= \frac{m}{i\hbar}\langle b|\mathbf{k} \cdot [\mathbf{rr}, H_0] \cdot \boldsymbol{\epsilon}^{\lambda}|a\rangle \\
&= -im\omega \mathbf{k} \cdot \langle b|\mathbf{rr}|a\rangle \cdot \boldsymbol{\epsilon}^{\lambda}. \qquad (17.132)
\end{aligned}
$$

We see that first term of Eq. (17.130) gives rise to electric quadrupole ($E2$) transition. The selection rules for $E2$ transition are $\Delta l \leq 2$, no change in parity of atomic states (this excludes $\Delta l = 1$). Thus the selection rule for $E2$ is $\Delta l = 2$. The matrix elements for $E2$ transition are suppressed by a factor $Z\alpha$ as compared with $E1$ transition as is the case for $M1$ transition.

17.8   Decay Width

The decay rate for the radiative transition from a state a to state b is given by

$$\frac{1}{\hbar}\Gamma_{a \to b} = \int \sum_{\lambda=1}^{2} W_{ba} d\Omega. \qquad (17.133)$$

Γ is called the decay width. The mean life time for this decay is given by

$$\tau_{a \to b} = \hbar/\Gamma_{a \to b}. \qquad (17.134)$$

17.8.1 E_1 Transition

For $E1$ transition, we have from Eq. (17.119)

$$\Gamma_{a \to b} = \frac{\alpha E_{\gamma}^3}{2\pi\hbar^2 c^2} \int \sum_{\lambda} |\langle b|\mathbf{r}|a\rangle \cdot \boldsymbol{\epsilon}^{\lambda *}|^2 d\Omega. \qquad (17.135)$$

The sum over two polarisation states gives, using Eq. (17.121e)

$$\sum_\lambda |\langle b|\mathbf{r}|a\rangle \cdot \boldsymbol{\epsilon}^{\lambda*}|^2 = \sum_\lambda \langle b|r_i|a\rangle\langle b|r_j|a\rangle^* \epsilon_i^{\lambda*} \epsilon_j^\lambda$$

$$= \langle b|r_i|a\rangle\langle b|r_j|a\rangle^* (\delta_{ij} - \frac{k_i k_j}{k^2})$$

$$= |\langle b|\mathbf{r}|a\rangle|^2 - \frac{|\mathbf{k} \cdot \langle b|\mathbf{r}|a\rangle|^2}{k^2}.$$

If we take $\mathbf{k}$ along the z-axis, this gives

$$\sum_\lambda |\langle b|\mathbf{r}|a\rangle \cdot \boldsymbol{\epsilon}^{\lambda*}|^2 = |\langle b|\mathbf{r}|a\rangle|^2 (1 - \cos^2\theta_k). \qquad (17.136)$$

Integrating over $d\Omega$, we have from Eq. (17.135),

$$\Gamma_{a\to b} = \frac{\alpha E_\gamma^3}{2\pi\hbar^2 c^2} |\langle b|\mathbf{r}|a\rangle|^2 \int_0^\pi \int_0^{2\pi} (1 - \cos^2\theta_k) \sin\theta_k d\theta_k d\phi_k$$

$$= \frac{4\alpha}{3c^2\hbar^2} E_\gamma^3 |\langle b|\mathbf{r}|a\rangle|^2. \qquad (17.137)$$

This rate depends on E_γ^3 which is typical of $E1$ emission. To get an order of magnitude for τ, for hydrogen-like atom (a_0 is Bohr radius, $\frac{1}{2} \times 10^{-8}$cm)

$$\langle b|\mathbf{r}|a\rangle| \sim a_0, E_\gamma \simeq \frac{e^2}{2a_0} \to \frac{\alpha}{2}\left(\frac{c\hbar}{a_0}\right),$$

giving

$$\tau = \frac{\hbar}{\Gamma} \simeq \frac{6}{\alpha^4}\frac{a_0}{c} \simeq 3.5 \times 10^{-10}\text{sec},$$

which gives time of a typical spontaneous $E1$ atomic transition.

As a concrete example, we calculate the mean life-time for the transition $2p \to 1s$ for a hydrogen-like atom. From Eq. (17.137), we have

$$\Gamma_{2p\to 1s} = \frac{4\alpha}{3c^2\hbar^2} E_\gamma^3 \frac{1}{3} \sum_{m=-1,0,1} |\int u_{100}^*(\mathbf{r})\mathbf{r}u_{21m}(\mathbf{r})d^3 r|^2, \qquad (17.138)$$

where we have averaged over the three possible values of m for the $2p$ state. Now

$$\langle \mathbf{r}\rangle \equiv \langle 100|\mathbf{r}|21m\rangle = \left(\int_0^\infty R_{10}(r)r R_{21}(r)r^2 dr\right)$$

$$\times \int_0^\pi \int_0^{2\pi} Y_{00}^* \begin{pmatrix} \sin\theta\cos\phi \\ \sin\theta\sin\phi \\ \cos\theta \end{pmatrix} Y_{1m} \sin\theta d\theta d\phi.$$

Then using $Y_{00} = \sqrt{\frac{1}{4\pi}}$, $\cos\theta = \sqrt{(4\pi/3)}Y_{10}$, $e^{\pm i\phi}\sin\theta = \mp\sqrt{\frac{8\pi}{3}}Y_{1,\pm 1}$,

$$\langle z \rangle = (\int_0^\infty R_{10}(r)R_{21}(r)r^3 dr)\frac{1}{\sqrt{3}}\delta_{m0},$$

$$\langle x \pm iy \rangle = (\int_0^\infty R_{10}(r)R_{21}(r)r^3 dr)(\mp\sqrt{\frac{2}{3}}\delta_{m,\pm 1}).$$

Thus

$$\frac{1}{3}\sum_m |\langle \mathbf{r} \rangle|^2$$

$$= \frac{1}{3}(\int_0^\infty R_{10}(r)R_{21}(r)r^3 dr)^2 \sum_m \{\frac{2}{3}\frac{2}{4}(\delta_{m1}\delta_{m1} + \delta_{m-1}\delta_{m-1})$$

$$+ \frac{1}{3}\delta_{m0}\delta_{m0}\},$$

and we have from Eq. (17.138)

$$\Gamma_{2p\to 1s} = \frac{4\alpha}{9c^2\hbar^2}E_\gamma^3(\int_0^\infty R_{10}(r)R_{21}(r)r^3 dr)^2), \tag{17.139a}$$

where

$$R_{10}(r) = 2\left(\frac{Z}{a_0}\right)^{3/2} e^{-zr/a_0}, \tag{17.139b}$$

$$R_{21}(r) = \frac{1}{\sqrt{24}}\left(\frac{Z}{a_0}\right)^{5/2} re^{-zr/2a_0}, \tag{17.139c}$$

and a_0 is the Bohr radius

$$a_0 = \frac{\hbar^2}{me^2} = \frac{\hbar}{mc\alpha}. \tag{17.139d}$$

Now

$$\int_0^\infty R_{10}^*(r)R_{21}(r)r^3 dr = \frac{1}{\sqrt{6}}\left(\frac{Z}{a_0}\right)^4\left(\frac{2a_0}{3Z}\right)^5 \int_0^\infty dx\, x^4 e^{-x}$$

$$= \frac{24}{\sqrt{6}}\left(\frac{2}{3}\right)^5 Z^{-1}a_0. \tag{17.140}$$

Hence

$$\Gamma_{2p\to 1s} = \frac{4\alpha}{9c^2\hbar^3}E_\Gamma^3 \frac{(24)^2}{6}\left(\frac{2}{3}\right)^{10} Z^{-2}a_0^2. \tag{17.141}$$

Now

$$E_\gamma = \hbar\omega = -\frac{mc^2 Z^2\alpha^2}{2}\left(\frac{1}{4} - 1\right)$$

$$= \frac{3m}{8}c^2 Z^2 \alpha^2. \tag{17.142}$$

Thus [using Eq. (17.141)]

$$\Gamma_{2p \to 1s} = \left(\frac{2}{3}\right)^8 \frac{mc^2}{\hbar} \alpha (Z\alpha)^4,$$

$$\tau_{2p \to 1s} = \left(\frac{3}{2}\right)^8 \frac{\hbar}{mc^2} \frac{1}{\alpha (Z\alpha)^4}. \tag{17.143}$$

For the hydrogen atom, $Z = 1$, therefore

$$\tau_{2p \to 1s} = \left(\frac{3}{2}\right)^8 \frac{\hbar}{mc^2} \frac{1}{\alpha^5} \tag{17.144}$$

$$\simeq 1.6 \times 10^{-9} \text{ sec.} \tag{17.145}$$

Finally, we note that if we multiply Eq. (17.119) for the decay rate, by the photon frequency ω, we get a formula for the intensity of radiation

$$I_{ba} = \frac{e^2 \omega^4}{2\pi c^3} \int \sum_{\lambda=1}^{2} |\langle b|\mathbf{r}|a\rangle \cdot \boldsymbol{\epsilon}^{\lambda*}|^2 d\Omega, \tag{17.146}$$

i.e. the classical formula for the intensity of light emitted by an oscillating dipole.

17.8.2 *Magnetic Dipole Transition due to Intrinsic Spin*

Here all one has to do is to replace the orbital magnetic moment $\boldsymbol{\mu}_L = \frac{e}{2mc}\mathbf{L}$ by the intrinsic magnetic moment of spin $\frac{1}{2}$ particle of charge e, namely [see Sec. 11.5 and 20.12]

$$\boldsymbol{\mu}_s = g_s \left(\frac{e}{2mc}\right)\mathbf{s}$$

where

$$\mathbf{s} = \frac{\hbar}{2}\boldsymbol{\sigma}$$

and $g_s = 2$ for an elementary particle like the electron. Then the right hand side of Eq. (17.131) is replaced by

$$\frac{g_s}{2}\hbar\frac{1}{2}\langle b|\boldsymbol{\sigma} \cdot \mathbf{k} \times \boldsymbol{\epsilon}^{*\lambda}|a\rangle. \tag{17.147}$$

Then from Eq. (17.111)

$$\frac{1}{\hbar}\Gamma_{ab}(M_1) = W_{ba} = \left(\frac{g_s}{2}\right)^2 \frac{\hbar^2}{4}\frac{e^2\omega}{2\pi m^2 \hbar c^3}$$

$$\times \int \sum_{\lambda} |\langle b|\boldsymbol{\sigma} \cdot (\mathbf{k} \times \boldsymbol{\epsilon}^{\lambda*})|a\rangle|^2 d\Omega. \tag{17.148}$$

This is the decay rate for the radiative transition from the state a to b due to intrinsic magnetic dipole moment μ_s. Now, using Eq. (17.121)

$$\sum_\lambda |\boldsymbol{\sigma} \cdot (\mathbf{k} \times \boldsymbol{\epsilon}^{*\lambda})|^2 = \sigma^i \sigma^j \sum_\lambda (\mathbf{k} \times \boldsymbol{\epsilon}^{*\lambda})_i (\mathbf{k} \times \boldsymbol{\epsilon}^\lambda)_j$$

$$= \sum_\lambda (\mathbf{k} \times \boldsymbol{\epsilon}^{*\lambda}) \cdot (\mathbf{k} \times \boldsymbol{\epsilon}^\lambda) = 2k^2 = 2\frac{\omega^2}{c^2}.$$

Hence $E_\gamma = \hbar\omega$

$$\Gamma_{ab}(M_1) = (\frac{g_s}{2})^2 \frac{\alpha}{(me^2)^2} E_\gamma^3. \tag{17.149}$$

We now apply the formula (17.149) to the radiative transition from a baryon Σ^0 to Λ^0:

$$\overset{0}{\Sigma} \to \Lambda^0 \gamma. \tag{17.150}$$

The transition magnetic moment [m_p is the proton mass used in nucleon magneton]

$$\mu(\overset{0}{\Sigma} - \Lambda^0) = \frac{g_s}{2} \left(\frac{e\hbar}{2m_p c} \right), \quad \frac{g_s}{2} \equiv 1.61. \tag{17.151}$$

Hence from Eq. (17.150):

$$\Gamma(\overset{0}{\Sigma} \to \Lambda^0 \gamma) = (1.61)^2 \frac{\alpha}{(m_p c^2)^2} E_\gamma^3. \tag{17.152}$$

Here $E_\gamma \equiv 74\text{MeV}$ so that Eq. (17.152) gives

$$\Gamma(\overset{0}{\Sigma} \to \Lambda^0 \gamma) \equiv 8.71 \times 10^{-3}\text{MeV}$$

$$\tau = \frac{\hbar}{\Gamma} \equiv 7.5 \times 10^{-20}s \quad (\text{Exp} : 7.4 \pm 0.07 \times 10^{-20}s) \tag{17.153}$$

17.9 Dipole Sum Rules, Two Point Function

Consider a particle in quantum mechanics described by dynamical variables $\hat{\mathbf{x}}$ and $\hat{\mathbf{p}}$. We want to discuss this dynamical system without assuming any form for the Hamiltonian and for the ground state as far as possible. This method is particularly suitable for the systems in which the detailed form of the Hamiltonian or dynamics is not known.

Let us start with the basic commutation relation of quantum mechanics,

$$[\hat{p}_i, \hat{x}_j] = -i\hbar\delta_{ij}. \tag{17.154}$$

In the Heisenberg picture

$$\hat{x}_i(t) = e^{iHt/\hbar}\hat{x}_i e^{-iHt/\hbar}, \tag{17.155a}$$

$$\hat{p}_i(t) = e^{iHt/\hbar}\hat{p}_i e^{-iHt/\hbar}. \tag{17.155b}$$

Define the two point function

$$\langle 0|\hat{x}_i(t)\hat{x}_j(t')|0\rangle = \int d^3 r u_0^*(\mathbf{r})x_i(t)x_j(t')u_0(\mathbf{r}), \tag{17.156}$$

where $u_0(\mathbf{r})$ is the ground state wave function.

We now want to obtain a "spectral representation" for this function (see below). For this purpose we use the following steps:

1. Using the completeness

$$\hat{1} = \sum_n |n\rangle\langle n| \tag{17.157a}$$

$$H|n\rangle = E_n|n\rangle, \tag{17.157b}$$

we have

$$\langle 0|\hat{x}_i(t)\hat{x}_j(t')|0\rangle = \sum_n \langle 0|\hat{x}_i(t)|n\rangle\langle n|\hat{x}_j(t')|0\rangle. \tag{17.158}$$

2. Using the translational invariance viz. Eq. (17.155), we have

$$\langle n|\hat{x}_i(t)|0\rangle = \langle n|e^{iHt/\hbar}\hat{x}_i e^{-iHt/\hbar}|0\rangle$$
$$= e^{-i(E_0-E_n)t/\hbar}\langle n|\hat{x}_i|0\rangle. \tag{17.159}$$

3. Using Eq. (17.159), we have

$$\langle 0|\hat{x}_i(t)\hat{x}_j(t')|0\rangle = \sum_n e^{i(E_0-E_n)(t-t')/\hbar}\langle 0|\hat{x}_i|n\rangle\langle n|\hat{x}_j|0\rangle$$
$$= \int_0^\infty dE e^{-iE(t-t')/\hbar}\sum_n \langle 0|\hat{x}_i|n\rangle\langle n|\hat{x}_j|0\rangle$$
$$\times \delta(E - E_n + E_0). \tag{17.160}$$

Let us define

$$\sum_n \langle 0|\hat{x}_i|n\rangle\langle n|\hat{x}_j|0\rangle\delta(E - E_n + E_0) = \rho_{ij}(E). \tag{17.161}$$

Thus we get the following representation of the two point function

$$\langle 0|\hat{x}_i(t)\hat{x}_j(t')|0\rangle = \int_0^\infty dE e^{-iE(t-t')/\hbar}\rho_{ij}(E). \tag{17.162}$$

This is known as the spectral representation of the two point function defined in Eq. (17.157) and $\rho_{ij}(E)$ is called the spectral function.

4. The spectral function $\rho_{ij}(E)$ has the following properties: (C's are complex numbers)

$$
\begin{aligned}
C_i^* \rho_{ij} C_j &= \sum_n \langle 0|C_i^* \hat{x}_i|n\rangle\langle n|\hat{x}_j C_j|0\rangle\delta(E - E_n + E_0) \\
&= \sum_n |\langle n|x_i C_i|0\rangle|^2 \delta(E - E_n + E_0) \\
&\geq 0,
\end{aligned}
\tag{17.163}
$$

for arbitrary constants $C_i \cdot \rho_{ij}$ is a positive matrix

$$
\rho_{11}, \rho_{22}, \rho_{33} > 0.
$$

5. Similarly, we have

$$
\langle 0|\hat{p}_i(t)\hat{p}_j(t')|0\rangle = \int_0^\infty dE e^{-iE(t-t')/\hbar}\tilde{\rho}_{ij}(E).
\tag{17.164}
$$

So far we have not used any dynamics. In order to relate $\tilde{\rho}_{ij}$ with ρ_{ij}, we have to put some dynamical information. Now we have

$$
\begin{aligned}
\frac{d}{dt}\hat{x}_i(t) &= \frac{d}{dt}(e^{iHt/\hbar}\hat{x}_i e^{-iHt/\hbar}) \\
&= \frac{i}{\hbar}e^{iHt/\hbar}[H, \hat{x}_i]e^{-iHt/\hbar}.
\end{aligned}
\tag{17.165}
$$

Let us now assume that the Hamiltonian H has the form

$$
H = \frac{\hat{\mathbf{p}}^2}{2m} + V(\mathbf{r}),
\tag{17.166}
$$

so that

$$
\begin{aligned}
[H, \hat{x}_i] &= \frac{1}{2m}[\hat{\mathbf{p}}^2, \hat{x}_i] \\
&= \frac{1}{2m}(\hat{p}_j[\hat{p}_j, \hat{x}_i] + [\hat{p}_j, \hat{x}_i]\hat{p}_j) \\
&= \frac{1}{2m}(-2i\hbar\delta_{ij}\hat{p}_j) = -\frac{i\hbar}{m}\hat{p}_i.
\end{aligned}
\tag{17.167}
$$

Hence

$$
\begin{aligned}
\frac{d}{dt}\hat{x}_i(t) &= \frac{1}{m}e^{iHt/\hbar}\hat{p}_i e^{-iHt/\hbar} \\
&= \frac{1}{m}\hat{p}_i(t).
\end{aligned}
\tag{17.168}
$$

Using Eq. (17.168), we obtain

$$
\begin{aligned}
\langle 0|\hat{p}_i(t)\hat{p}_j(t')|0\rangle &= m^2\langle 0|\frac{d}{dt}\hat{x}_i(t)\frac{d}{dt'}\hat{x}_j(t')|0\rangle \\
&= \frac{m^2}{\hbar^2}\int_0^\infty dE e^{-iE(t-t')/\hbar}E^2\rho_{ij}(E).
\end{aligned}
\tag{17.169}
$$

Comparing Eq. (17.169) with Eq. (17.164), we get

$$\tilde{\rho}_{ij}(E) = \frac{m^2}{\hbar^2} E^2 \rho_{ij}(E). \tag{17.170}$$

6. We also note from Eq. (17.168) and Eq. (17.169)

$$\langle 0|\hat{p}_i(t)\hat{x}_j(t')|0\rangle = -\frac{im}{\hbar}\int_0^\infty dE e^{-iE(t-t')/\hbar} E\rho_{ij}(E),$$

$$\langle 0|\hat{x}_i(t)\hat{p}_j(t')|0\rangle = \frac{im}{\hbar}\int_0^\infty dE e^{-iE(t-t')/\hbar} E\rho_{ij}(E). \tag{17.171a}$$

Changing $t \to t'$, $i \to j$, we have

$$\langle 0|\hat{x}_j(t')\hat{p}_i(t)|0\rangle = \frac{im}{\hbar}\int_0^\infty dE e^{iE(t-t')/\hbar} E\rho_{ji}(E). \tag{17.171b}$$

Under time reversal $t \to -t$, $\hat{\mathbf{r}} \to \hat{\mathbf{r}}$, $\hat{\mathbf{p}} \to -\hat{\mathbf{p}}$ and we see that commutation relation (17.154) is not invariant under time reversal unless either we change the order of operators in a product, or take the complex conjugate of complex number [see chapter 10]. Therefore, if we assume time reversal invariance, we get

$$\langle 0|\hat{x}_j(t')\hat{p}_i(t)|0\rangle \to -\langle 0|\hat{p}_i(t)\hat{x}_j(t')|0\rangle. \tag{17.172}$$

Therefore, from Eqs. (17.171a,b), we have

$$\frac{im}{\hbar}\int_0^\infty dE e^{-iE(t-t')/\hbar} E\rho_{ji}(E) = \frac{im}{\hbar}\int_0^\infty dE e^{-iE(t-t')/\hbar} E\rho_{ij}(E). \tag{17.173}$$

Hence we have

$$\rho_{ji}(E) = \rho_{ij}(E). \tag{17.174}$$

7. Now from Eqs. (17.171) and (17.174)

$$\langle 0|[\hat{p}_i(t), \hat{x}_j(t')]|0\rangle$$
$$= -\frac{im}{\hbar}\int_0^\infty dE (e^{-iE(t-t')/\hbar} + e^{iE(t-t')/\hbar}) E\rho_{ij}(E). \tag{17.175}$$

At equal time $t = t'$

$$[\hat{p}_i(t), \hat{x}_j(t)] = e^{iHt/\hbar}[\hat{p}_i, \hat{x}_j]e^{-iHt/\hbar}$$
$$= -i\hbar\delta_{ij}. \tag{17.176}$$

Hence from Eqs. (17.175) and (17.176), we have the sum rule

$$\int_0^\infty dE E\rho_{ij}(E) = \frac{\hbar^2}{2m}\delta_{ij}. \tag{17.177}$$

Note that in deriving Eq. (17.177), we have not used any detailed form of potential in the Hamiltonian—except that it does not involve the momentum p.

The ground state $|0\rangle$ is arbitrary. Consider the simple case when the ground state $|0\rangle$ is spherically symmetric, i.e. it is an s-state. Then we have rotational invariance so that

$$\rho_{ij}(E) = \delta_{ij}\rho(E). \tag{17.178}$$

If the ground state has non-zero spin, then we sum over spin and again we have rotational invariance. In this case the sum rule (17.177) reduces to

$$\int_0^\infty dE\, E\rho(E) = \frac{\hbar^2}{2m}. \tag{17.179a}$$

This is known as the spectral function sum rule.

Two additional sum rules are obtained from Eqs. (17.162) and (17.164), by putting $t = t'$, $i = j$ and using Eqs. (17.170) and (17.178)

$$\langle \mathbf{r}^2 \rangle_0 = 3 \int_0^\infty dE\, \rho(E), \tag{17.179b}$$

$$\langle \mathbf{p}^2 \rangle_0 = \frac{3m^2}{\hbar^2} \int_0^\infty dE\, E^2 \rho(E). \tag{17.179c}$$

To sum up, sum rules (17.179) have been obtained on very general grounds, namely, completeness, translational invariance, existence of the ground state, and the basic commutation relation of quantum mechanics. In addition, we have used time reversal invariance and the form (17.166) of the Hamiltonian to derive Eq. (17.168). Equation (17.168) follows essentially from the fact that the potential in the Hamiltonian does not depend upon the momentum.

We can easily show that Eqs. (17.179) lead to the uncertainty principle. For this purpose, we use the following mathematical result.

Define an inner product

$$(f, g) = \int dE\, \rho(E) f(E) g(E) \tag{17.180a}$$

in a certain function space. The Schwartz inequality gives

$$|(f, g)| \leq ||f||\, ||g||, \tag{17.180b}$$

where

$$||f|| = \sqrt{(f, f)}. \tag{17.180c}$$

Choose $f(E) = 1$, $g(E) = E$. Then we have

$$\int_0^\infty dE E |\rho(E)|^2 \le \left(\int_0^\infty dE \rho(E)\right) \left(\int_0^\infty dE E^2 \rho(E)\right). \quad (17.181)$$

Now using Eqs. (17.179), we have

$$\frac{\hbar^4}{4m^2} \le \frac{1}{3}\langle \mathbf{r}^2\rangle \frac{\hbar^2}{3m^2}\langle \mathbf{p}^2\rangle \quad (17.182a)$$

or

$$\langle \mathbf{r}^2\rangle\langle \mathbf{p}^2\rangle \ge \frac{9}{4}\hbar^2, \quad (17.182b)$$

which is the uncertainty relation.

17.10 Application of Dipole Sum Rules

Consider the reaction of the type

$$\gamma + |0\rangle \to |n\rangle, \quad (17.183)$$

where $|0\rangle$ is the ground state and $|n\rangle$ denotes all possible final states consistent with the conservation laws. Then to the lowest order in e, the transition probability is given by

$$|C_n(\infty)|^2,$$

where (cf. Eq. (17.96a) with $n_\lambda(\mathbf{k}) = 1$)

$$C_n(\infty) = -\frac{ie}{m\hbar}\sqrt{\frac{4\pi\hbar}{2\omega V}}2\pi\hbar\delta(E_n - E_0 - \hbar\omega)\langle n|e^{i\mathbf{k}\cdot\mathbf{r}}\hat{\mathbf{p}}\cdot\boldsymbol{\epsilon}^\lambda|0\rangle. \quad (17.184)$$

Using the dipole approximation, the total transition rate is given by (average over initial polarisation)

$$W = \frac{|C_n(\infty)|^2}{t} = \frac{e^2}{m^2\hbar^2}\frac{4\pi\hbar}{2\omega V}2\pi\hbar\left(\frac{1}{2}\sum_{\lambda=1}^2 \epsilon_i^\lambda\epsilon_j^\lambda\right)$$

$$\times \sum_n \delta(E_n - E_0 - \hbar\omega)\langle 0|\hat{\mathbf{p}}_i|n\rangle\langle n|\hat{\mathbf{p}}_j|0\rangle. \quad (17.185)$$

Now

$$\frac{1}{2}\sum_{\lambda=1}^2 \epsilon_i^\lambda\epsilon_j^\lambda = \frac{1}{3}\delta_{ij}, \quad (17.186a)$$

$$\tilde{\rho}_{ij}(\omega) = \sum_n \langle 0|\hat{p}_i|n\rangle\langle n|\hat{p}_j|0\rangle\delta(E_n - E_0 - \hbar\omega)$$

$$= m^2 \omega^2 \rho_{ij}(\omega)$$
$$= m^2 \omega^2 \delta_{ij} \rho(\omega). \tag{17.186b}$$

Hence

$$W = \frac{e^2}{m^2 \hbar^2} \frac{4\pi\hbar}{2\omega V} 2\pi\hbar \frac{1}{3} (\delta_{ij}\delta_{ij}) m^2 \omega^2 \rho(\omega)$$
$$= \frac{e^2}{\hbar^2} \frac{4\pi\hbar}{2V} 2\pi\hbar\omega\rho(\omega). \tag{17.187}$$

The total cross-section $\sigma(\omega)$ for the process (17.183) is given by

$$\sigma(\omega) = \frac{W}{\text{Flux}}, \tag{17.188a}$$

$$\text{Flux} = \rho v = \frac{1}{V}c. \tag{17.188b}$$

Hence

$$\sigma(\omega) = \pi(4\pi\frac{e^2}{c})\omega\rho(\omega)$$
$$= 4\pi^2 \hbar\alpha\omega\rho(\omega) \tag{17.189a}$$

or

$$\rho(\omega) = \frac{1}{4\pi^2\hbar\alpha} \frac{\sigma(\omega)}{\omega}. \tag{17.189b}$$

Rewriting the sum rules (17.179) in terms of ω, we have

$$\int_0^\infty d\omega\,\omega\rho(\omega) = \frac{1}{2m}, \tag{17.190}$$

$$\int_0^\infty d\omega\,\rho(\omega) = \langle \mathbf{r}^2 \rangle_0 \frac{1}{3\hbar}, \tag{17.191}$$

$$\int_0^\infty d\omega\,\omega^2 \rho(\omega) = \frac{1}{3m^2} \hbar\langle \mathbf{p}^2 \rangle_0. \tag{17.192}$$

Now using Eq. (17.189b), we have

$$\frac{1}{4\pi^2\alpha} \int_0^\infty d\omega\,\sigma(\omega) = \frac{\hbar}{2m}, \tag{17.193}$$

$$\frac{1}{4\pi^2\alpha} \int_0^\infty d\omega\,\frac{\sigma(\omega)}{\omega} = \frac{1}{3}\langle \mathbf{r}^2 \rangle_0, \tag{17.194}$$

$$\frac{1}{4\pi^2\alpha} \int_0^\infty d\omega\,\omega\sigma(\omega) = \hbar^2 \frac{1}{3} \frac{\langle \mathbf{p}^2 \rangle_0}{m^2}. \tag{17.195}$$

Since $\sigma(\omega)$ is the total cross-section and can be determined experimentally, the above sum rules can be tested. Since we have used the dipole approximation, the third sum rule (17.195) cannot be trusted as in this sum rule because of ω factor in the numerator, large values of ω become important, but for large values of ω, the dipole approximation breaks down. In this respect, the sum rule (17.194) is the best one, as ω is in the denominator. $\langle r^2 \rangle_0$ is related to the radius of the atomic system. This sum rule relates the radius of atomic system with the total cross-section.

17.11   Dispersion Relation

We have seen that $\langle 0|\hat{x}_i(t)\hat{x}_j(t')\rangle_0$ has the following spectral representation (cf. Eqs. (17.162) and (17.178)), $E = \hbar\omega$

$$\langle 0|\hat{x}_i(t)\hat{x}_j(t')|0\rangle = \delta_{ij}\hbar \int_0^\infty d\omega e^{-i\omega(t-t')}\rho(\omega). \qquad (17.196)$$

Let us consider its analytical continuation for complex $(t - t')$. Put

$$t - t' = u + iv. \qquad (17.197)$$

Then right-hand side of Eq. (17.197) becomes

$$\delta_{ij}\hbar \int_0^\infty d\omega e^{-i\omega u + \omega v}\rho(\omega).$$

If $v < 0$ or $\mathrm{Im}(t - t') < 0$, $e^{\omega v}$ acts as a damping factor. We can then take derivative of the above expression with respect to v, since the spectral representation of

$$\langle 0|\hat{p}_i(t)\hat{p}_j(t')|0\rangle$$

contains the spectral function

$$\tilde{\rho}_{ij}(\omega) = m^2\omega^2\rho_{ij}(\omega). \qquad (17.198)$$

Further, the quantity of physical interest is the time ordered product of two point functions and their Fourier transformations. We define time ordered product as follows:

$$T(\hat{x}_i(t)\hat{x}_j(t')) = \hat{x}_i(t)\hat{x}_j(t') \quad t > t',$$
$$= \hat{x}_j(t')\hat{x}_i(t) \quad t < t'. \qquad (17.199)$$

Let us consider the Fourier transform of

$$\langle 0|T(\hat{p}_i(t)\hat{p}_j(t'))|0\rangle,$$

$$G_{ij}(\omega) = \int_{-\infty}^\infty dt e^{-i\omega(t-t')}\langle 0|T(\hat{p}_i(t)\hat{p}_j(t'))|0\rangle$$

$$= \int_{t'}^\infty dt e^{-i\omega(t-t')}\langle 0|\hat{p}_i(t)\hat{p}_j(t')|0\rangle + \int_{-\infty}^{t'} dt e^{-i\omega(t-t')}$$

$$\times \langle 0|\hat{p}_j(t')\hat{p}_i(t)|0\rangle. \qquad (17.200)$$

Using Eq. (17.169), we have

$$G_{ij}(\omega) = m^2\hbar\delta_{ij}\Big(\int_{t'}^\infty dt e^{-i\omega(t-t')}\int_0^\infty d\omega' e^{-i\omega'(t-t')}\omega'^2\rho(\omega')$$

$$+ \int_{-\infty}^{t'} dt e^{-i\omega(t-t')}\int_0^\infty d\omega' e^{i\omega'(t-t')}\omega'^2\rho(\omega')\Big). \qquad (17.201)$$

Interchanging the order of integration (this is justified if the integrals exist), we get

$$G_{ij}(\omega) = m^2 \hbar \delta_{ij} \left(\int_0^\infty d\omega' \omega'^2 \rho(\omega') \int_{t'}^\infty dt e^{-i(\omega'+\omega)(t-t')} \right.$$
$$\left. + \int_0^\infty d\omega' \omega'^2 \rho(\omega') \int_{-\infty}^{t'} dt e^{-i(\omega-\omega')(t-t')} \right). \qquad (17.202)$$

In order to give meaning to time integration, we use the following prescription:

$$\int_{t'}^\infty dt e^{-i(\omega'+\omega-i\epsilon)(t-t')} = \frac{-i}{\omega'+\omega-i\epsilon},$$
$$\int_{-\infty}^{t'} dt e^{-i(\omega-\omega'+i\epsilon)(t-t')} dt = \frac{i}{\omega-\omega'+i\epsilon}, \qquad (17.203)$$

where we take the limit $\epsilon \to 0$ at the end. Thus

$$G_{ij}(\omega) = \int_{-\infty}^\infty dt e^{-i\omega(t-t')} \langle 0|T(\hat{p}_i(t)\hat{p}_j(t'))|0 \rangle$$
$$= -i\delta_{ij}m^2\hbar \int_0^\infty d\omega' \omega'^2 \rho(\omega') \left(\frac{1}{\omega'+\omega-i\epsilon} - \frac{1}{\omega-\omega'+i\epsilon} \right). \qquad (17.204)$$

Hence

$$G_{ij}(\omega) = i\delta_{ij}m^2\hbar \int_0^\infty d\omega' \frac{2\omega'^3 \rho(\omega')}{\omega^2 - \omega'^2 + i\epsilon}. \qquad (17.205)$$

Now using Eq. (17.190b), we have

$$G_{ij}(\omega) = i\delta_{ij} \frac{m^2}{2\pi^2\alpha} \int_0^\infty d\omega' \frac{\omega'^2 \sigma(\omega')}{\omega^2 - \omega'^2 + i\epsilon}. \qquad (17.206)$$

We note that

$$\frac{1}{\omega^2 - \omega'^2 + i\epsilon} = P \frac{1}{\omega^2 - \omega'^2} - i\pi \delta(\omega^2 - \omega'^2), \qquad (17.207a)$$

$$\delta(\omega^2 - \omega'^2) = \frac{1}{2\omega} (\delta(\omega - \omega') + \delta(\omega + \omega')), \qquad (17.207b)$$

where P denotes the principal value integral. Thus we obtain

$$\text{Im}(iG_{ij}(\omega)) = \delta_{ij} \frac{m^2}{2\pi^2\alpha} \pi \frac{\omega^2 \sigma(\omega)}{2\omega}$$
$$= \delta_{ij} \frac{m^2 c}{\alpha} \frac{\omega}{4\pi c} \sigma(\omega), \qquad (17.208)$$

$$\text{Re}(iG_{ij}(\omega)) = \delta_{ij} \frac{m^2}{2\pi^2\alpha} P \int_0^\infty \frac{\omega'^2 \sigma(\omega')}{\omega'^2 - \omega^2} d\omega'. \qquad (17.209)$$

By optical theorem

$$\text{Im } f(\omega, 0) \equiv \text{Im } f(\omega) = \frac{\omega}{4\pi c}\sigma(\omega), \qquad (17.210)$$

we see that

$$\left(\frac{\alpha}{m^2 c}iG_{ij}(\omega)\right) \equiv \delta_{ij}f(\omega) \qquad (17.211)$$

represents the forward scattering amplitude for the photon scattering. Hence we have from Eqs. (17.207) and (17.210)

$$f(\omega) = \frac{2}{\pi}\int_0^\infty d\omega'\omega'\frac{\text{Im } f(\omega')}{\omega'^2 - \omega^2 - i\epsilon}, \qquad (17.212a)$$

$$\text{Re } f(\omega) = \frac{2}{\pi}P\int_0^\infty d\omega'\omega'\frac{\text{Im } f(\omega')}{\omega'^2 - \omega^2}, \qquad (17.212b)$$

$$\text{Re } f(\omega) = \frac{1}{2\pi^2 c}P\int_0^\infty \frac{\omega'^2\sigma(\omega')}{\omega'^2 - \omega^2}d\omega'. \qquad (17.212c)$$

The above equations are known as dispersion relations. In actual physical cases, the forward scattering amplitude $f(\omega)$ does not vanish as $\omega \to \infty$. Therefore the above integrals do not converge. In such cases, we make a subtraction at $\omega = 0$, so that we get

$$f(\omega) - f(0) = \frac{2\omega^2}{\pi}\int_0^\infty d\omega'\frac{\text{Im } f(\omega')}{\omega'(\omega'^2 - \omega^2 - i\epsilon)}, \qquad (17.213a)$$

$$\text{Re }(f(\omega) - f(0)) = \frac{2\omega^2}{\pi}P\int_0^\infty d\omega'\frac{\text{Im } f(\omega')}{\omega'(\omega'^2 - \omega^2)}, \qquad (17.213b)$$

$$\text{Re }(f(\omega) - f(0)) = \frac{\omega^2}{2\pi^2 c}P\int_0^\infty \frac{\sigma(\omega')d\omega'}{\omega'^2 - \omega^2}. \qquad (17.213c)$$

From the above equations, we see that $f(\omega)$ is an analytic function of ω in the cut plane starting from 0 to ∞. Since $\sigma(\omega)$ can be determined experimentally, by the dispersion relation (17.214c) we can determine the forward scattering amplitude.

17.12   Problems

17.1 Let **B** be a uniform magnetic field. Verify that $\mathbf{A} = \frac{1}{2}\mathbf{B} \times \mathbf{r}$ satisfies the relation $\mathbf{B} = \nabla \times \mathbf{A}$ and the gauge condition $\nabla \cdot \mathbf{A} = 0$. Take $\mathbf{B} \equiv (0, 0, B)$ and show that $\mathbf{A} \equiv \frac{1}{2}(-By, Bx, 0)$.

17.2 Consider the motion of a particle of charge e and mass m in $x - y$ plane with a uniform magnetic field along z-axis, $\mathbf{B} = \nabla \times \mathbf{A}$, $\mathbf{A} = \frac{1}{2}\mathbf{B} \times \mathbf{r}$.

Show that

$$\hat{Q} = \frac{1}{m\omega}(\hat{p}_x - \frac{e}{c}A_x) \qquad (17.214)$$

and

$$\hat{P} = \frac{1}{m\omega}(\hat{P}_y - \frac{e}{c}A_y) \qquad (17.215)$$

satisfy

$$[\hat{Q}, \hat{p}] = i\hbar. \qquad (17.216)$$

Show that Hamiltonian

$$H = \frac{1}{2m}(\hat{\mathbf{p}} - \frac{e}{c}\mathbf{A})^2 \qquad (17.217)$$

can be written as

$$H = \frac{\hat{P}^2}{2m} + \frac{1}{2}m\omega^2\hat{Q}^2 \qquad (17.218)$$

and hence the energy eigenvalues, called Landau levels are given by

$$E = (n + \frac{1}{2})\hbar\omega. \qquad (17.219)$$

Show that the other canonical pair

$$\hat{Q}' = (\hat{p}_y - \frac{e}{c}A_y)$$

$$\hat{P}' = \frac{1}{m\omega}(\hat{p}_x + \frac{e}{c}A_x) \qquad (17.220)$$

are cyclic i.e. they commute with $\hat{Q}$, $\hat{P}$ and do not occur in the Hamiltonian.

17.3 For a particle of charge e and mass m, in an electromagnetic field $(\phi, \mathbf{A})$:

$$H = -\frac{\hbar^2}{2m}\nabla^2 + \frac{i\hbar}{mc}e\mathbf{A}\cdot\nabla + \frac{e^2}{2mc^2}\mathbf{A}^2 + e\phi.$$

The expectation value of H is

$$\langle \psi|H|\psi \rangle = \langle \psi|H|\psi \rangle^*.$$

Show that

$$2\langle \psi|H|\psi \rangle = \int \Big[-\frac{\hbar^2}{2m}(\psi^*\nabla^2\psi + \psi\nabla^2\psi^*)$$

$$+ \Big(\frac{ie\hbar}{mc}\Big)(\psi^*(\mathbf{A}\cdot\nabla)\psi - \psi(\mathbf{A}\cdot\nabla)\psi^*)$$

$$+ \frac{e^2}{mc^2}\mathbf{A}^2 + 2e\phi\Big]d^3x.$$

Using the relation

$$\nabla \cdot (\nabla\psi^*\psi) = \psi\nabla^2\psi^* + \psi^*\nabla^2\psi + 2\nabla\psi^* \cdot \nabla\psi,$$

and the fact that by Gauss's theorem, the left hand can be written as surface integral for $\nabla(\psi^*\psi)$, which vanishes for a large surface, one gets

$$\langle\psi|H|\psi\rangle = \int d^3x \Big[\frac{\hbar^2}{2m}\nabla\psi^*\nabla\psi + \frac{ie\hbar}{2mc}(\psi^*(\mathbf{A} \cdot \nabla)\psi - \psi(\mathbf{A} \cdot \nabla)\psi^*)$$
$$+ \frac{e^2}{2mc^2}\psi^*\psi\mathbf{A}^2 + e\psi^*\psi\phi \Big].$$

Hence the Hamiltonian density is

$$\mathcal{H} = \frac{\hbar^2}{2m}\nabla\psi^* \cdot \nabla\psi + \frac{ie\hbar}{2mc}(\psi^*\nabla\psi - \psi\nabla\psi^*)$$
$$+ \frac{e^2}{2mc^2}\mathbf{A}^2\psi^*\psi - e\psi^*\psi\phi$$
$$= -\frac{\hbar^2}{2m}\nabla\psi^* \cdot \nabla\psi + e(\rho\phi - \mathbf{j} \cdot \mathbf{A})$$

where

$$\rho = \psi^*\psi, \quad \mathbf{j} = \frac{\hbar}{2mci}(\psi^*\nabla\psi - \psi\nabla\psi^*) = \frac{e}{2mc^2}\psi^*\psi\mathbf{A}^2.$$

The Lagrangian density $\mathcal{L} = -\mathcal{H}$. Verify that

$$\frac{\partial\rho}{\partial t} + \nabla \cdot \mathbf{j} = 0.$$

17.4 Show that for a hydrogen-like atom, parity operator $\hat{P}$ gives

$$\langle n_f l_f m_f|\mathbf{r}|n_i l_i m_i\rangle = (-1)^{l_i+l_f-1}\langle n_f l_f m_f|\mathbf{r}|n_i l_i m_i\rangle.$$

Hence parity conservation implies that $(-1)^{l_i} \neq (-1)^{l_f}$ i.e. atomic state must change parity for E_1 transition. This is the parity selection rule. Note also that $(l_f - l_i)$ must be odd.

17.5 By using the relation

$$[L^2, [L^2, \mathbf{r}]] = 2\hbar(\mathbf{r}L^2 + L^2\mathbf{r}),$$

show that for a hydrogen-like atom

$$l_f - l_i = \pm 1$$

or $l_f = -l_i$. However $l_f = -l_i$ only if $l_f = 0 = l_i$ but then

$$\langle n_f 00|\mathbf{r}|n_i 00\rangle.$$

vanishes.

17.6 Photoelectric Effect: Consider the ejection of an electron from the ground state of hydrogen like atom by absorption of a single photon viz. the reaction

$$\text{Photon} + \text{Atom} = e + \text{Atom}'.$$

Calculate the cross-section $\frac{d\sigma}{d\Omega}$ for this process by neglecting the recoil of atom.

Hint: The number of states when the electron has momentum between $\mathbf{p}_e$ and $\mathbf{p}_e + d\mathbf{p}_e = \frac{V}{(2\pi\hbar)^3}d^3p_e = \frac{V}{(2\pi\hbar)^3}2mp_edE_ed\Omega$, $E_e = p_e/2m$. The ground state of hydrogen like atom is represented by the state function

$$\langle \mathbf{r}|a \rangle = \langle r|100 \rangle = u_{100}(r)$$
$$= \frac{1}{\pi}(Z/a_0)^{3/2}e^{-zr/a_0}$$

and for free electron the wave function is given by

$$\langle \mathbf{r}|b \rangle = \frac{1}{\sqrt{V}}e^{i\mathbf{p}_e \cdot \mathbf{r}/\hbar}.$$

Using these results, we can express the total transition probability per unit time for the absorption of photons of frequency ω as

$$W_{ba} = \frac{e^2}{m^2\hbar^2}\frac{2\pi\hbar}{V\omega}\frac{V}{(2\pi\hbar)^3}\int 2mp_edE_ed\Omega 2\pi\hbar$$
$$\times \delta(E_e + E_B - \hbar\omega)|\langle b|e^{i\mathbf{k}\cdot\mathbf{r}}\boldsymbol{\epsilon}^\lambda \cdot \hat{\mathbf{p}}|a\rangle|^2.$$

Now we can write

$$\langle a|e^{-i\mathbf{k}\cdot\mathbf{r}}\boldsymbol{\epsilon}^\lambda \cdot \hat{\mathbf{p}}|b \rangle = \boldsymbol{\epsilon}^\lambda \cdot \frac{1}{\sqrt{V}}\int d^3ru_{100}^*(r)e^{-i\mathbf{k}\cdot\mathbf{r}}\mathbf{p}_ee^{i\mathbf{p}_e \cdot \mathbf{r}/\hbar}$$
$$= \frac{1}{\sqrt{V}}\boldsymbol{\epsilon}^\lambda \cdot \mathbf{p}_e\frac{1}{\sqrt{\pi}}(Z/a_0)^{3/2}\int d^3re^{-i(\mathbf{k}-\frac{\mathbf{p}_e}{\hbar})\cdot\mathbf{r}}e^{-zr/a_0}.$$

The flux of incoming photon is $\frac{1}{V}c$ and $\frac{d\sigma}{d\Omega} = \frac{1}{\text{Flux}}\frac{W_{ba}}{d\Omega}$. Note E_B is the binding energy of the electron in the ground state. In evaluating $(\boldsymbol{\epsilon}^\lambda \cdot \mathbf{p}_e)^2$, sum over polarisations; for this purpose it is convenient to select as unit vectors $\boldsymbol{\epsilon}^1$, $\boldsymbol{\epsilon}^2$ and $\frac{\mathbf{k}}{|\mathbf{k}|}$ and take $\mathbf{k}$ along Z-axis. The cross-section can be expressed as

$$\frac{d\sigma}{d\Omega} = \frac{32\sqrt{2}Z^5\alpha^8a^2(E_e/mc^2)^{1/2}\sin^2\theta}{\left\{(\alpha Z)^2 + \frac{2E_e}{mc^2}(1 - v_e/c\cos\theta)\right\}^4},$$

where we have neglected the binding energy and have put $\hbar\omega \approx p_e^2/2m$, $\alpha = e^2/\hbar c$ and v_e is the electron velocity.

17.7 Show that $\langle 0|E_i(\mathbf{r},t)E_j(\mathbf{r}',t)|0\rangle$, the expectation value of electric field at 2 points in vacuum (i.e. no photons), is given by

$$4\pi\hbar c^2 \left(\delta_{ij}\mathbf{n}\cdot\mathbf{n}' - \frac{\partial}{\partial x_i}\frac{\partial}{\partial x_j'} \right) D_1\mathbf{R},$$

where

$$D_1(\mathbf{R}) = \frac{1}{(2\pi)^3} \int \frac{d^3k}{2\omega} e^{i\mathbf{k}\cdot(\mathbf{r}-\mathbf{r}')}, \quad \mathbf{R} = \mathbf{r} - \mathbf{r}'$$

$$= \frac{1}{4\pi^2|\mathbf{R}|c} \int_0^\infty dk\,\sin k|\mathbf{R}|.$$

By defining the last integral as

$$\lim_{\alpha\to 0}\int_0^\infty e^{-\alpha k}\sin k|\mathbf{R}|dk,$$

show that

$$D_1(R) = \frac{1}{4\pi^2 c|\mathbf{R}|^2}$$

and $D_1(R) \to \infty$ when $\mathbf{R} \to 0$, i.e. $\mathbf{r} = \mathbf{r}'$. Since what we measure by a test body is the field strength averaged over some region in space, it may be more realistic to consider the average field operator about some point, e.g. defined as

$$\bar{\mathbf{E}} = \frac{1}{\Delta V}\int_{\Delta V}\mathbf{E}d^3r,$$

where ΔV is a small volume containing the position in question. Then show that

$$\langle 0|\bar{\mathbf{E}}\cdot\bar{\mathbf{E}}|0\rangle \sim \hbar c/(\Delta\ell)^4, \tag{I}$$

where $\Delta\ell$ is the linear dimension of volume ΔV. The above expression characterizes the fluctuations in the electric field when no photons are present. Compare it with the square of the field strength for a classical electromagnetic wave of wavelength $2\pi\lambda$ where the time average of $\mathbf{E}^2$ can be equated with the energy density of electromagnetic wave so that

$$(\mathbf{E}^2)_{\text{average}} = \bar{n}\hbar(c/\lambda), \tag{II}$$

where $\bar{n}$ stands for the number of photons per unit volume. Thus, for the validity of classical description, purely quantum effects such as (I) (with $\Delta\ell \sim \lambda$) must be completely negligible in comparison with (II). For this, we must have

$$\bar{n} \gg \frac{1}{\lambda^3},$$

i.e. the description of physical phenomenon based on classical electrodynamics is reliable when number of photons per volume λ^3 is $\gg 1$. Show that this condition is satisfied by number of photons per volume λ^3 at a distance of 10 km from an antenna with a power of 135,000 Watts giving waves of $\lambda \simeq 50$ cm.

17.7 For quantised radiation field, show that

$$[E_i(\mathbf{r}, t), B_j(\mathbf{r}', t)] = ic(4\pi)\hbar E_{ijk} \frac{\partial}{\partial x_k} \delta(\mathbf{r} - \mathbf{r}').$$

What does this say about the simultaneous measureability of **B** and **E**?

Chapter 18

Formal Theory of Scattering

18.1   Introduction

We have discussed scattering theory before, but we restricted ourselves to potential scattering involving elastic collisions of two simple structureless particles. We used coordinate representation explicitly in our discussion, whereas in some problems, other representations may be more advantageous.

We now discuss the general but formal theory of scattering because it has the following advantages:

(i) It is not restricted to any particular representation.

(ii) It can be easily generalised to complex systems such as nuclear reactions where there are many channels, elastic as well as inelastic.

(iii) Symmetry properties of the system can be easily incorporated in this formalism and the selection rules for the scattering process can be easily derived.

It is useful and natural to look at a scattering process as a transition from one unperturbed state (initial state) to another (final state). The scattering process corresponds to the situation where the energy spectrum is continuous and the perturbation is such that

$$V = 0 \quad \text{for} \quad t \ll 0 \quad \text{and} \quad t \gg 0$$

and independent of time in between.

If the scattering region is of finite extent, the initial and final states can be taken, to a good degree of approximation, as plane wave eigenstates of definite momentum of the unperturbed Hamiltonian

$$H_0 = \mathbf{p}^2 / 2\mu$$

(μ is the reduced mass for scattering of a system of two particles) and the scattering interaction causes transitions from an initial state with momentum $\mathbf{p}_i$ to final states characterised by momentum $\mathbf{p}_f$.

We write the Hamiltonian as

$$H = H_0 + V, \tag{18.1}$$

where H_0 is the kinetic energy operator of two colliding particles. For this simple case in the centre of mass system, $H_0 = \mathbf{p}^2/2\mu$. For the case of two colliding systems H_0 also contains their separate internal energies. V is the interaction between the colliding particles responsible for their scattering.

H_0 is so simple that solutions to the eigenvalue equation

$$H_0|n\rangle = E_n|n\rangle \tag{18.2}$$

are known.

The Schrödinger equation is

$$i\hbar\frac{d}{dt}|\psi(t)\rangle = H|\psi(t)\rangle. \tag{18.3}$$

The eigenstates $|n\rangle$ of H_0 form a complete set so that we can write the solution $|\psi(t)\rangle$ in terms of them, i.e.

$$|\psi(t)\rangle = \sum_n C_n(t)e^{-iE_nt/\hbar}|n\rangle. \tag{18.4}$$

Substituting Eq. (18.4) in Eq. (18.3) and using the orthonormality of eigenvectors

$$\langle b|n\rangle = \delta_{bn},$$

we get

$$i\hbar\frac{dC_b(t)}{dt} = \sum_n V_{bn}e^{i\omega_{bn}t}C_n(t), \tag{18.5}$$

where

$$V_{bn} = \langle b|V|n\rangle, \tag{18.6}$$

$$\hbar\omega_{bn} = E_b - E_n. \tag{18.7}$$

Now for $t \ll 0$, $V = 0$ so that our system is in an eigenstate of H_0

$$\begin{aligned}|\psi(t \ll 0)\rangle &= |a, t\rangle \\ &= |a\rangle e^{-iE_at/\hbar}. \end{aligned} \tag{18.8}$$

It is convenient to introduce 'in' and 'out' state. Thus $|a\rangle_{in} = |a, t \to -\infty\rangle$, $|a\rangle_{out} = |a, t \to \infty\rangle$. Comparing Eq. (18.8) with Eq. (18.4), we can write the initial conditions of the problem:

$$C_a(-\infty) = 1, C_n(-\infty) = 0, \quad n \neq a. \tag{18.9}$$

We are interested in calculating $C_b(t)$ for large t, so that

$$\frac{d}{dt}|C_b(t)|^2 \quad \text{for large } t$$

gives the transition probability per unit time from initial state $|a\rangle$ with propagation vector $\mathbf{k}_i(\mathbf{p}_i = \hbar\mathbf{k}_i)$ to final state $|b\rangle$ characterised by propagation vectors $\mathbf{k}_f(\mathbf{p}_f = \hbar\mathbf{k}_f)$. As we shall see the above quantity is related to the cross section.

To first order in perturbation theory (i.e. put $C_n = 0, n \neq a, C_a = 1$ on the right-hand side of Eq. (18.5), we get a solution of Eq. (18.5)

$$C_b(t) = -\frac{i}{\hbar} \int_{-\infty}^{t} dt' e^{i\omega_{ba}t'} V_{ba}(t') + \delta_{ba} \tag{18.10a}$$

with the initial condition of Eq. (18.9).

If V is independent of time as we shall assume, then

$$C_b(t) = -\frac{i}{\hbar} V_{ba} \int_{-\infty}^{t} e^{i\omega_{ba}t'} dt' + \delta_{ba}. \tag{18.10b}$$

The integral in Eq. (18.10b) is not defined for $t' = -\infty$. To give a meaning to the integral, we use the following prescription. Introduce a factor $e^{\eta t'}$ in the integrand and write

$$C_b(t) = -\lim_{\eta \to 0} \frac{i}{\hbar} V_{ba} \int_{-\infty}^{t} e^{i\omega_{ba}t' + \eta t'} dt' + \delta_{ba}, \tag{18.11}$$

where the limit $\eta \to 0$ is taken at the end, i.e. $\lim \eta \to 0$ is taken after the $\lim t_0 \to -\infty$. The above prescription will hold for a time t which satisfies

$$|t| \ll \frac{1}{\eta}.$$

18.2 Interaction Picture

To get a general solution of Eq. (18.3) or Eq. (18.5), it is convenient to go to the interaction picture. In the Schrödinger picture, the state vector is time dependent:

$$i\hbar\frac{d}{dt}|\psi(t)\rangle_S = H|\psi(t)\rangle_S \tag{18.12}$$

$$H = H_0 + V,$$

where H_0 is independent of time, but V may in general depend on time, although most of the time we shall take it to be independent of time.

We define a state vector

$$|\psi(t)\rangle_I = e^{iH_0 t/\hbar}|\psi(t)\rangle_S, \tag{18.13}$$

$$i\hbar\frac{d}{dt}|\psi(t)\rangle_I = -H_0 e^{iH_0 t/\hbar}|\psi(t)\rangle_S + e^{iH_0 t/\hbar}i\hbar\frac{d}{dt}|\psi(t)\rangle_S$$

$$= -H_0|\psi(t)\rangle_I + e^{iH_0 t/\hbar}H e^{-iH_0 t/\hbar}|\psi(t)\rangle_I. \tag{18.14}$$

Now define

$$H_0^I(t) = e^{iH_0 t/\hbar}H_0 e^{-iH_0 t/\hbar}$$

$$= H_0 \tag{18.15a}$$

and

$$H_I(t) = e^{iH_0 t/\hbar}H e^{-iH_0 t/\hbar}$$

$$= H_0 + V_I(t), \tag{18.15b}$$

where

$$V_I(t) = e^{iH_0 t/\hbar}V e^{-iH_0 t/\hbar}. \tag{18.15c}$$

From Eq. (18.14), we have

$$i\hbar\frac{d}{dt}|\psi(t)\rangle_I = V_I(t)|\psi(t)\rangle_I. \tag{18.16}$$

We see that $V_I(t)$ is the effective Hamiltonian for the state vector $|\psi(t)\rangle_I$. This is the reason it is called an interaction picture. Since the transformation is effected by a unitary operator $e^{iH_0 t/\hbar}$, $|\psi(t)\rangle_I$ describes the same quantum state $|\psi(t)\rangle_S$ but in a different picture. The unitary transformations Eq. (18.13) and Eq. (18.15) define the interaction picture.

An arbitrary operator $\hat{A}$ in the Schrödinger picture is transformed into

$$\hat{A}_I(t) = e^{iH_0 t/\hbar}\hat{A}e^{-iH_0 t/\hbar}, \tag{18.17a}$$

$$i\hbar\frac{d\hat{A}_I(t)}{dt} = [\hat{A}_I(t), H_0]. \tag{18.17b}$$

To summarise, we compare the three pictures

Schrödinger picture	Interaction picture	Heisenberg picture
$i\hbar\frac{d}{dt}|\psi(t)\rangle_S = H|\psi(t)\rangle_S$	$i\hbar\frac{d}{dt}|\psi(t)\rangle_I$	$i\hbar\frac{d}{dt}|\psi(t)\rangle_H = 0$
$i\hbar\frac{d\hat{A}}{dt} = 0$	$= V_I(t)|\psi(t)\rangle_I$	$|\psi(t)\rangle_H = |\psi(0)\rangle_S$
$\hat{A} = \hat{A}(0)$	$\hat{A}_I(t) = e^{iH_0 t/\hbar}\hat{A}e^{-iH_0 t/\hbar}$	$\hat{A}_H(t) = e^{iHt/\hbar}\hat{A}e^{-iHt/\hbar}$
	$i\hbar\frac{d}{dt}\hat{A}_I(t) = [\hat{A}_I(t), H_0]$	$i\hbar\frac{d}{dt}\hat{A}_H(t) = [\hat{A}_H(t), H]$
$H = H_0 + V$	$H_I(t) = H_0 + V_I(t)$	$H_H = H$.

The interaction picture is intermediate between the Schrödinger and Heisenberg pictures. Here both the state vectors and operators change with time but the time development is governed by different portions of the Hamiltonian, the state vectors by the interaction V and the operators by H_0.

18.3 Formal Solution of Schrödinger Equation in the Interaction Picture

Assume that $|\psi(t)\rangle_I$ is generated from $|\psi(t_0)\rangle_I$ by a linear operator $U(t, t_0)$

$$|\psi(t)\rangle_I = U(t, t_0)|\psi(t_0)\rangle_I, \tag{18.18}$$

where

$$U(t_0, t_0) = 1, \tag{18.19}$$

by definition. Substituting Eq. (18.18) in Eq. (18.16), we have

$$i\hbar \frac{\partial U(t, t_0)}{\partial t}|\psi(t_0)\rangle_I = V_I(t)U(t, t_0)|\psi(t_0)\rangle_I,$$

which implies that

$$i\hbar \frac{\partial U(t, t_0)}{\partial t} = V_I(t)U(t, t_0). \tag{18.20}$$

The important point is that the operator $U(t, t_0)$ depends only on the structure of the physical system and not on the particular choice of initial state $|\psi(t_0)\rangle_I$.

We see that

$$\begin{aligned}|\psi(t)\rangle_I &= U(t, t_0)|\psi(t_0)\rangle_I \\ &= U(t, t')|\psi(t')\rangle_I \\ &= U(t, t')U(t', t_0)|\psi(t_0)\rangle_I.\end{aligned}$$

Therefore

$$U(t, t')U(t', t_0) = U(t, t_0). \tag{18.21a}$$

This is known as the group property of the operator U. In particular,

$$1 = U(t_0, t_0) = U(t_0, t)U(t, t_0).$$

Hence

$$U(t_0, t) = U^{-1}(t, t_0). \tag{18.21b}$$

Sometimes it is convenient to replace the differential equation by an integral equation. Thus integrating Eq. (18.20) from t_0 to t and making use of the condition (18.19), we have

$$U(t, t_0) = 1 - \frac{i}{\hbar} \int_{t_0}^{t} V_I(t')U(t', t_0)dt'. \tag{18.22}$$

Substituting back, the expression $U(t,t_0)$ from Eq. (18.22), we have an iteration procedure:

$$U(t,t_0) = 1 - \frac{i}{\hbar}\int_{t_0}^{t} dt_1 V_I(t_1)[1 - \frac{i}{\hbar}\int_{t_0}^{t_1} V_I(t_2)U(t_2,t_0)dt_2]$$

$$= 1 - \frac{i}{\hbar}\int_{t_0}^{t} dt_1 V_I(t_1) + \left(-\frac{i}{\hbar}\right)^2 \int_{t_0}^{t} dt' V_I(t_1)\int_{t_0}^{t_1} V_I(t_2)dt_2 + \cdots.$$

$$(18.23)$$

This can be used as a basis for a perturbation expansion, each successive term giving one higher order of expansion. For $t \to \infty$, $t_0 \to -\infty$

$$U(\infty,-\infty) = 1 - \frac{i}{\hbar}\int_{-\infty}^{\infty} dt_1 V_I(t_1) + (\frac{-i}{\hbar})^2 \int_{-\infty}^{\infty} dt_1 \int_{-\infty}^{t_1} dt_2 V_I(t_1)V_I(t_2) + \cdots.$$

We can write it in more compact form. By interchanging $t_1 \to t_2$, we can write the second term as $(\frac{-i}{\hbar})^2 \int_{-\infty}^{\infty} dt_2 \int_{-\infty}^{t_2} dt_1 V_I(t_2)V_I(t_1)$. It can be shown this term is equally well written as $(\frac{-i}{hbar})^2 \int_{-\infty}^{\infty} dt_1 \int_{-\infty}^{t_1} dt_2$ $V_I(t_2)V_I(t_1)$. Using this identity we can write the second term as $(\frac{-i}{\hbar})^2 \frac{1}{2!} \int_{-\infty}^{\infty} dt_1 \int_{-\infty}^{\infty} dt_2 T(V_I(t_1)V_I(t_2))$, where the time ordered product is defined as

$$T(V_I(t_1)V_I(t_2)) = V_I(t_1)V_I(t_2) \qquad t_2 < t_1,$$
$$= V_I(t_2)V_I(t_1) \qquad t_2 > t_1.$$

Thus

$$U(\infty,-\infty)$$
$$= 1 - \frac{i}{\hbar}\int_{-\infty}^{\infty} dt_1 V_I(t_1) + (\frac{-i}{\hbar})^2(\frac{1}{2!}) \int_{-\infty}^{\infty} dt_1 \int_{-\infty}^{\infty} dt_2 T(V_I(t_1)V_I(t_2))$$
$$+ \cdots + (\frac{-i}{\hbar})^n(\frac{1}{n!}) \int_{-\infty}^{\infty} dt_1 \int_{-\infty}^{\infty} dt_2 \cdots \int_{-\infty}^{\infty} dt_n T(V_I(t_1)V_I(t_2)\cdots V_I(t_n))$$

$$(18.24)$$

Now from Eqs. (18.4) and (18.13)

$$|\psi(t)\rangle_I = e^{iH_0t/\hbar}|\psi(t)\rangle$$
$$= \sum_n C_n(t)e^{-iE_nt/\hbar}e^{iH_0t/\hbar}|n\rangle$$
$$= \sum_n C_n(t)|n\rangle,$$

---

[1]see J. J. Sakurai, Advanced quantum mechanics, Addison-Wesley (1973) p.186

so that

$$\langle b|\psi(t)\rangle_I = \sum_n C_n(t)\delta_{bn}$$
$$= C_b(t). \tag{18.25}$$

If the experimental arrangement is such that the system is known to be in an eigenstate $|a\rangle$ of H_0 at $t = t_0 \to -\infty$, then the transition amplitude to some eigenstate $|b\rangle$ of H_0 is given by

$$C_b(t) = \langle b|\psi(t)\rangle_I$$
$$= \lim_{t_0 \to -\infty} \langle b|U(t, t_0)|\psi(t_0)\rangle_I. \tag{18.26}$$

But

$$|\psi(t_0)\rangle_I = e^{iH_0 t_0/\hbar}|\psi(t_0)\rangle.$$

Therefore,

$$\lim_{t_0 \to -\infty} |\psi(t_0)\rangle_I = \lim_{t_0 \to -\infty} e^{iH_0 t_0/\hbar} e^{-iE_a t_0/\hbar}|a\rangle$$
$$= \lim_{t_0 \to -\infty} e^{-iE_a t_0/\hbar} e^{iE_a t_0/\hbar}|a\rangle$$
$$= |a\rangle. \tag{18.27}$$

Thus

$$C_b(t) = \lim_{t_0 \to -\infty} \langle b|U(t, t_0)|a\rangle = \langle b|U(t, -\infty)|a\rangle. \tag{18.28}$$

From Eq. (18.23), we see that to first order in perturbation

$$U(t, -\infty) = 1 - \frac{i}{\hbar} \int_{-\infty}^{t} dt' V_I(t')dt'.$$

Therefore

$$C_b(t) = \delta_{ab} - \frac{i}{\hbar} \langle b| \int_{-\infty}^{t} dt' V_I(t')dt'|a\rangle$$
$$= \delta_{ab} - \frac{i}{\hbar} \langle b| \int_{-\infty}^{t} dt' e^{iH_0 t'/\hbar} V e^{-iH_0 t'/\hbar}|a\rangle$$
$$= \delta_{ab} - \frac{i}{\hbar} \int_{-\infty}^{t} dt' e^{i\omega_{ba} t'} V_{ba}(t'),$$

which is a re-derivation of Eq. (18.11) if V is independent of time.

We shall now assume that V is independent of time. We note that the formal solution of Eq. (18.3) or (18.12) is

$$|\psi(t)\rangle = e^{-iH(t-t_0)/\hbar}|\psi(t_0)\rangle \tag{18.29}$$

or

$$|\psi(t)\rangle_I = e^{iH_0t/\hbar}e^{-iH(t-t_0)/\hbar}|\psi(t_0)\rangle. \qquad (18.30a)$$

But

$$|\psi(t)\rangle_I = U(t,t_0)|\psi(t_0)\rangle_I$$
$$= U(t,t_0)e^{iH_0t_0/\hbar}|\psi(t_0)\rangle. \qquad (18.30b)$$

Comparing Eqs. (18.30a) and (18.30b), we have

$$U(t,t_0)e^{iH_0t_0/\hbar} = e^{iH_0t/\hbar}e^{-iH(t-t_0)/\hbar}$$

or

$$U(t,t_0) = e^{iH_0t/\hbar}e^{-iH(t-t_0)/\hbar}e^{-iH_0t_0/\hbar}. \qquad (18.31a)$$

Hence

$$|\psi(t)\rangle_I = e^{iH_0t/\hbar}e^{-iH(t-t_0)/\hbar}e^{-iH_0t_0/\hbar}|\psi(t_0)\rangle_I. \qquad (18.31b)$$

From Eq. (18.31a), we have

$$U^\dagger(t,t_0) = e^{iH_0t_0/\hbar}e^{iH(t-t_0)/\hbar}e^{-iH_0t/\hbar}$$
$$= e^{iH_0t_0/\hbar}e^{-iH(t_0-t)/\hbar}e^{-iH_0t/\hbar}$$
$$= U(t_0,t)) \qquad (18.32a)$$

or

$$U^\dagger(t,t_0)U(t,t_0) = U(t_0,t)U(t,t_0)$$
$$= U(t_0,t_0)$$
$$= 1. \qquad (18.32b)$$

Hence

$$U^\dagger(t,t_0) = U^{-1}(t,t_0). \qquad (18.32c)$$

Thus $U(t,t_0)$ is an unitary operator.

Note that H_0 and H do not in general commute and therefore we cannot add up the exponents in Eq. (18.31). Since ultimately we are interested in

$$\lim_{t_0\to-\infty} U(t,t_0) \quad \text{and} \quad \lim_{t\to+\infty}\lim_{t_0\to-\infty} U(t,t_0),$$

we first define the above limiting procedure. We shall use the following prescription. Consider first a function $f(t)$. Then we define

$$\lim_{t\to-\infty} f(t) = \lim_{\eta\to0} \eta \int_{-\infty}^{0} e^{\eta t'} f(t')dt' \qquad (18.33a)$$

and

$$\lim_{t \to +\infty} f(t) = \lim_{\eta \to 0} \eta \int_0^\infty e^{-\eta t'} f(t') dt'. \qquad (18.33b)$$

If $f(t)$ is a smooth function having well defined limits $f(\mp\infty)$ as $t \to \mp\infty$, then the above are identities as can be easily seen by integrating by parts:

$$\lim_{t \to -\infty} f(t) = \lim_{\eta \to 0} \left(\frac{\eta e^{\eta t'}}{\eta} f(t')|^0_{-\infty} - \int_{-\infty}^0 e^{\eta t'} f'(t') dt' \right)$$

$$= f(0) - f(t')|^0_{-\infty}$$

$$= f(-\infty).$$

If $f(t)$ oscillates at $t = \mp\infty$, then at large $|t|$, the corresponding exponential factors $e^{\pm\eta t}$ will average out such oscillations to zero. In such cases, we get finite η dependent expressions. We can then use these expressions for further calculation and take the limit $\eta \to 0$ in the final result.

Example: (i)

$$\lim_{t \to -\infty} e^{i\omega t} = \lim_{\eta \to 0} \eta \int_{-\infty}^0 e^{i\omega t'} e^{\eta t'} dt'$$

$$= \frac{\eta}{\eta + i\omega}.$$

Similarly

$$\lim_{t \to +\infty} e^{i\omega t} = \frac{-\eta}{-\eta + i\omega}.$$

Thus we have a finite but η-dependent expression.

(ii) $\int_{-\infty}^\infty e^{i\omega t} dt$. This function oscillates at $t = \mp\infty$. Now

$$\int_{-\infty}^\infty e^{i\omega t} dt = \frac{1}{i\omega} \left(\lim_{t \to \infty} e^{i\omega t} - \lim_{t \to -\infty} e^{i\omega t} \right)$$

$$= \lim_{\eta \to 0} \frac{1}{i\omega} \left(-\frac{\eta}{-\eta + i\omega} - \frac{\eta}{\eta + i\omega} \right)$$

$$= \lim_{\eta \to 0} \frac{2\eta}{\eta^2 + \omega^2}$$

$$= 2\pi\delta(\omega).$$

We now extend the above prescription to the operator (18.31a) and write

$$U(t, -\infty) = \lim_{\eta \to 0} \eta \int_{-\infty}^0 e^{\eta t'} e^{iH_0 t/\hbar} e^{-iHt/\hbar} e^{iHt'/\hbar} e^{-iH_0 t'/\hbar} dt'. \qquad (18.34a)$$

In particular,

$$U(0, -\infty) = \lim_{\eta \to 0} \eta \int_{-\infty}^0 e^{\eta t'} e^{iHt'/\hbar} e^{-iH_0 t'/\hbar} dt'. \qquad (18.34b)$$

18.4   Scattering States — The S-Matrix

We have

$$H_0|a\rangle = E_a|a\rangle,$$
$$e^{-iH_0t/\hbar}|a\rangle = e^{-iE_at/\hbar}|a\rangle.$$

Thus

$$U(0,-\infty)|a\rangle = \lim_{\eta\to 0}\eta\int_{-\infty}^{0} e^{[\eta+(i/\hbar)(H-E_a)]t'}\,dt'\,|a\rangle$$

$$= \lim_{\eta\to 0}\frac{\eta}{\eta+(i/\hbar)(H-E_a)}|a\rangle$$

$$= \lim_{\eta\to 0}\frac{i\hbar\eta}{E_a-H+i\hbar\eta}|a\rangle$$

$$= \lim_{\varepsilon\to 0}\frac{i\varepsilon}{E_a-H+i\varepsilon}|a\rangle, \qquad (18.35)$$

where $\varepsilon = \hbar\eta$. The meaning of Eq. (18.35) is as follows

$$(E_a - H + i\varepsilon)U(0,-\infty)|a\rangle = i\varepsilon|a\rangle,$$

i.e. in the limit $\varepsilon \to 0$,

$$(E_a - H)U(0,-\infty)|a\rangle = 0$$

or

$$HU(0,-\infty)|a\rangle = E_aU(0,-\infty)|a\rangle. \qquad (18.36)$$

Thus $U(0,-\infty)|a\rangle$ is an eigenstate of the total Hamiltonian H belonging to the eigenvalue E_a which is also the eigenvalue of H_0. The above procedure can succeed if in the limit $\varepsilon \to 0$, E_a is an eigenstate of both H_0 and H. In other words, it will be assumed that the continuum part of the energy spectrum of H_0 and H extend from 0 to ∞ and to each state $|a\rangle$ in the continuum part of the H_0-spectrum which has energy E_a, there belongs a corresponding state $U(0,-\infty)|a\rangle$ in the continuum part of the spectrum of the H having the same energy E_a. This assumption of no 'level shifts' in the continuum is not satisfied in some problems and then we have to drop this restriction. However, we shall not consider such cases.

We denote the state $U(0,-\infty)|a\rangle$ by

$$|a\rangle_{in} = |a^{(+)}\rangle = U(0,-\infty)|a\rangle$$

$$= \frac{i\varepsilon}{E_a-H+i\varepsilon}|a\rangle, \qquad (18.37)$$

or

$$(E_a - H + i\varepsilon)|a^+\rangle = i\varepsilon|a\rangle$$
$$= (E_a - H_0 + i\varepsilon)|a\rangle - V|a\rangle + V|a\rangle$$
$$= (E_a - H + i\varepsilon)|a\rangle + V|a\rangle. \tag{18.38a}$$

Alternatively, noting that $H = H_0 + V$, we have

$$(E_a - H_0 + i\varepsilon)|a^+\rangle = i\varepsilon|a\rangle + V|a^+\rangle$$
$$= (E_a - H_0 + i\varepsilon)|a\rangle + V|a^+\rangle. \tag{18.38b}$$

We can write Eqs. (18.38a, 18.38b) as

$$|a^{(+)}\rangle = |a\rangle + \frac{1}{E_a - H + i\varepsilon}V|a\rangle, \tag{18.39a}$$

$$|a^{(+)}\rangle = |a\rangle + \frac{1}{E_a - H_0 + i\varepsilon}V|a^{(+)}\rangle. \tag{18.39b}$$

The notation $|a^{(+)}\rangle$ is intended to emphasise that the solution of the above equation will tend to the 'unperturbed' state $|a\rangle$ as $t \to -\infty$.

In exactly the same manner, we can show that

$$U(0,\infty)|b\rangle = \lim_{\eta \to 0} \eta \int_0^\infty e^{[-\eta + (i/\hbar)(H - E_b)]t'} dt' |b\rangle$$

$$= \lim_{\varepsilon \to 0} \frac{-i\varepsilon}{E_b - H - i\varepsilon}|b\rangle. \tag{18.40}$$

We denote $U(0,\infty)|b\rangle$ by $|b^{(-)}\rangle$, so that

$$|b\rangle_{out} = |b^{(-)}\rangle = U(0,\infty)|b\rangle$$

$$= -\frac{i\varepsilon}{E_b - H - i\varepsilon}|b\rangle. \tag{18.41}$$

Equivalently

$$|b^{(-)}\rangle = |b\rangle + \frac{1}{E_b - H - i\varepsilon}V|b\rangle, \tag{18.42a}$$

$$|b^{(-)}\rangle = |b\rangle + \frac{1}{E_b - H_0 - i\varepsilon}V|b^{(-)}\rangle. \tag{18.42b}$$

The notation $|b^{(-)}\rangle$ is intended to emphasise that the solution of the above equation will tend to an 'unperturbed' state $|b\rangle$ as $t \to +\infty$.

The Eqs. (16.39) and (16.42) are the fundamental equations of the formal theory of scattering and are known as Lippmann–Schwinger equations.

Our purpose is to calculate the transition amplitude

$$\langle b|U(t, -\infty)|a\rangle \quad \text{for large } t, \text{ viz.}$$

$$\lim_{t \to \infty} \langle b|U(t, -\infty)|a\rangle \equiv S_{ba}. \tag{18.43}$$

Now using the group property of U,

$$S_{ba} = \langle b|U(\infty, -\infty)|a\rangle$$
$$\langle b|U(\infty, 0)U(0, -\infty)|a\rangle.$$

Now

$$U(0, -\infty)|a\rangle = |a^{(+)}\rangle$$
$$|b^{(-)}\rangle = U(0, \infty)|b\rangle,$$

so that

$$\langle b^{(-)}| = \langle b|U^\dagger(0, \infty)$$
$$= \langle b|U(\infty, 0).$$

Therefore

$$S_{ba} = \langle b|U(\infty, -\infty)|a\rangle$$
$$= \langle b^{(-)}|a^{(+)}\rangle = \,_{out}\langle b|a\rangle_{in}. \qquad (18.44)$$

The operator $S = U(\infty, -\infty)$, with matrix elements given above is called the scattering matrix or S-matrix. Note that this operator is unitary.

Now we want to calculate $\langle b^{(-)}|a^{(+)}\rangle$. For this purpose, we first calculate $\langle b|a^{(+)}\rangle$. Now, on using Eq. (18.39b)

$$\langle b|a^{(+)}\rangle = \langle b|a\rangle + \langle b|\frac{1}{E_a - H_0 + i\varepsilon}V|a^{(+)}\rangle. \qquad (18.45a)$$

But

$$\langle b|(E_a - H_0 + i\varepsilon) = \langle b|(E_a - E_b + i\varepsilon).$$

Operating on both sides by $\frac{1}{E_a - H_0 + i\varepsilon}$, we have

$$\langle b| = (E_a - E_b + i\varepsilon)\langle b|\frac{1}{E_a - H_0 + i\varepsilon}$$

or

$$\frac{1}{E_a - E_b + i\varepsilon}\langle b| = \langle b|\frac{1}{E_a - H_0 + i\varepsilon}. \qquad (18.45b)$$

Hence

$$\langle b|a^{(+)}\rangle = \langle b|a\rangle + \frac{1}{E_a - E_b + i\varepsilon}\langle b|V|a^{(+)}\rangle. \qquad (18.45c)$$

Now from Eq. (18.42a), we also have

$$\langle b^{(-)}| = \langle b| + \langle b|V\frac{1}{E_b - H + i\varepsilon},$$

therefore

$$\langle b^{(-)}|a^{(+)}\rangle = \langle b|a^{(+)}\rangle + \langle b|V\frac{1}{E_b - H + i\varepsilon}|a^{(+)}\rangle.$$

The argument is similar to that used in the derivation of Eq. (18.45b) gives

$$\frac{1}{E_b - H + i\varepsilon}|a^{(+)}\rangle = \frac{1}{E_b - E_a + i\varepsilon}|a^{(+)}\rangle.$$

Therefore

$$\langle b^{(-)}|a^{(+)}\rangle = \langle b|a^{(+)}\rangle + \frac{1}{E_b - E_a + i\varepsilon}\langle b|V|a^{(+)}\rangle$$

$$= \langle b|a\rangle + \left(\frac{1}{E_a - E_b + i\varepsilon} + \frac{1}{E_b - E_a + i\varepsilon}\right)\langle b|V|a^{(+)}\rangle$$

$$= \delta_{ba} + \lim_{\varepsilon\to 0}\frac{-2i\varepsilon}{(E_b - E_a)^2 + \varepsilon^2}\langle b|V|a^{(+)}\rangle$$

$$= \delta_{ba} - 2\pi i\delta(E_b - E_a)\langle b|V|a^{(+)}\rangle. \tag{18.46}$$

We define an operator T with matrix elements

$$T_{ba} = \langle b|T|a\rangle = -\langle b|V|a^{(+)}\rangle, \tag{18.47}$$

so that

$$S_{ba} = \langle b|S|a\rangle$$

$$= \langle b^{(-)}|a^{(+)}\rangle$$

$$= \delta_{ba} + 2\pi i\delta(E_b - E_a)T_{ba}. \tag{18.48}$$

T is called the transition matrix or T-matrix. This gives the amplitude for the transition from state $|a\rangle$ to $|b\rangle$.

Now from Eq. (18.39b), introducing a complete set of states $|c\rangle$, we have

$$|a^{(+)}\rangle = |a\rangle + \sum_c |c\rangle\langle c|\frac{1}{E_a - H_0 + i\varepsilon}V|a^{(+)}\rangle$$

$$= |a\rangle - \sum_c \frac{1}{E_a - E_c - i\varepsilon}|c\rangle T_{ca}$$

$$= \left(|a\rangle - \sum_c \frac{1}{E_a - H_0 + i\varepsilon}|c\rangle\langle c|T|a\rangle\right)$$

$$= \left(1 - \frac{1}{E_a - H_0 + i\varepsilon}T\right)|a\rangle. \tag{18.49}$$

Furthermore, in order to obtain perturbative approximation for T_{ba}, we need an equation for T_{ba} in terms of V_{ba}. This can be easily obtained as follows:

$$T_{ba} = -\langle b|V|a^+\rangle$$

$$= -\sum_c \langle b|V|c\rangle\langle c|a^+\rangle$$

$$= -\sum_c V_{bc}\left\{\delta_{ca} + \frac{1}{E_a - E_c + i\varepsilon}\langle c|V|a^+\rangle\right\},$$

where we have used Eq. (18.45a). Hence

$$T_{ba} = -V_{ba} + \sum_c \frac{V_{bc}T_{ca}}{E_a - E_c + i\varepsilon}. \tag{18.50}$$

These give a system of simultaneous equations from which perturbative approximations to the transition amplitude T_{ba} can be obtained. Equivalently in operator form Eq. (18.50) can be written as

$$T = -V + V\frac{1}{E_a - H_0 + i\varepsilon}T. \tag{18.51}$$

The iterative solution to this equation is given by

$$T = -V - V\frac{1}{E_a - H_0 + i\varepsilon}V \cdots . \tag{18.52}$$

Finally we want to relate $|T_{ba}|^2$ with the transition probability per unit time from the state $|a\rangle$ to the state $|b\rangle$. For this purpose we note that the probability of transition from the state $|a\rangle$ to the state $|b\rangle$ is given by $|C_b(t)|^2$, where

$$C_b(t) = \langle b|U(t, -\infty)|a\rangle. \tag{18.53}$$

Now from Eq. (18.31a),

$$e^{-iH_0t/\hbar}U(t, t_0) = e^{-iHt/\hbar}e^{iHt_0/\hbar}e^{-iH_0t_0/\hbar},$$

therefore

$$e^{-iH_0t/\hbar}U(t, -\infty)|a\rangle = e^{-iHt/\hbar}\lim_{\eta\to 0}\eta\int_{-\infty}^{0}dt'e^{\eta t'}e^{(i/\hbar)(H-E_a)t'}|a\rangle$$

$$= e^{-iHt/\hbar}\frac{i\varepsilon}{E_a - H + i\varepsilon}|a\rangle$$

$$= e^{-iHt/\hbar}|a^{(+)}\rangle$$

$$= e^{-iE_at/\hbar}|a^{(+)}\rangle. \tag{18.54}$$

Hence

$$\langle b|e^{-iH_0t/\hbar}U(t, -\infty)|a\rangle = e^{-iE_at/\hbar}\langle b|a^+\rangle. \tag{18.55}$$

But

$$\langle b|e^{-iH_0 t/\hbar} = e^{-iE_b t/\hbar}\langle b|.$$

Therefore

$$C_b(t) = e^{i(E_b - E_a)t/\hbar}\langle b|a^{(+)}\rangle$$

$$= e^{i(E_b - E_a)t/\hbar}\left(\delta_{ba} - \frac{1}{E_a - E_b + i\varepsilon}T_{ba}\right)$$

on using (18.45b) and (18.49). Hence

$$C_b(t) = -T_{ba}\frac{e^{i(E_b - E_a)t/\hbar}}{(E_a - E_b + i\varepsilon)} + \delta_{ba}. \tag{18.56}$$

Note that equivalently it can be written as

$$C_b(t) = -\frac{i}{\hbar}(-T_{ba})\int_{-\infty}^{t} dt'\, e^{i(E_b - E_a)t'/\hbar + \eta t'} + \delta_{ba}, \tag{18.57}$$

where the limit $\eta \to 0$ is taken in the end. The perturbative expression (18.11) can be obtained from the exact expression (18.57) by replacing T_{ba} by $-V_{ba}$.

For the state $b \neq a$, the probability of transition from state $|a\rangle$ to $|b\rangle$ is given by [on using Eq. (18.56)]

$$|C_b(t)|^2 = \frac{1}{\hbar^2}\left|\lim_{\eta\to 0}\frac{e^{i\omega t}}{-\omega + i\eta}\right|^2 |T_{ba}|^2. \tag{18.58}$$

We are interested in this probability of transition for large t. For large t

$$\lim_{\eta\to 0}\frac{e^{i\omega t}}{\omega - i\eta} = 2\pi i\delta(\omega).$$

Thus for large t, we have

$$|C_b(t)|^2 = (2\pi)^2\delta(\omega)\delta(\omega)|T_{ba}|^2$$

$$= (2\pi)^2\delta(0)\delta\left(\frac{E_b - E_a}{\hbar}\right)\frac{1}{\hbar^2}|T_{ba}|^2$$

$$= (2\pi)^2\delta(0)\delta(E_b - E_a)\frac{1}{\hbar}|T_{ba}|^2.$$

Now

$$\delta(\omega) = \frac{1}{2\pi}\lim_{t\to\infty}\int_{-t/2}^{t/2} e^{i\omega t'}\, dt'$$

$$\delta(0) = \frac{1}{2\pi}\int_{-t/2}^{t/2} dt$$

$$= \frac{1}{2\pi}\lim_{t\to\infty} t.$$

Therefore, for large t,

$$|C_b(t)|^2 = \frac{(2\pi)^2}{(2\pi)}\delta(E_b - E_a)\frac{1}{\hbar}|T_{ba}|^2 t \tag{18.59}$$

or

$$\frac{d}{dt}|C_b(t)|^2 = 2\pi\delta(E_b - E_a)\frac{1}{\hbar}|T_{ba}|^2. \tag{18.60}$$

The same result is obtained from Eq. (18.57) if the limit $\eta \to 0$ is taken at the very end. It is assumed that T_{ba} has no singularity as a function of energy at $E_b = E_a$. The above result implies a constant transition rate. This is what we expect to be the effect of the the scattering causing transition from state $|a\rangle$ to state $|b\rangle$ and the above transition rate is closely related to the scattering cross-section as we shall see.

Equation (18.60) explicitly exhibits the fact that energy is conserved in the transitions. If the transition is to a continuum of states about state b, then $\rho_f(E_b)dE_b$ is the number of final states with energies between E_b and $E_b + dE_b$, where $\rho_f(E_b)$ is the density of final states, i.e. the number of states per unit energy. In this case, the transition probability per unit time is given by

$$\begin{aligned}
W &= \int \frac{d}{dt}|C_b(t)|^2\rho_f(E_b)dE_b \\
&= \frac{2\pi}{\hbar}\int \delta(E_b - E_a)|T_{ba}|^2\rho_f(E_b)dE_b \\
&= \frac{2\pi}{\hbar}|T_{ba}|^2\rho_f(E_b).
\end{aligned} \tag{18.61}$$

If we use the first order perturbation theory, then

$$W = \frac{2\pi}{\hbar}|V_{ba}|^2\rho_f(E_b). \tag{18.62}$$

This is the famous Fermi Golden Rule.

18.5 The Scattering Cross-section

In this case, we take the unperturbed states $|a\rangle$ and $|b\rangle$ to be normalised momentum eigenstates with initial and final momentum to be $\mathbf{p}_i = \hbar\mathbf{k}_i$, $\mathbf{p}_f = \hbar\mathbf{k}_f$ so that

$$|a\rangle = |\mathbf{p}_i\rangle, |b\rangle = |\mathbf{p}_f\rangle.$$

We note that the number of states between $\mathbf{k}_f$ and $\mathbf{k}_f + d\mathbf{k}_f$ in volume V is given by

$$\frac{V}{(2\pi)^3}d^3\mathbf{k}_f = \frac{V}{(2\pi\hbar)^3}d^3\mathbf{p}_f.$$

Now

$$u_{\mathbf{p}}(\mathbf{r}) = \langle \mathbf{r}|\mathbf{p}\rangle = \frac{1}{(2\pi\hbar)^{3/2}} e^{(i\mathbf{p}/\hbar)\cdot\mathbf{r}},$$

$$|u_{\mathbf{p}}(\mathbf{r})|^2 = \frac{1}{(2\pi\hbar)^3},$$

and

$$\int_V |u_{\mathbf{p}}(\mathbf{r})|^2 d^3r = \frac{V}{(2\pi\hbar)^3} = 1.$$

Thus we take $V = (2\pi\hbar)^3$. Hence the number of final states between $\mathbf{p}_f$ and $\mathbf{p}_f + d\mathbf{p}_f$

$$\rho_f(E_f)dE_f = d^3\mathbf{p}_f$$
$$= |\mathbf{p}_f|^2 d|\mathbf{p}_f| d\Omega$$
$$= |\mathbf{p}_f|^2 \frac{d|\mathbf{p}_f|}{dE_f} dE_f d\Omega. \qquad (18.63a)$$

Thus

$$\rho_f(E_f) = |\mathbf{p}_f|^2 \frac{d|\mathbf{p}_f|}{dE_f} d\Omega \qquad (18.63b)$$

and

$$W = \frac{2\pi}{\hbar} \int |T_{fi}|^2 d^3 p_f (E_f - E_i)$$
$$= \frac{2\pi}{\hbar} |T_{fi}|^2 \rho_f(E_f). \qquad (18.64)$$

Now the incident flux $= \rho_{in} v$ where ρ_{in} is the incident particle density and v is the incident velocity. Therefore,

$$\text{Flux} = \rho_{in} v$$
$$= \frac{1}{(2\pi\hbar)^3} v \qquad (18.65)$$

is the probability that a particle is incident on a unit area perpendicular to the beam per unit time. Let $d\sigma$ denote the differential cross-section. Then

$$d\sigma(\text{Flux}) = \text{Transition probability per unit time } W.$$

Hence

$$d\sigma = \frac{W}{\text{Flux}}$$
$$= \frac{1}{\text{Flux}} \frac{2\pi}{\hbar} |T_{fi}|^2 \rho_f(E_f), \qquad (18.66)$$

where $\rho_f(E_f)$ is given in Eq. (18.63b).

For the case of non-relativistic elastic scattering in the centre of mass system

$$E_i = \frac{p_i^2}{2\mu}, \quad E_f = \frac{p_f^2}{2\mu}, \quad \frac{dE_f}{d|\mathbf{p}_f|} = \frac{|\mathbf{p}_f|}{\mu},$$

$$|\mathbf{p}_i| = |\mathbf{p}_f| = p \quad \text{and} \quad v = \frac{p}{\mu}.$$

Therefore

$$\rho_f(E_f) = p^2 \frac{\mu}{p} d\Omega$$

$$= \mu p d\Omega \tag{18.67}$$

and

$$\text{Flux} \quad = \left(\frac{1}{2\pi\hbar}\right)^3 \frac{p}{\mu}.$$

Hence

$$d\sigma = \left((2\pi)^2 \mu\hbar\right)^2 |T_{fi}|^2 d\Omega. \tag{18.68}$$

18.6   Properties of the Scattering States

We have derived

$$|a^{(\pm)}\rangle = |a\rangle + \frac{1}{E_a - H \pm i\varepsilon} V|a\rangle, \tag{18.69}$$

$$|a^{(\pm)}\rangle = |a\rangle + \frac{1}{E_a - H_0 \pm i\varepsilon} V|a^{\pm}\rangle. \tag{18.70}$$

We may consider Eq. (18.69) to be a formal solution of Eq. (18.70) in the sense that $|a^{(\pm)}\rangle$ are given in terms of the known states $|a\rangle$. If we substitute the formal solution in Eq. (18.47), we get the transition matrix elements

$$T_{ba} = -\langle b|V|a^+\rangle$$

$$= -\langle b|V|a\rangle - \langle b|V \frac{1}{E_a - H + i\varepsilon} V|a\rangle. \tag{18.71}$$

In this way, $|T_{ba}|^2$ or the scattering cross-section for a scattering process can in principle be obtained. However, in practice, not much is gained since the effect of the operator $\frac{1}{E_a - H + i\varepsilon}$ is not known unless eigenvalues of H have already been determined but this is the problem we wish to solve. Hence, in practice, one has to resort to approximate methods to solve Eq. (18.70).

For this purpose, it is convenient to introduce the Green's operators

$$G_0^{\pm}(E_a) \equiv \frac{1}{E_a - H_0 \pm i\varepsilon} \tag{18.72a}$$

$$G^{\pm}(E_a) = \frac{1}{E_a - H \pm i\varepsilon}, \tag{18.72b}$$

so that

$$|a^{\pm}\rangle = |a\rangle + G_0^{\pm}(E_a)V|a^{\pm}\rangle \tag{18.73a}$$

with solutions

$$|a^{\pm}\rangle = |a\rangle + G^{\pm}(E_a)V|a\rangle. \tag{18.73b}$$

Using the identity

$$\frac{1}{A} - \frac{1}{B} = \frac{1}{B}(B - A)\frac{1}{A}$$

with

$$A = E_a - H \pm i\varepsilon,$$
$$B = E_a - H_0 \pm i\varepsilon,$$

we have

$$G^{\pm}(E_a) = G_0^{\pm}(E_a) + G_0^{\pm}(E_a)VG^{\pm}(E_a). \tag{18.74}$$

We can solve $G^{\pm}(E_a)$ by the perturbative expansion

$$G^{\pm}(E_a) = G_0^{\pm}(E_a) + G_0^{\pm}(E_a)VG_0^{\pm}(E_a) + G_0^{\pm}(E_a)VG_0^{\pm}(E_a)VG_0^{\pm}(E_a) + \cdots, \tag{18.75}$$

so that

$$|a^{\pm}\rangle = |a\rangle + G_0^{\pm}(E_a)V|a\rangle + G_0^{\pm}(E_a)VG_0^{\pm}(E_a)V|a\rangle + \cdots. \tag{18.76}$$

This can be regarded as an expansion of

$$|a^{\pm}\rangle = \frac{1}{1 - G_0^{\pm}(E_a)V}|a\rangle, \tag{18.77}$$

where $\frac{1}{1 - G_0^{\pm}(E_a)V}$ is expanded in power series.

The formal solution Eq. (18.69) can be used to demonstrate the orthogonality of the scattering states. Now

$$\langle b^{(+)}|a^{(+)}\rangle = \langle b|a^+\rangle + \langle b|V\frac{1}{E_b - H - i\varepsilon}|a^{(+)}\rangle$$

$$= \langle b|a\rangle + \frac{1}{E_a - E_b + i\varepsilon}\langle b|V|a^+\rangle + \langle b|V\frac{1}{E_b - E_a - i\varepsilon}|a^{(+)}\rangle$$

$$= \langle b|a\rangle = \delta_{ab}, \tag{18.78a}$$

where we have used Eq. (18.45b). Similarly

$$\langle b^{(-)} | a^{(-)} \rangle = \delta_{ba}. \tag{18.78b}$$

Corresponding to the orthonormal set $|a\rangle$, we have obtained two sets $|a^{(+)}\rangle$ and $|a^{(-)}\rangle$ of orthonormal eigenvectors of H. Are these sets complete? The answer in general is no, since H may also have discrete eigenvalues corresponding to bound states produced by the interaction V. These discrete states, which have no counterpart in the spectrum of H, are never found among the solutions of Eq. (18.70), as these solutions refer to the continuum part of H spectrum. These bound states are orthogonal to the scattering states as they have energies lower than energies in the scattering process and must be added to the scattering states $|a^{(+)}\rangle$ [or $|a^{(-)}\rangle$] for the completeness relation. Thus we have the completeness relation

$$\sum_c |c^{(+)}\rangle\langle c^{(+)}| + \sum_B |B\rangle\langle B| = 1, \tag{18.79}$$

where $|B\rangle$ refer to the bound states and

$$\langle c^{(+)} | B \rangle = 0, \tag{18.80a}$$
$$\langle c^{(-)} | B \rangle = 0. \tag{18.80b}$$

18.7   The Coordinate Representation

Let us consider the Green's operators

$$G_0^{\pm}(E_a) = \frac{1}{E_a - H_0 \pm i\varepsilon}. \tag{18.81a}$$

We can represent them as follows:

$$G_0^{\pm}(E_a) = \sum_c \frac{1}{E_a - H_0 \pm i\varepsilon} |c\rangle\langle c|$$
$$= \sum_c \frac{|c\rangle\langle c|}{E_a - E_c \pm i\varepsilon}. \tag{18.81b}$$

We now take the **r**-representation of our fundamental equation (18.73a)

$$\langle \mathbf{r} | a^{(\pm)} \rangle = \langle \mathbf{r} | a \rangle + \langle \mathbf{r} | G_0^{\pm}(E_a) V | a^{(\pm)} \rangle$$
$$= \langle \mathbf{r} | a \rangle + \int \int \langle \mathbf{r} | G_0^{\pm}(E_a) | \mathbf{r}'' \rangle$$
$$\times \langle \mathbf{r}'' | V | \mathbf{r}' \rangle \langle \mathbf{r}' | a^{(\pm)} \rangle d^3 r'' d^3 r'. \tag{18.82}$$

We can write

$$\langle \mathbf{r} | G_0^\pm(E_a) | \mathbf{r}'' \rangle = \lim_{\varepsilon \to 0} G_{\pm\varepsilon}(\mathbf{r}, \mathbf{r}'')$$
$$= G_\pm(\mathbf{r}, \mathbf{r}'') \tag{18.83a}$$

and

$$\langle \mathbf{r}'' | V | \mathbf{r}' \rangle = \delta(\mathbf{r}' - \mathbf{r}'') V(\mathbf{r}'), \tag{18.83b}$$

so that Eq. (18.82) becomes

$$\psi_a^{(\pm)}(\mathbf{r}) = \psi_a(\mathbf{r}) + \int G_\pm(\mathbf{r}, \mathbf{r}') V(\mathbf{r}') \psi_a^{(\pm)}(\mathbf{r}') d^3 r'. \tag{18.84}$$

On the other hand, the **r**-representation of the transition matrix is given by

$$T_{ba} = -\langle b | V | a^{(+)} \rangle$$
$$= -\int \int \langle b | \mathbf{r}'' \rangle \langle \mathbf{r}'' | V | \mathbf{r}' \rangle \langle \mathbf{r}' | a^+ \rangle d^3 r'' d^3 r'$$
$$= -\int \psi_b^*(\mathbf{r}') V(\mathbf{r}') \psi_a^{(+)}(\mathbf{r}') d^3 r'. \tag{18.85}$$

In the Born approximation, we can replace in the integrals above $\psi_a^{(+)}(\mathbf{r}')$ by $\psi_a(\mathbf{r}')$, so that

$$\psi_a^{(+)}(\mathbf{r}) = \psi_a(\mathbf{r}) + \int G_+(\mathbf{r}, \mathbf{r}') V(\mathbf{r}') \psi_a(\mathbf{r}') d^3 r', \tag{18.86a}$$

$$T_{ba}^B = -\int \psi_b^*(\mathbf{r}') V(\mathbf{r}') \psi_a(\mathbf{r}') d^3 r'. \tag{18.86b}$$

We consider two applications of the above considerations.

Application I: Consider two spinless and distinguishable particles α and β which interact via a potential

$$V(\mathbf{r}) \quad \text{with} \quad \mathbf{r} = \mathbf{r}_\alpha - \mathbf{r}_\beta.$$

No other potential is present in the problem. Thus

$$H = H_0 + V(\mathbf{r}),$$

$$H_0 = \frac{\hat{p}_\alpha^2}{2m_\alpha} + \frac{\hat{p}_\beta^2}{2m_\beta} = \frac{\hat{p}^2}{2\mu},$$

in the c.m. system and μ is the reduced mass. Hence $|a\rangle$ and $|b\rangle$ are normalised momentum eigenstates with initial and final momenta

$$\mathbf{p}_i = \hbar \mathbf{k}_i, \quad \mathbf{p}_f = \hbar \mathbf{k}_f,$$

so that

$$\langle \mathbf{r}|a\rangle = \langle \mathbf{r}|\mathbf{p}_i\rangle = \frac{1}{(2\pi\hbar)^{3/2}} e^{i\mathbf{p}_i \cdot \mathbf{r}/\hbar}, \qquad (18.87a)$$

$$\langle \mathbf{r}|b\rangle = \langle \mathbf{r}|\mathbf{p}_f\rangle = \frac{1}{(2\pi\hbar)^{3/2}} e^{i\mathbf{p}_f \cdot \mathbf{r}/\hbar}, \qquad (18.87b)$$

$$\langle \mathbf{p}_f|\mathbf{p}_i\rangle = \frac{1}{(2\pi\hbar)^3} \int e^{i(\mathbf{p}_i - \mathbf{p}_f)\cdot \mathbf{r}/\hbar}$$

$$= \delta(\mathbf{p}_f - \mathbf{p}_i). \qquad (18.87c)$$

Now

$$\langle \mathbf{r}|G_0^{\pm}(E_a)|\mathbf{r}'\rangle = \sum_{\mathbf{q}} \frac{\langle \mathbf{r}|\mathbf{q}\rangle\langle \mathbf{q}|\mathbf{r}'\rangle}{E_p - E_q \pm i\varepsilon}$$

$$= \int d^3 q\, e^{i\mathbf{q}\cdot(\mathbf{r}-\mathbf{r}')/\hbar} \frac{1}{E_p - E_q \pm i\varepsilon}. \qquad (18.88)$$

For elastic scattering

$$\mu_i = \mu_f = \mu, \quad E_i = E_f = E = E_p, \quad E_p = \frac{p^2}{2\mu},$$

$$E_q = \frac{q^2}{2\mu},$$

and Eq. (18.88) gives

$$\lim_{\varepsilon \to 0} G_{\pm}(\mathbf{r}, \mathbf{r}') = \langle \mathbf{r}|G_0^{\pm}(E)|\mathbf{r}'\rangle = \frac{2\mu}{(2\pi\hbar)^3} \int e^{i\mathbf{q}\cdot(\mathbf{r}-\mathbf{r}')/\hbar}$$

$$\times \frac{1}{p^2 - q^2 \pm i\varepsilon} d^3 q = \frac{2\mu}{\hbar^2}\left(-\frac{1}{4\pi}\right)\frac{1}{|\mathbf{r}-\mathbf{r}'|} e^{\pm i\frac{p}{\hbar}|\mathbf{r}-\mathbf{r}'|}. \quad (18.89)$$

(See Eqs. (8.171) for $d^3 q$ integration, here $\mathbf{p} = \hbar\mathbf{k}$.) Then Eq. (18.84) becomes

$$\psi_p^{(\pm)}(\mathbf{r}) = \frac{1}{(2\pi\hbar)^{3/2}} e^{i\mathbf{p}\cdot\mathbf{r}/\hbar} - \frac{1}{4\pi}\left(\frac{2\mu}{\hbar^2}\right)\int \frac{e^{\pm i\frac{p}{\hbar}|\mathbf{r}-\mathbf{r}'|}}{|\mathbf{r}-\mathbf{r}'|}$$

$$\times V(\mathbf{r}')\psi_p^{(\pm)}(\mathbf{r}')d^3 r'. \qquad (18.90a)$$

Now

$$|\mathbf{r} - \mathbf{r}'| = (\mathbf{r}^2 - 2\mathbf{r}\cdot\mathbf{r}' + \mathbf{r}'^2)^{1/2} \simeq r\left[1 - \frac{\hat{\mathbf{r}}\cdot\mathbf{r}'}{r} + O\left(\frac{1}{r^2}\right)\right],$$

so that

$$\psi_p^{(+)}(\mathbf{r}) \underset{r \to \infty}{\sim} \frac{1}{(2\pi\hbar)^{3/2}} e^{i\mathbf{p}\cdot\mathbf{r}/\hbar} - \frac{1}{4\pi}\frac{e^{i\frac{p}{\hbar}r}}{r}$$

$$\times \left(\frac{2\mu}{\hbar^2}\right)\int e^{-i\frac{p}{\hbar}(\hat{\mathbf{r}}\cdot\mathbf{r}')} V(\mathbf{r}')\psi_p^{(+)}(\mathbf{r}')d^3 r'. \qquad (18.90b)$$

This shows that the solutions $\psi_p^{(+)}(\mathbf{r})$ of the above integral equation asymptotically represent outgoing spherical waves in addition to incident plane waves. Hence $|a^{(+)}\rangle$ is usually called the outgoing eigenstate of H, if $|a\rangle$ is a momentum eigenvector $|\mathbf{p}\rangle$ while $|a^{(-)}\rangle$ is called the incoming eigenstate of H.

Transition Amplitude in Born approximation: We have derived Eq. (18.86b), which on using Eqs. (18.87) gives

$$T_{ba}^B = -V_{ba} = -\langle \mathbf{p}_f | V | \mathbf{p}_i \rangle,$$

i.e.

$$T_{p_f p_i}^B = -V_{p_f p_i} = -\int \frac{1}{(2\pi\hbar)^3} e^{-i\mathbf{p}_f \cdot \mathbf{r}'/\hbar} V(\mathbf{r}')$$
$$\times e^{i\mathbf{p}_i \cdot \mathbf{r}'/\hbar} d^3 r'$$
$$= -\frac{1}{(2\pi\hbar)^3} \tilde{V}(\Delta), \tag{18.91a}$$

where

$$\tilde{V}(\Delta) = \int e^{i\Delta \cdot \mathbf{r}/\hbar} V(\mathbf{r}) d^3 r, \tag{18.91b}$$

with

$$\Delta = \mathbf{p}_i - \mathbf{p}_f \quad = \text{momentum transfer.}$$

Now the scattering cross-section is given by

$$d\sigma = W/(\text{Flux})_{\text{in}}, \tag{18.92a}$$

where the golden rule gives

$$W = \frac{2\pi}{\hbar} |V_{p_f p_i}|^2 \rho_f(E_f), \tag{18.92b}$$

while $\rho_f(E_f)$ is given in Eq. (18.63b), i.e.

$$\rho_f(E_f) = \left(|\mathbf{p}_f|^2 \frac{d|\mathbf{p}_f|}{dE_f}\right) d\Omega$$
$$= \mu_f |\mathbf{p}_f| d\Omega. \tag{18.93a}$$

Now

$$(\text{Flux})_{\text{in}} = \rho_{in} v = \frac{1}{(2\pi\hbar)^3} \frac{|\mathbf{p}_i|}{\mu_i}. \tag{18.93b}$$

Hence Eqs. (18.91), (18.92) and (18.93) give

$$\left(\frac{d\sigma}{d\Omega}\right)^B = \frac{1}{4\pi^2} \frac{1}{\hbar^4} \frac{|\mathbf{p}_f|}{|\mathbf{p}_i|} \mu_i \mu_f |\tilde{V}(\Delta)|^2. \tag{18.94a}$$

For elastic scattering

$$\mu_i = \mu_f = \mu, \quad |\mathbf{p}_f| = |\mathbf{p}_i| = p,$$

we have

$$(\frac{d\sigma}{d\Omega})_{el}^B = \frac{1}{4\pi^2}\frac{\mu^2}{\hbar^4}|\tilde{V}(\Delta)|^2$$

$$= (\frac{\mu}{2\pi\hbar^2})^2|\tilde{V}(\Delta)|^2. \tag{18.94b}$$

Application II: 2 spinless and distinguishable particles α and β interacting via $V(\mathbf{r}_\alpha - \mathbf{r}_\beta) = V(\mathbf{r})$, with $\mathbf{r} = \mathbf{r}_\alpha - \mathbf{r}_\beta$ but there is also a fixed attractive potential $U(\mathbf{r}_\beta)$ acting on β only: Thus

$$H = H_0 + V(\mathbf{r}),$$

$$H_0 = \frac{\hat{p}_\alpha^2}{2m_\alpha} + \frac{\hat{p}_\beta^2}{2m_p} + U(\mathbf{r}_\beta). \tag{18.95a}$$

$U(\mathbf{r}_\beta)$ is strong enough so as to possess a number of bound states

$$(\frac{\hat{p}_\beta^2}{2m_\beta} + U(\mathbf{r}_\beta))\langle\mathbf{r}_\beta|n\rangle = E_n\langle\mathbf{r}_\beta|n\rangle. \tag{18.95b}$$

In other words, the particle β is bound in the potential $U(\mathbf{r}_\beta)$. Let Y_n stand for the nth bound state. We consider the following processes:

$$\alpha + Y_0 \to \alpha + Y_0 \qquad \text{elastic scattering}$$
$$\alpha + Y_0 \to \alpha + Y_n, \quad n \neq 0, \quad \text{inelastic scattering,}$$

for example

$$\alpha = e^-, \qquad Y_0 \text{ nucleus } (A, Z),$$
$$\beta = \text{proton}, \quad Y_n \text{ excited nucleus } (A, Z)^*.$$

In this case, states $|a\rangle$ and $|b\rangle$:

$$|a\rangle = |0\rangle|\mathbf{p}_\alpha\rangle, \quad |b\rangle = |n\rangle|\mathbf{p}_\alpha'\rangle$$

are eigenstates of H_0 with energies

$$E_a = E_\alpha + E_0,$$

$$E_b = E_\alpha' + E_n.$$

Then from Eq. (18.86b)

$$T_{ba}^B = -\langle b|V|a\rangle$$
$$= -\int\int\int\int \langle b|\mathbf{r}_\alpha''\mathbf{r}_\beta''\rangle\langle\mathbf{r}_\alpha''\mathbf{r}_\beta''|V|\mathbf{r}_\alpha'\mathbf{r}_\beta'\rangle$$

$$\times \langle \mathbf{r}'_\alpha \mathbf{r}'_\beta | a \rangle d^3 r''_\alpha d^3 r''_\beta d^3 r'_\alpha d^3 r'_\beta, \qquad (18.96a)$$

where

$$\langle \mathbf{r}_\alpha \mathbf{r}_\beta | a \rangle = \langle \mathbf{r}_\alpha | \mathbf{p}_\alpha \rangle \langle \mathbf{r}_\beta | 0 \rangle$$

$$= \phi_0(\mathbf{r}_\beta) \frac{1}{(2\pi\hbar)^{3/2}} e^{i\mathbf{p}_\alpha \cdot \mathbf{r}/\hbar}, \qquad (18.96b)$$

$$\langle \mathbf{r}_\alpha \mathbf{r}_\beta | b \rangle = \langle \mathbf{r}_\alpha | \mathbf{p}'_\alpha \rangle \langle \mathbf{r}_\beta | n \rangle$$

$$= \phi_n(\mathbf{r}_\beta) \frac{1}{(2\pi\hbar)^{3/2}} e^{i\mathbf{p}'_\alpha \cdot \mathbf{r}'/\hbar}, \qquad (18.96c)$$

and

$$\langle \mathbf{r}''_\alpha \mathbf{r}''_\beta | V | \mathbf{r}'_\alpha \mathbf{r}'_\beta \rangle = \delta(\mathbf{r}'_\alpha - \mathbf{r}''_\alpha)(\mathbf{r}'_\beta - \mathbf{r}''_\beta) V(\mathbf{r}'), \qquad (18.97)$$

with $\mathbf{r}' = \mathbf{r}'_\alpha - \mathbf{r}'_\beta$. Then

$$T^B_{ba} = -\frac{1}{(2\pi\hbar)^3} \int \int e^{-i\mathbf{p}'_\alpha \cdot \mathbf{r}'_\alpha/\hbar} \phi_n^*(\mathbf{r}'_\beta) V(\mathbf{r}') \phi_0(\mathbf{r}'_\beta)$$

$$\times e^{i\mathbf{p}_\alpha \cdot \mathbf{r}'_\alpha/\hbar} d^3 r'_\alpha d^3 r'_\beta. \qquad (18.98a)$$

We change the variable of integration $\mathbf{r}'_\alpha = \mathbf{r}' + \mathbf{r}'_\beta$ and treat $\mathbf{r}'$ and $\mathbf{r}'_\beta$ as independent variables. Thus we can write:

$$T^B_{ba} = -\frac{1}{(2\pi\hbar)^3} \{ \int e^{i\mathbf{q}\cdot\mathbf{r}'/\hbar} V(\mathbf{r}') d^3 r' \cdot \int e^{i\mathbf{q}\cdot\mathbf{r}'_\beta} \phi_n^*(\mathbf{r}'_\beta) \times \phi_0(\mathbf{r}'_\beta) d^3 r'_\beta \},$$

$$(18.98b)$$

where $\mathbf{q} = \mathbf{p}_\alpha - \mathbf{p}'_\alpha$ is the momentum transfered to particle α. Now define

$$\tilde{V}(\mathbf{q}) = \int e^{i\mathbf{q}\cdot\mathbf{r}/\hbar} V(\mathbf{r}) d^3 r, \qquad (18.99a)$$

$$F_n(\mathbf{q}) = \int e^{i\mathbf{q}\cdot\mathbf{r}/\hbar} \phi_n^*(\mathbf{r}) \phi_0(\mathbf{r}) d^3 r. \qquad (18.99b)$$

Equation (18.98b) becomes

$$T^B_{ba} = -\frac{1}{(2\pi\hbar)^3} \tilde{V}(\mathbf{q}) F_n(\mathbf{q}). \qquad (18.100)$$

We have the following picture that the particle α is scattered, with initial momentum $\mathbf{p}_\alpha$ and final momentum $\mathbf{p}'_\alpha$. Therefore if we compare Eq. (18.100) with Eq. (18.91a), we get from an analogue of our formula (18.94b), i.e. with $\mu_i = \mu_f = m_\alpha$, $\mathbf{p}_i = \mathbf{p}_\alpha$, $\mathbf{p}_f = \mathbf{p}'_\alpha$,

$$\frac{d\sigma^B_n}{d\Omega'_\alpha} = \frac{1}{4\pi^2} \frac{1}{\hbar^4} \frac{|\mathbf{p}'_\alpha|}{|\mathbf{p}_\alpha|} m_\alpha^2 |\tilde{V}(\mathbf{q})|^2 |F_n(\mathbf{q})|^2$$

$$= \frac{|\mathbf{p}'_\alpha|}{|\mathbf{p}_\alpha|} \left(\frac{m_\alpha}{2\pi\hbar^2}\right)^2 |\tilde{V}(\mathbf{q})|^2 |F_n(\mathbf{q})|^2. \qquad (18.101)$$

For the elastic case, we have $n = 0$, $\mathbf{p}'_\alpha = \mathbf{p}_\alpha$ and formula (18.101) becomes

$$\frac{d\sigma^B_{el}}{d\Omega'_\alpha} = (\frac{m_\alpha}{2\pi\hbar^2})^2 |\tilde{V}(\mathbf{q})|^2 |F_0(\mathbf{q})|^2, \tag{18.102}$$

where $F_0(\mathbf{q})$ is the elastic form factor

$$F_0(\mathbf{q}) = \int e^{i\mathbf{q}\cdot\mathbf{r}} |\phi_0(\mathbf{r})|^2 d^3r. \tag{18.103}$$

We can rewrite (18.102) as

$$\frac{d\sigma^B_{el}}{d\Omega'_\alpha} = \sigma^B(q)|F_0(\mathbf{q})|^2, \tag{18.104a}$$

where

$$\sigma^B(q) = (\frac{m_\alpha}{2\pi\hbar^2})^2 |\tilde{V}(\mathbf{q})|^2 \tag{18.104b}$$

is the usual Born approximation to the *differential cross-section* for scattering through momentum transfer $\mathbf{q}$ for a fixed target. A better approximation is to replace σ^B by σ_{el}, the exact elastic differential cross-section off a fixed β-particle target:

$$\frac{d\sigma^B_{el}}{d\Omega'_\alpha} = \sigma_{el}(q)|F_0(\mathbf{q})|^2, \tag{18.105}$$

the so called impulse approximation.

18.8   Unitarity of the S-Matrix

We have defined the S-matrix

$$S = \lim_{\substack{t_0 \to -\infty \\ t \to +\infty}} U(t, t_0), \tag{18.106}$$

where

$$U(t, t_0) = e^{iH_0 t/\hbar} e^{-iH(t-t_0)/\hbar} e^{-iH_0 t_0/\hbar}. \tag{18.107}$$

Then the unitarity of $U(t, t_0)$ and hence of S-matrix follows from the hermiticity of the Hamiltonian.

The unitarity property can also be seen as follows

$$\begin{aligned}
(SS^\dagger)_{ba} &= \langle b|SS^\dagger|a\rangle \\
&= \sum_c \langle b|S|c\rangle\langle c|S^\dagger|a\rangle \\
&= \sum_c \langle b|S|c\rangle\langle a|S|c\rangle^* \\
&= \sum_c \langle b^{(-)}|c^{(+)}\rangle\langle a^{(-)}|c^{(+)}\rangle^* \\
&= \sum_c \langle b^{(-)}|c^{(+)}\rangle\langle c^{(+)}|a^{(-)}\rangle + \sum_B \langle b^{(-)}|B\rangle\langle B|a^{(-)}\rangle,
\end{aligned} \tag{18.108}$$

since the bound states $|B\rangle$ are orthogonal to $|b^{(-)}\rangle$ and $|a^{(-)}\rangle$.

But

$$\sum_c |c^{(+)}\rangle\langle c^{(+)}| + \sum_B |B\rangle\langle B| = 1. \tag{18.109}$$

Therefore

$$(SS^\dagger)_{ba} = \langle b^{(-)}|a^{(-)}\rangle$$
$$= \delta_{ba}, \tag{18.110a}$$

or

$$SS^\dagger = 1. \tag{18.110b}$$

We now write the same relation in terms of the transition matrix T. We denote the initial and final states by $|i\rangle$ and $|f\rangle$ instead of $|a\rangle$ and $|b\rangle$.

Now

$$T_{fi} = \langle f|T|i\rangle,$$
$$S_{fi} = \delta_{fi} + 2\pi i\delta(E_f - E_i)T_{fi}.$$

Thus

$$(SS^\dagger)_{fi} = \delta_{fi}$$

gives

$$\sum_c \langle f|S|c\rangle\langle c|S^\dagger|i\rangle = \delta_{fi}. \tag{18.111}$$

Now

$$\langle c|S^\dagger|i\rangle = \langle i|S|c\rangle^*$$
$$= \delta_{ci} - 2\pi i\delta(E_c - E_i)\langle i|T|c\rangle^*$$
$$= \delta_{ci} - 2\pi i\delta(E_c - E_i)T_{ic}^*, \tag{18.112}$$

where

$$T_{ic}^* = \langle i|T|c\rangle^* = \langle c|T^\dagger|i\rangle = T_{ci}^\dagger. \tag{18.113}$$

From Eqs. (18.111) and (18.112), we have

$$\sum_c (\delta_{fc} + 2\pi i\delta(E_f - E_c)T_{fc})(\delta_{ci} - 2\pi i\delta(E_c - E_i)T_{ic}^*) = \delta_{fi}$$

or

$$\delta_{fi} + 2\pi i\delta(E_f - E_i)\left(T_{fi} - T_{if}^*\right) + (2\pi)^2 \sum_c \delta(E_f - E_c)\delta(E_c - E_i)T_{fc}T_{ic}^*$$
$$= \delta_{fi}.$$

Therefore

$$i\delta(E_f - E_i)\left(T^*_{if} - T_{fi}\right) = (2\pi)\delta(E_f - E_i)\sum_c T^*_{ic}T_{fc}\delta(E_f - E_c).$$

$$(18.114)$$

If $E_i \neq E_f$, this is a trivial identity. However, if $E_i = E_f$, we can cancel $\delta(E_f - E_i)$ on both sides and get the relation

$$i(T^*_{if} - T_{fi}) = 2\pi\sum_c T^*_{ic}T_{fc}\delta(E_f - E_c). \qquad (18.115a)$$

If we had started with

$$(SS^\dagger)_{fi} = \delta_{fi},$$

we would have obtained

$$i(T^*_{if} - T_{fi}) = 2\pi\sum_c T^*_{cf}T_{ci}\delta(E_f - E_c), \qquad (18.115b)$$

for $E_f = E_i$.

Equations (18.115) are frequently used unitarity relations. In the above relations, $\sum_c$ means summation over c extended to all possible states.

We now show that conservation of probability is a consequence of the unitarity relation, which in turn is a consequence of the hermiticity of the Hamiltonian. We recall that the probability of transition from a state $|i\rangle$ to any state $|b\rangle$ is given by

$$|C_b(t)|^2 = |\langle b|U(t, -\infty)|i\rangle|^2.$$

We sum over all possible states b

$$\sum_b |C_b(t)|^2 = \sum_b \langle b|U(t, -\infty)|i\rangle^*\langle b|U(t, -\infty)|i\rangle$$

$$= \sum_b \langle i|U^\dagger(t, -\infty)|b\rangle\langle b|U(t, -\infty)|i\rangle$$

$$= \langle i|U^\dagger(t, -\infty)U(t, -\infty)|i\rangle$$

because the states $|b\rangle$ form a complete set. Thus

$$\sum_b |C_b(t)|^2 = \langle i|U^\dagger(t, -\infty)U(t, -\infty)|i\rangle$$

$$= \langle i|U(-\infty, t)U(t, -\infty)|i\rangle$$

$$= \langle i|\hat{1}|i\rangle$$

$$= 1, \qquad (18.116)$$

which is the statement for the conservation of probability.

18.9 Optical Theorem

Let us now specialise to the case $f = i$. Then

$$T_{ii}^* - T_{ii} = -2i\mathrm{Im}\,T_{ii}. \tag{18.117}$$

Therefore, from Eq. (18.115), we have

$$\mathrm{Im}\,T_{ii} = \pi \sum_c |T_{ic}|^2 \delta(E_i - E_c). \tag{18.118}$$

Let us now take the index i to refer to momentum $\mathbf{p}_i$ and f to the momentum $\mathbf{p}_f$. There may be other quantum numbers necessary to specify the states $|i\rangle$ and $|f\rangle$ but we do not write them explicitly. For $\mathbf{p}_i = \mathbf{p}_f$, we have forward scattering ($\theta = 0$, or momentum transfer $\Delta^2 = (\mathbf{p}_i - \mathbf{p}_f)^2 = \mathbf{p}_i^2 + \mathbf{p}_f^2 - 2|\mathbf{p}_i||\mathbf{p}_f|\cos\theta = 2p^2(1 - \cos\theta) = 0$). Thus we have

$$T_{ii} = \frac{1}{\hbar^3} T_{kk} = \frac{1}{\hbar^3} \frac{2\pi\hbar^2}{(2\pi)^3 \mu} f_k(0), \tag{18.119}$$

where $f_k(0)$ is the elastic (since $i = f$) scattering amplitude for an energy $E_i = E = \frac{p^2}{2\mu} = \frac{\hbar^2 k^2}{2\mu}$ at the momentum transfer $\Delta = 0$.

Now

$$\frac{2\pi}{\hbar} \sum_c |T_{ci}|^2 \delta(E_c - E_i) = \frac{2\pi}{\hbar} \int \sum |T_{ci}|^2 \delta(E_c - E_i) d^3 p_c, \tag{18.120}$$

where $\sum$ denotes summation over all other quantum numbers necessary to specify the state $|c\rangle$. From Eqs. (18.120) and (18.64), the quantity

$$\frac{2\pi}{\hbar} \sum_c |T_{ci}|^2 \delta(E_c - E_i) = \frac{p}{(2\pi\hbar)^3 \mu} \sigma_{\text{total}}, \tag{18.121}$$

where σ_{total} denotes the total cross-section of the process starting from the initial state i, since we have not only summed over angles but all possible states consistent with the energy conservation. Thus Eq. (18.118) gives

$$\mathrm{Im}\,T_{ii} = \frac{p}{2\hbar^2 (2\pi)^3 \mu} \sigma_{\text{total}}$$

or

$$\mathrm{Im}\,f_k(0) = \frac{k}{4\pi} \sigma_{\text{total}}. \tag{18.122}$$

This is the generalised form of the optical theorem. It should be emphasised that on the right-hand side of Eq. (18.122), we have σ_{total} rather than σ_{el} because of the summation over c which involves all possible channels, both elastic and otherwise.

18.10 Problems

16.1 A system is prepared in the state $|a'\rangle$ at $t = 0$. Show that the probability of finding it in the state $|b'\rangle$ at time t is given by

$$|\langle b'|e^{-iH_0 t/\hbar}U(t,0)|a'\rangle|^2,$$

where U is the time displacement operator in the interaction picture

$$|a(t)\rangle_I = U(t,t_0)|a(t_0)\rangle_I.$$

16.2 Show that the unitarity of S-matrix follows from the conservation of probability.

16.3 If the S-matrix is represented as

$$S = \frac{1 - i/2 \, K}{1 + i/2 \, K},$$

where K is a hermitian operator, show that it satisfies $SS^\dagger = 1$. Derive the relation (analogue of unitarity relation)

$$T_{fi} = -K_{fi} - i\pi \sum_a K_{fa}T_{ai}\delta(E_a - E_i),$$

where

$$\langle f|K|i\rangle = 2\pi\delta(E_f - E_i)K_{fi}.$$

16.4 Show that for the scattering process

$$\alpha + Y_0 \rightarrow \alpha + Y_n,$$

if the potential $U(r)$, which possesses the bound states Y_n, is central, the inelastic form factor ($n \neq 0$) given in Eq. (18.99b) is

$$F_{lm}(q) = 4\pi i^l Y_{lm}^*(\hat{\mathbf{q}}) \int_0^\infty u_n^*(r)j_l(qr)u_0(r)r^2 dr,$$

where $l(l+1)\hbar^2$ and $m\hbar$ are eigenvalues of L^2 and L_z. Show further that
(i) $F_{lm}(q) \sim q^l$ as $q \rightarrow 0$, and (ii) the partial wave cross-section is

$$\frac{d\sigma_l}{d\Omega'_\alpha} = (2l+1)\frac{p'_\alpha}{p_\alpha}\sigma^B(q)\left|\int_0^\infty u_n^*(r)j_l(qr)u_0(r)r^2 dr\right|^2,$$

where $u_n(r)$ are the solutions of the radial Schrödinger equation.
Hint: Write

$$\phi_n(\mathbf{r}) = \langle \mathbf{r}|n\rangle = Y_{lm}(\hat{\mathbf{r}})u_n(r)$$

$$e^{i\mathbf{q}\cdot\mathbf{r}} = 4\pi \sum_{l'm'} i^{l'} j_{l'}(qr)Y_{l'm'}^*(\hat{\mathbf{q}})Y_{l'm'}(\hat{\mathbf{r}})$$

and note

$$j_l(qr) \sim (qr)^l \quad \text{as} \quad q \to 0.$$

In calculating the partial wave cross-section, remember that we do not distinguish final states which only differ in magnetic quantum number m so that the observed cross-section involves the sum

$$\sum_m |Y_{lm}(\hat{\mathbf{q}})|^2 = \frac{2l+1}{4\pi}.$$

16.5 Show that for

$$V = V_0 \quad r \le a$$
$$= 0 \quad r > a$$

the Born approximation (B.A.) is valid if

$$V_0 \ll \frac{p\hbar}{\mu a}, \quad \text{when} \quad \frac{pa}{\hbar} \gg 1$$

and

$$V_0 \ll \frac{\hbar^2}{\mu a^2}, \quad \text{when} \quad \frac{pa}{\hbar} \ll 1.$$

Show further that the first condition is satisfied if

$$\mu \frac{V_0 a^2}{\hbar^2} \ll 1$$

which is also the second condition. Thus B.A. is justified at all energies if $V_0 \ll \frac{\hbar^2}{\mu a^2}$.

Hint: In the first approximation

$$\psi_p^+(\mathbf{r}) = \frac{1}{(2\pi\hbar)^{3/2}} e^{i\mathbf{p} \cdot \mathbf{r}/\hbar}.$$

In the second approximation

$$\psi_p^+(\mathbf{r}) = \frac{1}{(2\pi\hbar)^{3/2}} \left\{ e^{i\mathbf{p} \cdot \mathbf{r}/\hbar} - \frac{1}{4\pi} \frac{2\mu}{\hbar^2} \int \frac{e^{i\frac{p}{\hbar}|\mathbf{r}-\mathbf{r}'|}}{|\mathbf{r}-\mathbf{r}'|} V(r') e^{i\mathbf{p} \cdot \mathbf{r}'/\hbar} d^3 r' \right\}.$$

For B.A. to be valid, the modulus square of the 2nd term in { } should be $\ll 1$. Then calculate { } for the potential given for $\mathbf{r} = 0$. Put $\mathbf{R} = \mathbf{r}' - \mathbf{r}$ and take $\mathbf{p}$ along the z-axis.

16.6 Starting from the equation

$$G^{(+)}(E) = \frac{1}{E - H + i\varepsilon},$$

show that

i

$$G^+(\mathbf{r}, \mathbf{r}', E) = \langle \mathbf{r}|G^+(E)|\mathbf{r}'\rangle = \sum_n \frac{u_n(\mathbf{r})u_n^*(\mathbf{r}')}{E - E_n + i\varepsilon}$$

ii

$$G^+(t) = \frac{1}{2\pi}\int e^{-iEt/\hbar}G^+(E)dE = -i\theta(t)K(t),$$

where

$$K(t) = \sum_n e^{-iE_n t/\hbar}|E_n\rangle\langle E_n|$$

is called the propagator.

iii

$$K(E) = \int e^{iEt/\hbar}K(t)dt = 2\pi\sum_n |E_n\rangle\langle E_n|\delta(E - E_n),$$

where $|E_n\rangle$ are eigenstates of H, $H|E_n\rangle = E_n|E_n\rangle$.

iv

$$K(\mathbf{r}, \mathbf{r}', t) = \sum_n e^{-iE_n t/\hbar}u_n(\mathbf{r})u_n^*(\mathbf{r}'),$$

$$K(\mathbf{r}, \mathbf{r}', E) = 2\pi\sum_n u_n(\mathbf{r})u_n^*(\mathbf{r}')\delta(E - E_n).$$

v For a free particle

$$K_0(\mathbf{r}, \mathbf{r}', t) = \left(\frac{m}{2\pi i\hbar t}\right)^{3/2}e^{-\frac{m}{2i\hbar t}(\mathbf{r}-\mathbf{r}')^2},$$

$$K_0(0,0,E) = \frac{m\sqrt{2mE}}{\pi\hbar^3}.$$

Hint:

$$\sum_n |E_n\rangle\langle E_n| = 1$$

$$\theta(t) = \frac{1}{2\pi i}\int_{-\infty}^{\infty}\frac{e^{i\omega t}}{\omega - i\varepsilon}d\omega.$$

For a free particle

$$u_n(\mathbf{r}) = u_p(\mathbf{r}) = \frac{1}{(2\pi\hbar)^{3/2}}e^{i\mathbf{p}\cdot\mathbf{r}/\hbar}. \tag{18.123}$$

Chapter 19

S-Matrix and Invariance Principles

19.1 Introduction

We have seen in Sec. 10.5 that an invariance principle is connected with a conservation law. By invariance we mean the following.

If the result of any experiment on some system is unchanged by a *physical transformation* of the apparatus, then the system is said to be invariant with respect to that transformation. This is not an empty statement but gives information about that system.

In quantum mechanics, a transformation is described by a unitary transformation U. The result of an experiment involving a transition from an initial state $|i\rangle$ to a final $|f\rangle$ is described by the matrix elements of an operator S called the scattering matrix (S-matrix) between states $|i\rangle$ and $|f\rangle$. Invariance means

$$\langle f|S|i\rangle = \langle f^u|S|i^u\rangle$$
$$= \langle f|U^\dagger S U|i\rangle \qquad (19.1)$$

or

$$S = U^\dagger S U \qquad (19.2)$$

or

$$[S, U] = 0. \qquad (19.3)$$

Here

$$|i^u\rangle = U|i\rangle \qquad (19.4)$$
$$|f^u\rangle = U|f\rangle \qquad (19.5)$$

are the transformed states. Equation (19.3) shows that invariance under a unitary transformation U means that a S-matrix commutes with it. Equation (19.3) also follows from the result $[H, U] = 0$ (see Sec. 10.5), by noting that the transition matrix T can be formally written as

$$T = -V - V \frac{1}{E_0 - H + i\varepsilon} V \tag{19.6}$$

where

$$V = H - H_0. \tag{19.7}$$

We consider two cases:

(a) U continuous: U can be built out of infinitesimal transformations. We consider an infinitesimal unitary transformation

$$U = 1 - i\varepsilon\hat{F}, \tag{19.8}$$

where $\hat{F}$ is hermitian. $\hat{F}$ can often be identified with an observable of the system. For example, corresponding to an infinitesimal rotation of coordinates, the unitary operator U is given by

$$U = 1 - \frac{i}{\hbar}\epsilon \cdot \mathbf{J} \tag{19.9}$$

where $\mathbf{J}$ is the angular momentum operator. From Eqs. (19.3) and (19.8), it follows that

$$[S, \hat{F}] = 0. \tag{19.10}$$

In case $\hat{F}$ is one of the observables of the system, we have

$$\hat{F}|i\rangle = F'_i|i\rangle \tag{19.11}$$
$$\hat{F}|f\rangle = F'_f|f\rangle, \tag{19.12}$$

i.e. $|i\rangle$ and $|f\rangle$ are eigenstates of $\hat{F}$, with eigenvalues F'_i and F'_f.

The transition from the state $|i\rangle$ to state $|f\rangle$ is given by the S-matrix $\langle f|S|i\rangle \neq 0$. Invariance under transformations (19.4) and (19.5) imply

$$\langle f|[S, \hat{F}]|i\rangle = 0. \tag{19.13}$$

Hence

$$(F'_i - F'_f)\langle f|S|i\rangle = 0 \tag{19.14}$$

or

$$F'_i = F'_f, \tag{19.15}$$

that is, the eigenvalues of $\hat{F}$ are conserved in the transition. $\hat{F}$ is then said to be a constant of motion. Operators like $\hat{F}$, whose eigenvalues are conserved, are called 'good operators'.

(b) U discrete (for example space reflection) U is of course unitary. If

$$U^2 = 1 \qquad (19.16)$$

then the eigenvalues of U are given by

$$U' = \pm 1. \qquad (19.17)$$

U is also hermitian. Therefore U itself can be regarded as one of the observables of the system, for example, the parity operator, provided that the system is invariant under U

$$[S, U] = 0. \qquad (19.18)$$

Then $U_i' = U_f'$ in any transition from $|i\rangle$ to $|f\rangle$, provided that $\langle f|S|i\rangle \neq 0$.

From symmetry or invariance principles, (i) we can derive selection rules viz. which processes are allowed and which are forbidden. (ii) We can relate different processes

$$\langle f|S|i\rangle = \langle fU^\dagger|S|Ui\rangle. \qquad (19.19)$$

(iii) We can derive restrictions on the form of the transition amplitudes.

19.2 Rotational Invariance

In a scattering process, we specify the initial and final states by the initial and final momenta $\mathbf{p}_i$ and $\mathbf{p}_f$ in the centre of mass frame. If rotational invariance holds (as for example, is the case for central forces $V = V(r)$), then

$$\langle \mathbf{p}_f|S|\mathbf{p}_i\rangle = \langle \mathbf{p}_f|U_R^\dagger SU_R|\mathbf{p}_i\rangle$$
$$= \langle \mathbf{p}_f U_R^\dagger|S|U_R\mathbf{p}_i\rangle. \qquad (19.20)$$

Thus the scattering matrix can depend only on those quantities which are invariant under rotations, i.e. on the scalars

$$\mathbf{p}_i^2, \mathbf{p}_f^2 \quad \text{and} \quad \mathbf{p}_i \cdot \mathbf{p}_f. \qquad (19.21)$$

If we specialise to two body elastic scattering,

$$|\mathbf{p}_i| = |\mathbf{p}_f| = p$$

$$E_i = \frac{p_i^2}{2\mu} = \frac{p_f^2}{2\mu} = E_f = E \qquad (19.22)$$

$$\mathbf{p}_i \cdot \mathbf{p}_f = p^2 \cos\theta, \qquad (19.23)$$

then we have two independent variables (E, θ) or (p, θ), i.e. energy E and scattering angle θ

$$\langle \mathbf{p}_f | S | \mathbf{p}_i \rangle = \langle p_f, \theta, \phi | S | p_i, 0, 0 \rangle$$
$$= (p_i - p_f) \langle \theta, \phi | S(p) | 0, 0 \rangle \tag{19.24}$$
$$= (p_i - p_f) A(p, \theta) \tag{19.25}$$

because the S-matrix has non-vanishing matrix elements only on the 'energy shell', i.e. between states of same energy $E_i = E_f$. We note

$$\langle \theta, \phi | lm \rangle = Y_{lm}(\theta, \phi) = (-1)^m \left(\frac{(2l+1)(l-m)!}{4\pi(l+m)!} \right)^{1/2} P_l^m(\cos \theta) e^{im\phi},$$

$$\langle 0, 0 | lm \rangle = Y_{lm}(0, 0). \tag{19.26}$$

Because of rotational invariance, there is no dependence on m and we can put $m = 0$ and thus we have

$$\langle \theta, \phi | l, 0 \rangle = \langle \theta, \phi | l \rangle = \left(\frac{2l+1}{4\pi} \right)^{1/2} P_l(\cos \theta), \tag{19.27}$$

$$\langle 0, 0 | l \rangle = \left(\frac{2l+1}{4\pi} \right)^{1/2}. \tag{19.28}$$

From Eq. (19.10), we have

$$\langle \mathbf{p}_f | S | \mathbf{p}_i \rangle = \delta(p_f - p_i) \langle \theta, \phi | S(p) | 0, 0 \rangle$$
$$= \delta(p_f - p_i) \sum_l \sum_{l'} \langle \theta, \phi | l \rangle \langle l | S(p) | l' \rangle \langle l' | 0, 0 \rangle$$
$$= \delta(p_f - p_i) \sum_l \frac{2l+1}{4\pi} F_l(p) P_l(\cos \theta), \tag{19.29}$$

where we have put

$$\langle l | S(p) | l' \rangle = \delta_{ll'} F_l(p), \tag{19.30}$$

because due to rotational invariance, angular momentum is conserved.

Now from the unitarity of the S-matrix

$$SS^\dagger = 1 \tag{19.31}$$

$$\langle \mathbf{p}_f | SS^\dagger | \mathbf{p}_i \rangle = \delta(\mathbf{p}_i - \mathbf{p}_f) \tag{19.32}$$

or

$$\int d^3 p' \langle \mathbf{p}_f | S | \mathbf{p}' \rangle \langle \mathbf{p}_i | S | \mathbf{p}' \rangle^* = \delta(\mathbf{p}_i - \mathbf{p}_f). \tag{19.33}$$

Using Eq. (19.29), we have from Eq. (19.33)

$$\sum_l \sum_{l'} \frac{2l+1}{4\pi} \frac{2l'+1}{4\pi} \int_0^{2\pi} \int_0^\pi \int p'^2 dp' \delta(p'-p_f)\delta(p'-p_i)F_l(p')F_{l'}^*(p')$$

$$\times P_l(\hat{p}' \cdot \hat{p}_f)P_{l'}(\hat{p}' \cdot \hat{p}_i) \sin\theta' d\theta' d\phi' = \delta(\mathbf{p}_i - \mathbf{p}_f), \quad (19.34)$$

$$p^2 \frac{1}{(4\pi)^2}\delta(p_i - p_f)\sum_l \sum_{l'}(2l+1)(2l'+1)F_l(p)F_{l'}^*(p)I_{ll'} = \delta(\mathbf{p}_i - \mathbf{p}_f),$$
$$(19.35)$$

where

$$I_{ll'} = \int_0^{2\pi} \int_0^\pi P_l(\cos\alpha)P_{l'}(\cos\theta') \sin\theta' d\theta' d\phi'. \quad (19.36)$$

Now

$$\hat{p}' \equiv (\theta', \phi'), \quad \hat{p}_i \equiv (0,0),$$

$$\hat{p}_f \equiv (\theta, \phi),$$

$$\hat{p}' \cdot \hat{p}_i = \cos\theta', \quad \hat{p}_f \cdot \hat{p}_i = \cos\theta,$$

$$\hat{p}' \cdot \hat{p}_f = \cos\alpha. \quad (19.37)$$

Using the identity,

$$P_l(\cos\alpha) = \frac{4\pi}{2l+1}\sum_{m=-l}^l Y_l^m(\theta', \phi')Y_{lm}(\theta, \phi), \quad (19.38)$$

we have

$$I_{ll'} = \frac{4\pi}{2l+1}\sum_{m=-l}^l \int_0^{2\pi} \int_0^\pi P_{l'}(\cos\theta')Y_l^{*m}(\theta', \phi')Y_{lm}(\theta, \phi) \sin\theta' d\theta' d\phi'$$

$$= \frac{4\pi}{(2l+1)}\sum_{m=-l}^l \sqrt{\frac{4\pi}{2l'+1}}\delta_{ll'}\delta_{m0}Y_{lm}(0, \psi)$$

$$= \frac{4\pi}{(2l+1)}P_l(\cos\theta)\delta_{ll'}. \quad (19.39)$$

Therefore, from Eq. (19.35), we have

$$\delta(p_i - p_f)\frac{p^2}{4\pi}\sum_l(2l+1)|F_l(p)|^2 P_l(\cos\theta) = \delta(\mathbf{p}_i - \mathbf{p}_f). \quad (19.40)$$

Using the completeness property of Legendre's polynomials, we can write

$$\delta(\mathbf{p}_i - \mathbf{p}_f) = \sum_l a_l(p)P_l(\cos\theta). \quad (19.41)$$

Now

$$d^3 p_f = p^2 dp d(\cos\theta) d\phi \tag{19.42}$$

and we can write

$$\delta(\mathbf{p}_i - \mathbf{p}_f) = \frac{1}{p^2} \delta(p_i - p_f)\delta(\cos\theta)\delta(\phi). \tag{19.43}$$

Therefore

$$\frac{1}{p^2}\delta(p_i - p_f)\delta(\cos\theta)\delta(\phi) = \sum_l a_l(p) P_l(\cos\theta). \tag{19.44}$$

Multiplying both sides by $P_{l'}(\cos\theta)$, integrating over angles and using the orthogonality of Legendre polynomials, we have

$$\frac{1}{p^2}\delta(p_i - p_f)\int_0^{2\pi}\int_{-1}^{+1}\delta(\cos\theta)\delta(\phi)P_{l'}(\cos\theta)d(\cos\theta)d\phi$$

$$= \sum_l a_l(p)\int_0^{2\pi}\int_{-1}^{+1}P_l(\cos\theta)P_{l'}(\cos\theta)d(\cos\theta)d\phi$$

$$= \sum_l a_l(p)\frac{2}{2l+1}\delta_{ll'}2\pi. \tag{19.45}$$

Therefore

$$a_l(p) = \frac{2l+1}{4\pi}\frac{1}{p^2}\delta(p_i - p_f). \tag{19.46}$$

Hence

$$\delta(\mathbf{p}_i - \mathbf{p}_f) = \frac{\delta(p_i - p_f)}{p^2}\sum_{l=0}^{\infty}\frac{2l+1}{4\pi}P_l(\cos\theta). \tag{19.47}$$

Therefore from Eqs. (19.40) and (19.47), we have

$$\delta(p_i - p_f)\frac{p^2}{4\pi}\sum_l(2l+1)|F_l(p)|^2 P_l(\cos\theta) = \frac{\delta(p_i - p_f)}{p^2}\sum_{l=0}^{\infty}\frac{2l+1}{4\pi}P_l(\cos\theta) \tag{19.48}$$

or

$$|F_l(p)|^2 = \frac{1}{p^4}. \tag{19.49}$$

Hence

$$F_l(p) = \frac{1}{p^2}e^{2i\delta_l(p)}, \tag{19.50}$$

where $\delta_l(p)$ are real functions of momentum or energy.

Therefore from Eq. (19.29), we have

$$\langle \mathbf{p}_f | S | \mathbf{p}_i \rangle = \delta(p_i - p_f) \frac{1}{4\pi p^2} \sum_l (2l+1) e^{2i\delta_l(p)} P_l(\cos\theta). \qquad (19.51)$$

Now

$$\langle \mathbf{p}_f | S | \mathbf{p}_i \rangle = \delta(\mathbf{p}_i - \mathbf{p}_f) + 2\pi i \delta(E_i - E_f) T_{fi}$$
$$= \delta(\mathbf{p}_i - \mathbf{p}_f) + 2\pi i \frac{\mu}{p} \delta(p_i - p_f) T_{fi}. \qquad (19.52)$$

But [see Eq. (18.119)]

$$T_{fi} = \frac{4\pi\hbar^2}{2\mu(2\pi)^3} f_{\mathbf{p}_i}(\hat{p}_f) \qquad (19.53)$$

therefore

$$\langle \mathbf{p}_f | S | \mathbf{p}_i \rangle = \delta(\mathbf{p}_i - \mathbf{p}_f) + 2\pi i \frac{\mu}{p} \frac{4\pi\hbar^2}{2\mu(2\pi)^3} f_{\mathbf{p}_i}(\hat{p}_f) \delta(p_i - p_f). \qquad (19.54)$$

Hence

$$\delta(p_i - p_f) \frac{1}{4\pi p^2} \sum_l (2l+1) e^{2i\delta_l(p)} P_l(\cos\theta) = \frac{\delta(p_i - p_f)}{p^2} \sum_{l=0}^{\infty} \frac{2l+1}{4\pi} P_l(\cos\theta)$$
$$+ \frac{i}{2\pi} \frac{\hbar^2}{p} f_{\mathbf{p}_i}(\hat{p}_f) \delta(p_i - p_f)$$
$$(19.55)$$

or

$$\frac{i}{2\pi} \frac{\hbar^2}{p} f_{\mathbf{p}_i}(\hat{p}_f) = \frac{1}{4\pi p^2} \sum_l (2l+1) \left(e^{2i\delta_l(p)} - 1 \right) P_l(\cos\theta) \qquad (19.56)$$

or

$$f_{\mathbf{k}_i}(\hat{k}_f) = \frac{1}{2ik} \sum_l (2l+1) \left(e^{2i\delta_l(k)} - 1 \right) P_l(\cos\theta), \qquad (19.57)$$

where

$$\hbar^3 f_{\mathbf{p}_i}(\hat{p}_f) = f_{\mathbf{k}_i}(\hat{k}_f). \qquad (19.58)$$

Equation (19.57) is the main result of the partial wave analysis. Here it has been obtained by using rotational invariance and unitarity of the S-matrix.

19.3   Parity

If invariance under a parity transformation $\hat{P}$ holds, then

$$\langle f|S|i\rangle = \langle f|\hat{P}^\dagger S\hat{P}|i\rangle$$
$$= \langle f\hat{P}^\dagger|S|\hat{P}i\rangle. \qquad (19.59)$$

Now

$$\hat{P}|i\rangle = \eta_P^i|i\rangle \qquad (19.60a)$$

$$\langle f|\hat{P}^\dagger = \eta_P^f\langle f| \qquad (19.60b)$$

where

$$\eta_P^i = \pm 1, \quad \eta_P^f = \pm 1 \qquad (19.61)$$

denote the parity of the states $|i\rangle$ and $|f\rangle$, respectively. Thus invariance under $\hat{P}$ implies

$$\langle f|S|i\rangle = \eta_P^i\eta_P^f\langle f|S|i\rangle. \qquad (19.62)$$

Hence

$$\eta_P^i\eta_P^f = 1, \qquad (19.63)$$

that is

$$\eta_P^i = \eta_P^f. \qquad (19.64)$$

This means that the S-matrix connects states of the same parity. In other words, the invariance of the S-matrix under space reflection requires that the parity of the initial state be the same as that of the final state.

Explicit Examples:

(i) Consider two body elastic scattering, one of the particles with spin 0 while the other with spin 1/2.

$$a + b \to a + b,$$
$$\mathbf{p}_1, \ (\mathbf{p}_2, \boldsymbol{\sigma}) \quad \mathbf{p}_1', \ (\mathbf{p}_2', \boldsymbol{\sigma}). \tag{19.65}$$

In the centre of mass system

$$\mathbf{p}_1 = -\mathbf{p}_2 = \mathbf{p}_i,$$
$$\mathbf{p}_1' = -\mathbf{p}_2' = \mathbf{p}_f,$$
$$|\mathbf{p}_i| = |\mathbf{p}_f| = p. \tag{19.66}$$

The initial and final states can be labelled as

$$|i\rangle = |\mathbf{p}_i, \boldsymbol{\sigma}\rangle, \tag{19.67a}$$

$$|f\rangle = |\mathbf{p}_f, \boldsymbol{\sigma}\rangle. \tag{19.67b}$$

Under the parity operation $\hat{P}$

$$\hat{P}|i\rangle = \eta_P^i |-\mathbf{p}_i, \boldsymbol{\sigma}\rangle, \tag{19.68a}$$

$$\hat{P}|f\rangle = \eta_P^f |-\mathbf{p}_f, \boldsymbol{\sigma}\rangle. \tag{19.68b}$$

Now

$$\langle \mathbf{p}_f, \boldsymbol{\sigma}|S|\mathbf{p}_i, \boldsymbol{\sigma}\rangle = \langle \mathbf{p}_f, \boldsymbol{\sigma}|\hat{P}^\dagger \hat{P} S \hat{P}^\dagger \hat{P}|\mathbf{p}_i, \boldsymbol{\sigma}\rangle$$
$$= \eta_P^f \eta_P^i \langle -\mathbf{p}_f, \boldsymbol{\sigma}|\hat{P} S \hat{P}^\dagger| -\mathbf{p}_i, \boldsymbol{\sigma}\rangle. \tag{19.69}$$

Invariance under $\hat{P}$ implies

$$\hat{P} S \hat{P}^\dagger = S \tag{19.70}$$

and because of elastic scattering $\eta_P^i = \eta_P^f$ so that $\eta_P^f \eta_P^i = 1$. Thus

$$\langle \mathbf{p}_f, \boldsymbol{\sigma}|S|\mathbf{p}_i, \boldsymbol{\sigma}\rangle = \langle -\mathbf{p}_f, \boldsymbol{\sigma} S| -\mathbf{p}_i, \boldsymbol{\sigma}\rangle. \tag{19.71}$$

If we assume only rotational invariance, then $\langle \mathbf{p}_f, \boldsymbol{\sigma}|S|\mathbf{p}_i, \boldsymbol{\sigma}\rangle$ can depend only on the scalars p, $\mathbf{p}_i \cdot \mathbf{p}_f$, $\mathbf{p}_i \cdot \boldsymbol{\sigma}$, $\mathbf{p}_f \cdot \boldsymbol{\sigma}$, $\boldsymbol{\sigma} \cdot (\mathbf{p}_i \times \mathbf{p}_f)$. We need not consider $\boldsymbol{\sigma}^2$ and higher powers since $\boldsymbol{\sigma}^2 = 3$ and higher powers can be reduced to either a constant or to $\boldsymbol{\sigma}$. In other words, in spin space, we can write

$$\langle \mathbf{p}_f, \boldsymbol{\sigma}|S|\mathbf{p}_i, \boldsymbol{\sigma}\rangle = \delta(p_i - p_f) \left(A(p, \theta) + A_2(p, \theta)\boldsymbol{\sigma} \cdot \mathbf{p}_i + A_3(p, \theta)\boldsymbol{\sigma} \cdot \mathbf{p}_f \right.$$
$$\left. + B(p, \theta)\boldsymbol{\sigma} \cdot (\mathbf{p}_i \times \mathbf{p}_f) \right). \tag{19.72}$$

This is a 2×2 matrix in spin space and it is understood that the above matrix elements are to be taken between spin wave functions χ_f^+ and χ_i for the final and initial states. Thus with rotational invariance alone, 4 independent amplitudes in spin space are possible. If, in addition, we assume invariance under $\hat{P}$ also, then Eq. (19.71) implies that

$$A_2 = 0, \ A_3 = 0 \tag{19.73}$$

since $\boldsymbol{\sigma} \cdot \mathbf{p}_i$ and $\boldsymbol{\sigma} \cdot \mathbf{p}_f$ change sign under $\hat{P}$.

Thus invariance under rotation and space inversion gives

$$\langle \mathbf{p}_f, \boldsymbol{\sigma} | S | \mathbf{p}_i, \boldsymbol{\sigma} \rangle = \delta(p_i - p_f) \left(A(p, \theta) + B(p, \theta) \boldsymbol{\sigma} \cdot (\mathbf{p}_i \times \mathbf{p}_f) \right). \tag{19.74}$$

This is an example of how a symmetry principle restricts the form of the transition amplitude.

(ii) Consider the decay of a state with spin 0 and even parity into three spinless particles, each having odd parity

$$A \to P_1 + P_2 + P_3. \tag{19.75}$$

Take the state A at rest, so that we have

$$\mathbf{p}_1 + \mathbf{p}_2 + \mathbf{p}_3 = 0, \tag{19.76}$$

where $\mathbf{p}_1, \mathbf{p}_2$ and $\mathbf{p}_3$ are the momenta of particles P_1, P_2 and P_3, respectively.

The transition matrix elements for the decay (19.75) are given by

$$M(\mathbf{p}_1, \mathbf{p}_2, \mathbf{p}_3) = \langle P_1(\mathbf{p}_1) P_2(\mathbf{p}_2) P_3(\mathbf{p}_3) | T | A \rangle. \tag{19.77}$$

Under parity

$$\hat{P} | A(\mathbf{0}) \rangle = | A(\mathbf{0}) \rangle \hat{P} | P_i(\mathbf{p}_i) \rangle = -| P_i(-\mathbf{p}_i) \rangle, \quad i = 1, 2, 3 \tag{19.78}$$

Now

$$M(\mathbf{p}_1, \mathbf{p}_2, \mathbf{p}_3) = \langle P_1(\mathbf{p}_1) P_2(\mathbf{p}_2) P_3(\mathbf{p}_3) | \hat{P}^\dagger \hat{P} T \hat{P}^\dagger \hat{P} | A(\mathbf{0}) \rangle \tag{19.79}$$

$$= (-1)^3 \langle P_1(-\mathbf{p}_1) P_2(-\mathbf{p}_2) P_3(-\mathbf{p}_3) | \hat{P} T \hat{P}^\dagger | A \rangle. \tag{19.80}$$

If parity is conserved in the above decay,

$$\hat{P} T \hat{P}^\dagger = T \tag{19.81}$$

and we have

$$M(\mathbf{p}_1, \mathbf{p}_2, \mathbf{p}_3) = -M(-\mathbf{p}_1, -\mathbf{p}_2, -\mathbf{p}_3). \tag{19.82}$$

Because of rotational invariance, M can be a function of scalars $\mathbf{p}_1 \cdot \mathbf{p}_2$, $\mathbf{p}_2 \cdot \mathbf{p}_3$, $\mathbf{p}_3 \cdot \mathbf{p}_1$ only; it cannot be a function of $\mathbf{p}_1 \cdot (\mathbf{p}_2 \times \mathbf{p}_3)$ because of Eq. (19.83). Hence rotational and space inversion invariance imply

$$M(\mathbf{p}_1 \cdot \mathbf{p}_2, \mathbf{p}_2 \cdot \mathbf{p}_3, \mathbf{p}_3 \cdot \mathbf{p}_1) = -M(\mathbf{p}_1 \cdot \mathbf{p}_2, \mathbf{p}_2 \cdot \mathbf{p}_3, \mathbf{p}_3 \cdot \mathbf{p}_1) \quad (19.83)$$

or

$$M = 0. \quad (19.84)$$

Thus the above decay is forbidden if we assume invariance under space inversion. In other words, a spinless particle with even parity cannot decay into three spinless particles each having odd parity if parity invariance holds. This is an example of how an invariance principle leads to selection rules.

19.4 Time Reversal

Under time reversal

$$t \to -t, \qquad r \to r. \quad (19.85)$$

Thus under time reversal

$$\mathbf{p} \to -\mathbf{p}, \quad \mathbf{L} = \mathbf{r} \times \mathbf{p} \to \mathbf{r} \times (-\mathbf{p}) = -\mathbf{L}, \quad (19.86)$$

$$\boldsymbol{\sigma} \to -\boldsymbol{\sigma}. \quad (19.87)$$

Let Π denote the operator which transforms quantum mechanical states and operators under the above transformation $t \to -t$.

First we show that Π cannot be a unitary operator. Under Π, the commutation relation

$$[q_i, p_j] = i\hbar\delta_{ij}$$
$$\to -i\hbar\delta_{ij}. \quad (19.88)$$

Hence transformation generated by Π cannot be unitary. But we want the above commutation relation to be invariant under Π. A way out of this difficulty is as follows.

(i) All c-numbers are simultaneously transformed into their complex conjugates, or

(ii) If more than one operator is involved, then order of factors is inverted and ket vectors go into bra vectors. Thus

$$\Pi \hat{A} \Pi^{-1} = \hat{A}^t, \tag{19.89}$$

$$\Pi \hat{B} \Pi^{-1} = \hat{B}^t, \tag{19.90}$$

$$\Pi \hat{A} \hat{B} \Pi^{-1} = \hat{B}^t \hat{A}^t, \tag{19.91}$$

$$\Pi |\psi\rangle = \langle \psi^t|, \tag{19.92}$$

$$\hat{A} |\psi\rangle = \hat{B} \hat{C} |\phi\rangle \hat{C}^t \hat{B}^t \tag{19.93}$$

$$\Rightarrow \langle \psi^t| \hat{A}^t = \langle \phi^t| \hat{C}^t \hat{B}^t. \tag{19.94}$$

The transformation (i) or (ii) is called antiunitary.

Using alternative (i)

$$q_i \rightarrow q_i, p_j \rightarrow -p_j, i \rightarrow -i \tag{19.95}$$

hence the commutation relation remains invariant. Similarly using alternative (ii)

$$\begin{aligned}
\Pi[q_i, p_j]\Pi^{-1} &= \Pi q_i p_j \Pi^{-1} - \Pi p_j q_i \Pi^{-1} \\
&= (-p_j)q_i - q_i(-p_j) \\
&= [q_i, p_j].
\end{aligned} \tag{19.96}$$

We also note that

$$\Pi \mathbf{J} \Pi^{-1} = -\mathbf{J} \tag{19.97}$$

and the commutation relation

$$[J_i, J_j] = i\hbar \varepsilon_{ijk} J_k \tag{19.98}$$

is preserved either in (i) or (ii).

Let us now discuss how the transition matrix T transforms under Π. If H_0 and V are invariant under time reversal, then

$$\Pi H \Pi^{-1} = H,$$

$$\Pi V \Pi^{-1} = V.$$

Now

$$T = -V - V \frac{1}{E_a - H + i\varepsilon} V \tag{19.99}$$

and we have

$$\begin{aligned}
\Pi T \Pi^{-1} &= -V - V \frac{1}{E_a - H - i\varepsilon} V \\
&= T^\dagger,
\end{aligned} \tag{19.100}$$

if we use definition (i).

Using definition (ii),

$$\Pi T \Pi^{-1} = T. \tag{19.101}$$

Now invariance under Π implies

$$
\begin{aligned}
\langle f|T|i\rangle &= \langle f|\Pi^{-1}\Pi T\Pi^{-1}\Pi|i\rangle \\
&= \langle f^t|T^\dagger|i^t\rangle^* \\
&= \langle i^t|T|f^t\rangle.
\end{aligned} \tag{19.102}
$$

In deriving Eq. (19.105), we have used definition (i). The same result also follows from definition (ii)

$$
\begin{aligned}
\langle f|T|i\rangle &= \langle f|\Pi^{-1}\Pi T\Pi^{-1}\Pi|i\rangle \\
&= \langle i^t|T|f^t\rangle.
\end{aligned} \tag{19.103}
$$

Hence we see that the two definitions are completely equivalent.

We now consider the scattering states under Π. We specify a state $|a\rangle$ by its momentum $\mathbf{p}_a$, z component of spin m_a and α, which denotes other quantum numbers which may be necessary to specify the state

$$
\begin{aligned}
|a\rangle &= |\alpha, \mathbf{p}_a, m_a\rangle \\
|a^{(+)}\rangle &= |\alpha, \mathbf{p}_a, m_a\rangle_{out}
\end{aligned}
$$

where

$$|a^{\pm}\rangle = |a\rangle + \frac{1}{E_a - H \pm i\varepsilon}V|a\rangle. \tag{19.104}$$

Using definition (i)

$$\Pi|\alpha, \mathbf{p}_a, m_a\rangle = |\alpha, -\mathbf{p}_a, -m_a\rangle, \tag{19.105}$$

$$
\begin{aligned}
\Pi|\alpha, \mathbf{p}_a, m_a\rangle_{out} &= |\alpha, -\mathbf{p}_a, -m_a\rangle + \frac{1}{E_a - H - i\varepsilon}V|\alpha, -\mathbf{p}_a, m_a\rangle \\
&= |\alpha, -\mathbf{p}_a, -m_a\rangle_{in}.
\end{aligned} \tag{19.106}
$$

If we use definition (ii)

$$\Pi|\alpha, \mathbf{p}_a, m_a\rangle = \langle\alpha, -\mathbf{p}_a, -m_a|. \tag{19.107}$$

$$
\begin{aligned}
\Pi|\alpha, \mathbf{p}_a, m_a\rangle_{out} &= \langle\alpha, -\mathbf{p}_a, -m_a| + \langle\alpha, -\mathbf{p}_a, -m_a|V\frac{1}{E_a H + i\varepsilon} \\
&= {}_{in}\langle\alpha, -\mathbf{p}_a, -m_a|.
\end{aligned} \tag{19.108}
$$

From now on, we confine ourselves to definition (i). The derivation using definition (ii) is left as an exercise.

Let us specify the initial and final states as

$$|i\rangle = |\alpha, \mathbf{p}_i, m_i\rangle, \tag{19.109a}$$

$$|f\rangle = |\beta, \mathbf{p}_f, m_f\rangle, \tag{19.109b}$$

then

$$|i^t\rangle = |\alpha, -\mathbf{p}_i, -m_i\rangle, \tag{19.110a}$$

$$|f^t\rangle = |\beta, -\mathbf{p}_f, -m_f\rangle. \tag{19.110b}$$

From Eq. (19.100), using Eqs. (19.107), we have

$$\langle\beta, \mathbf{p}_f, m_f|T|\alpha, \mathbf{p}_i, m_i\rangle = \langle\alpha, -\mathbf{p}_i, -m_i|T|\beta, -\mathbf{p}_f, -m_f\rangle. \tag{19.111}$$

This expresses the equality of two scattering processes obtained by reversing the momenta and spin components. This is known as the reciprocity relation. It is a consequence of the invariance under time reversal.

Further, if we assume that T is invariant under space inversion, i.e. if parity is conserved in the scattering process, then

$$\hat{P}T\hat{P}^\dagger = T\hat{P}|\beta, -\mathbf{p}_f, -m_f\rangle$$
$$= |\beta, \mathbf{p}_f, -m_f\rangle. \tag{19.112}$$

From Eq. (19.115), we have

$$\langle\beta, \mathbf{p}_f, m_f|T|\alpha, \mathbf{p}_i, m_i\rangle = \langle\alpha, -\mathbf{p}_i, -m_i|\hat{P}^\dagger\hat{P}T\hat{P}^\dagger\hat{P}|\beta, -\mathbf{p}_f, -m_f\rangle$$
$$= \langle\alpha, \mathbf{p}_i, -m_i|T|\beta, \mathbf{p}_f, -m_f\rangle. \tag{19.113}$$

If the spins are summed, then we can write Eq. (19.117) as

$$\sum_{spins} |\langle\beta, \mathbf{p}_f, m_f|T|\alpha, \mathbf{p}_i, m_i\rangle|^2 = \sum_{spins} |\langle\alpha, \mathbf{p}_i, m_i|T|\beta\mathbf{p}_f, m_f\rangle|^2. \tag{19.114}$$

This is called the 'semi detailed balance principle'. We apply the above result to two body scattering $a + b \to c + d$.

Now

$$d\sigma = \frac{2\pi}{\hbar(\text{Flux})_i}\rho(E_f)|T_{f_i}|^2$$
$$\rho(E_f) = \mu_f p_f d\Omega$$
$$= \frac{p_f^2}{v_f}d\Omega \tag{19.115}$$

and

$$(\text{Flux})_{in} = \frac{v_i}{(2\pi\hbar)^3}. \qquad (19.116)$$

If the particles are with spin, then in calculating the cross-section, we average over the initial spins and sum over the final spins. Hence

$$\frac{d\sigma}{d\Omega}(a+b \rightarrow c+d) = (2\pi)^4 \hbar^2 \frac{1}{(2s_a+1)(2s_b+1)} \frac{p_{cd}^2}{v_{ab}v_{cd}}$$
$$\times \sum_{spin} |T_{ab \rightarrow cd}|^2. \qquad (19.117)$$

$$\frac{d\sigma}{d\Omega}(c+d \rightarrow a+b) = (2\pi)^4 \hbar^2 \frac{1}{(2s_c+1)(2s_d+1)} \frac{p_{ab}^2}{v_{cd}v_{ab}} \sum_{spin} |T_{cd \rightarrow ab}|^2. \qquad (19.118)$$

But from Eq. (19.118)

$$\sum_{spin} |T_{ab \rightarrow cd}|^2 = \sum_{spin} |T_{cd \rightarrow ab}|^2. \qquad (19.119)$$

Hence

$$\frac{d\sigma}{d\Omega}(a+b \rightarrow c+d) = \frac{(2s_c+1)(2s_d+1)}{(2s_a+1)(2s_b+1)} \frac{p_{cd}^2}{p_{ab}^2} \frac{d\sigma}{d\Omega}(c+d \rightarrow a+b). \qquad (19.120)$$

This is known as the principle of detailed balance.

We now use the above principle to determine the spin of pion. Consider the process

$$p + p \rightleftharpoons \pi^+ + d \qquad (19.121)$$

where d denotes the deuteron. Now the spin of a proton $= 1/2$ and the spin of a deuteron $= 1$. Let s_π denote the spin of a pion. Then from Eq. (17.62) we have

$$\frac{d\sigma}{d\Omega}(p+p \rightarrow \pi^+ + d) = \frac{3(2s_\pi+1)}{4} \frac{p_\pi^2}{p_p^2} \frac{d\sigma}{d\Omega}(\pi^+ + d \rightarrow p+p). \qquad (19.122)$$

For total cross-sections, we have

$$\sigma(p+p \rightarrow \pi^+ + d) = \frac{3}{4}(2s_\pi+1)\frac{p_\pi^2}{p_p^2}\sigma(\pi^+ + d \rightarrow p+p). \qquad (19.123)$$

From the experimentally determined cross-sections, we find $s_\pi = 0$, i.e. the spin of the pion is zero.

19.5 Problems

19.1 Show that a state with spin 0 and negative parity cannot decay into 2 spinless particles each having negative parity, if parity conservation is assumed (rotation invariance is of course assumed).

19.2 A spinless state decays into two spinless particles, show that (assuming rotational invariance) the matrix elements must be real if time reversal invariance is used.

19.3 Using either the first or the second definition of time reversal transformation, show that the Schrödinger equation is invariant under time reversal.

Chapter 20

Relativistic Quantum Mechanics: Dirac Equation

"Dirac's equation can be written in one line as

$$i\hbar \frac{\partial}{\partial t}|\psi\rangle = (c\boldsymbol{\alpha}\mathbf{p} + \beta mc)|\psi\rangle$$

From the solution of this equation come details about the hydrogen atom, the spin of the electron, and the existence of antimatter. Poets bring us fresh insights with the right sequence of words, Dirac brought us fresh insights with the right sequence of symbols[1]".

20.1 A Brief Review of the Theory of Relativity

The special theory of relativity is based on two postulates:

(i) Laws of physics take the same form in all inertial frames (Principle of relativistic invariace).

(ii) In any given inertial frame, the speed of light c is the same whether light is emitted by a source at rest or in uniform motion (light principle).

From these postulates, it follows that the "length" $(c^2t^2 - x^2 - y^2 - z^2)$ is invariant.

In relativity, space and time are treated on equal footing and as such we deal with four vectors in four [1+3] dimensional space:

$$x^\mu = (x^o, x^i) = (ct, \boldsymbol{x}), \qquad \text{(contravariant vector)}$$

---

[1] John S Rigden, Hydrogen: The Essential Element (Cambridge University Press) Cambridge, Massachusetts, 2005, p. 94.

where $\mu = 0, 1, 2, 3;$ $\mu = 0, i = 1, 2, 3,$ and

$$x_\mu = (x_o, x_i) = (ct, -\boldsymbol{x}), \qquad \text{(covariant vector)}$$

so that

$$x^2 = x^\mu x_\mu = x^o x_o + x^i x_i = c^2 t^2 - (x^2 + y^2 + z^2).$$

Introduce metric tensors $g^{\mu\nu}$, $g_{\mu\nu}$ to raise or lower the indices:

$$x^\mu = g^{\mu\nu} x_\nu, \quad x_\mu = g_{\mu\nu} x^\nu \tag{20.1}$$

where

$$g^{\mu\nu} = \begin{pmatrix} 1 & 0 & 0 & 0 \\ 0 & -1 & 0 & 0 \\ 0 & 0 & -1 & 0 \\ 0 & 0 & 0 & -1 \end{pmatrix} = g_{\mu\nu} \tag{20.2}$$

$$g^{00} = g_{00} = 1, \quad g^{ij} = -\delta^{ij} = -\delta_{ij} = g_{ij}.$$

From (i) and (ii), it follows that the length of the vector x is invariant:

$$x'^\mu x'_\mu = x^\mu x_\mu. \tag{20.3}$$

The vectors x'^μ and x^μ are related to each other by a linear transformation (called Lorentz Transformation)

$$x'^\mu = \Lambda^\mu_\nu x^\nu \tag{20.4}$$

$$x'_\mu = \Lambda^\nu_\mu x_\nu$$

which leaves $x^\mu x_\mu$ invariant:

$$
\begin{aligned}
x^\mu g_{\mu\nu} x^\nu &= x'^\alpha g_{\alpha\beta} x'^\beta \\
&= \Lambda^\alpha_\mu x^\mu g_{\alpha\beta} \\
&= \Lambda^\beta_\nu x^\nu.
\end{aligned} \tag{20.5}
$$

The above equation gives

$$g_{\mu\nu} = \Lambda^\alpha_\mu g_{\alpha\beta} \Lambda^\beta_\nu. \tag{20.6}$$

In matrix form Lorentz transformation [LT] is

$$x' = \Lambda x, \qquad x = \Lambda^T x'$$

and the matrix

$$
\begin{aligned}
g &= (\Lambda^T)_{\mu\alpha} g_{\alpha\beta} (\Lambda)_{\beta\gamma} \\
&= \Lambda^T g \Lambda
\end{aligned} \tag{20.7}
$$

where Λ is a 4×4 matrix. Further

$$\det g = (\det \Lambda^T)\det(g\Lambda)$$
$$= (\det \Lambda)^2 \det g$$

$$(\det \Lambda)^2 = 1, \qquad \det \Lambda = \pm 1. \qquad (20.8)$$

If $\det \Lambda = 1$, LT is called proper, denoted by L_+. This excludes spatial reflection for which $\det \Lambda = -1$. Further from Eq. (20.6)

$$1 = (\Lambda_o^o)^2 - (\Lambda_o^i)^2$$

i.e. $(\Lambda_o^o)^2 \geq 1$, which implies $\Lambda_o^o \geq 1$ or $\Lambda_o^o \leq -1$. If $\Lambda_o^o \geq 1$, the time direction is unaltered and the LT is called orthochronus, written as $L^\dagger$. We shall confine ourselves to $L_+^\dagger$, which excludes space reflection and time reversal.

Lorentz group is a six parameter group; three parameters correspond to rotation in three dimensional space(x, y, z) and the remaining three parameters correspond to Lorentz velocity transformation (Lorentz boost).

In order to see this, first consider the transformation

$$x'^o = x^o,$$
$$x'^i = \Lambda_j^i x^j. \qquad (20.9)$$

Consider for simplicity rotation around $z-$axis by an angle ω:

$$\Lambda = \begin{pmatrix} 1 & 0 & 0 & 0 \\ 0 & \cos\omega & \sin\omega & 0 \\ 0 & -\sin\omega & \cos\omega & 0 \\ 0 & 0 & 0 & 1 \end{pmatrix} = \begin{pmatrix} 1 & 0 \\ 0 & \Lambda_R \end{pmatrix}. \qquad (20.10)$$

The matrix

$$\Lambda_R = \begin{pmatrix} \cos\omega & \sin\omega & 0 \\ -\sin\omega & \cos\omega & 0 \\ 0 & 0 & 1 \end{pmatrix}$$

giving

$$x'^o = x^o,$$
$$x' = x\cos\omega + y\sin\omega,$$
$$y' = -x\sin\omega + y\cos\omega,$$
$$z' = z. \qquad (20.11)$$

For Lorentz boost, for simplicity consider Lorentz velocity transformation along x-axis ($\mathbf{v} = v, 0, 0$). For this case

$$\Lambda = \begin{pmatrix} \gamma & -\dfrac{\gamma v}{c} & 0 & 0 \\ -\dfrac{\gamma v}{c} & \gamma & 0 & 0 \\ 0 & 0 & 1 & 0 \\ 0 & 0 & 0 & 1 \end{pmatrix} = \begin{pmatrix} \Lambda(v) & 0 \\ 0 & 1 \end{pmatrix} \tag{20.12}$$

giving

$$t' = \gamma \left(t - \frac{v}{c^2} x \right)$$
$$x' = \gamma(x - vt)$$
$$y' = y$$
$$z' = z \tag{20.13}$$

where

$$\gamma = \frac{1}{\sqrt{1 - \frac{v^2}{c^2}}}.$$

This is the Lorentz transformation, when the primed frame is moving with uniform velocity v along x-axis with respect to unprimed system.

For infinitesimal Lorentz transformation:

$$\Lambda^\mu_\nu = \delta^\mu_\nu + \epsilon^\mu_\nu,$$
$$x'^\mu = x^\mu + \epsilon^\mu_\nu x^\nu$$
$$= x^\mu + g^{\mu\lambda} \epsilon_{\lambda\nu} x^\nu. \tag{20.14}$$

Then Eq.(20.6) gives

$$g_{\mu\nu} = (\delta^\alpha_\mu + \epsilon^\alpha_\mu) g_{\alpha\beta} (\delta^\beta_\nu + \epsilon^\beta_\nu)$$
$$= g_{\mu\nu} + \epsilon_{\mu\nu} + \epsilon_{\nu\mu},$$
$$\epsilon_{\mu\nu} = -\epsilon_{\nu\mu}. \tag{20.15}$$

Thus

$$\delta x^\lambda = x'^\lambda - x^\lambda$$
$$= \frac{1}{2} \epsilon_{\mu\nu} (g^{\lambda\mu} x^\nu - g^{\lambda\nu} x^\mu)$$
$$= \frac{1}{2} \epsilon_{\mu\nu} (x^\nu \partial^\mu - x^\mu \partial^\nu) x^\lambda$$
$$= \frac{i}{2} \epsilon_{\mu\nu} L^{\mu\nu} x^\lambda$$

where the hermitian operators
$$L^{\mu\nu} = i(x^\mu \partial^\nu - x^\nu \partial^\mu) \tag{20.16}$$
are six generators of the Lorentz group corresponding to six parameters $\epsilon_{\mu\nu}$. They satisfy the commutation relations
$$[L^{\mu\nu}, L^{\rho\sigma}] = i(g^{\mu\rho}L^{\nu\sigma} - g^{\nu\sigma}L^{\mu\rho} + \mu \leftrightarrow \nu). \tag{20.17}$$
In relativistic mechanics, energy and momentum are treated on equal footing just like the space and time. Accordingly we introduce energy-momentum 4-vector
$$p^\mu = (p^o, p^i) = \left(\frac{E}{c}, \boldsymbol{p}\right)$$
$$p_\mu = (p_o, p_i) = \left(\frac{E}{c}, -\boldsymbol{p}\right) \tag{20.18}$$
$$p^2 = p^\mu p_\mu = \frac{E^2}{c^2} - \boldsymbol{p}^2 = m^2 c^2 \tag{20.19}$$
so that for a particle on mass shell
$$p^\mu p_\mu = \frac{E^2}{c^2} - \boldsymbol{p}^2 = m^2 c^2.$$
In terms of γ,
$$E = m\gamma c^2, \quad \boldsymbol{p} = m\gamma \mathbf{v} = E\frac{\mathbf{v}}{c^2} \tag{20.20}$$
or
$$\mathbf{v} = \frac{c^2 \boldsymbol{p}}{E}. \tag{20.21}$$
Thus in analogy with space-time transformation (20.13),
$$E' = \gamma[E - \mathbf{v}.\boldsymbol{p}]$$
$$\boldsymbol{p}' = \boldsymbol{p} + [(\gamma - 1)\mathbf{v}\frac{\mathbf{v}.\boldsymbol{p}}{v^2} - \gamma\mathbf{v}\frac{E}{c^2}]$$
where the primed quantities are in the frame moving with velocity $\mathbf{v}$ relative to one in which $\mathbf{E}$ and $\boldsymbol{p}$ are measured. It is convenient to write the vector $\boldsymbol{p}$ in terms of its longitudinal and transverse components
$$\boldsymbol{p} = (\boldsymbol{p}_{11}, \boldsymbol{p}_\perp). \tag{20.22}$$
Under pure Lorentz transformation (Lorentz boost)
$$\mathbf{p}_{11} = \gamma \left[\mathbf{p}_{11} - \frac{\mathbf{v}}{c^2}E\right],$$
$$\boldsymbol{p}'_\perp = \boldsymbol{p}_\perp,$$
$$E' = \gamma \left[E - \mathbf{v}.\mathbf{p}_{11}\right]. \tag{20.23}$$
Finally
$$\partial_\mu = \frac{\partial}{\partial x^\mu} = \left(\frac{1}{c}\frac{\partial}{\partial t}, \boldsymbol{\nabla}\right),$$
$$\partial^\mu = \frac{\partial}{\partial x_\mu} = \left(\frac{1}{c}\frac{\partial}{\partial t}, -\boldsymbol{\nabla}\right),$$
$$\partial^\mu \partial_\mu = \partial_\mu \partial^\mu = \frac{1}{c^2}\frac{\partial^2}{\partial t^2} - \boldsymbol{\nabla}^2 = \Box. \tag{20.24}$$

20.2 Relativistic Quantum Mechanics: Introduction

After the discovery of quantum mechanics in 1926–27 the big problem was that the Schrödinger equation

$$i\hbar\frac{\partial\Phi}{\partial t} = -\frac{\hbar^2}{2m}\nabla^2\Phi \qquad (20.25)$$

was not consistent with the special theory of relativity as time and space coordinates are not treated on equal footing.

Now for a single free particle in non-relativistic classical mechanics

$$E = \frac{p^2}{2m}. \qquad (20.26)$$

To go over to quantum mechanics replace the dynamical variables E and p by differential operators

$$E \to i\hbar\frac{\partial}{\partial t}, \quad p \to -ih\nabla, \qquad (20.27)$$

which act on the state function Φ. We get the Schrödinger equation given in Eq. (20.25)

In order to get the relativistic equation, it is natural to replace the non-relativistic energy equation by the corresponding relativistic energy equation

$$\frac{E^2}{c^2} = p^2 + m^2c^2. \qquad (20.28)$$

Replacing E and p by operators given in Eq.(20.27), we obtain the relativistic equation in quantum mechanics

$$-\frac{\hbar^2}{c^2}\frac{\partial^2\Phi}{\partial t^2} = -\hbar^2\nabla^2\Phi + m^2c^2\Phi. \qquad (20.29)$$

Now

$$\partial^\mu\partial_\mu = \frac{1}{c^2}\frac{\partial^2}{\partial t^2} - \nabla^2. \qquad (20.30)$$

Hence, we can write the differential equation (20.29), in the form

$$\left(\partial^\mu\partial_\mu + \frac{m^2c^2}{\hbar^2}\right)\Phi = 0,$$

$$\left(\partial^\mu\partial_\mu + \frac{m^2c^2}{\hbar^2}\right)\Phi^* = 0. \qquad (20.31)$$

The above equation is known as the Klein–Gordon equation for a free particle of mass m. In this equation, space and time coordinates appear on the same footing as required by the theory of relativity. But this equation as a single particle equation faces difficulty.

The equation of continuity viz.

$$\partial^\mu j_\mu = 0 \quad \text{or} \quad \frac{\partial \rho}{\partial t} + \boldsymbol{\nabla} \cdot \mathbf{j} = 0 \tag{20.32}$$

must be satisfied as it is equivalent to conservation of probability. A natural generalization for j_μ from the Schrödinger theory [Eq.(2.46)] is

$$j_\mu = -\frac{\hbar}{2im} \left[\Phi^* \partial_\mu \Phi - \Phi \partial_\mu \Phi^*\right] \tag{20.33}$$

giving

$$j_0 = \frac{\rho}{c} = -\frac{\hbar}{2imc} \left[\Phi^* \frac{\partial \Phi}{\partial t} - \Phi \frac{\partial \Phi^*}{\partial t}\right], \tag{20.34}$$

$$\mathbf{j} = \frac{\hbar}{2im} \left[\Phi^* \boldsymbol{\nabla} \Phi - \Phi \boldsymbol{\nabla} \Phi^*\right]. \tag{20.35}$$

It is easy to see from Eqs. (20.29) and (20.33) that

$$\partial^\mu j_\mu = 0. \tag{20.36}$$

The requirement that probability current density ρ must be positive definite is not satisfied as the expression for ρ is not absolute square of the wave function, but contains the time derivative of the wave function. Thus for a mode with energy E which should oscillate as $e^{-iEt/\hbar}$, we have

$$\Phi(\mathbf{x}, t) \sim \phi(\mathbf{x}) e^{-iEt/\hbar}$$

so that

$$\rho = \frac{E}{mc^2} |\phi(\mathbf{x})|^2 \tag{20.37}$$

where

$$E = \pm\sqrt{c^2 \mathbf{p}^2 + m^2 c^4} = \pm E_p,$$

Thus for $E > 0$, ρ is positive but for $E < 0$, it is negative and thus cannot be interpreted as probability. We may arbitrarily omit all solutions of the form $\phi(\mathbf{x}) e^{-iEt/\hbar}$ for $E < 0$, but this is not allowed since the solutions with $E > 0$ do not form a complete set.

The Klein–Gordon equation cannot be considered as a quantum mechanical equation for a single particle. One encounters two problems: Energy can be negative for which probability density is negative. It is important to realise that two problems are interlinked.

We may note that, Klein–Gordon (K–G) equation

$$\left[\left(\frac{1}{c^2}\frac{\partial^2}{\partial t^2} - \nabla^2\right) + \frac{m^2 c^2}{\hbar^2}\right]\Phi(t,\mathbf{x}) = 0 \tag{20.38}$$

for $m = 0$, is similar to the second order differential equation

$$\left[\left(\frac{1}{c^2}\frac{\partial^2}{\partial t^2} - \nabla^2\right)\right]\mathbf{E}, \mathbf{B} \tag{20.39}$$

which the electromagnetic fields $\mathbf{E}, \mathbf{B}$ satisfy. Pauli and Weisskopf interpreted K–G equation as a field equation, where Φ is a field, the quantum of which describes the spin zero particle such as π-mesons.

Dirac made a breakthrough in 1928. He did it from pure logic by introducing spinors in addition to more familiar scalar, vector and tensor quantities.

By doing so the two problems are decoupled: ρ is always positive definite but negative energy problem remains. Dirac solved it by proposing Hole theory, thereby predicting antimatter.

20.3 Dirac Equation

We notice that in the Klein–Gordon equation the problem with ρ arises due to time derivative in Eq. (20.34) which in turn arises because of Klein–Gordon equation being second order in time derivative. Dirac started with a linear equation both with respect to time and space derivatives. Here an analogy with the Maxwell's equations, which are fully relativistic, will be helpful. One can write basic equations of electromagnetism in terms of electromagnetic fields $\mathbf{E}$ and $\mathbf{B}$ which have six components and satisfy Maxwell's equations, which are linear in time and space coordinates, or in terms of vector potential A^μ which have four components but satisfy a second order Klein–Gordon equation

$$\partial^\mu \partial_\mu A^\nu = 0.$$

As noticed previously each component of $\mathbf{E}$ and $\mathbf{B}$ also satisfies such a second order equation (20.39).

The above analogy suggests that

(1) We should expect several components of wave function ψ_n.
(2) Each component should be related to other by first order linear equations both in space and time derivatives.
(3) Each component should satisfy the Klein–Gordon-equation.

The most general equation one can write satisfying conditions (1) and (2) [we put $c = 1$ and $\hbar = 1$] is

$$\left(i\delta_{\ell n}\frac{\partial}{\partial t} + i(\alpha^i)_{\ell n}\frac{\partial}{\partial x^i} - \beta_{\ell n}m \right)\psi_n = 0 \qquad (20.40)$$

where $\ell, n = 1, 2, ...N$ and $i = 1, 2, 3$. In order to satisfy condition (3) operate by

$$\left(-i\delta_{m\ell}\frac{\partial}{\partial t} + i(\alpha^j)_{m\ell}\frac{\partial}{\partial x^j} - m\beta_{m\ell} \right).$$

One obtains, since $\dfrac{\partial}{\partial t}$ and $\dfrac{\partial}{\partial x}$ commute,

$$\left[\delta_{mn}\frac{\partial^2}{\partial t^2} - \frac{1}{2}\left[(\alpha^j)_{m\ell}(\alpha^i)_{\ell n} + i \leftrightarrow j\right]\frac{\partial}{\partial x^i}\frac{\partial}{\partial x^j} \right.$$
$$\left. -im\left((\alpha^j)_{m\ell}\beta_{\ell n}\frac{\partial}{\partial x^j} + \beta_{m\ell}(\alpha^i)_{\ell n}\frac{\partial}{\partial x^i}\right) + m^2\beta_{m\ell}\beta_{\ell n} \right]\psi_n = 0 \quad (20.41)$$

where the term involving $\alpha^i\alpha^j$ has been symmetrized which is permissible since $\dfrac{\partial}{\partial x^i}$ and $\dfrac{\partial}{\partial x^j}$ commute. The above equation reduces to the Klein–Gordon-like equation,

$$\left(\frac{\partial^2}{\partial t^2} - \nabla^2 - m^2 \right)\psi_m = 0, \qquad (20.42)$$

provided that

$$\frac{1}{2}\left[(\alpha^i)_{m\ell}(\alpha^j)_{\ell n} + (\alpha^j)_{m\ell}(\alpha^i)_{\ell n}\right] = \delta_{mn}\delta^{ij},$$
$$(\alpha^i)_{m\ell}(\beta)_{\ell n} + (\beta)_{m\ell}(\alpha^i)_{\ell n} = 0,$$
$$\beta_{m\ell}\beta_{\ell n} = \delta_{mn}. \qquad (20.43)$$

It is clear that α's and β's are not algebraic numbers. They are $N \times N$ matrices with the properties

$$\alpha^i\alpha^j + \alpha^j\alpha^i = 2\delta^{ij},$$
$$\alpha^i\beta + \beta\alpha^i = 0,$$
$$\beta^2 = 1. \qquad (20.44)$$

Thus the matrices $\alpha^1, \alpha^2, \alpha^3, \beta$ anticommute in pairs and their square is unity.

Thus the Eq. (20.40) takes the matrix form [inserting back $\hbar$ and c]

$$\left(i\hbar \frac{\partial}{c\partial t} + i\hbar \boldsymbol{\alpha}.\boldsymbol{\nabla} - \beta m \right) \psi = 0 \qquad (20.45)$$

where ψ is a column matrix with N components. Or

$$i\hbar \frac{\partial \psi}{\partial t} = H\psi = (c\boldsymbol{\alpha}.(-i\hbar\boldsymbol{\nabla}) + \beta mc^2)\psi,$$

$$i\hbar \frac{\partial \psi}{\partial t} = (c\boldsymbol{\alpha}.\hat{p} + \beta mc^2)\psi, \qquad (20.46)$$

where $\hat{p} = -i\hbar\nabla$ is the momentum operator. Since the Hamiltonian operator

$$H = \boldsymbol{\alpha}.\hat{p} + \beta m, \qquad (20.47)$$

should be hermitian, each of the four matrices α^i and β must be hermitian:

$$\alpha^\dagger = \alpha, \qquad \beta^\dagger = \beta. \qquad (20.48)$$

From the anticommutation relations $\alpha^i\beta = -\beta\alpha^i$, $\det(\alpha^i\beta) = (-1)^N\det(\beta\alpha^i)$ i.e. $\det(\alpha^i)\det\beta = (-1)^N(\det\beta)\det(\alpha^i)$. Thus $(-1)^N = 1$ so that N must be even. Further

$$\beta\alpha^i\beta = -\alpha^i \qquad (20.49)$$

so that

$$\text{Tr}(\beta\alpha^i\beta) = -\text{Tr}(\alpha^i). \qquad (20.50)$$

But

$$\text{Tr}(\beta\alpha^i\beta) = \text{Tr}(\beta\beta\alpha^i) = \text{Tr}(\alpha^i). \qquad (20.51)$$

Thus $\text{Tr}(\alpha^i) = 0$; similarly $\text{Tr}(\beta) = 0$. For $N = 2$ matrices 1 and σ^i form a complete set but since the unit matrix commutes with σ^i it cannot be identified with β. Thus the lowest possible rank is $N = 4$. One can choose a particular representation

$$\alpha^i = \begin{pmatrix} 0 & \sigma^i \\ \sigma^i & 0 \end{pmatrix}, \qquad \begin{pmatrix} I & 0 \\ 0 & -I \end{pmatrix}, \qquad \text{with} \quad I = \begin{pmatrix} 1 & 0 \\ 0 & 1 \end{pmatrix} \qquad (20.52)$$

where σ^i are 2×2 Pauli matrices, which satisfy

$$\sigma^i\sigma^j + \sigma^j\sigma^i = 2\delta ij, \qquad [\sigma^i, \sigma^j] = 2i\epsilon^{ijk}\sigma^k.$$

It is easy to show that the representation given in Eq. (20.51), known as the Pauli representation, satisfy the relations (20.44). Then ψ is a column matrix:

$$\psi = \begin{pmatrix} \psi_1 \\ \psi_2 \\ \psi_3 \\ \psi_4 \end{pmatrix} = \begin{pmatrix} \psi_A \\ \psi_B \end{pmatrix}, \tag{20.53}$$

where each of ψ_A and ψ_B has two components: when written in full

$$\beta = \begin{pmatrix} 1 & 0 & 0 & 0 \\ 0 & 1 & 0 & 0 \\ 0 & 0 & -1 & 0 \\ 0 & 0 & 0 & -1 \end{pmatrix}, \quad \alpha^1 = \alpha_x = \begin{pmatrix} 0 & 0 & 0 & 1 \\ 0 & 0 & 1 & 0 \\ 0 & 1 & 0 & 0 \\ 1 & 0 & 0 & 0 \end{pmatrix},$$

$$\alpha^2 = \alpha_y = \begin{pmatrix} 0 & 0 & 0 & -i \\ 0 & 0 & i & 0 \\ 0 & -i & 0 & 0 \\ i & 0 & 0 & 0 \end{pmatrix}, \quad \alpha^3 = \alpha_z = \begin{pmatrix} 0 & 0 & 1 & 1 \\ 0 & 0 & 1 & -1 \\ 1 & 0 & 0 & 0 \\ 0 & -1 & 0 & 0 \end{pmatrix}. \tag{20.54}$$

20.4 Covariant Form of Dirac Equation

We rewrite Dirac equation given in Eq. (20.45) by multiplying on the left by β:

$$i\hbar\beta \frac{\partial \psi}{\partial t} - [c\beta\alpha.(-i\hbar\nabla + \beta mc^2)]\psi = 0. \tag{20.55}$$

Noting that

$$\partial_\mu = \frac{\partial}{\partial x^\mu} = \left(\frac{1}{c}\frac{\partial}{\partial t}, \nabla \right), \qquad \beta^2 = 1,$$

we can write Eq. (20.55) as

$$\left[i\hbar \left(\gamma^o \frac{\partial}{\partial x^o} + \gamma^i \partial_i \right) - mc \right] \psi = 0$$

or

$$(i\hbar\gamma^\mu \partial_\mu - mc)\psi = 0 \tag{20.56}$$

where

$$\beta = \gamma^o, \qquad \beta\alpha = \gamma; \qquad \beta\alpha^i = \gamma^i,$$

$$\gamma^\mu = (\gamma^o, \gamma^i) = (\gamma^o, \gamma),$$

$$\gamma_\mu = (\gamma_o, \gamma_i) = (\gamma_o, -\boldsymbol{\gamma}),$$

$$\gamma^o = \gamma_o. \tag{20.57}$$

Hence, the Dirac equation in covariant form is given in Eq. (20.56), which can also be written as

$$[\gamma^\mu \hat{p}_\mu + mc]\psi(x) = 0, \qquad \hat{p}_\mu = -i\hbar\partial_\mu. \tag{20.58}$$

We note that

$$(\gamma^i)^\dagger = (\beta\alpha^i)^\dagger = \alpha^i\beta = -\beta\alpha^i = -\gamma^i$$

i.e. γ^i is antihermitian,

$$(\gamma^i)^2 = \beta\alpha^i\beta\alpha^i = -\beta^2(\alpha^i)^2 = -1, \qquad i = 1, 2, 3.$$

Noting that $g^{oo} = 1, g^{ij} = -\delta^{ij}$, the anticommutation relations (20.44) can be written in the compact form

$$\{\gamma^\mu, \gamma^\nu\} = (\gamma^\mu\gamma^\nu + \gamma^\nu\gamma^\mu) = 2g^{\mu\nu}. \tag{20.59}$$

Thus there are four independent γ-matrices $\gamma^o, \gamma^1, \gamma^2, \gamma^3$. The fifth one is defined as

$$\gamma^5 = i\gamma^o\gamma^1\gamma^2\gamma^3, \qquad \gamma^{5\dagger} = \gamma^5,$$

$$\gamma_5 = i\gamma_o\gamma_1\gamma_2\gamma_3, \qquad \gamma_5^\dagger = \gamma_5,$$

$$(\gamma^5)^2 = \gamma_5^2 = 1. \tag{20.60}$$

It is convenient to define

$$[\gamma^\mu, \gamma^\nu] = \gamma^\mu\gamma^\nu - \gamma^\nu\gamma^\mu = -2i\sigma^{\mu\nu}$$

or

$$\sigma^{\mu\nu} = \frac{i}{2}(\gamma^\mu\gamma^\nu - \gamma^\nu\gamma^\mu). \tag{20.61}$$

From the properties of γ-matrices, it is easy to see that there are sixteen independent γ-matrices listed in Table 1. The following identities are useful

$$\gamma^\mu\gamma^\nu\gamma^\lambda = g^{\mu\nu}\gamma^\lambda - g^{\mu\lambda}\gamma^\nu + g^{\nu\lambda}\gamma^\mu + i\epsilon^{\mu\nu\lambda\rho}\gamma^5\gamma_\rho,$$

$$[\gamma^\mu, \sigma^{\nu\lambda}] = 2i(g^{\mu\nu}\gamma^\lambda - g^{\mu\lambda}\gamma^\nu), \tag{20.62}$$

$$\{\gamma^\mu, \sigma^{\nu\lambda}\} = -2\epsilon^{\mu\nu\lambda\rho}\gamma^5\gamma_\rho,$$

$$\gamma^5 = -\frac{i}{4!}\epsilon^{\mu\nu\lambda\rho}\gamma_\mu\gamma_\nu\gamma_\lambda\gamma_\rho = \gamma_5,$$

where we see the convention

$$\epsilon^{0\nu\nu\lambda} = \epsilon^{ijk}, \text{e.g.} \qquad \epsilon^{0123} = 1 = -\epsilon_{0123}, \qquad \epsilon^{ijk} = -\epsilon_{ijk}.$$

The properties of the γ-matrices discussed above are independent of any representation of γ-matrices. Two representations of γ-matrices are useful:

Type	No. of independent components
1	1
γ^μ	4
$\sigma^{\mu\nu}$	6
$\gamma^\mu\gamma^\nu\gamma^\lambda$	4
γ^5	1

(i) Pauli representation:

$$\gamma^o = \beta = \begin{pmatrix} 1 & 0 \\ 0 & -1 \end{pmatrix}, \gamma^i = \beta\alpha^i = \begin{pmatrix} 0 & \sigma^i \\ -\sigma^i & 0 \end{pmatrix},$$

$$\boldsymbol{\gamma} = \beta\boldsymbol{\alpha} = \begin{pmatrix} 0 & \boldsymbol{\sigma} \\ -\boldsymbol{\sigma} & 0 \end{pmatrix},$$

$$\gamma^5 = \begin{pmatrix} 0 & 1 \\ 1 & 0 \end{pmatrix} = \gamma_5. \tag{20.63}$$

This representation is particularly suitable to go to the non-relativistic limit.

(ii) Weyl or Chiral Representation:

$$\gamma^o = \begin{pmatrix} 0 & 1 \\ 1 & 0 \end{pmatrix}, \gamma^i = \begin{pmatrix} 0 & \sigma^i \\ -\sigma^i & 0 \end{pmatrix}, \gamma^5 = \begin{pmatrix} -1 & 0 \\ 0 & 1 \end{pmatrix}. \tag{20.64}$$

Define

$$\sigma^\mu = (1, \sigma^i) = (1, \boldsymbol{\sigma}),$$

$$\sigma_\mu = (1, \sigma^i) = (1, -\boldsymbol{\sigma}),$$

$$\bar{\sigma}^\mu = (1, -\boldsymbol{\sigma}), \qquad \bar{\sigma}_\mu = (1, \boldsymbol{\sigma}).$$

Then we can write

$$\gamma^\mu = \begin{pmatrix} 0 & \sigma^\mu \\ \bar{\sigma}^\mu & 0 \end{pmatrix}. \tag{20.65}$$

This representation is specially suitable in taking the limit m→ 0 i.e. for chiral fermions.

In the Pauli representation, one can write the 4-component wave function ψ as $\psi = \psi_A + \psi_B$, where

$$\psi_A = \frac{1+\gamma^o}{2}\psi = \begin{pmatrix} \phi_A \\ 0 \end{pmatrix}, \qquad \psi_B = \frac{1-\gamma^o}{2}\psi = \begin{pmatrix} 0 \\ \phi_B \end{pmatrix}$$

so that

$$\psi = \begin{pmatrix} \phi_A \\ \phi_B \end{pmatrix}, \tag{20.66}$$

where each of ϕ_A and ϕ_B has two components. In the chiral representation

$$\psi = \psi_L + \psi_R,$$

where

$$\psi_L = \frac{1-\gamma^5}{2}\psi = \begin{pmatrix} \phi_L \\ 0 \end{pmatrix}$$

$$\psi_R = \frac{1+\gamma^5}{2}\begin{pmatrix} 0 \\ \phi_R \end{pmatrix} \tag{20.67}$$

where each of ϕ_L and ϕ_R has two components. Thus $(1\mp\gamma_5)\psi$ respectively pick ϕ_L and ϕ_R. These are known as chiral projections of ψ.

20.5 Relativistic Invariance of Dirac Equation

Under Lorentz transformation

$$x'^\mu = \Lambda^\mu_\nu x^\nu$$
$$\partial'_\mu = \Lambda^\nu_\mu \partial_\nu$$

and suppose that

$$\psi(x) \rightarrow \psi'(x') = S\psi(x). \tag{20.68}$$

Then

$$(\gamma^\mu \partial'_\mu - m)\psi'(x) = (\gamma^\mu \Lambda^\nu_\mu \partial_\nu - m)S\psi(x) = 0$$

or multiplying left by S^{-1},

$$[(S^{-1}\gamma^\mu S)\Lambda^\nu_\mu \partial_\nu - m]\psi(x) = 0$$

which reduces to

$$[\gamma^\nu \partial_\nu - m]\psi(x) = 0$$

as required by the invariance, provided that

$$S^{-1}\gamma^\mu S\Lambda^\nu_\mu = \gamma^\nu$$

or

$$S^{-1}\gamma^\mu S\Lambda^\nu_\mu g_{\nu\beta}\Lambda^\beta_\alpha = \gamma^\nu g_{\nu\beta}\Lambda^\beta_\alpha.$$

Therefore using Eq. (20.6),

$$S^{-1}\gamma^\mu S g_{\mu\alpha} = \Lambda^\beta_\alpha \gamma_\beta.$$

Equivalently

$$S^{-1}\gamma^\mu S = \Lambda^\mu_\lambda \gamma^\lambda. \tag{20.69}$$

For infinitesimal Lorentz transformation

$$\Lambda^\mu_\lambda = \delta^\mu_\lambda + \epsilon^\mu_\lambda$$
$$= \delta^\mu_\lambda + g^{\mu\nu}\epsilon_{\nu\lambda}$$

and the above condition becomes

$$S^{-1}\gamma^\mu S = \gamma^\mu + g^{\mu\nu}\epsilon_{\nu\mu}\gamma_\nu. \tag{20.70}$$

We can write

$$S = 1 - \frac{i}{4}\epsilon_{\mu\nu}\sigma^{\mu\nu} \tag{20.71}$$

since $\sigma^{\mu\nu}$ is the only antisymmetric (in $\mu \leftrightarrow \nu$) combination we can form out of γ-matrices.

We can verify that [using Eq. (20.62)]

$$S^{-1}\gamma^\mu S = \left(1 + \frac{i}{4}\epsilon_{\lambda\nu}\sigma^{\lambda\nu}\right)\gamma^\mu\left(1 - \frac{i}{4}\epsilon_{\lambda\nu}\sigma^{\lambda\nu}\right)$$

$$= \gamma^\mu + \frac{i}{4}\epsilon_{\lambda\nu}\left[\sigma^{\lambda\nu}\gamma^\mu - \gamma^\mu\sigma^{\lambda\nu}\right]$$

$$= \gamma^\mu - \frac{i}{4}\epsilon_{\lambda\nu}2i(g^{\mu\lambda}\gamma^\nu - g^{\mu\nu}\gamma^\lambda)$$

$$= \gamma^\mu + \frac{1}{2}(g^{\mu\lambda}\epsilon_{\lambda\nu}\gamma^\nu + g^{\mu\nu}\epsilon_{\nu\lambda}\gamma^\lambda)$$

$$= \gamma^\mu + g^{\mu\nu}\epsilon_{\nu\lambda}\gamma^\lambda$$

which agrees with Eq. (20.70).

We have seen that under infinitesimal Lorentz transformation

$$x'^\mu = x^\mu + g^{\mu\lambda}\epsilon_{\lambda\nu}x^\nu, \qquad x^\mu = x'^\mu - g^{\mu\lambda}\epsilon_{\lambda\nu}x'^\nu$$

$$\psi'(x') = S\psi(x) = \left(1 - \frac{i}{4}\epsilon_{\mu\nu}\sigma^{\mu\nu}\right)\left(\psi(x'^\mu - g^{\mu\lambda}\epsilon_{\lambda\nu}x'^\nu)\right)$$

or

$$\psi'(x) = \left(1 - \frac{i}{4}\epsilon_{\mu\nu}\sigma^{\mu\nu}\right)\left(\psi(x^\mu - g^{\mu\lambda}\epsilon_{\mu\nu}x^\nu)\right)$$

$$= \left(1 - \frac{i}{4}\epsilon_{\mu\nu}\sigma^{\mu\nu}\right)\left(\psi(x) - g^{\mu\lambda}\epsilon_{\lambda\nu}x^\nu\partial_\mu\psi(x)\right)$$

$$= \left(1 - \frac{i}{4}\epsilon_{\mu\nu}\sigma^{\mu\nu}\right)\left(1 - \frac{1}{2}\epsilon_{\lambda\nu}(x^\nu\partial^\lambda - x^\lambda\partial^\nu)\right)\psi(x)$$

$$= \left(1 - \frac{i}{2}\epsilon_{\mu\nu}\left(\frac{1}{2}\sigma^{\mu\nu} + L^{\mu\nu}\right)\right)\psi(x). \tag{20.72}$$

where

$$L^{\mu\nu} = i(x^\mu \partial^\nu - x^\nu \partial^\mu)$$

are the generators of the Lorentz group as previously noted. The most general representation of the generators of the Lorentz group are given by

$$M^{\mu\nu} = L^{\mu\nu} + S^{\mu\nu} \tag{20.73}$$

where the hermitian $S^{\mu\nu}$ satisfy the same commutation relations as in Eq. (20.17) and commute with $L^{\mu\nu}$. Thus

$$[M^{\mu\nu}, M^{\rho\sigma}] = -i[g^{\mu\rho}M^{\nu\sigma} - g^{\nu\sigma}M^{\mu\rho} + \mu \leftrightarrow \nu] \tag{20.74}$$

which form the Lie algebra of the Lorentz group. For a spinor ψ,

$$S^{\mu\nu} = \frac{1}{2}\sigma^{\mu\nu} = \frac{i}{4}[\gamma^\mu, \gamma^\nu]$$

as is clear from Eq. (20.72). Thus the relation (20.72) can be written as

$$\psi'(x) = \left[1 - \frac{i}{2}\epsilon_{\mu\nu}M^{\mu\nu}\right]\psi(x) = U(\Lambda(\epsilon))\psi(x) \tag{20.75}$$

where on exponentiation

$$U(\Lambda(\epsilon)) = e^{-\frac{i}{2}\epsilon_{\mu\nu}M^{\mu\nu}}. \tag{20.76}$$

For the rotation subgroup of Lorentz group

$$\epsilon_{\mu\nu} \rightarrow \epsilon_{ij}, \epsilon_{\mu\nu}\sigma^{\mu\nu} \rightarrow \epsilon_{ij}\sigma^{ij}, \epsilon_{\mu\nu}L^{\mu\nu} \rightarrow \epsilon_{ij}L^{ij}, \qquad M^{\mu\nu} \rightarrow \epsilon_{ij}M^{ij}.$$

In particular, for rotation about x^3–axis,

$$\epsilon_{12} = \epsilon, \qquad \sigma^{12} = i\gamma^1\gamma^2, \qquad L^{12} = i\left(x^1\frac{\partial}{\partial x^2} - x^2\frac{\partial}{\partial x^1}\right) = -\frac{1}{\hbar}L^3,$$

L^3 is the third component of the orbital angular momentum along x^3-axis. Now

$$(i\gamma^1\gamma^2)^2 = 1.$$

Thus the operator $i\gamma^1\gamma^2$ has eigenvalues ±1. Let us write

$$S^3 = -\frac{i\hbar}{2}\gamma^1\gamma^2. \tag{20.77}$$

Hence the eigenvalues of S^3 are $\pm\frac{\hbar}{2}$. Thus

$$M^{12} = L^{12} + \frac{1}{2}\sigma^{12} = -\frac{1}{\hbar}(L^3 + S^3) \tag{20.78}$$

and

$$\psi'(x) = \left[1 + \frac{i}{\hbar}(L^3 + S^3)\right]\psi(x). \tag{20.79}$$

Thus we see that change in the fundamental form of ψ induced by an infinitesimal rotation about x-axis is determined by the operator

$$1 + \frac{i}{\hbar}(L^3 + S^3) = 1 + \frac{i}{\hbar}J^3 \qquad (20.80)$$

where $J^3 = L^3 + S^3$ should correspond to the third component of the angular momentum operator, since it generates an infinitesimal rotation around the third axis. As already seen L^3 corresponds to the orbital angular momentum while $S^3 = -\frac{i\hbar}{2}\gamma^1\gamma^2$ which has eigenvalues $\pm\frac{\hbar}{2}$ represents an additional angular momentum associated with the particle and is called the spin angular momentum. Thus the Dirac equation represents a spin $\frac{1}{2}$ particle.

In general

$$\sigma^{ij} = \frac{i}{2}(\gamma^i\gamma^j - \gamma^j\gamma^i) = -\frac{i}{2}\begin{pmatrix} \sigma^i\sigma^j - \sigma^j\sigma^i & 0 \\ 0 & \sigma^i\sigma^j - \sigma^j\sigma^i \end{pmatrix}$$

where we have used

$$\gamma^i = \begin{pmatrix} 0 & \sigma^i \\ -\sigma^i & 0 \end{pmatrix},$$

so that

$$\sigma^{ij} = \frac{i}{2}(-i)2\epsilon^{ijk}\sigma^k = \epsilon^{ijk}\sigma^k. \qquad (20.81)$$

But we can write

$$\epsilon_{ij} = -\epsilon^{ijn}\omega^n$$

so that

$$\epsilon_{ij}\sigma^{ij} = -\epsilon^{ijn}\omega^n\epsilon^{ijk}\sigma^k = -2\boldsymbol{\sigma}.\boldsymbol{\omega}. \qquad (20.82)$$

Also

$$\epsilon_{ij}L^{ij} = -\frac{2}{\hbar}\boldsymbol{\omega}.\mathbf{L}. \qquad (20.83)$$

Hence from Eq. (20.72), we have

$$\psi'(x) = \left[1 + \frac{i}{\hbar}\boldsymbol{\omega}.\left(\mathbf{L} + \frac{\hbar}{2}\boldsymbol{\sigma}\right)\right]\psi(x)$$

$$= \left[1 + \frac{i}{\hbar}\boldsymbol{\omega}.\mathbf{J}\right]\psi(x). \qquad (20.84)$$

The angular momentum $\mathbf{J}$ is a vector sum of the orbital angular momentum $\mathbf{L}$ and $\mathbf{S} = \frac{\hbar}{2}\boldsymbol{\sigma}$, where $\mathbf{S}$ is spin angular momentum, with eigenvalues $S = \frac{1}{2}, S_3 = \pm\frac{1}{2}$.

Hence Dirac spinor ψ represents a spin $\frac{1}{2}$ particle.

Remarks: **J** is the generator of rotation group

$$U_R = e^{\frac{i}{\hbar}\boldsymbol{\omega}\cdot\mathbf{J}}.$$

Here U_R is a unitary operator, which rotates the co-ordinate system by an angle $\boldsymbol{\omega}$.

The usual operator

$$U_R = e^{-\frac{i}{\hbar}\boldsymbol{\omega}\cdot\mathbf{J}}$$

refers to the rotation of vector by an angle $\boldsymbol{\omega}$.

20.6   Transformation Properties of Dirac Bilinears

We have seen that under Lorentz transformation the Dirac spinor satisfies the transformation law

$$\psi(x) \to \psi'(x') = S\psi(x) \tag{20.85}$$

where for infinitesimal transformation

$$S = 1 - \frac{i}{4}\epsilon_{\mu\nu}\sigma^{\mu\nu} = 1 + \frac{1}{4}\epsilon_{\mu\nu}\gamma^{\mu}\gamma^{\nu}$$

$$S^{\dagger} = 1 + \frac{1}{4}\epsilon_{\mu\nu}(\gamma^{\nu})^{\dagger}(\gamma^{\mu})^{\dagger} \tag{20.86}$$

so that

$$\gamma^{0}S^{\dagger}\gamma^{0} = 1 + \frac{1}{4}\epsilon_{\mu\nu}\gamma^{\nu}\gamma^{\mu}$$

$$= 1 - \frac{1}{4}\epsilon_{\mu\nu}\gamma^{\mu}\gamma^{\nu}$$

$$= S^{-1}. \tag{20.87}$$

Then

$$\bar{\psi}(x) \to \bar{\psi}'(x') = \psi^{\dagger}(x)S^{\dagger}\gamma^{0}S^{\dagger}\gamma^{0} = \bar{\psi}(x)S^{-1}. \tag{20.88}$$

Thus

$$\bar{\psi}(x)\psi(x) \to \bar{\psi}'(x')\psi'(x') = \bar{\psi}(x)S^{-1}S\psi(x) = \bar{\psi}(x)\psi(x) \tag{20.89}$$

i.e. $\bar{\psi}(x)\psi(x)$ is a scalar.

We want to find the transformation properties of the bilinears $\bar{\psi}\Gamma\psi$, where Γ is any of the independent products of γ^{μ}, given in Table 1. Thus

$$\bar{\psi}\Gamma\psi \to \bar{\psi}'(x')\Gamma\psi'(x') = \bar{\psi}(x)S^{-1}\Gamma S\psi(x). \tag{20.90}$$

(i)

$$\Gamma = \gamma^\mu$$

Using Eq. (20.69), we see that

$$J^\mu(x) = \bar\psi(x)\gamma^\mu\psi(x) \rightarrow$$

$$J'_\mu(x') = \Lambda^\mu_\nu\bar\psi(x)\gamma^\nu\psi(x) = \Lambda^\mu_\nu J^\nu(x),$$

i.e. it transforms like a vector.

(ii)

$$\Gamma = \gamma^5$$

We note that

$$S^{-1}\gamma^5 S = -\frac{i}{4!}\epsilon_{\mu\nu\lambda\rho}S^{-1}\gamma^\mu S S^{-1}\gamma^\nu S S^{-1}\gamma^\lambda S S^{-1}\gamma^\rho S$$

$$= -\frac{i}{4!}\epsilon_{\mu\nu\lambda\rho}\Lambda^\mu_\alpha\gamma^\alpha\Lambda^\nu_\beta\gamma^\beta\Lambda^\lambda_\sigma\gamma^\sigma\Lambda^\rho_\delta\gamma^\delta$$

$$= -\frac{i}{4!}\epsilon_{\mu\nu\lambda\rho}\Lambda^\mu_\alpha\Lambda^\nu_\beta\Lambda^\lambda_\sigma\Lambda^\rho_\delta\gamma^\alpha\gamma^\beta\gamma^\sigma\gamma^\delta.$$

But

$$\epsilon_{\mu\nu\lambda\rho}\Lambda^\mu_\alpha\Lambda^\nu_\beta\Lambda^\lambda_\sigma\Lambda^\rho_\delta = \epsilon_{\alpha\beta\sigma\delta}\det(\Lambda).$$

Therefore

$$S^{-1}\gamma^5 S = \det(\Lambda)\left(-\frac{i}{4}\right)\epsilon_{\alpha\beta\sigma\delta}\gamma^\alpha\gamma^\beta\gamma^\sigma\gamma^\delta$$

$$= \det(\Lambda)\gamma^5. \tag{20.91}$$

Thus for proper Lorentz transformation L_+

$$S^{-1}\gamma^5 S = \gamma^5. \tag{20.92}$$

(iii)

$$\Gamma = \gamma^\mu\gamma^5$$

$$S^{-1}\gamma^\mu\gamma^5 S = S^{-1}\gamma^\mu S S^{-1}\gamma^5 S$$

$$= \det(\Lambda)\Lambda^\mu_\nu\gamma^\mu\gamma^5. \tag{20.93}$$

(iv)

$$\Gamma = \sigma^{\mu\nu}$$

Now

$$S^{-1}\sigma^{\mu\nu} = \frac{i}{2}[S^{-1}\gamma^\mu S S^{-1}\gamma^\nu S - S^{-1}\gamma^\nu S S^{-1}\gamma^\mu S]$$
$$= \frac{i}{2}[\Lambda^\mu_\lambda \gamma^\lambda \Lambda^\nu_\rho \gamma^\rho - \Lambda^\nu_\rho \gamma^\nu \Lambda^\mu_\lambda \gamma^\lambda]$$
$$= \Lambda^\mu_\lambda \Lambda^\nu_\rho \sigma^{\lambda\rho}. \tag{20.94}$$

Hence using Eq. (20.90) under proper Lorentz transformation L_+ the bi-linears transform as

$$\begin{array}{ll} \bar\psi\psi : & \text{Scalar} \\ \bar\psi\gamma^5\psi : & \text{Scalar} \\ \bar\psi\gamma^\mu\psi : & \text{Vector} \\ \bar\psi\gamma^\mu\gamma^5\psi : & \text{Vector} \\ \bar\psi\sigma^{\mu\nu}\psi : & \text{Tensor} \end{array}$$

The proper Lorentz transformation L_+, does not distinguish between $\bar\psi\psi$ and $\bar\psi\gamma^5\psi$ and between $\bar\psi\gamma^\mu\psi$ and $\bar\psi\gamma^\mu\gamma^5\psi$. The appearance of $\det(\Lambda)$ suggests that we should consider space reflection to distinguish between these bilinears.

20.7 Discrete Transformations

(1) Space Reflection:

$$\mathbf{x} \to -\mathbf{x}, \qquad t \to t. \tag{20.95}$$

Under space reflection

$$x^\mu \to x'^\mu = \Lambda^\mu_\nu x^\nu$$

$$\Lambda^\mu_\nu = \begin{pmatrix} 1 & 0 & 0 & 0 \\ 0 & -1 & 0 & 0 \\ 0 & 0 & -1 & 0 \\ 0 & 0 & 0 & -1 \end{pmatrix}, \qquad \det(\Lambda) = -1. \tag{20.96}$$

Under space reflection

$$x \to x^P = (x^o, -\mathbf{x}) \tag{20.97}$$

and suppose that

$$\psi(x) \to \psi^P(x^P) = S_P \psi(x), \tag{20.98}$$

where S_P is a 4×4 matrix. Invariance under space reflection means that for free particle Dirac Eq. (20.56) we should have

$$(i\hbar\gamma^\mu\partial_\mu^P - m)\psi^P(x^P) = 0. \qquad (20.99)$$

Let us write Eq. (20.56) in terms of x^P :

$$\left(i\hbar(\gamma^o\partial_o^P - \gamma^i\partial_i^P - m)\right) S_P^{-1}\psi^P(x^P) = 0.$$

Multiplying by S_P on the left

$$\left(i\hbar(S_P\gamma^o S_P^{-1}\partial_o^P - S_P\gamma^i S_P^{-1}\partial_i^P) - m\right)\psi^P(x^P) = 0$$

i.e. it reduces to Eq. (20.99) as required by the invariance if

$$S_P\gamma^o S_P^{-1} = \gamma^o,$$

$$S_P\gamma^i S_P^{-1} = -\gamma^i. \qquad (20.100)$$

These can be satisfied if we choose

$$S_P = \eta_P\gamma^o$$

where η_P is a multiplicative constant. We note that

$$\bar{\psi}(x) \rightarrow \bar{\psi}^P(x^P)\gamma^o\eta_P. \qquad (20.101)$$

Thus $|\eta_P|^2 = 1$ if we want the probability density $\psi^\dagger(x)\psi(x)$ to be invariant i.e. to be the same in both coordinate systems. Thus under space reflection.

$$\psi^P(x^P) = \eta_P\gamma^o\psi(x), \qquad \bar{\psi}^P(x^P) = \bar{\psi}(x)\gamma^o\eta_P. \qquad (20.102)$$

Since γ^5 anticommutes with γ^o, it follows then

$$\psi_L^P(x^P) = \psi_R(x),$$

$$\psi_R^P(x^P) = \psi_L(x). \qquad (20.103)$$

Thus the bilinears, under space reflections transform as

$$\bar{\psi}(x)\psi(x) \rightarrow \bar{\psi}'(x)\psi'(x') = \bar{\psi}(x)\gamma^o\gamma^o\psi(x)$$
$$= \bar{\psi}(x)\psi(x): \qquad \text{(Scalar)}$$

$$\bar{\psi}(x)\gamma^5\psi(x) \rightarrow \bar{\psi}'(x)\gamma^5\psi'(x') = \bar{\psi}(x)\gamma^o\gamma^5\gamma^o\psi(x)$$
$$= -\bar{\psi}\gamma^5(x)\psi(x): \qquad \text{(Pseudoscalar)}$$

$$\bar{\psi}(x)\gamma^\mu\psi(x) \rightarrow \bar{\psi}^P(x^P)\gamma^\mu\psi^P(x^P) = \bar{\psi}(x)\gamma^o\gamma^\mu\gamma^o\psi(x)$$
$$= (-1)^\mu\bar{\psi}\gamma^\mu\psi(x): \qquad \text{(Vector)}$$

where

$$(-1)^\mu = 1: \quad \text{for} \quad \mu = 0$$
$$= -1: \quad \text{for} \quad \mu = i = 1, 2, 3$$

since

$$\gamma^o \gamma^i \gamma^o = -\gamma^i = -1.$$

Further

$$\bar{\psi}(x)\gamma^\mu\gamma^5\psi(x) \rightarrow \bar{\psi}^P(x^P)\gamma^\mu\gamma^5\psi^P(x^P)$$
$$= \bar{\psi}(x)\gamma^o\gamma^\mu\gamma^5\gamma^o\psi(x)$$
$$= -(-1)^\mu \bar{\psi}\gamma^\mu\gamma^5\psi(x) : \text{(Axial Vector)}$$

using

$$\gamma^o\gamma^\mu\gamma^5\gamma^o = -\gamma^o\gamma^\mu\gamma^o\gamma^5.$$

Both $\bar{\psi}\gamma^\mu\psi$ and $\bar{\psi}\gamma^\mu\gamma^5\psi$ transform as vectors under proper Lorentz transformation L_+, but transform differently under space reflection. Now

$$\bar{\psi}(x)\sigma^{\mu\nu}\psi(x) \rightarrow \bar{\psi}^P(x^P)\sigma^{\mu\nu}\psi^P(x^P)$$
$$= \bar{\psi}(x)\gamma^o\sigma^{\mu\nu}\gamma^o\psi(x)$$
$$= \bar{\psi}(x)\sigma^{\mu\nu}\psi(x) \qquad \text{(Tensor)}.$$

This follows from the fact that

$$\gamma^o\sigma^{\mu\nu} = \frac{i}{2}\gamma^o[\gamma^\mu\gamma^\nu - \gamma^\nu\gamma^\mu]\gamma^o$$
$$= \frac{i}{2}[\gamma^o\gamma^\mu\gamma^o\gamma^o\gamma^\nu - (\mu \leftrightarrow \nu)]$$
$$= (-1)^\mu(-1)^\nu\sigma^{\mu\nu} = \sigma^{\mu\nu}.$$

(2) Time Reversal:

$$t \rightarrow -t, \qquad \mathbf{x} \rightarrow \mathbf{x}$$
$$x \rightarrow x^T = (-x^o, \mathbf{x}),$$

i.e.

$$x'^T\mu = \Lambda^\mu_\nu x^\nu,$$

$$\Lambda^\mu_\nu = \begin{pmatrix} -1 & 0 & 0 & 0 \\ 0 & 1 & 0 & 0 \\ 0 & 0 & 1 & 0 \\ 0 & 0 & 0 & 1 \end{pmatrix}. \qquad (20.104)$$

Under time reversal, suppose

$$\psi(x) \rightarrow \psi_T(x^T) = S_T\psi^*(x) \qquad (20.105)$$

and all the complex numbers go to their complex conjugates [see Chap. 10]. Invariance under time reversal implies

$$(i\hbar\gamma^\mu\partial_\mu^T - m)\psi_T(x^T) = 0. \tag{20.106}$$

Taking the complex conjugate of free particle Dirac equation (20.56),

$$[-i\hbar\left(\gamma^{o*}(\partial_o) + (\gamma^i)^*\partial_i\right) - mc]\psi^*(x) = 0.$$

Writing it in terms of x^T, it becomes

$$[-i\hbar\left(\gamma^{o*}(-\partial_o^T) + (\gamma^i)^*\partial_i^T\right) - mc]S_T^{-1}\psi_T(x^T) = 0.$$

Multiplying on the left by S_T,

$$[i\hbar\left(S_T\gamma^{o*}S_T^{-1}(\partial_o^T) - S_T(\gamma^i)^*S_T^{-1}\right) - mc]\psi_T(x^T) = 0,$$

which reduces to Eq. (20.106) as required by invariance if

$$S_T\gamma^{o*}S_T^{-1} = \gamma^o,$$

$$S_T\gamma^{i*}S_T^{-1} = -\gamma^i. \tag{20.107}$$

However,

$$(\gamma^i)^* = \gamma^i, \qquad i = 1, 3$$
$$= -\gamma^i, \qquad i = 2$$

giving

$$S_T\gamma^i S_T^{-1} = -\gamma^i, \quad i = 1, 3$$
$$= \gamma^i. \qquad i = 2$$

The matrix which satisfies these conditions is

$$S_T = \eta_T B$$

where

$$B = \gamma^1\gamma^3 \tag{20.108}$$

and η_T is a phase factor

$$|\eta_T|^2 = 1.$$

Now

$$B^\dagger = (\gamma^1\gamma^3)^\dagger = \gamma^{3\dagger}\gamma^{1\dagger} = \gamma^3\gamma^1 = -\gamma^1\gamma^3 = -B,$$

$$BB^\dagger = (\gamma^1\gamma^3)(\gamma^3\gamma^1) = 1,$$

$$B^* = \gamma^{1*}\gamma^{3*} = \gamma^1\gamma^3 = B,$$

$$B^T = (\gamma^1\gamma^3)^T = (\gamma^3)^T(\gamma^1)^T = \gamma^3\gamma^1 = B^\dagger = B^{-1}. \qquad (20.109)$$

From Eq. (20.105):

$$\psi_T(x^T) = \eta_T B\psi^*(x)$$

$$\psi_T^\dagger(x^T) = \eta_T^*(\psi^*)^\dagger B^\dagger = \eta_T^*(\psi^\dagger(x))^* B^{-1}$$

or

$$\bar{\psi}_T(x^T) = \eta_T^*(\psi^\dagger(x))^* B^{-1}\gamma^o = \eta_T^*(\psi^\dagger(x))^*\gamma^o B^{-1}.$$

Thus

$$[\bar{\psi}_T(x^T)]^* = \eta_T\psi^\dagger(x)\gamma^o B^{-1}$$
$$= \eta_T\bar{\psi}(x)B^{-1}. \qquad (20.110)$$

On the other hand from Eq. (20.105)

$$[\psi_T(x^T)]^* = \eta_T^* B\psi(x).$$

Hence from Eq. (20.110) invariance of probability density $\psi^\dagger\psi$ gives

$$|\eta_T|^2 = 1.$$

Thus, using Eqs. (20.105), (20.109) and (20.110), for bilinears:

$$\bar{\psi}(x)\psi(x) \rightarrow \bar{\psi}_T^*(x^T)\psi_T^*(x^T) = \bar{\psi}(x)B^{-1}B\psi(x)$$
$$= \bar{\psi}(x)\psi(x). \qquad \text{(Scalar)}$$

$$\bar{\psi}(x)\gamma^5\psi(x) \rightarrow \bar{\psi}_T^*\gamma^{5*}(x^T)\psi_T^*(x^T)$$
$$= \bar{\psi}(x)B^{-1}\gamma^5 B\psi(x)$$
$$= \bar{\psi}(x)\gamma^5\psi(x). \qquad \text{(Scalar)}$$

$$\bar{\psi}(x)\gamma^\mu\psi(x) \rightarrow \bar{\psi}_T^*(x^T)\gamma^{\mu*}\psi_T^*(x^T)$$
$$= (-1)^\mu\bar{\psi}(x)B^{-1}\gamma^\mu B\psi(x)$$
$$= (-1)^\mu\bar{\psi}(x)\gamma^\mu\psi(x). \qquad \text{(Vector)}$$

$$\bar{\psi}(x)\gamma^\mu\gamma^5\psi(x) \rightarrow \bar{\psi}_T^*(x^T)(\gamma^\mu)^*(\gamma^5)^*\psi_T^*(x^T)$$
$$= \bar{\psi}(x)B^{-1}(\gamma^\mu)^*(\gamma^5)^* B\psi(x)$$
$$= \bar{\psi}(x)B^{-1}(\gamma^\mu)^* BB^{-1}(\gamma^5)^* B\psi(x)$$
$$= (-1)^\mu\bar{\psi}(x)\gamma^\mu\gamma^5\psi(x). \qquad \text{(Vector)}$$

20.8 Generators of the Lorentz Group and their Representations

20.8.1 *Matrix Representation of the Generators of the Lorentz Group*

We can construct the matrix representation of Lorentz group as follows. For an infinitesimal Lorentz transformation viz

$$x'^{\mu} = \Lambda^{\mu}_{\nu} x^{\nu}, \text{ with}$$
$$\Lambda^{\mu}_{\nu} = \delta^{\mu}_{\nu} + \epsilon^{\mu}_{\nu} \tag{20.111}$$

let us introduce an operator (matrix) $U(\Lambda)$

$$U(\Lambda) = 1 + \frac{1}{2}\epsilon_{\alpha\beta} A^{\alpha\beta}. \tag{20.112}$$

Then A gives the matrix representation (adjoint) of Lorentz transformation. From Eqs. (20.111) and (20.112), we have

$$\frac{1}{2}\epsilon_{\alpha\beta}(A^{\alpha\beta})^{\mu}_{\nu} = \epsilon^{\mu}_{\nu}$$
$$= g^{\mu\lambda}\epsilon_{\lambda\nu}$$
$$= \frac{1}{2}g^{\mu\lambda}(\epsilon_{\lambda\nu} - \epsilon_{\nu\lambda})$$
$$= \frac{1}{2}g^{\mu\lambda}\left[\epsilon_{\alpha\beta}\delta^{\alpha}_{\lambda}\delta^{\beta}_{\nu} - \epsilon_{\alpha\beta}\delta^{\alpha}_{\nu}\delta^{\beta}_{\lambda}\right]$$
$$= \frac{1}{2}\epsilon_{\alpha\beta}\left[g^{\mu\lambda}\delta^{\alpha}_{\lambda}\delta^{\beta}_{\nu} - g^{\mu\lambda}\delta^{\alpha}_{\nu}\delta^{\beta}_{\lambda}\right],$$

so that

$$(A^{\alpha\beta})^{\mu}_{\nu} = [g^{\mu\alpha}\delta^{\beta}_{\nu} - g^{\mu\beta}\delta^{\alpha}_{\nu}] \tag{20.113}$$

$$A^{\alpha\beta} = -A^{\beta\alpha}.$$

Let us write

$$A^{\alpha\beta} = -iM^{\alpha\beta}$$

so that we can write

$$U(\Lambda) = 1 - \frac{i}{2}\epsilon_{\alpha\beta} M^{\alpha\beta}. \tag{20.114}$$

Then six matrices

$$M^{ij} = M_{ij} = (M^{23}, M^{31}, M^{12})$$

$$M^{0i} = -M_{0i} = (M^{01}, M^{02}, M^{03})$$

are given by

$$(M^{12})^\mu_\nu = i(g^{\mu 1}\delta^2_\nu - g^{\mu 2}\delta^1_\nu)$$

$$= \begin{pmatrix} 0 & 0 & 0 & 0 \\ 0 & 0 & -i & 0 \\ 0 & i & 0 & 0 \\ 0 & 0 & 0 & 0 \end{pmatrix} \equiv S^3 = S_z, \qquad (20.115a)$$

$$(M^{23})^\mu_\nu = i(g^{\mu 2}\delta^3_\nu - g^{\mu 3}\delta^2_\nu)$$

$$= \begin{pmatrix} 0 & 0 & 0 & 0 \\ 0 & 0 & 0 & 0 \\ 0 & 0 & 0 & -i \\ 0 & 0 & i & 0 \end{pmatrix} \equiv S^1 = S_x, \qquad (20.115b)$$

$$(M^{31})^\mu_\nu = i(g^{\mu 3}\delta^1_\nu - g^{\mu 1}\delta^3_\nu)$$

$$= \begin{pmatrix} 0 & 0 & 0 & 0 \\ 0 & 0 & 0 & i \\ 0 & 0 & 0 & 0 \\ 0 & -i & 0 & 0 \end{pmatrix} \equiv S^2 = S_y. \qquad (20.115c)$$

Thus $\mathbf{S} = (S_x, S_y, S_z)$ gives the matrix representation of rotation subgroup of Lorentz group.

Now

$$(M^{01})^\mu_\nu = i(g^{\mu 0}\delta^1_\nu - g^{\mu 1}\delta^0_\nu)$$

$$= i\begin{pmatrix} 0 & 0 & 1 & 0 \\ 1 & 0 & 0 & 0 \\ 0 & 0 & 0 & 0 \\ 0 & 0 & 0 & 0 \end{pmatrix} \equiv \kappa^1, \qquad (20.116a)$$

$$(M^{02})^\mu_\nu = i(g^{\mu 0}\delta^2_\nu - g^{\mu 2}\delta^0_\nu)$$

$$= i\begin{pmatrix} 0 & 0 & 1 & 0 \\ 0 & 0 & 0 & 0 \\ 1 & 0 & 0 & 0 \\ 0 & 0 & 0 & 0 \end{pmatrix} \equiv \kappa^2, \qquad (20.116b)$$

$$(M^{03})^\mu_\nu = i(g^{\mu 0}\delta^3_\nu - g^{\mu 3}\delta^0_\nu)$$

$$= i\begin{pmatrix} 0 & 0 & 0 & 1 \\ 0 & 0 & 0 & 0 \\ 0 & 0 & 0 & 0 \\ 1 & 0 & 0 & 0 \end{pmatrix} \equiv \kappa^3. \qquad (20.116c)$$

$\kappa \equiv (\kappa_x, \kappa_y, \kappa_z)$ give the matrix representation of Lorentz boost. The matrices S_i, S_j, κ_i and κ_j satisfy the following commutation relations

$$[S_i, S_j] = i\epsilon_{ijk}S_k,$$
$$[S_i, \kappa_j] = i\epsilon_{ijk}\kappa_k, \qquad (20.117)$$
$$[\kappa_i, \kappa_j] = -i\epsilon_{ijk}\kappa_k.$$

We note that 4×4 matrices $S's$ are hermitian but matrices $\kappa's$ are antihermitian. Thus finite representation of Lorentz group is not unitary.

20.8.2 Representations of Lorentz Group

As already seen in Secs. 20.1 and 20.5, the most general representations of the Lorentz group are given by $M^{\mu\nu}$, which satisfy the commutation relations given in Eq. (20.74). Just as $L_+^\uparrow$ transformations has two subgroups, the six generators of the Lorentz group split into three generators M^{ij} which belong to the rotation group $O(3)$. This can be seen as follows. Define

$$M^{ij} = -\epsilon^{ijk}J_k = \epsilon^{ijk}J^k$$

$$M_{ij} = -\epsilon_{ijk}J^k = \epsilon_{ijk}J_k$$

so that

$$M^{12} = -J_3 = J^3, M_{12} = J^3 \quad \text{etc.} \qquad (20.118)$$

From Eq. (20.74), we can easily write down the commutation relations for J's

$$[J^i, J^j] = i\epsilon^{ijk}J^k. \qquad (20.119)$$

Thus the generators J^i satisfy the commutation relations of the rotation group $O(3)$. Hence out of 6 generators of Lorentz group, three generators M^{ij} belong to the subgroup $O(3)$. The other three generators M^{0i} give the Lorentz boosts. Define

$$M^{oi} = K^i. \qquad (20.120)$$

Then from Eq. (20.74), we get the commutation relations for K's:

$$[K^j, J^l] = i\epsilon^{jlm}K^m,$$
$$[K^i, K^j] = [M^{0i}, M^{0j}] = -ig^{00}M^{ij} = -i\epsilon^{ijk}J^k. \qquad (20.121)$$

Note that the minus sign in Eq. (20.121) is the manifestation of the non-compactness of the Lorentz group. Note also that K's are anti-hermitian. The negative sign originates from the Minkowski matrix, $g^{00} = 1$, $g^{ij} = -\delta_{ij}$. As a consequence the irreducible representations are radically difficult in nature from those of the rotation group. Further

$$\epsilon_{\mu\nu} M^{\mu\nu} = \epsilon_{0j} M^{0j} + \epsilon_{i0} M^{i0} + \epsilon_{ij} M^{ij}$$
$$= 2\epsilon_{0j} K^j + \epsilon_{ij}(-\epsilon^{ijk} J_k). \qquad (20.122)$$

Writing

$$\epsilon_{0j} = \eta_j = -\eta^j, \epsilon_{ij} = \epsilon^{ijk} \omega^k \qquad (20.123)$$

so that $U(\Lambda)$ given in Eq. (20.76) can be written

$$U(\Lambda) = e^{i\omega \cdot \mathbf{J} + i\eta \cdot \mathbf{K}}$$
$$= U_\Lambda(R) U_L(v) \qquad (20.124)$$

where

$$\boldsymbol{\omega} = \boldsymbol{\theta}\mathbf{n},$$

$$\boldsymbol{\eta} = \eta \frac{\mathbf{v}}{v}. \qquad (20.125)$$

explicitly showing that Lorentz group is a direct product of the rotation group in three dimensions and Lorentz boosts. It should be noted that $U_L(v)$ is not unitary unlike unitary operator $e^{i\boldsymbol{\omega} \cdot \mathbf{J}} : U_L(v) U_L^\dagger(v) \neq 1$ but is in fact hermitian $U_L^\dagger(v) = e^{i\eta \cdot \mathbf{K}} = U_L(v)$.

The mathematical designation for proper Homogeneous Lorentz group $L_+^\uparrow$ is $SO(3,1)$ which refers to the fact that signature of the metric $g^{\mu\nu}$ is 1, -1, -1, -1.

For the classification of irreducible representation of $L_+^\uparrow$, it is useful to introduce hermitian combinations

$$M^i = \frac{1}{2}(J^i + iK^i) \qquad (20.126)$$

$$N^i = \frac{1}{2}(J^i - iK^i)$$

which satisfy the commutation relation.

$$[M^i, M^j] = i\epsilon^{ijk} M^k,$$
$$[N^i, N^j] = i\epsilon^{ijk} N^k,$$
$$[N^i, M^j] = 0. \qquad (20.127)$$

u, v	Representation
0,0	Lorentz scalar
(1/2,0)	2-component Lorentz spinor of first kind
(0,1/2)	2-component Lorentz spinor of second kind
(1/2,1/2)	Lorentz four vector

This algebra is identical to the Lie algebra of group $SU(2)_M \otimes SU(2)_N$ with two Casimir operators

$$M^2 = M^i M^i,$$

$$N^2 = N^i N^i,$$

$$[M^2, M^i] = [N^2, N^i] = 0. \tag{20.128}$$

Since in Eq. (20.126) M^i and N^i are not linear combinations of basic elements, the commutators in Eq. (20.127) do not define the (real) Lie algebra of $L_+^\uparrow$. However, due to analogy with angular momentum, it is convenient to use the eigenvalues of M^2, M_3, N^2, N_3 which are respectively $u(u+1), v(v+1), u, v = 0, 1/2, 3/2, ...$ to label the elements of irreducible representations. This provides the basis $|kl\rangle$, where

$$|kl\rangle = \{|u, k\rangle |v, l\rangle\},$$

$$k = -u, ..., u, \qquad l = -v, ..., v.$$

Suppressing u, v on the basis vectors of the product space:

$$J^3 |kl\rangle = (M^3 + N^3)|kl\rangle$$

$$= M^3 |u, k\rangle |v, l\rangle + |u, k\rangle N^3 |v, l\rangle$$

$$= (k + l)|k, l\rangle,$$

$$J^\pm |k, l\rangle = |k \pm 1, l\rangle [u(u+1) - k(k \pm 1)]^{1/2}$$

$$+ |k, l \pm 1\rangle [v(v+1) - l(l \pm 1)]^{1/2},$$

$$K^3 |k, l\rangle = |k, l\rangle i(l - k),$$

$$K^\pm |k, l\rangle = |k, l \pm 1\rangle i[v(v+1) - l(l \pm 1)]^{1/2} - |k \pm 1, l\rangle [u(u+1) - k(k \pm 1)]^{1/2}. \tag{20.129}$$

We see that in the (u, v) representation matrix representation of K^3 is diagonal and imaginary. This implies that finite dimensional representations of the Lorentz groups are non-unitary. We can label the representation of

Lorentz group as follows.

Under space reflection $J^i \to J^i$, but $K^i = M^{0i}$ change sign so that $M^i \leftrightarrow N^i$ and thus $\left(0, \frac{1}{2}\right) \leftrightarrow \left(0, \frac{1}{2}\right)$. Hence if a left handed spinor belongs to representation $\psi_L\left(\frac{1}{2}, 0\right)$, the right handed spinor belongs to representation $\psi_R\left(\frac{1}{2}, 0\right)$.

Now for the fundamental spinor representation $\left(\frac{1}{2}, 0\right)$,

$$J^i = \frac{1}{2}\sigma^i, K^i = -\frac{i}{2}\sigma^i,$$

$$M^i = \frac{\sigma^i}{2}, N^i = 0. \tag{20.130}$$

Thus under rotation

$$\phi_L = \begin{pmatrix} \phi_1 \\ \phi_2 \end{pmatrix} \to e^{i\theta \mathbf{n} \cdot \frac{\sigma}{2}} \begin{pmatrix} \phi_1 \\ \phi_2 \end{pmatrix}$$

and under boost

$$\phi_L \to e^{\eta \frac{\mathbf{v}}{v} \cdot \sigma} \begin{pmatrix} \phi_1 \\ \phi_2 \end{pmatrix}. \tag{20.131}$$

For the spinor representation $(0, 1/2)$

$$J^i \to \frac{1}{2}\sigma^i, K^i \to \frac{i}{2}\sigma^i \tag{20.132}$$

which give

$$M^i = 0, N^i = \sigma^i,$$

$$\phi_R \xrightarrow{\text{rotation}} e^{i\theta \mathbf{n} \cdot \frac{\sigma}{2}} \phi_R,$$

$$\phi_R \xrightarrow{\text{boost}} e^{-\eta \frac{\mathbf{v}}{v} \cdot \mathbf{n}} \phi_R. \tag{20.133}$$

The Dirac spinor ψ_D transforms in $\left(\frac{1}{2}, 0\right) \oplus \left(0, \frac{1}{2}\right)$ representation of the Lorentz group, which is a reducible representation

$$\psi_D = \begin{pmatrix} \phi_L \\ \phi_R \end{pmatrix} \xrightarrow{\text{rotation}} \begin{pmatrix} e^{i\theta \mathbf{n} \cdot \frac{\sigma}{2}} & 0 \\ 0 & e^{i\theta \mathbf{n} \cdot \frac{\sigma}{2}} \end{pmatrix} \begin{pmatrix} \phi_L \\ \phi_R \end{pmatrix},$$

$$\psi_D = \begin{pmatrix} \phi_L \\ \phi_R \end{pmatrix} \xrightarrow{\text{boost}} \begin{pmatrix} e^{\eta \frac{\mathbf{v}}{v} \cdot \frac{\sigma}{2}} & 0 \\ 0 & e^{-\eta \frac{\mathbf{v}}{v} \cdot \frac{\sigma}{2}} \end{pmatrix} \begin{pmatrix} \phi_L \\ \phi_R \end{pmatrix}. \tag{20.134}$$

20.9 Gauge Invariance: Dirac Equation in the Presence of Electromagnetic Field

The Dirac equation

$$(i\hbar\gamma^\mu\partial_\mu - mc)\psi(x) = 0 \tag{20.135}$$

is clearly not invariant under local gauge transformation [Λ is a function of x]:

$$\psi(x) \to e^{ie\Lambda/c\hbar}\psi(x), \tag{20.136}$$

since

$$\partial_\mu\psi(x) \to \frac{ie}{c\hbar}e^{ie\Lambda/c\hbar}\psi(x)\partial_\mu\Lambda + e^{ie\Lambda/c\hbar}\partial_\mu\psi(x).$$

To compensate the first term on the right hand side, we must introduce a vector gauge field, in this case electromagnetic potential $A_\mu(x)$, and replace $\partial_\mu\psi$ by

$$\partial_\mu\psi \to \left(\partial_\mu + \frac{ie}{c\hbar}A_\mu\right)\psi \equiv D_\mu\psi.$$

Then the resulting Dirac equation

$$\left[i\hbar\left(\partial_\mu + \frac{ie}{c\hbar}A_\mu\right) - mc\right]\psi(x) = 0$$

or

$$(i\hbar\gamma^\mu D_\mu - mc)\psi(x) = 0 \tag{20.137}$$

is invariant under the combined transformation (20.136) and

$$A_\mu(x) \to A_\mu(x) + \partial_\mu\Lambda. \tag{20.138}$$

It is easy to see that, under gauge transformations (20.136) and (20.138)

$$D_\mu\psi \to e^{ie\Lambda/c\hbar}D_\mu\psi(x) \tag{20.139}$$

ensuring the gauge invariance of Eq. (20.137). Now

$$[D_\mu, D_\nu]f(x) = \left[\partial_\mu + \frac{ie}{c\hbar}A_\mu, \partial_\nu + \frac{ie}{c\hbar}A_\nu\right]f(x)$$

$$= [\partial_\mu, \partial_\nu]f(x) + \frac{ie}{c\hbar}[\partial_\mu, A_\nu]f(x) + \frac{ie}{c\hbar}[A_\mu, \partial_\nu]f(x)$$

$$+ \frac{e^2}{c^2\hbar^2}[A_\mu, A_\nu]f(x)$$

so that

$$[D_\mu, D_\nu]f(x) = \frac{ie}{c\hbar}[\partial_\mu, A_\nu]f(x) + \frac{ie}{c\hbar}[A_\mu, \partial_\nu]f(x)$$

$$= \frac{ie}{c\hbar}f(x)[\partial_\mu A_\nu - \partial_\nu A_\mu] = \frac{ie}{c\hbar}f(x)F_{\mu\nu}.$$

Hence we have (since $f(x)$ is arbitrary)

$$[D_\mu, D_\nu] = \frac{ie}{c\hbar} F_{\mu\nu}. \tag{20.140}$$

In terms of momentum operator $\hat{p}_\mu$

$$\hat{P}_\mu \equiv i\hbar D_\mu = i\hbar\partial_\mu + i\hbar\left(\frac{ie}{c\hbar}\right)A_\mu$$

$$= i\hbar\partial_\mu - \frac{e}{c}A_\mu = \hat{p}_\mu - \frac{e}{c}A_\mu. \tag{20.141}$$

The Dirac Equation (20.135) is

$$\left[\gamma^\mu\left(\hat{p}_\mu - \frac{e}{c}A_\mu\right) - mc\right]\psi = 0. \tag{20.142}$$

Let us take the hermitian conjugate of Eq. (20.135):

$$(-i\hbar(\partial_\mu\psi^\dagger)(\gamma^\mu)^\dagger - mc\psi^\dagger(x)) = 0.$$

Multiplying on the right by γ^0 and using $\gamma^{\mu\dagger} = \gamma^0\gamma^\mu\gamma^0$,

$$(-i\hbar(\partial_\mu\bar{\psi})\gamma^\mu - mc\bar{\psi}) = 0$$

which is the Dirac equation for the adjoint spinor $\bar{\psi}(x)$. The above equation can be written in the form

$$\bar{\psi}[-i\hbar\overleftarrow{\partial}_\mu\gamma^\mu - mc] = 0 \tag{20.143}$$

and in the presence of electromagnetic field:

$$\bar{\psi}\left[\left(-i\hbar\overleftarrow{\partial}_\mu - \frac{e}{c}A_\mu\right)\gamma^\mu - mc\right] = 0. \tag{20.144}$$

We now derive the equation of continuity. We have already seen that

$$j^\mu = c\bar{\psi}\gamma^\mu\psi \tag{20.145}$$

is a vector. Now if we multiply Eq. (20.135) on the left by $\bar{\psi}$ and Eq. (20.143) on the right by ψ and subtract, we obtain

$$\bar{\psi}(\gamma^\mu\partial_\mu\psi) + (\partial_\mu\bar{\psi}\gamma^\mu\psi) = 0$$

or

$$\partial_\mu(\bar{\psi}\gamma^\mu\psi) = 0 \quad\text{i.e.}\quad \partial_\mu j^\mu = 0 \tag{20.146}$$

i.e. the current $j^\mu = c\bar{\psi}\gamma^\mu\psi$ is conserved. Further

$$j^{\mu\dagger} = c\psi^\dagger(\gamma^\mu)^\dagger\gamma^0\psi = c\psi^\dagger\gamma^0\gamma^0(\gamma^\mu)^\dagger\gamma^0\psi = c\bar{\psi}\gamma^\mu\psi = j^\mu \tag{20.147}$$

i.e. j^μ is hermitian and j^0 is given by

$$j^0 = c\bar{\psi}\gamma^0\psi = c\psi^\dagger\psi = c\rho \tag{20.148}$$

where ρ is the probability density. The Eq. (20.146) is equivalent to the equation of continuity. This can be seen as follows:

$$\partial_\mu j^\mu = \partial_o j^o + \partial_i j^i = \frac{1}{c}\frac{\partial j^o}{\partial t} + \nabla \cdot \mathbf{j}$$

$$= \frac{1}{c}\frac{\partial \rho}{\partial t} + \nabla \cdot \mathbf{j} = 0 \qquad (20.149)$$

where

$$j^i = c\bar{\psi}\gamma^i\psi \qquad \text{or} \qquad \Rightarrow \mathbf{j} = c\bar{\psi}\boldsymbol{\gamma}\psi$$

and

$$j^o = c\bar{\psi}\gamma^o\psi = c\psi^\dagger\psi = c\rho \geq 0$$

i.e. the probability density is always positive and is delinked from the problem of negative energies.

20.10   Constants of Motion

With the help of the Heisenberg equation of motion, we can determine whether an observable is a constant of motion or not. Consider the Dirac Hamiltonian for a free spin $\frac{1}{2}$ particle:

$$H = c\boldsymbol{\alpha}.\hat{\boldsymbol{p}} + \beta mc^2,$$

$$\hat{\boldsymbol{p}} \to -i\hbar\nabla. \qquad (20.150)$$

The Heisenberg equation of motion for $\hat{p}_i$ is given by

$$i\hbar\frac{d\hat{p}_i}{dt} = [\hat{p}_i, H]$$

$$= [\hat{p}_i, c\alpha_j\hat{p}_j + \beta mc]$$

$$= 0$$

since $[\hat{p}_i, \hat{p}_j] = 0$ and α_j and β commute with $\hat{p}_i$. Thus for a free particle, momentum is a constant of motion and we can find simultaneous eigenvalues of both energy and momentum. For a spin $\frac{1}{2}$ particle in a centrally symmetric field, the Hamiltonian is

$$H = c\boldsymbol{\alpha}.\hat{\boldsymbol{p}} + \beta mc^2 + V(r). \qquad (20.151)$$

In a central field, the angular momentum should be conserved, i.e.

$$\frac{d\mathbf{J}}{dt} = 0.$$

Now the orbital angular momentum

$$\mathbf{L} = \mathbf{r} \times \hat{p},$$

$$L_i = \epsilon_{i\ell m} x_\ell \hat{p}_m.$$

The Heisenberg equation for angular momentum $\mathbf{L}$ is given by

$$i\hbar \frac{dL_i}{dt} = [L_i, H]$$

$$= [L_i, c\alpha_j \hat{p}_j + \beta mc^2 + V(r)]$$

$$= c\alpha_j [L_i, \hat{p}_j], \qquad (20.152)$$

since L_i commute with $\beta mc^2, \alpha_j$ and $V(r)$. But

$$[L_i, \hat{p}_j] = \epsilon_{i\ell m}[x_\ell \hat{p}_m, \hat{p}_j]$$

$$= \epsilon_{i\ell m}\{[x_\ell, \hat{p}_j]\hat{p}_m + x_\ell[\hat{p}_m, \hat{p}_j]\}$$

$$= i\hbar \epsilon_{i\ell m} \delta_{\ell j} \hat{p}_m$$

$$= i\hbar \epsilon_{ijm} \hat{p}_m.$$

Therefore

$$i\hbar \frac{dL_i}{dt} = i\hbar c \epsilon_{ijm} \alpha_j \hat{p}_m$$

or

$$\frac{d\mathbf{L}}{dt} = c\boldsymbol{\alpha} \times \hat{p}, \qquad (20.153)$$

i.e. the orbital angular momentum is not a constant of motion in sharp contrast to the corresponding situation in Schrödinger theory. Now let us consider

$$\boldsymbol{\sigma} = \begin{pmatrix} \boldsymbol{\sigma} & 0 \\ 0 & \boldsymbol{\sigma} \end{pmatrix}.$$

Then

$$i\hbar \frac{d\sigma_i}{dt} = [\sigma_i, H]$$

$$= [\sigma_i, c\alpha_j \hat{p}_j + \beta mc^2 + V(r)]$$

$$= c\hat{p}_j[\sigma_i, \alpha_j] + mc^2[\sigma_i, \beta].$$

We can easily calculate these commutators by using the Pauli representation of $\boldsymbol{\alpha}$ and β. Since β is a diagonal matrix

$$[\sigma_i, \beta] = 0.$$

On the other hand

$$\sigma_i \alpha_j - \alpha_j \sigma_i = \begin{pmatrix} \sigma_i & 0 \\ 0 & \sigma_i \end{pmatrix} \begin{pmatrix} 0 & \sigma_j \\ \sigma_j & 0 \end{pmatrix} - \begin{pmatrix} 0 & \sigma_j \\ \sigma_j & 0 \end{pmatrix} \begin{pmatrix} \sigma_i & 0 \\ 0 & \sigma_i \end{pmatrix}$$

$$= \begin{pmatrix} 0 & \sigma_i \sigma_j - \sigma_j \sigma_i \\ \sigma_i \sigma_j - \sigma_j \sigma_i & 0 \end{pmatrix}.$$

Therefore

$$[\sigma_i, \alpha_j] = 2i\epsilon_{ijk} \begin{pmatrix} 0 & \sigma_k \\ \sigma_k & 0 \end{pmatrix} = 2i\epsilon_{ijk}\alpha_k.$$

Hence

$$i\hbar \frac{d\sigma_i}{dt} = 2ic\epsilon_{ijk}\hat{p}_j\alpha_k$$

or

$$\frac{\hbar}{2} \frac{d\sigma}{dt} = -c\boldsymbol{\alpha} \times \hat{p}. \tag{20.154}$$

Thus the operator $\boldsymbol{\sigma}$ for an electron in a central field is not a constant of motion. However, it follows from Eqs. (20.153) and (20.154) that

$$\frac{d}{dt}\left(\mathbf{L} + \frac{1}{2}\hbar\boldsymbol{\sigma}\right) = 0,$$

i.e.

$$\frac{d}{dt}\mathbf{J} = 0, \tag{20.155}$$

where

$$\mathbf{J} = \mathbf{L} + \frac{1}{2}\hbar\boldsymbol{\sigma}.$$

Thus although $\mathbf{L}$ and $\frac{1}{2}\hbar\boldsymbol{\sigma}$ taken separately are not constants of motion, the sum

$$\mathbf{J} = \mathbf{L} + \frac{1}{2}\hbar\boldsymbol{\sigma}$$

is a constant of a motion. We can interpret this result by saying that the electron has a spin angular momentum $\mathbf{S} = \frac{1}{2}\hbar\boldsymbol{\sigma}$, which must be added to the orbital angular momentum $\mathbf{L}$, so as to get total angular momentum $\mathbf{J}$ which is a constant of motion for a spin $\frac{1}{2}$ particle in a central field of force.

20.11   'Velocity' in Dirac Theory

For a free particle, the Hamiltonian is given by

$$H = c\boldsymbol{\alpha}.\hat{\boldsymbol{p}} + \beta mc^2.$$

Using the Heisenberg equation of motion, we have

$$
\begin{aligned}
i\hbar \dot{x}_i &= [x_i, H] \\
&= [x_i, c\alpha_j \hat{p}_j + \beta mc^2] \\
&= c\alpha_j i\hbar \delta_{ij} \\
&= i\hbar c\alpha_i
\end{aligned}
$$

or

$$\dot{x}_i = c\alpha_i$$

which shows that $\boldsymbol{\alpha}$ is the velocity of particle in terms of c.

This result is rather surprising, as it means an altogether different relation between velocity and momentum from what one has in classical mechanics ($v = c^2 p/E$). Note, however, that the eigenvalues of α_i are ± 1 [as $\alpha_i^2 = 1$]. Thus $\dot{x}_i$ has eigenvalues $\pm c$ corresponding to the eigenvalues ± 1 of α_i. We conclude that a measurement of a component of the velocity of a free electron is certain to lead to the result $\pm c$. We may also note that because α_j and α_k do not commute for $j \neq k$, a measurement of the x component of the velocity is incompatible with a measurements of the y component of the velocity; this may appear strange as we know that $\hat{p}_1$ and $\hat{p}_2$ commute. In spite of these peculiarities, there is actually no contradiction. The theoretical velocity in the above conclusion is the velocity at one instant of time while the observed velocities are always average velocities through appreciable time intervals. We shall find upon further examination of the equation of motion that the velocity oscillates rapidly about a mean value which agrees with the observed value.

We have

$$
\begin{aligned}
i\hbar \dot{\alpha}_i &= [\alpha_i, H] \\
&= 2\alpha_i H - (\alpha_i H + H\alpha_i).
\end{aligned}
$$

But

$$
\begin{aligned}
\alpha_i H + H\alpha_i &= c[\alpha_i \alpha_j + \alpha_j \alpha_i]\hat{p}_j + mc^2(\alpha_i \beta + \beta \alpha_i) \\
&= 2c\delta_{ij}\hat{p}_j \\
&= 2c\hat{p}_i.
\end{aligned}
$$

Thus

$$i\hbar\dot{\alpha}_i = 2\alpha_i H - 2c\hat{p}_i. \tag{20.156}$$

We see that the velocity operator $\dot{x}_i = c\dot{\alpha}(t)$ is not a constant of motion in spite of the fact that the particle is free.

Noting that H and $\hat{p}_i$ are constants of motion, we can find the solution of the differential Eq. (20.156) as follows. First we note that

$$i\hbar\ddot{\alpha}_i(t) = 2\dot{\alpha}_i(t)H.$$

Therefore

$$\dot{\alpha}_i(t) = \dot{\alpha}_i(0)e^{-2iHt/\hbar}. \tag{20.157}$$

Substituting Eq. (20.157) in Eq. (20.156), we have

$$i\hbar\dot{\alpha}_i(0)e^{-2iHt/\hbar} = 2\alpha_i(t)H - 2c\hat{p}_i,$$

$$\alpha_i(t) = c\hat{p}_i H^{-1} + \frac{i\hbar}{2}\dot{\alpha}_i(0)e^{-2iHt/\hbar}H^{-1}.$$

In particular, for $t = 0$

$$\frac{i\hbar}{2}\dot{\alpha}_i(0)H^{-1} = (\alpha_i(0) - c\hat{p}_i H^{-1}). \tag{20.158}$$

Thus the solution of Eq. (20.157) is

$$\alpha_i(t) = c\hat{p}_i H^{-1} + (\alpha_i(0) - c\hat{p}_i H^{-1})e^{-2iHt/\hbar}.$$

Therefore

$$\dot{x}_i = c\alpha_i(t)$$
$$= c^2\hat{p}_i H^{-1} + c(\alpha_i(0) - c\hat{p}_i H^{-1})e^{-2iHt/\hbar}. \tag{20.159}$$

Thus, the ith component of velocity consists of two parts, a constant part $c^2\hat{p}_i H^{-1}$ and an oscillatory part of frequency $2H/\hbar$ that is known as zitter-bewegung term. This frequency is at least $2mc^2/h$ and hence high. The first part of the velocity is reasonable since, for an eigenstate of momentum and energy, it gives $c^2 p_i/E$, in agreement with the classical relativistic formula. Taken literally, the second part shows that the velocity of the electron has an additional term that fluctuates rapidly about its average value in the absence of any potential. Only the constant part would be observed in a practical measurement of velocity, such a measurement giving the average velocity through a time-interval much larger than $\hbar/2mc^2$. The oscillatory part insures that instantaneous value of $\dot{x}_i$ shall have the eigenvalues $\pm c$. As for the coordinate operator, the relation can be easily integrated to give

$$x_i(t) = x_i(0) + c^2\hat{p}_i H^{-1} + \frac{ic\hbar}{2}(\alpha_i(0) - c\hat{p}_i H^{-1})H^{-1}e^{-2iHt/\hbar}. \tag{20.160}$$

The first and second terms in Eq. (20.160) are understandable since their expectation values give the trajectory of the wave packet according to the classical law:

$$x_i^{\text{classical}}(t) = x_i^{\text{classical}}(0) + t(c^2 p_i/E)^{\text{classical}}.$$

The presence of the third term in Eq. (20.160), which is a consequence of the second term in Eq. (20.159) appears to imply that the free electron executes very rapid oscillation in addition to the uniform rectilinear motion.

20.12 Two Component Pauli Form of Dirac Equation: Existence of Magnetic Moment

The Dirac equation in the presence of electromagnetic field is given in Eq. (20.137):

$$(i\hbar\gamma^\mu D_\mu - mc)\psi = 0. \tag{20.161}$$

The Klein–Gordon equation in the presence of electromagnetic field is

$$(\hbar^2 D_\mu D^\mu + m^2 c^2)\phi = 0. \tag{20.162}$$

The equation (20.161) can be transformed into second order equation similar to Klein–Gordon equation (20.162) by multiplying on the left with

$$(i\hbar\gamma^\nu D_\nu + mc), \tag{20.163}$$

so that

$$(-\hbar^2\gamma^\mu\gamma^\nu D^\mu D_\nu - m^2 c^2)\psi = 0. \tag{20.164}$$

Using the anti commutation relations (20.159) and $\sigma^{\mu\nu}$ given in Eq. (20.161) One can reset the above equation as

$$[\hbar^2 D^\mu D_\mu - \frac{i}{2}\hbar^2\sigma^{\mu\nu}[D^\mu, D_\nu] + m^2]\psi = 0. \tag{20.165}$$

Using the relation

$$[D_\mu, D_\nu] = \frac{ie}{c\hbar}F_{\mu\nu}, \tag{20.166}$$

given in Eq. (20.140), the above equation becomes

$$\hbar^2 D^\mu D_\mu + \frac{e\hbar}{2c}\sigma^{\mu\nu}F_{\mu\nu} + m^2 c^2)\psi = 0. \tag{20.167}$$

This differers from the Klein–Gordon equation (20.162) in the term $\frac{e}{2}\sigma^{\mu\nu}F_{\mu\nu}$. Now

$$\sigma^{\mu\nu}F_{\mu\nu} = 2i[\gamma^\circ\gamma^i F_{oi} - \frac{i}{2}\sigma^{ij}F_{ij}].$$

But $\gamma^o \gamma^i = \alpha^i$ and

$$F_{oi} = (\partial_o A_i - \partial_i A_o)$$
$$= E^i$$

and $\sigma^{ij} = \epsilon^{ijk} \Sigma^k$ so that

$$\sigma^{ij} F_{ij} = \epsilon^{ijk} \Sigma^k (\partial_i) A_j - \partial_j A_i)$$
$$= -\Sigma.(\boldsymbol{\nabla} \times \mathbf{A})$$
$$= -\Sigma.\mathbf{B}$$

Thus in terms of electromagnetic field $\mathbf{E}$ and $\mathbf{B}$ the Eq. (20.167) is written as

$$(\hbar^2 D^\mu D_\mu + \frac{ie\hbar}{c}\boldsymbol{\alpha}.\mathbf{E} - \frac{e\hbar}{c}\Sigma.\mathbf{B} + m^2 c^2)\psi = 0 \qquad (20.168)$$

where

$$\Sigma = \begin{pmatrix} \boldsymbol{\sigma} & 0 \\ 0 & \boldsymbol{\sigma} \end{pmatrix}. \qquad (20.169)$$

Now in the Pauli representation $\boldsymbol{\alpha}$ is non-diagonal while every other term is diagonal. It is then preferable to use chiral representation for γ-matrices given in Eq. (20.64), where

$$\gamma^5 = \begin{pmatrix} -1 & 0 \\ 0 & 1 \end{pmatrix} \qquad (20.170)$$

and

$$\alpha^i = \gamma^o \gamma^i = \begin{pmatrix} -\sigma^i & 0 \\ 0 & \sigma^i \end{pmatrix} \qquad (20.171)$$

are both diagonal. Then in terms of two components wave functions ϕ_L, ϕ_R:

$$\psi = \begin{pmatrix} \phi_L \\ \phi_R \end{pmatrix} \qquad (20.172)$$

where $\psi_{L,R} = \dfrac{1 \mp \gamma_5}{2}\psi$, the equation (20.168) splits into two equations which are completely decoupled.

$$(\hbar^2 D^\mu D_\mu I \mp \frac{ie\hbar}{c}\sigma.\mathbf{E} - \frac{e\hbar}{c}\sigma.\mathbf{B} + m^2 c^2 I)\phi_{L,R} = 0 \qquad (20.173)$$

where $\mathbf{I}$ is a 2×2 unit matrix, which we will not write explicitly in what follows. The above equation differs from Klein–Gordon equation in the terms $\sigma.\mathbf{E}$ and $\boldsymbol{\sigma}.\mathbf{B}$.

In order to interpret these terms and see their physical significance we first consider only the radiation field:

$$A^o = 0, \qquad \mathbf{A} \neq 0, \qquad \mathbf{B} = \nabla \times \mathbf{A} \text{ and } \qquad \mathbf{E} = -\frac{\partial \mathbf{A}}{\partial t}. \qquad (20.174)$$

We take $\mathbf{A}$ independent of time t, so that $\mathbf{E} = 0$ and the equation (20.173) reduces to

$$\left(\hbar^2 \frac{\partial^2}{\partial t^2} - \hbar^2 \mathbf{D}^2 - \frac{e\hbar}{c}\sigma.\mathbf{B} + m^2 c^2 \right) \psi(\mathbf{r}, t) = 0 \qquad (20.175)$$

where $D^2 = (\nabla - \frac{ie}{\hbar c}\mathbf{A})^2$ and ψ is ϕ_L or ϕ_R.

Now a mode with energy $E = mc^2 + \epsilon$ should oscillate in time as $e^{-iEt/\hbar}$. In the non-relativistic limit, kinetic energy $\epsilon \ll mc^2$, so that for a slowly moving particle:

$$\psi(\mathbf{r}, t) = \psi(\mathbf{r}, t) e^{-imc^2 t/\hbar}. \qquad (20.176)$$

Thus

$$\frac{\partial}{\partial t}(e^{-imc^2 t/\hbar}\psi) = e^{-imc^2 t/\hbar}\left(-\frac{imc^2}{\hbar} + \frac{\partial}{\partial t} \right)\psi.$$

Using it again

$$\frac{\partial}{\partial t}[e^{-imc^2/\hbar}\left(-imc^2/\hbar + \frac{\partial}{\partial t} \right)\psi] = e^{-imc^2/\hbar}\left(\frac{-imc^2}{\hbar} + \frac{\partial}{\partial t} \right)^2 \psi$$

$$= e^{-imc^2/\hbar}\left(\frac{\partial^2}{\partial t^2} - 2im\frac{c^2}{\hbar}\frac{\partial}{\partial t} - \frac{mc^2}{\hbar^2} \right)\psi.$$

Dropping $\dfrac{\partial^2}{\partial t^2}$ compared to $2imc^2\dfrac{\partial}{\partial t}$, equation (20.175) becomes

$$i\hbar\frac{\partial \psi}{\partial t} = \left(-\frac{\hbar^2}{2m}\left(\nabla - \frac{ie}{\hbar c}\mathbf{A} \right)^2 - \frac{e\hbar}{2mc}\sigma.\mathbf{B} \right)\psi. \qquad (20.177)$$

Now

$$\left(\nabla - \frac{ie}{\hbar c}\mathbf{A} \right)^2 \psi = \left(\nabla^2 - \frac{ie}{\hbar c}\nabla.\mathbf{A} - \frac{ie}{\hbar c}\mathbf{A}.\nabla - \frac{e^2}{\hbar^2 c^2}\mathbf{A}^2 \right)\psi.$$

But

$$\nabla.\mathbf{A}\psi = (\nabla.\mathbf{A})\psi + \mathbf{A}.\nabla\psi$$

$$= \mathbf{A}.\nabla\psi$$

using $\nabla.\mathbf{A} = 0$. Further taking $\mathbf{B}$ to be uniform, and noting $\mathbf{A} = \dfrac{1}{2}\mathbf{B} \times \mathbf{r}$ satisfies $\mathbf{B} = \nabla \times \mathbf{A}$, we get $\mathbf{A}.\nabla = \dfrac{i}{2\hbar}\mathbf{B}.\mathbf{L}$, where $\mathbf{L} = -i\hbar(\mathbf{r} \times \nabla)$ is the orbital angular momentum. Thus finally

$$i\hbar\frac{\partial \psi}{\partial t} = \left[-\hbar^2\frac{\nabla^2}{2m} - \frac{e}{2mc}(\mathbf{L} + 2\mathbf{S}).\mathbf{B} + \frac{e^2}{2mc^2}\mathbf{A}^2\right]\psi \qquad (20.178)$$

where $\mathbf{S} = \dfrac{\hbar}{i}\sigma$ denotes spin $\dfrac{1}{2}$ operator. Now $\dfrac{e}{2mc}\mathbf{L}$ is the magnetic moment due to the orbital motion of a particle of charge e. Note the important factor of 2 in front of $\mathbf{S}$, known as magnetic spin anomaly which naturally comes from Dirac equation. The second term gives the intrinsic magnetic moment of a particle of charge e with spin $\dfrac{1}{2}$ i.e. the magnetic moment of electron is $\mu_e = 1$ or $g = 2$, in terms of Bohr's magnetron $\dfrac{e}{2mc}$ in agreement with experiment. This naturally comes out from Dirac equation while earlier it was calculated experimentally. This was regarded as a great success of Dirac equation.

20.13 Dirac Equation for the Hydrogen-like Atom

Now consider the Hydrogen atom for which

$$A^0 \neq 0 \qquad \mathbf{A} = 0, \qquad \mathbf{B} = 0$$

so that

$$D^\mu D_\mu = D^{o2} - \mathbf{D}^2 = \frac{1}{c^2}\left(\frac{\partial}{\partial t} + i\frac{e}{\hbar}A^o\right)^2 - \nabla^2. \qquad (20.179)$$

Here A^0 is the Coulomb potential $A^0 = -Z\dfrac{e}{r}$ so that

$$eA^0 = -Z\frac{e^2}{r} = -Z\frac{\alpha}{r},$$

and

$$\mathbf{E} = -\nabla A^0 = -\frac{dA^0}{dr}\hat{r}. \qquad (20.180)$$

Then equation (20.173), on using Eqs. (20.179) and (20.180), takes the form

$$\left[\frac{\hbar^2}{c^2}\left(\frac{\partial}{\partial t} + \frac{ie}{\hbar}A^0\right)^2 - \hbar^2\nabla^2 \pm \frac{ie\hbar}{c}\frac{dA^0}{dr}\boldsymbol{\sigma}.\hat{r} + m^2c^2\right]\phi_{L,R} = 0. \qquad (20.181)$$

For stationary states

$$\phi_{L,R} = e^{-iEt/\hbar}\phi_{L,R}(\mathbf{r}),$$

where

$$\phi_{L,R}(\mathbf{r}) = Y_{\ell m}(\theta, \phi) R_{L,R}(r).$$

Thus equation (20.181) [see Sec. 7.1] becomes

$$\left[2\frac{eEA^0}{c^2} - \frac{e^2 A_o^2}{c^2} - \hbar^2 \nabla^2 \pm i\frac{e\hbar}{c}\frac{dA^0}{dr}\boldsymbol{\sigma}.\hat{\mathbf{r}} - \left(\frac{E^2 - m^2 c^4}{c^2} \right) \right] R_{L,R}(r) = 0$$

$$(20.182)$$

where

$$\nabla^2 = \frac{\partial^2}{\partial r^2} + \frac{2}{r}\frac{\partial}{\partial r} - \frac{L^2}{\hbar r^2}, \qquad (20.183)$$

and L^2 is the square of orbital angular momentum $\mathbf{L}$. The total angular momentum

$$\mathbf{J} = \mathbf{L} + \mathbf{S} = \mathbf{L} + \frac{1}{2}\boldsymbol{\sigma}$$

commutes with L^2 as both $\mathbf{L}$ and $\boldsymbol{\sigma}$ commutes with L^2. It also commute with $\boldsymbol{\sigma}.\hat{\mathbf{r}} = \dfrac{\boldsymbol{\sigma}.\mathbf{r}}{r}$, since

$$[J^i, \boldsymbol{\sigma}.\mathbf{r}] = [J^i, \sigma^k x^k]$$

$$= [L^i, \sigma^k x^k] + \frac{1}{2}[\sigma^i, \sigma^k x^k]$$

$$= \sigma^k [L^i, x^k] + \frac{1}{2}x^k [\sigma^i, \sigma^k]$$

$$= \sigma^k i\epsilon^{ikj} x^j + ix^k \epsilon^{ikj}\sigma^j$$

$$= 0.$$

Thus $\mathbf{J}$ commutes with the operator acting on $\phi_{L,R}$. J^2 and J_z have eigenvalues $j(j+1)$ and $m\hbar$ respectively, where

$$m = -j, ..., +j; \qquad j = \frac{1}{2}, \frac{2}{3}, ..., \qquad (20.184)$$

and can be used to label the states. Now L^2 has eigenvalues $\ell(\ell+1)\hbar^2$ and the integer ℓ takes the values:

$$\ell = j + \frac{1}{2} = \ell_+,$$

$$\ell = j - \frac{1}{2} = \ell_-,$$

$\boldsymbol{\sigma} . \mathbf{r}$ is hermitian and has its square $(\boldsymbol{\sigma} . \hat{r})^2 = 1$ so that it has eigenvalues ± 1. In the two dimensional sub-space provided by $\ell_{\pm}, (\boldsymbol{\sigma} . \hat{r})$ has non-diagonal matrix elements (it is like dipole operator for which the selection rule $\Delta \ell \neq 0$ holds), so that[1],

$$\langle \ell_{\pm} | \boldsymbol{\sigma} . \hat{r} | \ell_{\pm} \rangle = 0. \tag{20.185}$$

Thus in the above subspace the operator

$$\hbar^2 \left(\frac{L^2}{\hbar^2 r^2} - \frac{e^2}{\hbar c^2} A_0^2 \pm \frac{ie}{\hbar c} \frac{dA_0}{dr} \boldsymbol{\sigma} . \hat{r} \right), \tag{20.186}$$

in Eq. (20.182) is represented by the 2×2 matrix

$$\hbar^2 \begin{pmatrix} \frac{\left(j+\frac{1}{2}\right)\left(j+\frac{3}{2}\right)}{r^2} - \frac{e^2}{\hbar c^2} A_0^2 & \pm \frac{ie}{\hbar^2 c} \frac{dA_0}{dr} \\ \pm \frac{ie}{\hbar c} \frac{dA_0}{dr} & \frac{\left(j-\frac{1}{2}\right)\left(j+\frac{1}{2}\right)}{r^2} - \frac{e^2}{\hbar^2 c^2} A_0^2 \end{pmatrix} \tag{20.187}$$

which has eigenvalues $\lambda \hbar^2$ given by

$$\left[\frac{\left(j+\frac{1}{2}\right)\left(j+\frac{3}{2}\right)}{r^2} - \frac{e^2}{\hbar^2 c^2} A_0^2 - \lambda \right] \left[\frac{\left(j-\frac{1}{2}\right)\left(j+\frac{1}{2}\right)}{r^2} - \frac{e^2}{\hbar c^2} A_0^2 - \lambda \right]$$
$$- \frac{e^2}{\hbar^2 c^2} \left(\frac{dA_0}{dr} \right)^2 = 0. \tag{20.188}$$

The solution of the quadratic equation gives

$$\lambda = \left[\frac{\left(j+\frac{1}{2}\right)^2}{r^2} - \frac{e^2}{\hbar c^2} A_0^2 \right] \pm \left[\frac{\left(j+\frac{1}{2}\right)^2}{r^4} - \frac{e^2}{\hbar^2 c^2} \left(\frac{dA_0}{dr} \right)^2 \right]^{1/2},$$
$$= \frac{1}{r^2} \left[\left(j+\frac{1}{2} \right)^2 - Z\alpha^2 \right]^{1/2} \left[\left[\left(j+\frac{1}{2} \right)^2 - Z\alpha^2 \right]^{1/2} \pm 1 \right], \tag{20.189}$$

where we have used $\left(A^0 = -Z\frac{e}{r} \right), \frac{e^2 A_0^2}{\hbar^2 c^2} = \frac{Z^2 \alpha^2}{r^2}$ and $\frac{e^2}{\hbar^2 c^2}$ $\left(\frac{dA^0}{dr} \right)^2 = \frac{Z^2 e^4}{r^4}$; where $\alpha = \frac{e^2}{\hbar c}$. Thus the eigenvalues of the operator

$$\hbar^2 \left(\frac{L^2}{r^2} - \frac{Z^2 \alpha^2}{r^2} \pm i \frac{Z\alpha}{r^2} \boldsymbol{\sigma} . \hat{r} \right) \tag{20.190}$$

are

$$\frac{\bar{\lambda}(\bar{\lambda}+1)\hbar^2}{r^2}, \tag{20.191}$$

---

[1] see C.Itzykson and J.Zuber, Quantum Field Theory, McGraw Hill, 1980. See, 2.3

where

$$\bar{\lambda} = \left[\left(j+\frac{1}{2}\right)^2 - Z^2\alpha^2\right]^{1/2}; \qquad \left[\left(j+\frac{1}{2}\right)^2 - Z^2\alpha^2\right]^{1/2} - 1, \quad (20.192)$$

which may be written as

$$\bar{\lambda} = \left(j+\frac{1}{2}\right) - \delta_j,$$

with

$$\delta_j = \left(j+\frac{1}{2}\right) - \sqrt{\left(j+\frac{1}{2}\right)^2 - Z^2\alpha^2}. \qquad (20.193)$$

Thus our equation (20.182) becomes

$$\left[\hbar^2\left(\frac{\partial^2}{\partial r^2} + \frac{2}{r}\frac{\partial}{\partial r}\right) + 2\frac{Z}{c^2}\frac{\hbar e^2}{r}E + (E^2 - m^2c^2) - \frac{\bar{\lambda}(\bar{\lambda}+1)\hbar^2}{r^2}\right]R(r) = 0. \qquad (20.194)$$

Writing

$$R(r) = \frac{\chi(r)}{r}, \qquad (20.195)$$

we have

$$\frac{\hbar^2}{2m}\frac{d^2\chi}{dr^2} + \left[\frac{E^2 - m^2c^2}{2m} + \frac{2Ze^2}{2mc^2r}E - \frac{\bar{\lambda}(\bar{\lambda}+1)\hbar^2}{2mr^2}\right]\chi(r) = 0. \qquad (20.196)$$

This is formally identical with the equation for Hydrogen-like atom in Schrödinger theory i.e. [see Eq. (7.16)]

$$\frac{\hbar^2}{2m}\frac{d^2\chi}{dr^2} + \left(E + \frac{Ze^2}{r} - \frac{\ell(\ell+1)\hbar^2}{2mr^2}\right)\chi = 0, \qquad (20.197)$$

with energy eigenvalues,

$$E = -m\frac{Z^2e^4}{2\hbar^2n^2}; \qquad n = \acute{n} + \ell + 1 \quad \text{is a positive integer,}$$

if

$$e^2 \to \frac{E}{mc^2}e^2,$$

$$E \to \frac{E^2 - m^2c^4}{2mc^2},$$

$$\ell \to \bar{\lambda} = \left(j+\frac{1}{2}\right) - \delta_j,$$

and

$$\acute{n} = n - (\ell + 1) \to n - (\bar{\lambda} + 1)$$
$$= n - \left(j + \frac{1}{2}\right) + \delta_j - 1.$$

Thus n must be shifted by δ_j.

$$n \to n = n - \delta_j.$$

Thus

$$\frac{E_{nj}^2 - m^2 c^4}{2mc^2} = -\frac{mZ^2\alpha^2}{c^2}\frac{E_{nj}^2}{(n - \delta_j)^2} \tag{20.198}$$

or

$$E_{nj} = \pm mc^2 \frac{1}{\left[1 + \frac{Z^2\alpha^2}{(n - \delta_j)^2}\right]^{1/2}}, \tag{20.199}$$

where $n = 1, 2, ... j$ and $j = \frac{1}{2}, \frac{3}{2}, ..., n - \frac{1}{2}$.

We note from Eq (20.193) that δ_j becomes complex for $Z\alpha > j + \frac{1}{2}$. This catastrophy first occurs for $j = \frac{1}{2}$ i.e. for $Z\alpha > 1$ or $Z > 137$. Since α is small therefore for $Z\alpha \ll 1$, we can expand equation (20.199) in powers of $Z\alpha$. This gives

$$(E_{nj} - mc^2) = -mc^2 \frac{Z^2\alpha^2}{2n^2}\left[1 + \frac{Z^2\alpha^2}{n}\left(\frac{1}{j + \frac{1}{2}} - \frac{3}{4n}\right)\right]. \tag{20.200}$$

For Hydrogen atom $Z = 1$,

$$E_n^0 = -\frac{m\alpha^2 c^2}{2n^2} = -\frac{me^4}{2\hbar^2 n^2} = -\frac{e^2}{2a_o n^2} \tag{20.201}$$

where a_o is the Bohr radius. Note the important fact that states with the same value of n but different j (e.g. $2p_{1/2}$ and $2p_{3/2}$ which were degenerate in Schrödinger theory) now split.:

$$[E(2p_{3/2}) - E(2p_{1/2})] = mc^2 \frac{\alpha^4}{32} = 4.53 \times 10^{-2} eV = 10.9\ GHz, \tag{20.202}$$

in agreement with experiment; on the other hand states with same n and j (e.g. $2s_{1/2}$ and $2p_{1/2}$) are still degenerate (Lamb shift). This requires interaction of electron with quantized radiation field.

20.14   Hole Theory and Existence of Positrons (Anti-matter)

We have already seen in Sec. 20.9 that there is no trouble with the probability density as it is always positive definite. But the interpretation of negative energy problem which the Dirac theory admits remains. The difficulty with the negative energy states is solved by Dirac's Hole theory. First we elaborate what is the difficulty? Consider the Hydrogen atom. An electron bound in an atom can emit radiation spontaneously and thus lose energy. It can make a radiative transition and fall into the continuum of energy of negative energy states by spontaneously emitting a photon of energy $\geq 2mc^2$. But the ground state of atom in stable. Clearly such a catastrophic transition must be prevented. Dirac solved it by giving a new picture of vacuum regarding it an infinite sea of electrons occupying negative energy states (see figure 20.1). The Pauli principle then forbids such a transition. We see that $E > 0$ is empty in the vacuum but $E < 0$ is filled

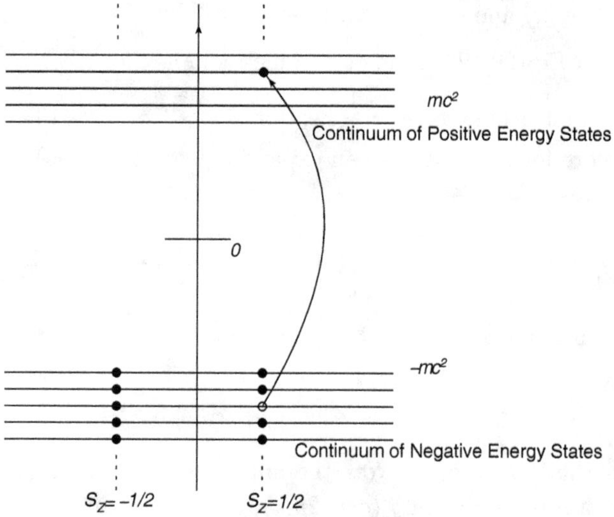

mc^2

Continuum of Positive Energy States

0

$-mc^2$

Continuum of Negative Energy States

$S_z = -1/2$ $S_z = 1/2$

Fig. 20.1   Dirac's picture of the vacuum

in the vacuum. The constant mc^2 produces energy gap between $E > 0$ and $E < 0$ states. The vacuum charge vanishes. A possibility exists that one of the electrons is lifted by a radiation quantum (photon) to a positive energy state, where it becomes observable as an ordinary electron, and the gap or

hole in the infinite sea is the positron (positive charge—opposite to electron due to charge conservation). This is what happens when a radiation
quantum of high energy ($\geq 2mc^2$) disappears with this process, giving rise
to an $e^- e^+$ pair. To describe the above phenomena, Feynman regarded the
positron and electron as essentially one particle, differing only in the sense
in which they move with respect to time (see Fig 20.2). A positron was
an electron moving backward in time. Interaction with radiation quantum
turns the path of e^+ around and then it appears as an electron.

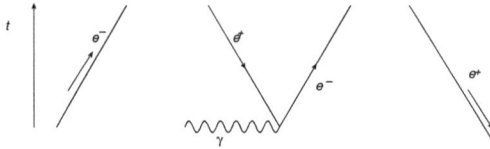

Fig. 20.2   Feynman's picture of electron and positron

20.15   Free Particle Solutions, Plane Wave Solutions of Dirac Equation

Dirac equation for a free particle is given in Eq. (20.56)

$$(i\hbar\gamma^\mu \frac{\partial}{\partial x^\mu} - mc)\psi(x) = 0. \tag{20.203}$$

Each component of Dirac equation satisfies the Klein–Gordon equation

$$\left(\frac{1}{c^2}\frac{\partial^2}{\partial t^2} - \nabla^2\right)\psi(x) = m^2 c^2 \psi(x). \tag{20.204}$$

The plane wave solution of Eq. (20.204) is

$$\psi(x) \sim e^{-\frac{i}{\hbar}p^\mu x_\mu} = e^{-\frac{i}{\hbar}(Et-\mathbf{p}.\mathbf{x})}. \tag{20.205}$$

Thus the plane wave solution of Eq. (20.203) can be written as

$$\psi(x) = u(\mathbf{p})e^{-\frac{i}{\hbar}p^\mu x_\mu} = u(\mathbf{p})e^{-i/\hbar(Et-\mathbf{p}.\mathbf{x})}. \tag{20.206}$$

Hence from Eq. (20.203):

$$(\gamma^\mu p_\mu - mc)u(\mathbf{p}) = 0$$

$$(/p - mc)u(\mathbf{p}) = 0 \tag{20.207}$$

where

$$\gamma^\mu p_\mu = /p. \tag{20.208}$$

Now $u(\boldsymbol{p})$ has four components:

$$u = \begin{pmatrix} u_1 \\ u_2 \\ u_3 \\ u_4 \end{pmatrix}. \tag{20.209}$$

Thus Eq. (20.207) is matrix equation:

$$(/p - mc)_{\alpha\beta} u_\beta(\boldsymbol{p}) = 0. \tag{20.210}$$

The hermitian conjugate of Eq. (20.207) is

$$u^\dagger \boldsymbol{p} [(\gamma^\mu)^\dagger p_\mu - mc] = 0.$$

By multiplying the above equation by γ^o on the right, writing $u^\dagger(\boldsymbol{p})\gamma^o$ as $\bar{u}(\boldsymbol{p})$ and using $\gamma^o \gamma^{\mu\dagger} \gamma^o = \gamma^\mu$, we get the equation for the adjoint spinor $\bar{u}(\boldsymbol{p})$:

$$\bar{u}(\boldsymbol{p}) [\gamma^\mu p_\mu - mc] \equiv \bar{u}(\boldsymbol{p}) [/p - mc] = 0. \tag{20.211}$$

First, we note that for a particle on the mass shell:

$$p^2 = p^\mu p_\mu = p_o^2 - \boldsymbol{p}^2 = m^2 c^2.$$

Thus

$$p_o^2 = \frac{E^2}{c^2} = \boldsymbol{p}^2 + m^2 c^2$$

or

$$E^2 = c^2 \boldsymbol{p}^2 + m^2 c^4.$$

Hence

$$E = \pm\sqrt{c^2 \boldsymbol{p}^2 + m^2 c^4} = \pm E_p$$

where

$$E_p = \sqrt{c^2 \boldsymbol{p}^2 + m^2 c^4}. \tag{20.212}$$

Let us write

$$u(\boldsymbol{p}) = \begin{pmatrix} u_A(\boldsymbol{p}) \\ u_B(\boldsymbol{p}) \end{pmatrix}. \tag{20.213}$$

Now in the Pauli representation of γ-matrices

$$/p = \gamma^\mu p_\mu = \gamma^o p_o - \boldsymbol{\gamma}.\boldsymbol{p} = \begin{pmatrix} p_o & -\boldsymbol{\sigma}.\boldsymbol{p} \\ \boldsymbol{\sigma}.\boldsymbol{p} & -p_o \end{pmatrix}. \tag{20.214}$$

Thus from Eqs. (20.207), (20.213) and (20.214), we obtain

$$(E - mc^2)u_A(\boldsymbol{p}) = c\boldsymbol{\sigma}.\boldsymbol{p}u_B(\boldsymbol{p}), \qquad (20.215a)$$

$$(E + mc^2)u_B(\boldsymbol{p}) = c\boldsymbol{\sigma}.\boldsymbol{p}u_A(\boldsymbol{p}). \qquad (20.215b)$$

In the rest frame of the particle $(\boldsymbol{p} = 0)$, we have

$$E = \pm mc^2$$

and from Eqs. (20.215), we get

$$(E - mc^2)u_A(0) = 0, \qquad (20.216)$$

$$(E + mc^2)u_B(0) = 0.$$

Thus for $E = mc^2$:

$$u_A(0) \neq 0, \qquad u_B(0) = 0 \qquad (20.217)$$

and for $E = -mc^2$:

$$u_B(0) \neq 0, \qquad u_A(0) = 0.$$

Hence there are four independent relations:

$$E = mc^2 = u^{(r)}(0) = \begin{pmatrix} u_A^{(r)}(0) \\ 0 \end{pmatrix} = \begin{pmatrix} \chi^{(r)} \\ 0 \end{pmatrix}$$

$$E = -mc^2 = u^{(r)}(0) = \begin{pmatrix} 0 \\ u_A^{(r)}(0) \end{pmatrix} = \begin{pmatrix} 0 \\ \chi^{(r)} \end{pmatrix} \qquad (20.218)$$

where $\chi^{(r)}$ are two component Pauli spinors: $r = 1, 2,$

$$\chi^{(1)} = \begin{pmatrix} 1 \\ 0 \end{pmatrix} : \text{Spin up} \qquad \sigma^3 \chi^{(1)} = \chi^{(1)},$$

$$\chi^{(2)} = \begin{pmatrix} 0 \\ 1 \end{pmatrix} : \text{Spin down} \qquad \sigma^3 \chi^{(2)} = \chi^{(2)}.$$

The four solutions in the rest frame are:

$$E = mc^2 :$$

$$u^{(1)}(0) = \begin{pmatrix} \chi^{(1)} \\ 0 \end{pmatrix} = \begin{pmatrix} 1 \\ 0 \\ 0 \\ 0 \end{pmatrix}, \text{Spin up},$$

$$u^{(2)}(0) = \begin{pmatrix} \chi^{(2)} \\ 0 \end{pmatrix} = \begin{pmatrix} 0 \\ 1 \\ 0 \\ 0 \end{pmatrix}, \text{Spin down}.$$

$$E = -mc^2 : \qquad\qquad (20.219)$$

$$u^{(3)}(0) = \begin{pmatrix} 0 \\ \chi^{(1)} \end{pmatrix} = \begin{pmatrix} 0 \\ 0 \\ 1 \\ 0 \end{pmatrix} , \text{Spin up,}$$

$$u^{(4)}(0) = \begin{pmatrix} 0 \\ \chi^{(2)} \end{pmatrix} = \begin{pmatrix} 0 \\ 0 \\ 0 \\ 1 \end{pmatrix} , \text{Spin down.}$$

Thus

$$\begin{pmatrix} \chi^{(1)} \\ 0 \end{pmatrix} e^{-imc^2 t/\hbar} \qquad \text{and} \qquad \begin{pmatrix} 0 \\ \chi^{(1)} \end{pmatrix} e^{imc^2 t/\hbar} \qquad (20.220)$$

are positive and negative energy solutions for spin up and

$$\begin{pmatrix} \chi^{(2)} \\ 0 \end{pmatrix} e^{-imc^2 t/\hbar} \qquad \text{and} \qquad \begin{pmatrix} 0 \\ \chi^{(2)} \end{pmatrix} e^{imc^2 t/\hbar} \qquad (20.221)$$

are positive and negative energy solutions for spin down for a particle at rest.

For a particle with momentum $\boldsymbol{p}$, we note that

$$u^r(\boldsymbol{p}) = N'(\slashed{p} + mc)u^r(0)$$

satisfies Eq. (20.207), since

$$(\slashed{p} - m)(\slashed{p} + m) = p^2 - m^2 c^2 = 0. \qquad (20.222)$$

Written in full in Pauli representation of γ-matrices, we have

$$u(\boldsymbol{p}) = N' \begin{pmatrix} p^o + mc & -\boldsymbol{\sigma}.\boldsymbol{p} \\ \boldsymbol{\sigma}.\boldsymbol{p} & -p^o + mc \end{pmatrix} u(0) \qquad (20.223)$$

where N' is a normalization constant. Thus the four independent solutions are:

$$E > 0, \qquad p^o = \frac{E}{c} = \frac{E_p}{c} = \sqrt{p^2 + m^2 c^2} \qquad u^{(r)}(\boldsymbol{p}) = N \begin{pmatrix} \chi^{(r)} \\ \frac{c\boldsymbol{\sigma}.\boldsymbol{p}}{E_p + mc^2} \chi^{(r)} \end{pmatrix},$$

$$r = 1, 2 \qquad\qquad (20.224)$$

$$E < 0, \qquad p^o = -\frac{|E|}{c} = \frac{E_p}{c} \qquad u^{(r)}(\boldsymbol{p}) = N \begin{pmatrix} \frac{-c\boldsymbol{\sigma}.\boldsymbol{p}}{E_p + mc^2} \chi^{(r)} \\ \chi^{(r)} \end{pmatrix}, \qquad r = 3, 4$$

$$(20.225)$$

where $N = |\sqrt{p^0 + mc}|N'$. It is easy to see that for $\boldsymbol{p} \neq 0$, the above spinors are not eigenstates of 4×4 matrix $\Sigma^3 = \begin{pmatrix} \sigma^3 & 0 \\ 0 & \sigma^3 \end{pmatrix}$ as is the case for the spinors with $\boldsymbol{p} = 0$. The reason is that $\boldsymbol{\sigma}.\mathbf{n}$ does not commute with one Dirac Hamiltonian, where $\mathbf{n}$ is an arbitrary direction. However, for $\mathbf{n} = \frac{\boldsymbol{p}}{|\boldsymbol{p}|}$, the operator

$$\mathcal{H} = \frac{\boldsymbol{\sigma}.\boldsymbol{p}}{|\boldsymbol{p}|} \tag{20.226}$$

called the helicity operator, commutes with the Dirac Hamiltonian. Now from Eq. (20.224)

$$u^{(r)\dagger}(\boldsymbol{p})u^{(s)}(\boldsymbol{p}) = |N|^2 \left(\chi^{(r)\dagger}, \frac{c\boldsymbol{\sigma}.\boldsymbol{p}}{E_p+mc^2} \chi^{(r)\dagger} \right) \begin{pmatrix} \chi^{(s)} \\ \frac{c\boldsymbol{\sigma}.\boldsymbol{p}}{E_p+mc^2} \chi^{(s)} \end{pmatrix}$$

$$= |N|^2 \left(1 + \frac{c^2(\boldsymbol{\sigma}.\boldsymbol{p})^2}{(E_p+mc^2)^2} \right) \chi^{(r)\dagger} \chi^{(s)}$$

$$= |N|^2 \frac{2E_p}{E_p + mc^2} \delta_{rs} \qquad r, s = 1, 2. \tag{20.227a}$$

Similarly

$$u^{(r)\dagger}(\boldsymbol{p})u^{(r)}(\boldsymbol{p}) = |N|^2 \frac{2E_p}{E_p + mc^2}, \quad r, s = 3, 4. \tag{20.227b}$$

Now

$$\bar{u}^{(r)}(\boldsymbol{p}) = u^{(r)\dagger}\gamma^0 = \left(\chi^{(r)\dagger}, -\chi^{(r)\dagger} \frac{c\boldsymbol{\sigma}.\boldsymbol{p}}{E_p+mc^2} \right). \tag{20.228}$$

Therefore

$$\bar{u}^{(r)}(\boldsymbol{p})u^{(s)}(\boldsymbol{p}) = |N|^2 \frac{2mc^2}{(E_p + mc^2)} \delta_{rs}, \qquad r = 1, 2$$

$$\bar{u}^{(r)}(\boldsymbol{p})u^{(s)}(\boldsymbol{p}) = |N|^2 \frac{-2mc^2}{(E_p + mc^2)} \delta_{rs}, \qquad r = 3, 4$$

We normalise the spinors, so that

$$\bar{u}^{(r)}(\boldsymbol{p})u^{(s)}(\boldsymbol{p}) = \delta_{rs} \qquad r = 1, 2$$

$$= -\delta_{rs} \qquad s = 3, 4 \tag{20.229}$$

This normalisation gives

$$|N|^2 = \frac{(E_p + mc^2)}{2mc^2}.$$

Selecting the phase, so the N is positive

$$N = \sqrt{\frac{E_p + mc^2}{2mc^2}}. \tag{20.230}$$

Hence the normalized spinors for positive and negative energy states are

$$u^{(r)}(\boldsymbol{p}) = \sqrt{\frac{E_p + mc^2}{2mc^2}} \begin{pmatrix} \chi^{(r)} \\ \frac{c\boldsymbol{\sigma}\cdot\boldsymbol{p}}{E_p+mc^2}\chi^{(r)} \end{pmatrix}, \qquad r = 1, 2, E > 0 \quad (20.231)$$

$$u^{(r)}(\boldsymbol{p}) = \sqrt{\frac{E_p + mc^2}{2mc^2}} \begin{pmatrix} \frac{-c\boldsymbol{\sigma}\cdot\boldsymbol{p}}{E_p+mc^2}\chi^{(r)} \\ \chi^{(r)} \end{pmatrix}, \qquad r = 3, 4, E < 0 \quad (20.232)$$

Finally, we have

$$\psi(x) \sim u^{(1,2)}e^{i/\hbar(-E_pt+\boldsymbol{p}\cdot\mathbf{x})}, \qquad E > 0 \qquad (20.233)$$
$$\sim u^{(3,4)}e^{i/\hbar(E_pt+\boldsymbol{p}\cdot\mathbf{x})}, \qquad E < 0 \qquad (20.234)$$

Change $\boldsymbol{p} \to -\boldsymbol{p}$ in the negative energy solutions and write

$$v^{(1)}(\boldsymbol{p}) = -u^{(4)}(-\boldsymbol{p})$$
$$v^{(2)}(\boldsymbol{p}) = u^{(3)}(-\boldsymbol{p})$$

then we can write the above solutions as

$$\psi(x) = u^{(r)}(\boldsymbol{p})e^{-i/\hbar p^{\mu}x_{\mu}}, \qquad r = 1, 2, E > 0,$$
$$\psi(x) = v^{(r)}(\boldsymbol{p})e^{i/\hbar p^{\mu}x_{\mu}}, \qquad r = 1, 2, E < 0. \qquad (20.235)$$

Without any ambiguity, we take E to be positive i.e. $E = E_p$, so that

$$u^{(r)}(\boldsymbol{p}) = \sqrt{\frac{E + mc^2}{2mc^2}} \begin{pmatrix} \chi^{(r)} \\ \frac{c\boldsymbol{\sigma}\cdot\boldsymbol{p}}{E+mc^2}\chi^{(r)} \end{pmatrix}, \qquad r = 1, 2. \qquad (20.236)$$

$$v^{(r)}(\boldsymbol{p}) = \sqrt{\frac{E + mc^2}{2mc^2}} \begin{pmatrix} \frac{c\boldsymbol{\sigma}\cdot\boldsymbol{p}}{E+mc^2}\chi^{(r)} \\ \chi^{(r)} \end{pmatrix}, \qquad r = 1, 2 \qquad (20.237)$$

Hence for positive energy solution i.e. for particle, the plane wave solution is

$$\psi(x) = u(\boldsymbol{p})e^{-i/\hbar p^{\mu}x_{\mu}} \qquad (20.238)$$

whereas for the negative energy solution

$$\psi(x) = v(\boldsymbol{p})e^{i/\hbar p^{\mu}x_{\mu}}. \qquad (20.239)$$

One way to interpret the negative energy solution [c.f. Sec. 20.14] is to regard a particle with negative energy as an antiparticle with positive energy but travelling backward in time. $u^{(r)}$ and $v^{(r)}(r = 1, 2)$ are called particle and anti-particle spinors. Substituting Eqs. (20.238) and (20.239) in Dirac equation (20.203), we obtain

$$(\slashed{p} - mc)u^{(r)} = 0, \qquad (20.240)$$

$$(\not{p} + mc)v^{(r)} = 0. \tag{20.241}$$

From Eq. (20.229), it follows that $u^{(r)}$ and $v^{(r)}$ are normalised as follows

$$\bar{u}^{(r)}(\boldsymbol{p})u^{(s)}(\boldsymbol{p}) = \delta_{rs},$$

$$\bar{v}^{(r)}(\boldsymbol{p})v^{(s)}(\boldsymbol{p}) = -\delta_{rs},$$

$$\bar{v}^{(r)}(\boldsymbol{p})u^{(s)}(\boldsymbol{p}) = \bar{u}^{(r)}(\boldsymbol{p})v^{(s)}(\boldsymbol{p}) = 0. \tag{20.242}$$

The spinors $u(\boldsymbol{p})$ and $v(\boldsymbol{p})$ also satisfy the completeness relation (in spinor space):

$$\Sigma_{r=1}^2 [u_\alpha^{(r)}(\boldsymbol{p})\bar{u}_\beta^{(r)}(\boldsymbol{p}) - v_\alpha^{(r)}(\boldsymbol{p})\bar{v}_\beta^{(r)}(\boldsymbol{p})] = \delta_{\alpha\beta}. \tag{20.243}$$

In matrix notation:

$$\Sigma_{r=1}^2 [u^{(r)}(\boldsymbol{p})\bar{u}^{(r)}(\boldsymbol{p}) - v^{(r)}(\boldsymbol{p})\bar{v}^{(r)}(\boldsymbol{p})] = 1. \tag{20.244}$$

Define the projection operators:

$$\Lambda_\pm(\boldsymbol{p}) \equiv \Lambda_\pm = \frac{mc \pm \not{p}}{2mc} \tag{20.245}$$

$$\Lambda_\pm^2 = \Lambda_\pm, \qquad \Lambda_+ + \Lambda_- = 1$$

$$\Lambda_+\Lambda_- = \frac{m^2c^2}{2mc} = \frac{m^2c^2 - p^2}{2mc} = \frac{m^2c^2 - m^2c^2}{2mc} = 0. \tag{20.246}$$

Now

$$\Lambda_+u(\boldsymbol{p}) = u(\boldsymbol{p}), \qquad \Lambda_+v(\boldsymbol{p}) = 0,$$

$$\Lambda_-v(\boldsymbol{p}) = v(\boldsymbol{p}), \qquad \Lambda_-u(\boldsymbol{p}) = 0. \tag{20.247}$$

Thus Λ_+ and Λ_- project out positive and negative energy solutions respectively. Further we note that

$$\Sigma_{r=1}^2 u^{(r)}(\boldsymbol{p})\bar{u}^{(r)}(\boldsymbol{p}) = \Lambda_+(\boldsymbol{p}),$$

$$-\Sigma_{r=1}^2 v^{(r)}(\boldsymbol{p})\bar{v}^{(r)}(\boldsymbol{p}) = \Lambda_-(\boldsymbol{p}). \tag{20.248}$$

To show this

$$\Sigma_{r=1}^2 u_\alpha^{(r)}(\boldsymbol{p})\bar{u}_\beta^{(r)} = \Sigma_{r=1}^2 (\Lambda_+)_{\alpha\sigma} u_\sigma^{(r)}(\boldsymbol{p})\bar{u}_\beta^{(r)}(\boldsymbol{p})$$

$$= (\Lambda_+)_{\alpha\sigma}\Sigma_{r=1}^2 u_\sigma^{(r)}(\boldsymbol{p})\bar{u}_\beta^{(r)}(\boldsymbol{p})$$

$$= (\Lambda_+)_{\alpha\sigma}\delta_{\alpha\beta} = (\Lambda_+)_{\alpha\beta}$$

where we have used Eq. (20.243) and the relations (20.247).

20.16   Charge Conjugation

The operation which takes a particle state function $\psi(x)$ to an antiparticle state function ψ^c is called charge conjugation. Let us consider the Dirac equation in the presence of electromagnetic field given in Eq. (20.137)

$$\left[i\hbar\gamma^\mu \left(\partial_\mu + \frac{ie}{\hbar c} A_\mu \right) - m \right] \psi(x) = 0. \tag{20.249}$$

We now take the complex conjugate of this equation:

$$\left[-i\hbar\gamma^{\mu*} \left(\partial_\mu - \frac{ie}{\hbar c} A_\mu \right) - mc \right] \psi^* = 0. \tag{20.250}$$

By multiplying Eq. (20.250) by $-i\gamma^2$ on the left, we get

$$-i\gamma^2 \left[(-i\hbar\gamma^{\mu*} \left(\partial_\mu - \frac{ie}{\hbar c} A_\mu \right) - mc \right] \psi^* = 0$$

or $[(\gamma^2)^2 = -1]$

$$-i\gamma^2 \left[i\hbar\gamma^{\mu*}\gamma^2\gamma^2 \left(\partial_\mu - \frac{ie}{\hbar c} A_\mu \right) - mc \right] \psi^* = 0. \tag{20.251}$$

First, we note that

$$\gamma^{0*} = \gamma^0, \qquad \gamma^{1*} = \gamma^1, \qquad \gamma^{2*} = -\gamma^2, \qquad \gamma^{3*} = \gamma^3$$

so that

$$\gamma^2\gamma^{\mu*}\gamma^2 = \gamma^\mu \tag{20.252}$$

and this is the reason for using $-i\gamma^2$ as above. Hence, from Eq. (20.251), we get

$$\left[i\hbar\gamma^\mu \left(\partial_\mu - \frac{ie}{\hbar c} A_\mu \right) - mc \right] (-i\gamma^2\psi^*) = 0. \tag{20.253}$$

Thus if the Dirac equation is to be invariant under the charge conjugation, then

$$\left[i\hbar\gamma^\mu \left(\partial_\mu + \frac{ie}{\hbar c} A_\mu^c \right) - mc \right] \psi^c = 0. \tag{20.254}$$

Now the Eq. (20.254) is identical to Eq. (20.253) provided that

$$\psi^c = -i\gamma^2\psi^*, \qquad A_\mu^c = -A_\mu. \tag{20.255}$$

Note that $A_\mu^c = -A_\mu$, as it should be, since

$$\partial^\nu \partial_\nu A_\mu = j_\mu \tag{20.256}$$

and electromagnetic current j_μ must change sign under charge conjugation. The matrix $-i\gamma^2\gamma^0$ is called the charge conjugate matrix and is written as [superscript T stands for transpose]:

$$C = -i\gamma^2\gamma^0, \qquad C^2 = -1,$$

$$C^\dagger = -C = C^T = C^{-1},$$

$$C\gamma^\mu C^{-1} = -(\gamma^\mu)^T. \tag{20.257}$$

Then we can write ψ^c given in Eq. (20.254) as

$$\psi^c = C\bar{\psi}^T. \tag{20.258}$$

Now under charge conjugation, $\psi \to \psi^c$, so that for plane wave solution

$$\psi^c = -i\gamma^2 u^*(\boldsymbol{p})e^{i/\hbar p^\mu x_\mu} = u^c(\boldsymbol{p})e^{i/\hbar p^\mu x_\mu} \tag{20.259}$$

implying

$$u^c(\boldsymbol{p}) = -i\gamma^2 u^*(\boldsymbol{p}) = v(\boldsymbol{p}). \tag{20.260}$$

In Section 20.7, we have discussed the discrete symmetries viz. space reflection and time reversal. Under space reflection

$$\psi(t, \mathbf{x}) \to \psi^P(t, -\mathbf{x}) = \gamma^o\psi(t, \mathbf{x})$$
$$= \gamma^o u(\boldsymbol{p})e^{-i/\hbar p^\mu x_\mu} \tag{20.261}$$

where

$$\gamma^o u(\boldsymbol{p}) = N \begin{pmatrix} 1 & 0 \\ 0 & -1 \end{pmatrix} \begin{pmatrix} \chi \\ \frac{c\sigma.\boldsymbol{p}}{E+mc^2}\chi \end{pmatrix}$$
$$= N \begin{pmatrix} \chi \\ -\frac{c\sigma.\boldsymbol{p}}{E+mc^2}\chi \end{pmatrix} = u(-\boldsymbol{p}).$$

Similarly

$$\gamma^o v(\boldsymbol{p}) = -v(-\boldsymbol{p}) \tag{20.262}$$

i.e. particle and antiparticle have opposite parities.
Under time reversal

$$\psi(t, \mathbf{x}) \to \psi^T(-t, \mathbf{x}) = -\gamma^1\gamma^3\psi^*(t, \mathbf{x})$$
$$= -\gamma^1\gamma^3 u^*(\boldsymbol{p})e^{i/\hbar p^\mu x_\mu}$$
$$= \begin{pmatrix} \sigma^1\sigma^3 & 0 \\ o & \sigma^1\sigma^3 \end{pmatrix} u^*(\boldsymbol{p})e^{i/\hbar p^\mu x_\mu}$$
$$= \begin{pmatrix} -i\sigma^2 & 0 \\ 0 & -i\sigma^2 \end{pmatrix} u^*(\boldsymbol{p})e^{i/\hbar p^\mu x_\mu}. \tag{20.263}$$

Therefore

$$-\gamma^1\gamma^3 u^*(\boldsymbol{p}) = \begin{pmatrix} -i\sigma^2 & 0 \\ 0 & -i\sigma^2 \end{pmatrix} u^*(\boldsymbol{p}) = N \begin{pmatrix} -i\sigma^2\chi \\ c\frac{-i\sigma^2\sigma^*\cdot\boldsymbol{p}}{E+mc^2}\chi \end{pmatrix}$$

$$= N \begin{pmatrix} -i\sigma^2\chi \\ c\frac{\sigma^*\cdot\boldsymbol{p}}{E+mc^2}(-i\sigma^2\chi) \end{pmatrix}. \qquad (20.264)$$

Now

$$-i\sigma^2\chi^{(1)} = \begin{pmatrix} 0 & -1 \\ 1 & 0 \end{pmatrix}\begin{pmatrix} 1 \\ 0 \end{pmatrix} = \begin{pmatrix} 0 \\ 1 \end{pmatrix} = \chi^{(2)},$$

$$-i\sigma^2\chi^{(2)} = \begin{pmatrix} 0 & -1 \\ 1 & 0 \end{pmatrix}\begin{pmatrix} 1 \\ 0 \end{pmatrix} = -\begin{pmatrix} 1 \\ 0 \end{pmatrix} = -\chi^{(1)}. \qquad (20.265)$$

Thus

$$-i\sigma^2\chi^{(r)} = r\chi^{(-r)}, \qquad (20.266)$$

where $r = \pm 1$ corresponds to $\frac{1}{2}$ (spin ↑) and $-\frac{1}{2}$ (spin ↓) respectively.
Hence

$$-\gamma^1\gamma^3 u^{(r)*}(\boldsymbol{p}) = r u^{(-r)}(-\boldsymbol{p}) \qquad (20.267)$$

so that under time reversal

$$u^{(r)}(\boldsymbol{p}) \to r u^{(-r)*}(-\boldsymbol{p}). \qquad (20.268)$$

20.17 Large and Small Bilinear Covariants

For not too relativistic spin $\frac{1}{2}$ particles, some of the covariant bilinears

$$\bar{\psi}\psi,\ \bar{\psi}\gamma_\mu\psi,\ \bar{\psi}\gamma_5\psi,\ \bar{\psi}i\gamma_\mu\gamma_5\psi \quad \text{and} \quad \bar{\psi}\sigma_{\mu\nu}\psi$$

are 'large' while others are 'small'. To see this recall from Sec. 20.4 that in
the Pauli representation we can write the four component wave function ψ
as

$$\psi = \begin{pmatrix} \phi_A \\ \phi_B \end{pmatrix},\ \psi_A = \begin{pmatrix} \phi_A \\ 0 \end{pmatrix},\ \psi_B = \begin{pmatrix} 0 \\ \phi_B \end{pmatrix} \qquad (20.269)$$

where each ϕ_A and ϕ_B has two components so that from the Dirac equation
(20.56), using Eq. (20.63)

$$i\hbar\left[\frac{1}{c}\frac{\partial}{\partial t}\begin{pmatrix} \phi_A \\ \phi_B \end{pmatrix} + \begin{pmatrix} 0 & \boldsymbol{\sigma}\cdot\boldsymbol{\nabla} \\ -\boldsymbol{\sigma}\cdot\boldsymbol{\nabla} & 0 \end{pmatrix}\begin{pmatrix} \phi_A \\ \phi_B \end{pmatrix}\right] - mc\begin{pmatrix} \phi_A \\ \phi_B \end{pmatrix} = 0$$

i.e.

$$\frac{i\hbar}{c}\frac{\partial\phi_A}{\partial t} = -i\hbar\boldsymbol{\sigma}.\boldsymbol{\nabla}\phi_B + mc\phi_A,$$

$$\frac{i\hbar}{c}\frac{\partial\phi_B}{\partial t} = -i\hbar\boldsymbol{\sigma}.\boldsymbol{\nabla}\phi_A + mc\phi_B.$$

For stationary states

$$i\hbar\boldsymbol{\sigma}.\boldsymbol{\nabla}\phi_B = -\frac{1}{c}(E - mc^2)\phi_A, \qquad i\hbar\boldsymbol{\sigma}.\boldsymbol{\nabla}\phi_A = -\frac{1}{c}(E + mc^2)\phi_B.$$

Thus for

$$E \approx mc^2, \tag{20.270}$$

$$\phi_B \approx \frac{-i\hbar\boldsymbol{\sigma}\cdot\boldsymbol{\nabla}}{2mc}\phi_A \approx \boldsymbol{\sigma}\cdot\frac{\hat{\mathbf{p}}}{2mc}\phi_A, \tag{20.271}$$

i.e. ϕ_B is of order v/c compared to ϕ_A. Using the Pauli representation of γ-matrices, we have

$$\bar{\psi}\psi = \psi^\dagger\gamma^0\psi$$

$$= (\phi_A^\dagger \phi_B^\dagger)\begin{pmatrix} I & 0 \\ 0 & -I \end{pmatrix}\begin{pmatrix} \phi_A \\ phi_B \end{pmatrix}$$

$$= \phi_A^\dagger\phi_A - \phi_B^\dagger\phi_B \simeq \phi_A^\dagger\phi_A = \psi_A^\dagger\psi_A + O(v^2/c^2), \tag{20.272}$$

$$\bar{\psi}\gamma^0\psi = \psi^\dagger\psi = \phi_A^\dagger\phi_A + \phi_B^\dagger\phi_B \simeq \phi_A{}^\dagger\phi_A = \psi_A^\dagger\psi_A + O(v^2/c^2), \tag{20.273}$$

$$\bar{\psi}\gamma^k\gamma_5\psi = \psi^\dagger\gamma^0\gamma^k\gamma_5\psi$$

$$= (\phi_A^\dagger \phi_B^\dagger)\begin{pmatrix} I & 0 \\ 0 & -I \end{pmatrix}\begin{pmatrix} \sigma^k & 0 \\ 0 & -\sigma^k \end{pmatrix}\begin{pmatrix} \phi_A \\ \phi_B \end{pmatrix}$$

$$= [\phi_A^\dagger\sigma^k\phi_A + \phi_B^\dagger\sigma^k\phi_B^\dagger]$$

$$= -\bar{\psi}\gamma_5\gamma^k\psi, \tag{20.274}$$

$$\bar{\psi}\sigma^{ij}\psi = \varepsilon^{ijk}(\phi_A^\dagger \phi_B^\dagger)\begin{pmatrix} I & 0 \\ 0 & -I \end{pmatrix}\begin{pmatrix} \sigma^k & 0 \\ 0 & \sigma^k \end{pmatrix}\begin{pmatrix} \phi_A \\ phi_B \end{pmatrix}$$

$$= \varepsilon^{ijk}[\phi_A^\dagger\sigma^k\phi_A - \phi_B^\dagger\sigma^k\phi_B]. \tag{20.275}$$

All the above bilinears are seen to be 'large' if we neglect terms of order $(v/c)^2$. Thus

$$\begin{pmatrix} \bar{\psi}\psi \\ \psi\gamma^0\psi \end{pmatrix} \approx \psi_A^\dagger\psi_A \tag{20.276}$$

$$\begin{pmatrix} \bar{\psi}\gamma_5\gamma^k\psi \\ \bar{\psi}\sigma^{ij}\psi \end{pmatrix} \approx \psi_A^\dagger \sigma_k \psi_A, \qquad (ijk \text{ cyclic}). \qquad (20.277)$$

In contrast to 1, γ_0, $i\gamma_5\gamma_k$ and σ_{ij} which connect $\psi_A^\dagger$ to ψ_A, the non-diagonal matrices γ_k, $i\gamma_5\gamma_0$, σ_{k0} and γ_5 connect $\psi_A^\dagger$ with ψ_B or $\psi_B^\dagger$ with ψ_A. Hence the corresponding bilinears are small and are of order v/c. For example

$$\psi\gamma_5\psi = \psi^\dagger\gamma^0\gamma_5\psi$$

$$= (\phi_A^\dagger \phi_B^\dagger) \begin{pmatrix} I & 0 \\ 0 & -I \end{pmatrix} \begin{pmatrix} 0 & I \\ I & 0 \end{pmatrix} \begin{pmatrix} \phi_A \\ \phi_B \end{pmatrix}$$

$$= -(\phi_A^\dagger \phi_B - \phi_B^\dagger \phi_A) \approx O(v/c). \qquad (20.278)$$

20.18   Weyl Equation

Recall from Sec. 20.4, that in the chiral representation of γ-matrices, we can write the four component wave function ψ as

$$\psi = \begin{pmatrix} \phi_L \\ \phi_R \end{pmatrix} \qquad (20.279)$$

so that using Eqs. (20.64), the Dirac Eq. (20.56) becomes

$$i\hbar\frac{\partial}{\partial x_o} \begin{pmatrix} \phi_R \\ \phi_L \end{pmatrix} + i\hbar \begin{pmatrix} 0 & \boldsymbol{\sigma}\cdot\boldsymbol{\nabla} \\ -\boldsymbol{\sigma}\cdot\boldsymbol{\nabla} & 0 \end{pmatrix} \begin{pmatrix} \phi_L \\ \phi_R \end{pmatrix} - mc \begin{pmatrix} \phi_L \\ \phi_R \end{pmatrix} = 0. \qquad (20.280)$$

Thus ϕ_L and ϕ_R satisfy the coupled equations

$$\left[i\boldsymbol{\sigma} \cdot \boldsymbol{\nabla} - i\frac{\partial}{\partial x_0} \right] \phi_L = -\frac{mc}{\hbar}\phi_R, \qquad (20.281a)$$

$$\left[-i\boldsymbol{\sigma} \cdot \boldsymbol{\nabla} - i\frac{\partial}{\partial x_0} \right] \phi_R = -\frac{mc}{\hbar}\phi_L. \qquad (20.281b)$$

It is the mass which links ϕ_L (or equivalently ψ_L) with ϕ_R (or ψ_R). Note the important fact that when the mass of a fermion is zero, Eqs. (20.281) are completely decoupled:

$$\left[i\boldsymbol{\sigma}\cdot\boldsymbol{\nabla} - i\frac{\partial}{\partial x_0} \right] \phi_L = 0, \qquad (20.282a)$$

$$\left[-i\boldsymbol{\sigma}\cdot\boldsymbol{\nabla} - i\frac{\partial}{\partial x_0} \right] \phi_R = 0. \qquad (20.282b)$$

It is easy to see that these equations are not disconnected. In fact using the usual representation of Pauli matrices $\sigma^i[\sigma^2* = -\sigma^2, \sigma^2\sigma^i\sigma^2 = -\sigma^i, i = 1, 2, 3]$ one verifies that, if $\phi_L(x)$ is a solution of Eq. (20.282a), $\sigma^2\phi_L^*$ is a solution of Eq. (20.282b). The above equations are known as the Weyl equations for massless spin 1/2 particles. Now $[(1\mp\gamma_5)/2]\psi$ respectively pick up ϕ_L and ϕ_R. These are known as the chiral projections of ψ. We further note, that for a massless spin 1/2 particle, the Dirac equation becomes

$$\gamma_\mu \frac{\partial}{\partial x_\mu}\psi = 0. \tag{20.283}$$

If we multiply the above equation on the left by γ_5 and use the fact that γ_5 anticommute with γ_μ, we see that $\gamma_5\psi$ is also a solution of the Dirac equation. In other words, the Dirac equation for a massless spin 1/2 particle is invariant under the transformation

$$\psi \longrightarrow \gamma_5\psi. \tag{20.284}$$

The above transformation is called the chiral transformation.

We now discuss the physical significance of the Weyl Eqs. (20.282). Consider first Eq. (20.282a). The plane wave solutions of this can be written as

$$\phi_L(x) = \omega(\mathbf{p})e^{(i/\hbar)(\mathbf{p}\cdot r - Et)}, \tag{20.285}$$

where $\omega(\mathbf{p})$ is a 2-component spinor. Substituting Eq. (20.285) in Eq. (20.282a) we find that $\omega(\mathbf{p})$ satisfies the equation

$$[\boldsymbol{\sigma}\cdot\mathbf{p} + E/c]\omega(\mathbf{p}) = 0. \tag{20.286}$$

Non-trivial solutions of this equation exist only when $E^2 = c^2\mathbf{p}^2$. There are, therefore, two solutions, one corresponding to positive energy $E = c|\mathbf{p}|$, and the other to negative energy $E = -c|\mathbf{p}|$. We denote the positive energy spinor by $u(\boldsymbol{p})$ and negative energy $(E = -m|\boldsymbol{p}|)$ spinor by $v(\boldsymbol{p})$. Thus we see

$$\frac{\boldsymbol{\sigma}\cdot\boldsymbol{p}}{|\boldsymbol{p}|}u(\boldsymbol{p}) = -u(\boldsymbol{p}),$$

$$\frac{\boldsymbol{\sigma}\cdot\boldsymbol{p}}{|\boldsymbol{p}|}v(\boldsymbol{p}) = v(\boldsymbol{p}).$$

Hence we conclude that a 2-component Weyl wave function $\phi_L(x)$ satisfying Eq. (20.282a) or equivalently positive chiral projection $[\frac{1}{2}(1 - \gamma_5)\psi(x)]$ of a 4-component wave function $\psi(x)$ satisfying the Dirac equation for a massless particle describes a left handed (negative helicity) spin 1/2 particle and a right handed (positive helicity) spin 1/2 antiparticle. In an entirely similar manner one verifies that the wave function ϕ_R or equivalently $[(1 - \gamma_5)/2]\psi(x)$ describes a right handed particle and a left handed antiparticle. We know that the neutrino mass is consistent with zero. If we take its mass to be exactly zero, the above results are relevant to the neutrino (antineutrino). A theory of neutrinos based on $\phi_L \neq 0$, $\phi_R = 0$ (or vice versa) is called the two component theory of the neutrino. Actually in nature, the helicity of neutrino is determined to be negative and thus we only have the first possibility.

20.19 Problems

20.1 In the presence of electromagnetic field, Dirac Hamiltonian is given by

$$H = c\boldsymbol{\alpha} \cdot (\hat{p} - \frac{e}{c}\mathbf{A}) + eA_0 + \beta mc^2.$$

Write

$$\hat{\pi} = (\hat{\mathbf{p}} - \frac{e}{c}\mathbf{A}).$$

Show that

$$\frac{d\hat{\pi}}{dt} = e(\mathbf{E} + \boldsymbol{\alpha} \times \mathbf{B})$$

where $\mathbf{B} = \boldsymbol{\nabla} \times \mathbf{A}$ is the magnetic field ($\mathbf{A}$ does not depend explicitly on time) and $\mathbf{E}$ is the electric field.

20.2 From the fundamental properties of γ matrices, derive the following relations:

$$\gamma^\mu (\gamma \cdot a)\gamma_\mu = -2\gamma \cdot a$$

$$\gamma^\mu (\gamma \cdot a)(\gamma \cdot b)\gamma_\mu = 4a \cdot b,$$

where summation over the μ-index is implied and where

$$\gamma \cdot a = \gamma_\nu a^\nu, \ a \cdot b = a_\nu b^\nu;$$

the summation convention is being used.

20.3 Using the anti-commutation relations for γ-matrices, show that

$$(i) \qquad \gamma^5 = -\frac{i}{4!}\varepsilon_{\mu\nu\lambda\rho}\gamma^\mu\gamma^n u\gamma^\lambda\gamma^\rho,$$

$$(ii) \qquad \gamma^5\gamma^\lambda = \frac{1}{3!}\varepsilon^{\mu\nu\rho\lambda}\gamma_\mu\gamma_\nu\gamma_\rho,$$

$$(iii) \qquad [\gamma^\mu, \sigma^{\nu\lambda}] = 2i(g^{\mu\nu}\gamma^\lambda - g^{\mu\lambda}\gamma^\nu),$$

$$(iv) \qquad \{\gamma^\mu, \sigma^{\nu\lambda}\} = -2i\varepsilon^{\mu\nu\lambda\rho}\gamma_5\gamma_\rho.$$

20.4 Show that in the rest frame of a spin $\frac{1}{2}$ particle, the vector

$$n_\mu = \frac{1}{2mc} \varepsilon_{\mu\nu\rho\lambda} \sigma^{\nu\rho} p^\lambda$$

has the value $n_i = \sigma_i$ i.e. $\mathbf{n} = \boldsymbol{\sigma}^i$.

20.5 The Dirac Equation is Lorentz invariant provided that ψ satisfies the transformation law $\psi'(x') = S\psi(x)$, where for the infinitesimal Lorentz transformation

$$S = 1 + \frac{1}{4} \varepsilon_{\mu\nu} \gamma^\mu \gamma^\nu.$$

Consider a 'Pure Lorentz Transformation' in which the primed system is moving with velocity v in the direction x^3-axis and that x'^1, x'^2 are parallel to x^1, x^2

$$x'^1 = x^1$$
$$x'^2 = x^2$$
$$x'^0 = x^0 \cosh\omega - x^3 \sinh\omega$$
$$x'^3 = x^0 \sinh\omega + x^3 \cosh\omega,$$

where

$$\cosh\omega = (1 - \beta^2)^{-1/2},$$
$$\sinh\omega = \frac{\beta}{(1 - \beta^2)^{1/2}}, \qquad \beta = \frac{v}{c}.$$

Take ω to be infinitesimal and find S for the above transformation. From this show that for finite ω, S can be written as

$$S = \cosh\omega/2 - i\gamma^3\gamma^0 \sinh\omega/2.$$

Choose the primed system in such a way that it coincides with the rest frame of the electron so that in the primed frame

$$\psi'(x') = \begin{pmatrix} 1 \\ 0 \\ 0 \\ 0 \end{pmatrix} e^{-imc^2t/\hbar}.$$

The above function for the same physical situation in the unprimed frame is given by

$$\psi(x) = S^{-1}\psi'(x').$$

Show that

$$S^{-1} \begin{pmatrix} 1 \\ 0 \\ 0 \\ 0 \end{pmatrix} = \sqrt{\frac{E+mc^2}{2mc^2}} \begin{pmatrix} 1 \\ 0 \\ \frac{p^3 c}{E+mc^2} \\ 0 \end{pmatrix}.$$

20.6 In Dirac theory of the electron, one identifies the electron intrinsic spin, **S**, as

$$\mathbf{S} = \hbar \frac{\boldsymbol{\alpha} \times \boldsymbol{\alpha}}{4i}.$$

Using only the anti-commutation rules for Dirac matrices α^1, α^2 and α^3, show that the electron has spin $1/2$ i.e. show $S^2 = \dfrac{3\hbar^2}{4}$

20.7 The large and small components of the solution of the Dirac equation

$$(i\gamma_\mu \hat{p}_\mu + mc)\psi = V\psi$$

are defined as $\psi_{\substack{A \\ B}} = \frac{1}{2}(1 \pm \gamma^0)\psi$.
(We choose a rep. in which $\gamma_0 = \beta$ is diagonal.)
Obtain the equation for ψ_A and show that when the kinetic energy, T, of the particle ($cp_0 = mc^2 + T$) is such that $T \ll mc^2$, ψ_A reduces to the non-relativistic Schrödinger wave function multiplied by $e^{-imc^2 t/\hbar}$.

20.8 Show that, using the Dirac equation,

$$\sum_s \bar{u}^s(p)\gamma^\mu u^s(p) = \frac{2}{m}p^\mu,$$

where $s = 1,2$ is the spin index. Show also that

$$p_\mu(\bar{u}(p)\sigma^{\mu\lambda}u(p)) = 0$$

while

$$p^\lambda(\bar{u}(p)\gamma_5\sigma^{\mu\nu}u(p)) = imc\bar{u}(p)\gamma^\mu\gamma^5 u(p).$$

20.9 A positive energy Dirac spinor (using the Pauli rep. of γ matrices) is given by

$$u^{(r)}(\mathbf{p}) = \begin{pmatrix} \chi^{(r)} \\ \frac{c}{E+mc^2}\boldsymbol{\sigma}\cdot\mathbf{p}\chi^{(r)} \end{pmatrix}.$$

As $m \to 0$, show that the γ_5 operator and the helicity $\boldsymbol{\sigma}\cdot\frac{\mathbf{p}}{|\mathbf{p}|}$ operators have the same effect on $u^{(r)}(\mathbf{p})$.

20.10 In the approximation $E \approx mc^2$, $|eA_0| \ll mc^2$, show the equation for ψ_A is given by

$$\frac{1}{2m}(\mathbf{p}-\frac{e}{c}\mathbf{A})^2 - \frac{e\hbar}{2mc}\boldsymbol{\sigma}\cdot\mathbf{B} + eA_0\psi_A = E^{(NR)}\psi_A,$$

where $E^{(NR)} = E - mc^2$ and $\mathbf{A} = \nabla \times \mathbf{A}$.
Take $\mathbf{A} = 0 = A_0$. Consider the probability current 4 vector

$$j^\mu = e\bar{\psi}\gamma^\mu\psi.$$

Show that in the lowest order ($E \approx mc^2$)

$$j^0 = c\psi_A^+\psi_A$$
$$j^i = c\psi^+\alpha^i\psi$$
$$= -\frac{i\hbar}{2m}\left\{\psi_A^+\frac{\partial\psi_A}{\partial x^i} - \frac{\partial\psi_A^\dagger}{\partial x^i}\psi_A\right\}$$

plus a term $i\varepsilon^{ijk}\frac{\partial}{\partial x^j}(\psi_A^+\sigma_k\psi_A)$ which can be ignored since when j^i is integrated over d^3x, it gives zero.

20.11 Show that under charge conjugation

$$\phi_L^c = -i\sigma^2\phi_R^*,$$

$$\phi_R^c = i\sigma^2\phi_L^*.$$

On the other hand under parity

$$\phi_L \leftarrow \phi_R.$$

Thus under CP

$$\phi_L \to -i\sigma^2\phi_L^*,$$

$$\phi_R \to i\sigma^2\phi_R^*.$$

20.12 Consider the coupled 2-component equations

$$i\sigma_L^\mu \partial_\mu \phi_L - \frac{mc}{\hbar c}\phi_R = 0$$

$$i\sigma_R^\mu \partial_\mu \phi_R - \frac{mc}{\hbar c}\phi_L = 0$$

where

$$\sigma_L^\mu = (1, -\boldsymbol{\sigma}) = \bar{\sigma}^\mu, \qquad \sigma_R^\mu = (1, \boldsymbol{\sigma}) = \sigma^\mu.$$

Write the above equation in the 4-component form

$$\left(i\gamma'^\mu \partial_\mu - \frac{mc}{\hbar}\right)\psi' = 0$$

where

$$\psi' = \begin{pmatrix} \phi_L \\ \phi_R \end{pmatrix}.$$

Obtain the equivalent form of γ'^μ and $\gamma'_5 = i\gamma'_0\gamma'_1\gamma'_2\gamma'_3$ and check that

$$(\gamma'^\mu\gamma'^\nu + \gamma'^\nu\gamma'^\mu) = 2g^{\mu\nu}.$$

Verify that the Weyl set $\{\gamma'^\mu\}$ and the Pauli set $\{\gamma^\mu\}$ are related by

$$\gamma'^\mu = S\gamma^\mu S^{-1}$$

where

$$S = \frac{1}{\sqrt{2}}\begin{pmatrix} i & i \\ i & -i \end{pmatrix}.$$

Chapter 21

Dirac Equation in (1+2) Dimensions: Application to Graphene

21.1   Dirac Equation in (1+2) Dimensions

The Dirac equation in $(1+2)$ dimensions has some interesting features. For one thing it has found applications in graphene, which we shall consider in some detail. Furthermore, being an equation in odd space-time dimensions, it has some novel features as we shall see. We start by writing the Dirac equation in $(1 + 2)$ dimensions: in covariant form $[x^\mu = (x^0, x^1, x^2)]$

$$(i\gamma^\mu \partial_\mu - mc)\psi = 0 \tag{21.1}$$

where $\partial_0 = \frac{1}{c}\frac{\partial}{\partial t}$. We may replace c by Fermi velocity v_f for a possible application to graphene. For the moment, we put $c = 1$ as we shall use natural units ($c = 1, \hbar = 1$). Here

$$\slashed{\partial} = \gamma^\mu \partial_\mu = \gamma^0 \partial_0 + \gamma^1 \partial_1 + \gamma^2 \partial_2 \tag{21.2}$$

and as usual [γ^0 is hermitian, γ^1, γ^2 are anti-hermitian]:

$$\{\gamma^\mu, \gamma^\nu\} = 2g^{\mu\nu}.$$

We note that the product of γ matrices $\gamma^0, \gamma^1, \gamma^2$: $\Gamma = \gamma^0\gamma^1\gamma^2$ commutes with each of $\gamma^0, \gamma^1, \gamma^2$ and is thus proportional to the unit matrix. On the other hand $\Gamma^2 = -1$ so that

$$\Gamma \equiv \gamma^0\gamma^1\gamma^2 = \pm iI_2$$

where I_2 is two dimensional identity matrix. Thus there exists two inequivalent representations for the γ matrices in (1+2) dimensions, which are given by the choice of sign. An explicit matrix realization of these two representations is given as follows:

$$\gamma^0 = \sigma^3, \ \gamma^1 = i\sigma^1, \gamma^2 = i\sigma^2, \tag{21.3}$$

$$\gamma^0 = \sigma^3, \ \gamma^1 = i\sigma^1, \gamma^2 = -i\sigma^2.$$

They are inequivalent because they cannot be related by a similarity transformation. This in fact is a feature of an arbitrary number of odd time space dimensions.

We now discuss the role of the parity operation vis a vis above two representations of the γ matrices. In (1+3) dimensions $x^p \longrightarrow (x^0, -x^1, -x^2, -x^3)$; recall for the Lorentz group, defined by the matrix Λ, for $L^\uparrow_+$, $\det \Lambda = +1$ and consists of three rotations and three boosts. For the parity transformation, which is a discrete transformation,

$$\Lambda = \begin{pmatrix} 1 & 0 & 0 & 0 \\ 0 & -1 & 0 & 0 \\ 0 & 0 & -1 & 0 \\ 0 & 0 & 0 & -1 \end{pmatrix}, \quad \det \Lambda = -1.$$

In (1+2) dimensions, there exist only one rotation and two boosts. The parity has a discrete symmetry (not rotation) which has $\det \Lambda = -1$, can be defined as

$$\Lambda = \begin{pmatrix} 1 & 0 & 0 \\ 0 & -1 & 0 \\ 0 & 0 & 1 \end{pmatrix} \quad or \quad \begin{pmatrix} 1 & 0 & 0 \\ 0 & 1 & 0 \\ 0 & 0 & -1 \end{pmatrix}. \tag{21.4}$$

We will make the second choice and thus define $x^p = (x^0, x^1, -x^2)$. Now using the first representation in Eq. (21.3), the Dirac equation given in Eq. (21.1) takes the form,

$$(i\slashed\partial - m)\psi_+(x) \equiv \left(i(\sigma^3\partial_0 + i\sigma^1\partial_1 + i\sigma^2\partial_2) - m\right)\psi_+(x) = 0. \tag{21.5}$$

If, however, we use the second representation in Eq. (21.3), we have

$$(i\slashed\partial - m)\psi_-(x) \equiv \left(i(\sigma^3\partial_0 + i\sigma^1\partial_1 - i\sigma^2\partial_2) - m\right)\psi_-(x) = 0. \tag{21.6}$$

Now if Eq. (21.5) is to be invariant under parity, we should have

$$\left(i(\sigma^3\partial_0^p + i\sigma^1\partial_1^p + i\sigma^2\partial_2^p) - m\right)\psi_+^p(x^p) = 0. \tag{21.7}$$

Now since $\partial_\mu^p \equiv (\partial_0, \partial_1, -\partial_2)$, there is no matrix M that simultaneously anticommutes with σ^2 and commutes with σ^1 and σ^3, therefore we cannot find a relation of the type

$$\psi_+^p(x^p) = M\psi_+(x) \tag{21.8}$$

so as to get back Eq. (21.5). One might therefore argue that the theory cannot be invariant under parity. However, comparing Eq. (21.7) with Eq. (21.6), one sees that a natural assumption would be

$$\psi_+^p(x^p) = -\eta_p \psi_-(x)$$

$$\psi_-^p(x^p) = -\eta_p \psi_+(x)$$
$$\tag{21.9}$$

where η_p is a phase factor, $|\eta_p|^2 = 1$. Thus to preserve parity, we have to take the two representations together. To put Eqs. (21.5) and (21.6) in symmetric form, we transform to new $\psi's$:

$$\psi_A = \psi_+,$$

$$\psi_B = i\gamma^2\psi_-. \tag{21.10}$$

Then Eqs. (21.5) and (21.6) take the form.

$$(i\slashed{\partial} - m)\psi_A = 0$$

$$(i\slashed{\partial} + m)\psi_B = 0 \tag{21.11}$$

where under the parity operation

$$\psi^p_{A,B}(x^p) = \eta_p\sigma^2\psi_{B,A}(x). \tag{21.12}$$

The peculiarity of the parity transformation (21.12) is that it changes A states into B states.

21.1.1   Dirac Equation in Hamiltonian Form and Plane Wave Solutions

We can write Eq. (21.1) in the form $[\hbar = 1, c = 1]$

$$i\gamma^0 \frac{\partial}{\partial x^0}\psi = (\boldsymbol{\gamma} \cdot \hat{\mathbf{p}} + m)\psi \tag{21.13}$$

where $\hat{\mathbf{p}}$ is the momentum operator $-i\boldsymbol{\nabla}$. Thus

$$i\frac{\partial}{\partial t}\psi = H\psi \tag{21.14a}$$

where the Hamiltonian is

$$H = \gamma^0(\boldsymbol{\gamma} \cdot \hat{\mathbf{p}} + m)$$
$$= (-\sigma^2\hat{p^1} + \sigma^1\hat{p^2} + m\sigma^3). \tag{21.14b}$$

Note that when $m = 0$, the H is not identical with $\boldsymbol{\sigma}\cdot\hat{\mathbf{p}}$, or $\boldsymbol{\sigma}\cdot\boldsymbol{\nabla}$, which is the case for 2-component Weyl spinor. Since in a plane, as already noticed we have only one rotation, therefore the orbital angular momentum operator has only one component

$$L^3 = x^1\hat{p^2} - x^2\hat{p^1}. \tag{21.15}$$

We see that

$$[H, L^3] = -\sigma^2[\hat{p}^1, x^1]\hat{p}^2 - \sigma^1[\hat{p}^2, x^2]$$
$$= i\boldsymbol{\sigma} \cdot \hat{\mathbf{p}} \neq 0.$$

On the other hand

$$[H, \frac{1}{2}\sigma^3] = -i\boldsymbol{\sigma} \cdot \hat{\mathbf{p}}.$$

Thus for $J^3 = L^3 + \frac{1}{2}\sigma^3$,

$$[H, J^3] = 0 \tag{21.16}$$

i.e. J^3 is a constant of motion.

For the plane wave solutions

$$\psi(x) \sim u(p)e^{-ip.x} \tag{21.17}$$

which when substituted in Eq. (21.11),

$$(\not{p} - m)u_A(p) = 0. \tag{21.18}$$

Writing

$$u_A(p) = \begin{pmatrix} \xi_1(p) \\ \xi_2(p) \end{pmatrix} \tag{21.19}$$

the above equation gives

$$\begin{pmatrix} p^0 - m & p^2 + ip^1 \\ -p^2 + ip^1 & -(p^0 + m) \end{pmatrix} \begin{pmatrix} \xi_1(p) \\ \xi_2(p) \end{pmatrix} = 0$$

which gives

$$\xi_2(p) = \frac{p^2 - ip^1}{p^0 + m}\xi_1(p) \quad , \quad \xi_1(p) = \frac{p^2 + ip^1}{p^0 - m}\xi_2(p).$$

Therefore we have two solutions,

$$u_A(p) = u_1(p) = \begin{pmatrix} 1 \\ \frac{p^2 - ip^1}{p^0 + m} \end{pmatrix}, \qquad p^0 = E > 0 \tag{21.20}$$

$$u_4(p) = \begin{pmatrix} \frac{p^2 + ip^1}{p^0 - m} \\ 1 \end{pmatrix}, \qquad p^0 = -|E| < 0 \tag{21.21}$$

Thus for antiparticle spinor

$$v_A(\mathbf{p}) = u_4(-\mathbf{p}) = \begin{pmatrix} \frac{p^2 + ip^1}{E + m} \\ 1 \end{pmatrix}. \tag{21.22}$$

We note that for $p^2 = 0 = p^1$,

$$\sigma^3 u_1(0) = \frac{1}{2}u_1(0),$$

$$\tag{21.23}$$

$$\sigma^3 v_1(0) = -\frac{1}{2}v_1(0).$$

Thus we have particle with spin clockwise (spin up) and antiparticle with spin anticlockwise (spin down). Finally

$$\psi_A = u_A(p)e^{-ip.x}$$
$$= v_A(p)e^{ip.x}$$

for particle (electron), antiparticle (hole) spinors. Since ψ_B is related to ψ_A by the parity operation (21.12), we have

$$\psi_B = u_B(p)e^{-ip.x}$$
$$= v_B(p)e^{ip.x},$$

$$u_B(p) = \begin{pmatrix} \frac{p^2+ip^1}{E+m} \\ 1 \end{pmatrix},$$

$$v_B(p) = \begin{pmatrix} 1 \\ \frac{p^2-ip^1}{E+m} \end{pmatrix}. \tag{21.24}$$

Now particle spinor has spin down and antiparticle has spin up.

21.2 Dirac Equation And Graphene

21.2.1 *Introduction*

Natural Carbon can exist in several forms; the most familiar are graphite and diamond. But there are other types.

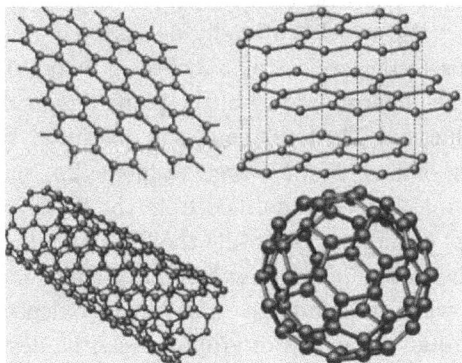

Fig. 21.1

Graphene is a single sheet of graphite and consists of a honeycomb lattice of carbon atoms. Graphite is a stack of graphene layers. Carbon nanotubes [see Fig 21.1] are rolled up cylinders of graphene. Fullerenes ($C60$) are molecules consisting of wrapped graphene by the introduction of pentagons on the hexagonal lattice. Recent progress in the experimental realization of a single layer of graphene has led to extensive exploration of electronic properties in this system. Experimental and theoretical studies have shown that the motion of quasiparticles in these 2D system are very different from those of conventional 2D electron gas systems realized in the semiconductor heterostructures.

In fact graphene is a unique system in many ways:

(A): It is a semimetal. Since it is a 2D system exposed to the environment, it can be easily modified chemically and/or structurally in order to change its functionality and hence its potential applications.

(B): Its low energy excitations are massless Dirac Fermions as we shall see: It mimics physics of quantum electrodynamics (QED) for massless fermions except that in graphene the Dirac Fermions move with the speed $1/300$ the velocity of light which implies many of the unusual properties of QED can show up in graphene but at much smaller speeds.

21.2.2 *Electronic Properties of Graphene*

The starting point in studying the electronic structure of graphene is the tight binding approach, which we briefly discuss. Graphite, a 3D carbon-based material, has 4 valence electrons: 3 of these in $2s, 2p_x, 2p_y$ states form tight bonds with neighboring atoms in the plane and as such will not play a part in the conduction. The fourth one is in $2p_z$ state. We therefore treat graphite as having one conduction electron in the $2p_z$ state. In graphene, the in-plane sigma(σ) bands are formed from the $2s, 2p_x$ and $2p_y$ orbitals, hybridized in a sp^2 configuration, while the $2p_z$ orbitals, perpendicular to the layer, form the out of plane pi (π) bands. The σ bands give rigidity to the structure, while π-bands give rise to the valence and conducting bands. The electronic properties of graphene can be described by a tight binding model with only one orbital per atom. Within this approximation a basis set is provided by the Bloch functions made up of $2p_z$ orbitals from the 2 inequivalent carbon atoms A and B which form the unit cell of the graphene hexagonal lattice [see Fig 21.2]. Graphene is made out of carbon atoms arranged in a hexagonal structure shown above. For a

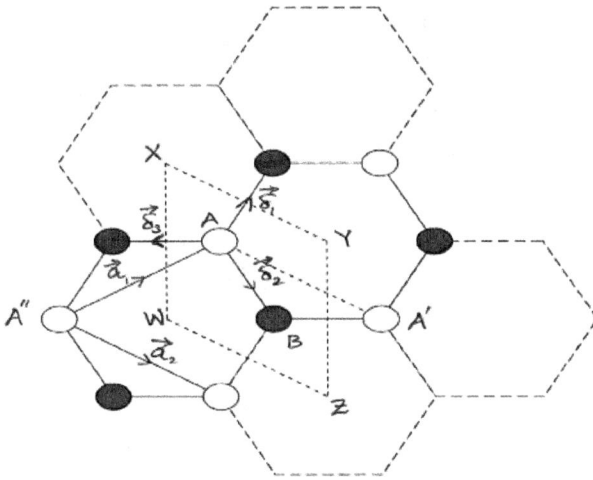

Fig. 21.2

hexagonal layer the unit cell contains two atoms A and B, belonging to the 2 triangular lattices. The lattice vectors are

$$\mathbf{a}_1 = \frac{a}{2}(3, \ \sqrt{3}), \ \mathbf{a}_2 = \frac{a}{2}(3, \ -\sqrt{3}) \tag{21.25}$$

$a = 1.42 A^0$ is the carbon–carbon distance. The reciprocal lattice vectors are given by

$$\mathbf{b}_1 = \frac{2\pi}{3a}(1, \ \sqrt{3}), \ \mathbf{b}_2 = \frac{2\pi}{3a}(1, \ -\sqrt{3}). \tag{21.26}$$

The first Brillouin zone is a hexagon [see Fig 21.3]. Of particular importance for the physics of graphene are 2 points K and K' at the corners of the graphene's Brillouin zone. Now

$$OM = b_{1x} = \frac{2\pi}{3a},$$

$$\frac{KM}{OM} = \tan 30° = \frac{1}{\sqrt{3}},$$

$$KM = \frac{2\pi}{3a}\frac{1}{\sqrt{3}}.$$

K and K' are called Dirac points. Their positions in momentum space are

$$\mathbf{K} = \frac{2\pi}{3a}(1, \ \frac{1}{\sqrt{3}}), \ \mathbf{K'} = \frac{2\pi}{3a}(1, \ -\frac{1}{\sqrt{3}}). \tag{21.27}$$

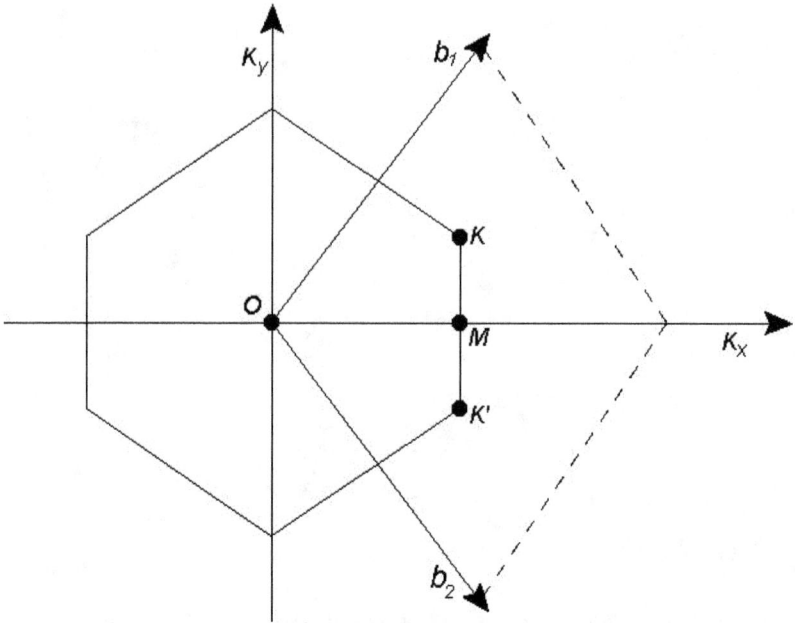

Fig. 21.3

The Dirac cones sit at the K, K' points. For an atom A of a sublattice, the 3 nearest neighbors B which belong to the other sublattice have position vectors [see fig 21.2]

$$\delta_1 = \frac{a}{2}(1, \ \sqrt{3}), \ \delta_2 = \frac{a}{2}(1, \ -\sqrt{3}), \ \ \delta_3 = -a(1, \ -0). \qquad (21.28)$$

The tight binding Hamiltonian of the electrons in graphene is

$$\mathcal{H} = \epsilon \sum_j (a^\dagger(\mathbf{R}_j^A)a(\mathbf{R}_j^A) + b^\dagger(\mathbf{R}_j^B)b(\mathbf{R}_j^B))$$

$$-t \sum_{\langle i,j \rangle} (a^\dagger(\mathbf{R}_i^A)b(\mathbf{R}_j^B) + h.c) \qquad (21.29)$$

where the $\langle i, j \rangle$ stands for summation over pairs of nearest neighbor atoms, i, j on the lattice. $a^\dagger(\mathbf{R}_i^A)(b^\dagger(\mathbf{R}_j^B))$ creates an electron on site $\mathbf{R}_i^A(\mathbf{R}_j^B)$ sublattice A(B) with a corresponding annihilation operator $a(\mathbf{R}_i^A)(b(\mathbf{R}_i^B))$. t ($\simeq 2.7$ eV) is the nearest neighbour hopping energy (hopping between different sublattices). Fourier transforming the annihilation and creation operators, we have

$$a(R_i^A) = \sum_k a_k\, e^{i\mathbf{k}\cdot\mathbf{R}_i^A},$$

$$b(R_j^B) = \sum_k b_k\, e^{i\mathbf{k}\cdot\mathbf{R}_j^B}. \tag{21.30}$$

Thus we can write the above Hamiltonian as

$$H = \sum_k \left(a_k^\dagger\ b_k^\dagger \right) \mathcal{H} \begin{pmatrix} a_k \\ b_k \end{pmatrix}.$$

where $\mathcal{H}$ is

$$\mathcal{H} = \begin{pmatrix} \varepsilon & -t\sum_i e^{i\mathbf{k}\cdot\mathbf{u}_i} \\ -t\sum_i e^{i\mathbf{k}\cdot\mathbf{v}_i} & \varepsilon \end{pmatrix} \tag{21.31}$$

Here $\mathbf{u}_i$ is a set of triad of vectors connecting an atom A with its B nearest neighbors, in our case $\boldsymbol{\delta}_1, \boldsymbol{\delta}_2, \boldsymbol{\delta}_3$ [see fig 21.2] and $\mathbf{v}_i$ the triad of their respective opposites. ε is the $2p_z$ energy level, taken as the origin of energy. Graphene's electronic states are described as a linear combination of atomic orbitals from the two atoms forming the primitive cell $O(A)$ and $\bullet(B)$ [see fig 21.2]

$$|\psi(\mathbf{r})\rangle = C_A|\phi_A\rangle + C_B|\phi_B\rangle. \tag{21.32}$$

The eigenfunctions and eigenvalues of the Hamiltonian are obtained from the equation

$$\mathcal{H}\begin{pmatrix} C_A \\ C_B \end{pmatrix} = E\begin{pmatrix} C_A \\ C_B \end{pmatrix} \tag{21.33}$$

where $\mathcal{H}$ is given in Eq. (21.31). The eigenfunctions are determined by the coefficients C_A and C_B in Eq. (21.32). The eigenvalues of this equation are obtained by diagonalization of $\mathcal{H}$ which gives the energy bands

$$E - \varepsilon = \pm t\{|e^{i\mathbf{k}\cdot\boldsymbol{\delta}_1} + e^{i\mathbf{k}\cdot\boldsymbol{\delta}_2} + e^{i\mathbf{k}\cdot\boldsymbol{\delta}_3}|^2\}^{1/2} \tag{21.34}$$

where ε and t are found by fitting experimental or the first-principles data. The most common practice is to adjust the tight-binding dispersion to a correct description of the pi-bands at the K point. This yields $\varepsilon = 0$, t ($\simeq 2.8$ eV). Thus, using Eqs. (21.28) and (21.34)

$$E_\pm(k) = \pm t\{3 + \cos k_y\sqrt{3}a + 4\cos k_x\frac{3a}{2}\cos k_y\frac{\sqrt{3}a}{2}\}^{1/2}$$

$$= \pm t\sqrt{3 + f(k)}. \tag{21.35}$$

It is instructive to calculate $f(k)$ close to one of the Dirac points: $\mathbf{k} = \mathbf{K} + \mathbf{p}$, $\mathbf{p}$ is the momentum measured relative to the Dirac points

$$f(k) \approx -3 + \frac{9}{4}a^2(p_x^2 + p_y^2).$$

Thus the energy band around the Dirac point is

$$E_{\pm}(p) \approx \pm v_f |\mathbf{p}| + O(|\mathbf{p}|^2) \tag{21.36}$$

where $v_f = \frac{3ta}{2}$ represents the Fermi velocity. It is seen that the dispersion is conical [fig 21.4]. The most striking difference between the result

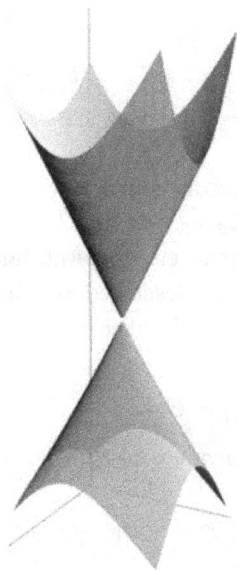

Fig. 21.4

(21.36) and the usual case $\varepsilon(p) = \frac{p^2}{2m}$, m electron mass, is that v_f changes substantially with energy whereas Fermi velocity in Eq. (21.36) does not depend on the energy or mass. The energy dispersion (21.38) resembles the energy of ultra-relativistic particles, (although v_f is 300 times smaller than the speed of light); these particles are quantum mechanically described by the massless Dirac Equation as we shall see. The most interesting aspect of graphene is that low energy excitations are massless, chiral Dirac Fermions. This difference in the nature of the quasiparticles in graphene from conventional 2D electron gas system has given rise to a lot of new and unusual phenomena such as anomalous Hall effect in graphene. The transport experiments have shown results in agreement with the presence of Dirac fermions. The 2D Dirac-like spectrum was confirmed by cyclotron resonance measurements, which we describe below.

21.2.3   Cyclotron Mass

The cyclotron mass is defined, within the classical approximation as

$$m^* = \frac{1}{2\pi} \frac{\partial A(E)}{\partial E} \tag{21.37}$$

where $A(E)$ is area in k-space enclosed by the orbit. Using the Energy dispersion $E^2 = v_f^2 p^2$

$$A(E) = \pi \frac{E^2}{v_f^2}$$

so that

$$m^* = \frac{p}{v_f}. \tag{21.38}$$

One may contrast it with Schrödinger dispersion $E = \frac{p^2}{2m}$ which would imply a constant cyclotron mass. The electronic density, n, related to the Fermi momentum is

$$n = \frac{k_F^2}{\pi}$$

which implies

$$m^* = \frac{\sqrt{\pi}}{v_f} \sqrt{n}. \tag{21.39}$$

Fitting the above equation provides an estimation to the Fermi velocity $[v_f = 3at]$ and the hopping parameter t respectively [see Fig 21.5]:

$$v_f \approx 10^6 ms^{-1} \quad and \quad t \approx 3 \ eV.$$

The experimental observation of the $\sqrt{n}$ dependence of the cyclotron mass, as shown in Fig 21.5, provides evidence for the existence of massless Dirac quasi-particles in graphene.

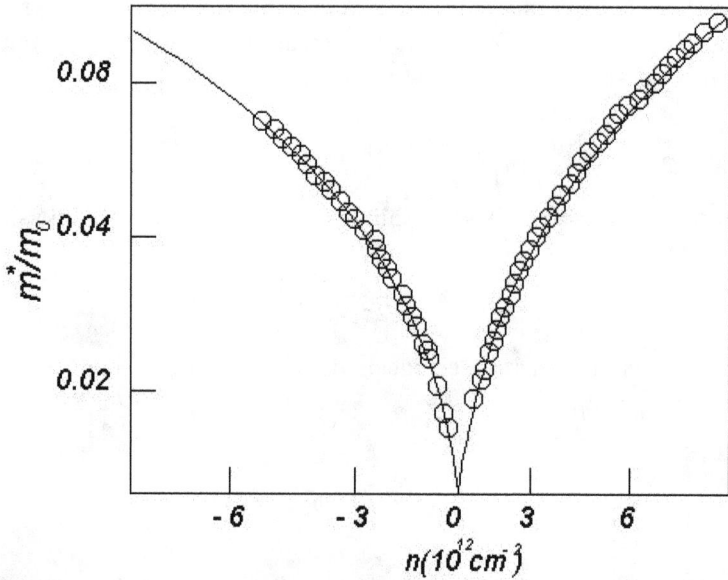

Fig. 21.5 Cyclotron mass of charge carriers as a function of their concentration n. Positive and negative n corresponds to electrons and holes. Symbols o are the experimental data.

21.2.4 *Dirac Hamiltonian and Dirac Equation for Quasiparticles in Graphene*

In Eq. (21.31), we may take $\mathbf{u}_i$ or $\mathbf{v}_j$ as two independent lattice vectors, e.g $\mathbf{a}_1$ and $\mathbf{a}_2$ as given in Eq. (21.25). Then Eq. (21.31) gives

$$H = \sum_k \left(a_k^\dagger \; b_k^\dagger \right) \mathcal{H}_D \begin{pmatrix} a_k \\ b_k \end{pmatrix}$$

where

$$\mathcal{H}_D = \begin{pmatrix} 0 & \mathcal{H}_{AB} \\ \mathcal{H}_{AB}^* & 0 \end{pmatrix}$$

with

$$\mathcal{H}_{AB} = -t[e^{i\mathbf{k}\cdot\mathbf{a}_1} + e^{i\mathbf{k}\cdot\mathbf{a}_2}]. \tag{21.40}$$

Let us expand around the Dirac point K' given in Eq. (21.27):

$$\mathbf{k} = \mathbf{K}' + \mathbf{p},$$

keeping terms linear in $|\mathbf{p}|$. Then

$$\mathcal{H}_{AB} = \frac{3at}{2}(ip_x + ip_y). \tag{21.41}$$

This gives

$$\mathcal{H}_D = \begin{pmatrix} 0 & ip_x + p_y \\ -ip_x + p_y & 0 \end{pmatrix}$$
$$= v_f[-p^1\sigma^2 + p^2\sigma^1] \qquad (21.42)$$

where $v_f = \frac{3at}{2}$ is the Fermi velocity, σ^1 and σ^2 are Pauli matrices, $p^i = (p^1, p^2)$ and in relativistic notation: $p^\mu = (p^0, p^1, p^2)$. This does not give $\sigma \cdot \mathbf{p}$ which is clear from Eq. (21.42). Eq. (21.42) can be put in the form

$$\mathcal{H}_D = v_f[i\sigma^3\sigma^1 p^1 + i\sigma^3\sigma^2 p^2]$$
$$= \gamma^0(\gamma \cdot \mathbf{p} v_f) \qquad (21.43)$$

where $\gamma^0 = \sigma^3$, $\gamma^1 = i\sigma^1$, $\gamma^2 = i\sigma^2$ are Dirac matrices in (1+2) dimensions. This is the Dirac Hamiltonian for massless fermion in (1+2) dimensions near the Dirac point K'. The corresponding Dirac equation is

$$\mathcal{H}_D \psi_\pm = E\psi_\pm, E = \pm v_f |\mathbf{p}| \qquad (21.44)$$

or in its time dependent form

$$i\frac{\partial \Psi}{\partial t} = \mathcal{H}_D \Psi$$

or, on using Eq. (21.43),

$$i\frac{\partial \psi}{\partial t} = \gamma^0(\gamma \cdot (-i\nabla v_f))\psi \qquad (21.45)$$

where $\mathbf{p}$ is the momentum operator $-i\nabla$. Writing $\partial_0 = \frac{1}{v_f}\frac{\partial}{\partial t}$, we can write Eq. (21.45) in covariant form

$$i(\gamma^0 \partial_0 + \gamma^1 \partial_1 + \gamma^2 \partial_2)\psi = 0$$

or

$$i(\gamma^\mu \partial_\mu)\psi = 0 \qquad (21.46)$$

which is the Dirac equation in (1+2) dimensions for a massless fermion. It is important to remark that if we expand around the Dirac point K, we obtain

$$\mathcal{H}_D = v_f[-p^1\sigma^2 - p^2\sigma^1]$$
$$= v_f[i\sigma^3\sigma^1 p^1\sigma^1 - i\sigma^3\sigma^2 p^2]. \qquad (21.47)$$

Now as discussed in section (21.1), in 3 space-time dimensions there exists two inequivalent representations for γ-matrices given in Eq. (21.3). We have used the first of these representations for the expansion around the

Dirac point K'. We take the second representation for the expansion around the Dirac point K, which is obtained from K' by the parity operation: $p \rightarrow p^p = (p^0, p^1, -p^2)$:

$$p^1 \longleftrightarrow p^1, p^2 \longleftrightarrow -p^2 \text{ and } \mathcal{H}_K \longleftrightarrow \mathcal{H}_{K'}.$$

Thus the parity operation connects Dirac Hamiltonian around the Dirac point K and K'.

21.2.5   *Tunneling and Klein Paradox*

We consider the scattering of massless Dirac fermions in two spatial dimensions by a square barrier shown in Fig (21.6). The Dirac equation is

$$\left(\frac{1}{v_f} \gamma^0 (i \frac{\partial}{\partial t} - V_0) + i\gamma^1 \partial_1 + i\gamma^2 \partial_2 \right) \Psi(x) = 0 \qquad (21.48)$$

where $\gamma^0 = \sigma^3$, $\gamma^1 = i\sigma^1, \gamma^2 = i\sigma^2$. Now

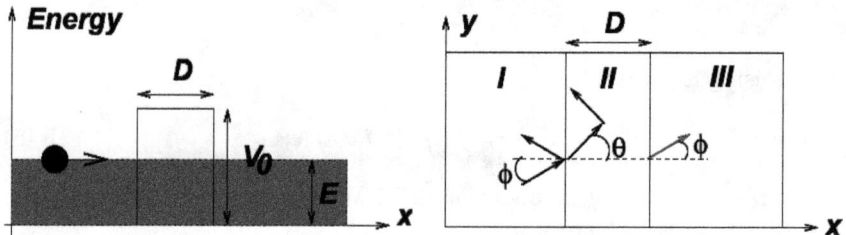

Fig. 21.6

$$\Psi(x) \sim e^{-ip \cdot x} u(p)$$
$$= e^{-iEt} e^{i \, \mathbf{p \cdot r}} u(p)$$
$$= e^{-iEt} \psi(x).$$

Then from Eq. (21.48), by replacing $p^0 \longrightarrow E - V_0$, the solution of Eq. (??) can be written

$$\Psi(x) = u_A(p) \, e^{i \, \mathbf{p \cdot r}}$$

where [see Eq. (21.20)]

$$u_A(p) = \begin{pmatrix} 1 \\ \frac{(-ip_x + p_y)v_f}{E - V_0} \end{pmatrix} \qquad (21.49)$$

where the energy dispersion is

$$E = \pm |\mathbf{p}| v_f.$$

We can write

$$p_x = p \cos \phi \quad , \quad p_y = \sin \phi$$

where

$$\tan \phi = \frac{p_y}{p_x}. \tag{21.50}$$

Then the wave functions in the three regions can be written as

$$\psi_I(\mathbf{r}) = \frac{1}{\sqrt{2}} \begin{pmatrix} 1 \\ ise^{i\phi} \end{pmatrix} e^{i(p_x x + p_y y)}$$

$$+ \frac{r}{\sqrt{2}} \begin{pmatrix} 1 \\ ise^{i(\pi - \phi)} \end{pmatrix} e^{i(-p_x x + p_y y)} \tag{21.51a}$$

where s is sign(E). For regions II and III:

$$\psi_{II}(\mathbf{r}) = \frac{a}{\sqrt{2}} \begin{pmatrix} 1 \\ is'e^{i\theta} \end{pmatrix} e^{i(q_x x + p_y y)}$$

$$+ \frac{b}{\sqrt{2}} \begin{pmatrix} 1 \\ is'e^{i(\pi - \theta)} \end{pmatrix} e^{i(-q_x x + p_y y)} \tag{21.51b}$$

$$\psi_{III}(\mathbf{r}) = \frac{t}{\sqrt{2}} \begin{pmatrix} 1 \\ ise^{i\phi} \end{pmatrix} e^{i(p_x x + p_y y)} \tag{21.51c}$$

where $s' = \text{sign}(E - V_0)$, $\tan \theta = \frac{q_x}{p_y}$, $q_x = \sqrt{\mathbf{q}^2 - p_y^2} = \sqrt{(V_0 - E)^2 \frac{1}{v_f^2} - p_y^2}$.
The coefficients r, a, b and t are determined from the continuity conditions

$$\psi_I(x = 0, y) = \psi_{II}(x = 0, y),$$
$$\psi_{II}(x = D, y) = \psi_{III}(x = D, y). \tag{21.52}$$

If we define $A_\pm = a \pm b$, then the above conditions give

$$1 + r = A_+,$$

$$e^{i\phi} + e^{-i\phi} - A_+ e^{-i\phi} = \frac{s'}{s}[iA_+ \sin \theta + A_- \cos \theta],$$

$$A_+ \cos(Dq_x) + iA_- \sin(Dq_x) = te^{iDp_x},$$

$$\frac{s'}{s}[iA_+ \sin(\theta + Dq_x) + A_- \cos(\theta + Dq_x)] = te^{i\phi}e^{iDp_x}. \tag{21.53}$$

The last three equations can be put in matrix form

$$M \begin{pmatrix} A_+ \\ A_- \\ t \end{pmatrix} = \begin{pmatrix} 2\cos\phi \\ 0 \\ 0 \end{pmatrix}$$

where

$$M = \begin{pmatrix} e^{-i\phi} + i\frac{s'}{s}\sin\theta & \frac{s'}{s}\cos\theta & 0 \\ \cos(Dq_x) & i\sin(Dq_x) & -e^{iDp_x} \\ i\frac{s'}{s}\sin(\theta + Dq_x) & \frac{s'}{s}\cos(\theta + Dq_x) & -e^{i\phi}e^{iDp_x} \end{pmatrix}. \qquad (21.54)$$

Then

$$\begin{pmatrix} A_+ \\ A_- \\ t \end{pmatrix} = M^{-1} \begin{pmatrix} 2\cos\phi \\ 0 \\ 0 \end{pmatrix}. \qquad (21.55)$$

Writing $M^{-1} = N$, to find t what we need is N_{31}, the cofactor of M_{13} which is

$$\frac{\frac{s'}{s}[\cos(Dq_x)\cos(\theta + Dq_x) + \sin(Dq_x)\sin(\theta + Dq_x)]}{\det M} = \frac{s'}{s}\frac{\cos\theta}{\det M}.$$

Hence

$$t = 2N_{31}\cos\phi/\det M$$
$$= 2\frac{s'}{s}\cos\theta\cos\phi\frac{1}{\det M}. \qquad (21.56)$$

and the transmitivity (tunneling probability) is given by

$$T(\phi) = tt^* = \frac{4\cos^2\theta\cos^2\phi}{|\det M|^2}$$

$$= \frac{\cos^2\theta\cos^2\phi}{[\cos\theta\cos\phi\cos(Dq_x)]^2 + \sin^2(Dq_x)[1 - \frac{s'}{s}\sin\phi\sin\theta]^2}. \qquad (21.57)$$

Notice the following:

(1) For $Dq_x = n\pi, n$ an integer, $T(\phi) = T(-\phi)$ and $T(\phi) = 1$, independent of the value of ϕ and the barrier becomes completely transparent.

(2) For normal incidence $\phi \to 0$, $\theta \to 0$, $T(0) = 1$ for any value of Dq_x. This result is the manifestation of Klein paradox and does not occur for non-relativistic particles. In this latter case for normal incidence, T is always smaller than one.

(3) In the limit $|V_0| \gg |E|$, $\theta \to 0$ and

$$T(\phi) \longrightarrow \frac{\cos^2\phi}{1 - \cos^2(Dq_x)\sin^2\phi}. \qquad (21.58)$$

The angular dependence of $T(\phi)$ is plotted in Fig (21.7), which shows that there are several directions for which transmission is one.

The above result has several applications, e.g. in filtering electron beams, and in transmission through graphene quantum dots.

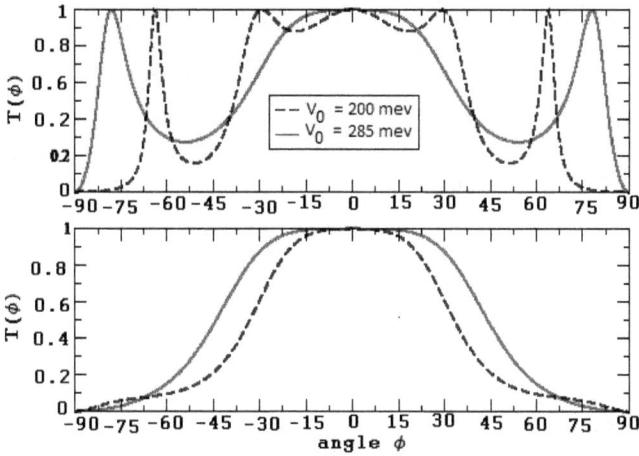

Fig. 21.7 Angular behaviour of $T(\phi)$ for two different values of V_0 : $V_0 = 200$ meV dashed line, $V_0 = 285$ meV solid line. The remaining parameters are $D = 110$ nm (top), $D = 50$ nm (bottom) $E = 80$ meV, $k_F = 2\pi/\lambda, \lambda = 50$ nm.

21.2.6 Dirac Fermions in an External Gauge Field (Electromagnetic): Landau Levels

Near the Dirac point K'; we have the Dirac equation for massless particles

$$i\gamma^\mu \partial_\mu \psi_A = 0.$$

Near the other Dirac Point K, the Dirac equation is

$$i\gamma^\mu \partial_\mu \psi_B = 0.$$

As already discussed ψ_A and ψ_B are related by parity transformation. We now introduce a gauge field A_μ in a gauge invariant way by replacing ∂_μ by covariant derivative D_μ:

$$\partial_\mu \to D_\mu = \partial_\mu + \frac{ie}{c} A_\mu$$

so that

$$i\gamma^\mu D_\mu \psi_A = 0. \tag{21.59}$$

Multiplying by $(-i\gamma^\nu D_\nu)$ on the left, we have

$$(-i\gamma^\nu D_\nu)(i\gamma^\mu D_\mu \psi_A) = 0.$$

Using Dirac algebra:

$$\sigma^{\mu\nu} = \frac{1}{2i}[\gamma^\mu\gamma^\nu - \gamma^\nu\gamma^\mu]$$

$$\gamma^\mu\gamma^\nu + \gamma^\nu\gamma^\mu = 2g^{\mu\nu}$$

we can write the above equation as

$$\{g^{\mu\nu}D_\mu D_\nu + \frac{i}{2}\sigma^{\mu\nu}[D_\mu, D_\nu]\}\psi_A = 0. \tag{21.60}$$

One can easily evaluate

$$[D_\mu, D_\nu] = \frac{ie}{c}F_{\mu\nu}, \tag{21.61}$$

$$F_{\mu\nu} = \partial_\mu A_\nu - \partial_\nu A_\mu.$$

Thus we obtain

$$[D_\mu D^\mu - \frac{e}{2c}\sigma^{\mu\nu}F_{\mu\nu}]\psi_A = 0, \quad \mu, \ \nu = 0, \ 1, \ 2$$

or

$$(\partial^\mu + \frac{ie}{c}A^\mu)(\partial_\mu + \frac{ie}{c}A_\mu)\psi_A - \frac{e}{2c}\sigma^{\mu\nu}F_{\mu\nu}\psi_A = 0. \tag{21.62}$$

This differs from the Klein–Gordon equation

$$D_\mu D^\mu \psi_A = 0 \tag{21.63}$$

in the term $\frac{e}{2c}\sigma^{\mu\nu}F_{\mu\nu}$. It is easy to verify that we can write

$$\sigma^{\mu\nu} = \epsilon^{\lambda\mu\nu}\gamma_\lambda. \tag{21.64}$$

Thus we can write

$$-\frac{e}{2c}\sigma^{\mu\nu}F_{\mu\nu} = -\frac{e}{2c}(\epsilon^{\lambda\mu\nu}F_{\mu\nu})\gamma_\lambda$$

$$= J^\lambda \gamma_\lambda \tag{21.65a}$$

where the induced current is

$$J^\lambda = -\frac{e}{2c}\epsilon^{\lambda\mu\nu}F_{\mu\nu} \tag{21.65b}$$

which gives the induced charge

$$Q = \int d^2x J^0(x) \tag{21.65c}$$

where

$$J^0(x) = -\frac{e}{c}\epsilon^{ij}F_{ij} = -\frac{e}{c}\epsilon^{ij}(\partial_i A_j - \partial_j A_i)$$

$$= 2\frac{e}{c}(\nabla \times \mathbf{A})^3 = 2\frac{e}{c}B \tag{21.65d}$$

so that

$$Q = 2\frac{e}{c}\int d^2x B$$

$$= 2\frac{e}{c}\Phi, \ \Phi \text{ is the magnetic flux.} \tag{21.65e}$$

We now return to Eq. (21.54) and explicitly evaluate

$$-\frac{e}{2c}\sigma^{\mu\nu}F_{\mu\nu} = -\frac{e}{2c}\epsilon^{\lambda\mu\nu}F_{\mu\nu}\gamma_\lambda$$

$$= -\frac{e}{c}\{\sigma^3(\partial_1 A^2 - \partial_2 A^1) + (i\sigma^1)(-\partial_0 A^2 - \partial_2 A^0)$$

$$-(i\sigma^2)(-\partial_0 A^1 - \partial_1 A^0)\}$$

$$= -\frac{e}{c}\{\boldsymbol{\sigma}\cdot\mathbf{B} + i\boldsymbol{\sigma}\times\mathbf{E}\} \qquad (21.66)$$

where

$$\mathbf{B} = \boldsymbol{\nabla}\times\mathbf{A} \quad, \quad \mathbf{E} = -\partial_0\mathbf{A} - \boldsymbol{\nabla}A^0.$$

Thus Eq. (21.54) becomes

$$[D^{0^2} - \mathbf{D}^2 - \frac{e}{c}\boldsymbol{\sigma}\cdot\mathbf{B} - i\frac{e}{c}\boldsymbol{\sigma}\times\mathbf{E}]\psi_A = 0 \qquad (21.67a)$$

where

$$D^0 = \partial_0 + \frac{ie}{c}A_0$$

$$= \frac{1}{v_f}\frac{\partial}{\partial t} + \frac{ie}{c}A_0. \qquad (21.67b)$$

For stationary states

$$\psi_A(t,\mathbf{x}) = \psi_A(\mathbf{x})e^{-i\epsilon t},$$

Eq. (21.5a) becomes

$$[(\frac{-i\epsilon}{v_f} + \frac{ie}{c}A_0)^2 - \mathbf{D}^2 - \frac{e}{c}\boldsymbol{\sigma}\cdot\mathbf{B} - i\frac{e}{c}\boldsymbol{\sigma}\times\mathbf{E}]\psi_A = 0 \qquad (21.68)$$

where

$$\mathbf{D}^2 = (\partial_1 + \frac{ie}{c}A_1)^2 + (\partial_2 + \frac{ie}{c}A_2)^2.$$

We now consider a static magnetic field:

$$A^0 = 0, \frac{\partial A^i}{\partial t} = 0, \mathbf{E} = 0.$$

We can select the gauge

$$A_1 = -A^1 = -A_x = By \qquad (21.69)$$

$$A_2 = 0 \qquad (21.70)$$

and we can write the wave function ψ_A as

$$\psi_A(x,y) = \varphi_A(y)e^{ikx} \qquad (21.71)$$

where the 2-component spinor $\varphi_A(y)$ is

$$\varphi_A = \begin{pmatrix} \varphi_+ \\ \varphi_- \end{pmatrix} \quad \begin{matrix} \sigma^3 = +1, \\ \sigma^3 = -1. \end{matrix} \tag{21.72}$$

Then Eq. (21.60) gives

$$\left[\frac{-\epsilon^2}{v_f^2} - i^2(k + \frac{e}{c}By)^2 - \frac{\partial^2}{\partial y^2} \right] \varphi_+ - \frac{eB}{c}\varphi_+ = 0. \tag{21.73}$$

By the change of variable $Y = k + \frac{e}{c}By$, we can write the above equation as

$$\left[(\frac{e}{c}B)^2 \partial_Y^2 + Y^2 \right] \varphi_+(y) = \left(\frac{\epsilon^2}{v_f^2} + \frac{e}{c}B \right) \varphi_+(y)$$

or

$$\left[-\frac{1}{2}\partial_Y^2 + \frac{1}{2}(\frac{e}{c}B)^2 Y^2 \right] \varphi_+(y) = \frac{1}{2}(\frac{c}{eB})^2 \left(\frac{\epsilon^2}{v_f^2} + \frac{e}{c}B \right) \varphi_+(y) \equiv E'\varphi_+, \tag{21.74}$$

which is the equation of 1-dimensional harmonic oscillator, so that energy levels are given by

$$E'_n = (n + \frac{1}{2})\frac{c}{eB} \tag{21.75}$$

or

$$\frac{1}{v_f^2}\epsilon_n^2 + \frac{e}{c}B = 2(n + \frac{1}{2})\frac{e}{c}B.$$

Thus

$$\epsilon_n = sgn|n| \, \omega_c |n|^{\frac{1}{2}}, \quad n = 0, \pm 1, \pm 2 \cdots \tag{21.76}$$

where

$$\omega_c = \sqrt{2}v_f\sqrt{\frac{e}{c}B}$$
$$= \sqrt{2}\frac{v_f}{l_B}, \tag{21.77}$$

$l_B = \sqrt{\frac{c}{eB}}$, l_B being the magnetic length: Thus the spectrum of Landau levels is given by [inserting back $\hbar$]

$$\epsilon_n = sgn|n|\sqrt{2|n|\hbar\frac{v_f^2}{c}eB} \quad , \quad n = 0 \pm 1, \pm 2 \cdots \tag{21.78}$$

These are Landau levels near the Dirac point K'. The Landau levels at the opposite Dirac point K have exactly the same spectrum and hence each Landau level is doubly degenerate. This particular Landau spectrum has been verified by several experimental groups. The result (21.69) is relevant to half integer (anomalous) quantum Hall effect in graphene.

21.3   Problems

21.1 Show that
 (i)
$$\sigma^{\mu\nu} = \epsilon^{\lambda\mu\nu}\gamma_\lambda$$

$\mu, \nu = 0, 1, 2.$ $\epsilon^{012} = 1,$ $\sigma^{\mu\nu} = \frac{1}{2i}(\gamma^\mu\gamma^\nu - \gamma^\nu\gamma^\mu)$
 (ii)
$$Tr[\gamma^\mu\gamma^\nu\gamma^\lambda] = -2i\epsilon^{\mu\nu\lambda}$$

21.2 For the static magnetic field:
$$A^0 = 0, \frac{\partial A^i}{\partial t} = 0, \ \mathbf{E} = 0$$

show that Eq. (21.60) of the text takes the form for $\sigma^3 = 1$
$$[-D_1^2 - D_2^2 - \frac{e}{c}B]\phi_+ = \frac{\epsilon^2}{v_f^2}\phi_+.$$

Show that
$$[D_1, D_2] = -\frac{ie}{c}B.$$

Define Canonical variables $\hat{Q}, \hat{P}$:
$$-i\frac{D_2}{\frac{eB}{c}} = \hat{Q}, \quad iD_1 = \hat{P},$$

so that
$$[\hat{Q}, \hat{P}] = i$$

and
$$[\frac{1}{2}\hat{P}^2 + \frac{1}{2}\omega^2\hat{Q}^2]\phi_+ = E\phi_+$$

where
$$E = \frac{1}{2}(\frac{\epsilon^2}{v_f^2} + \frac{eB}{c}), \quad \omega = \frac{eB}{c}.$$

The above is the equation of the one-dimensional harmonic oscillator, and give energy levels
$$E = (n + \frac{1}{2})\omega_c$$

$$\epsilon_\pm(n) = \pm\,\omega_c\,|n| \quad, \quad n = 0, \pm1, \pm2\cdots$$

$$\omega_c = \sqrt{2}v_f\sqrt{\frac{e}{c}B}.$$

21.4 Selected References and Further Reading

i. For 21.1: Lecture notes of a short course given on "Origin of Mass" by Adnan Bashir at National Center for Physics, Quaid-i-azam university, Islamabad, in Dec. 2005.

ii For the rest of the sections, the main reference is A. H. Castro Neto, F. Guinea, N. M. R. Peres, K. S. Novoselov and A. K. Geim, The electronic properties of graphene, *Rev. Mod. Phys.* 81, 109162 (2009) (arXiv:0709.1163 [cond-mat.other], Sep. 2007).

All figures, except 21.2, have been taken from this reference.

iii see also: Francisco Guinea, M. Pilar Lopez-Sancho and Maria A. H. Vozmediano, Interactions and disorder in 2D graphite sheets, arXiv:cond-mat/0511558 [cond-mat.str-el], Nov. 2005.

iv M. Mecklenburg, J. Woo and B. C. Regan, Electron–Photon Interactions in Graphene, *Phys. Rev. B* 81, 245401 (2010) (arXiv:1003.4419 [cond-mat.mes-hall], March. 2010).

v Carbon Nanotube and Graphene Device Physics by H.-S. Philip Wong and Deji Akinwande, Cambridge University Press, 2011.

SELECTED GENERAL REFERENCES

1. G. Bayem, *Lectures on Quantum Mechanics*, W. A. Benjamin, New York, 1969.

2. D. Bohm, *Quantum Theory*, Dover Publ., New York, 1989.

3. P. A. M. Dirac, *The Principle of Quantum Mechanics* (4th edition), Oxford University Press (Clarendon), Oxford, 1958.

4. S. Gasiorowicz, *Quantum Physics (Second Edition)*, John Wiley & Sons, Inc., 1996

5. K. Gottfried, *Quantum Mechanics, Vol 1, Fundamentals*, W. A. Benjamin, New York, 1966.

6. L. D. Landau and E. M. Lifshitz, *Quantum Mechanics (Nonrelativistic Theory)* (2nd edition), Addison-Wesley, Reading, Mass., 1965.

7. F. Mandl, *Quantum Mechanics*, John Wiley & sons, Inc., 1992.

8. P.T. Matthews, *Introduction to Quantum Mechanics (Third Edition)* McGraw-Hill Book Company (UK) Limited, 1974.

9. A. Messiah, *Quantum Mechanics (in 2 volumes)*, John Wiley & sons, New York, 1968.

10. E. Merzbacher, *Quantum Mechanics (2nd edition)* John Wiley & sons, New York, 1970.

11. J. J. Sakurai, *Advanced Quantum Mechanics*, Addision-Wesley Publishing Company, 1973; *Modern Quantum Mechanics* (S. F. Tuan, Editor), Addison-Wesley, Reading, Mass, 1994.
12. L. I. Schiff, *Quantum Mechanics (3rd edition)*, McGraw-Hill, New York, 1968.
13. R. Shankar, *Principle of Quantum Mechanics*, Plenum Press, New York, 1980.

Index

Index